Molecular Biology of the Flavivirus

Edited by Matthias Kalitzky and Peter Borowski

horizon bioscience

Horizon Bioscience
32 Hewitts Lane
Wymondham
Norfolk NR18 0JA
U.K.

www.horizonbioscience.com

British Library Cataloguing-in-Publication Data

A catalogue record for this book is available from the British Library
ISBN-10 1-904933-22-X
ISBN-13 978-1-904933-22-9

Printed and bound in Great Britain by Cromwell Press

Dedications

For Julia.
MK

This book is dedicated to my mentor and advisor, Prof. Dr. Helmuth Hilz.
PB

Contents

Contributors

Sven-Erik Behrens
Martin-Luther-University Halle-Wittenberg
Institute for Biotechnology
Department of Microbial Biotechnology
Halle/Saale
Germany

sven.behrens@biochemtech.uni-halle.de

Hubert E. Blum
Department of Medicine II
University of Freiburg
Freiburg
Germany

Hubert.Blum@uniklinik-freiburg.de

Peter Borowski
Department of Molecular Biology
John Paul II Catholic University in Lublin
Lublin
Poland

Xavier Deparis
Département d'Epidémiologie et de
Santé Publique
Ecole du Val-de Grâce
Saint Mande
France

xdeparis@pasteur-cayenne.fr

Ernest A. Gould
Centre for Ecology and Hydrology
Mansfield Road
Oxford
UK

eag@ceh.ac.uk

Tamara S. Gritsun
Centre for Ecology and Hydrology
Oxford
UK

tsg@ceh.ac.uk

Maria G. Guzman
Virology Department
PAHO/WHO Collaborating Center for the
Study of Dengue and its Vector
Pedro Kouri Tropical Medicine Institute
Havana City
Cuba

lupe@ipk.sld.cu

Olaf Isken
Institute for Cancer Research
Fox Chase Cancer Center
Philadelphia, PA
USA

Matthias Kalitzky
Institute of Medical Microbiology, Virology and Hygiene
Center of Clinical Pathology
University Hospital Hamburg-Eppendorf
Hamburg
Germany

mkalitzk@uke.uni-hamburg.de

Poonsook Keelapang
Department of Microbiology
Faculty of Medicine
Chiang Mai University
Chiang Mai
Thailand

Gustavo Kourí
Pedro Kouri Tropical Medicine Institute
Havana City
Cuba

Beate Kümmerer
Bernhard Nocht Institute for Tropical Medicine
Department of Virology
Hamburg
Germany

kuemmerer@bni.uni-hamburg.de

Young-Min Lee
Department of Microbiology
College of Medicine Chungbuk National University
Cheongju
South Korea

ymlee@chungbuk.ac.kr

Yoshihiro Makino
Division of Epidemiology
Department of Infectious Disease
Faculty of Medicine
Oita University
Oita
Japan

Prida Malasit
Medical Molecular Biology Center
Faculty of Medicine Siriraj Hospital
Mahidol University
Bangkok
Thailand

Darius Moradpour
Division of Gastroenterology and Hepatology
Centre Hospitalier Universitaire Vaudois
University of Lausanne
Lausanne
Switzerland

Ma Shao-Ping
Division of Epidemiology
Department of Infectious Disease
Faculty of Medicine
Oita University
Oita
Japan

masp@med.oita-u.ac.jp

Delfina Rosario
Virology Department
PAHO/WHO Collaborating Center for the Study of Dengue and its Vector
Pedro Kouri Tropical Medicine Institute
Havana City
Cuba

Nopporn Sittisombut
Department of Microbiology
Faculty of Medicine
Chiang Mai University
Chiang Mai
Thailand

nsittiso@mail.med.cmu.ac.th

Ryosuke Suzuki
Department of Virology II
National Institute of Infectious Diseases
Tokyo
Japan

Tetsuro Suzuki
Department of Virology II
National Institute of Infectious Diseases
Tokyo
Japan

tesuzuki@nih.go.jp

Kyuichi Tanikawa
International Institute for Liver Research
Kurume Research Center
Kurume
Japan

tanikawa@kurume.ktarn.or.jp

Andrew Tuplin
Centre for Ecology and Hydrology
Mansfield Road
Oxford
UK

Sang-Im Yun
Department of Microbiology
College of Medicine
Chungbuk National University
Cheongju
South Korea

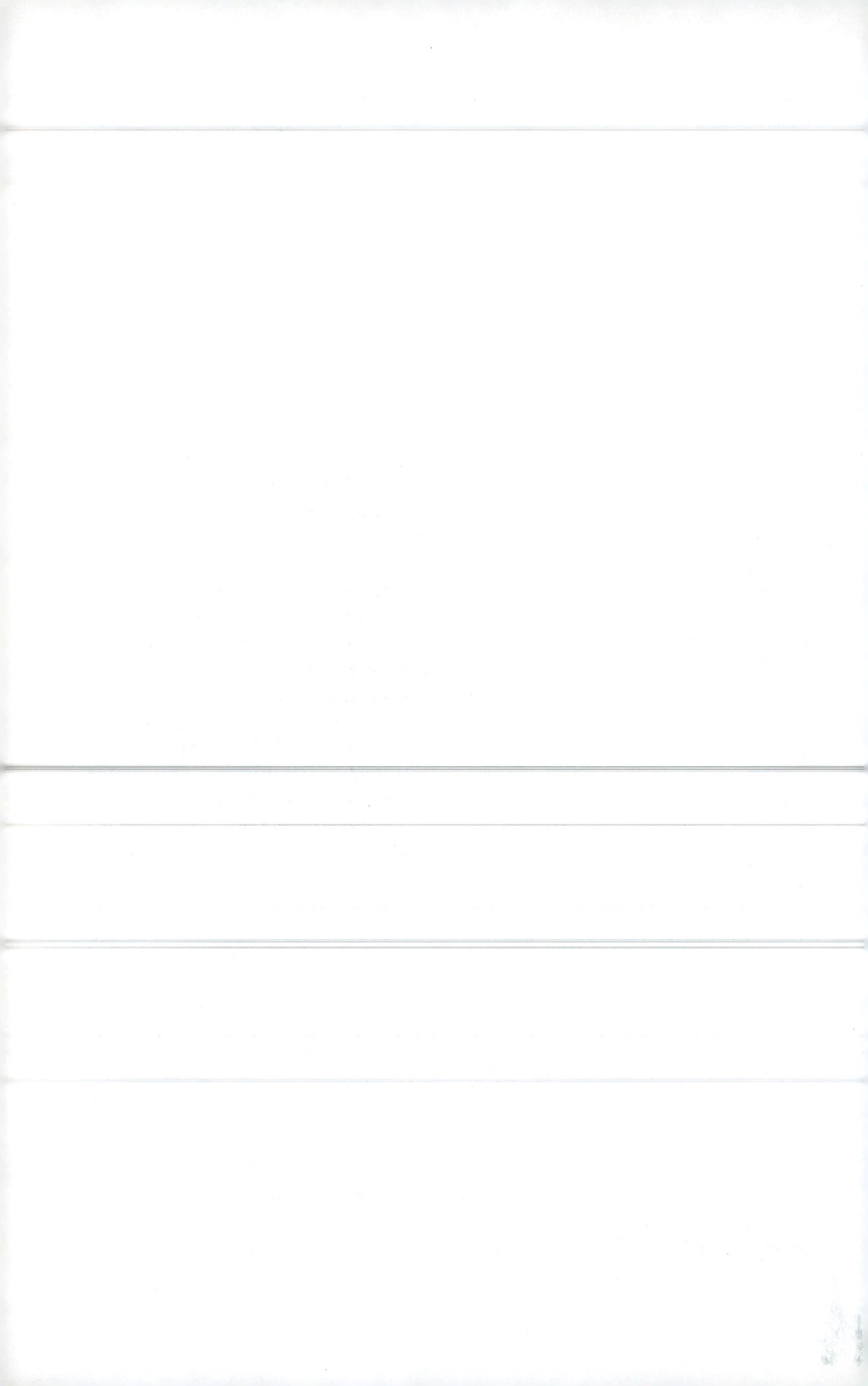

Preface

The Flaviviridae constitute an extremely interesting family of viruses. Whereas mosquito-borne flaviviruses such as dengue and yellow fever virus represent a challenge to health authorities worldwide, hepaciviruses, which are transmittable by a parenteral route, are responsible for a high percentage of cases of chronic hepatitis and hepatocellular carcinoma. In this way, infections by Flaviviridae contribute to a great extent to morbidity and mortality worldwide.

Our aim was primarily to collect together the latest knowledge about this family of viruses rather than to create another encyclopedic work on this subject. We did not want to limit the content of this book to molecular virologic issues but instead to combine both clinically relevant and scientifically new discoveries.

For this purpose, we have brought together an international community of experts, each one reporting on his or her field of expertise. The list of authors includes names such as Ernest Gould and Hubert Blum, known over the course of decades for their intensive scientific work.

This volume begins with a brief review of the molecular biology of yellow fever virus, followed by an introduction to the taxonomy and evolution of Flaviviridae. Next, our book focuses on three selected molecular genetic topics, before we come to two clinically important issues: the laboratory diagnosis of dengue virus, with a focus on the special circumstances in developing countries, and the pathogenesis of dengue hemorrhagic fever. This complication of dengue virus infection has for a long time been explained by the theory of "antibody-dependent enhancement" following secondary infections. Yet, this theory does not explain all hemorrhagic cases in dengue infection. In his review on dengue virus epidemics in the past, Xavier Deparis critically reflects this dogma.

Following two contributions from the Far East on vaccine development and the epidemiology of Japanese encephalitis virus, we elucidate on another major global health problem: the molecular biology, treatment and pathogenesis of hepatitis C virus.

As editors of this volume, we wish to thank everyone who has contributed to this book, in particular all of the authors, who originate from Cuba, Thailand, South Korea, Japan, France, Great Britain, the USA, and Germany, and also the staff at Horizon Scientific Press, who have continuously supported our work.

Hamburg, January 2006
Matthias Kalitzky, MD
Peter Borowski, MD

The Molecular Biology of Yellow Fever Virus

1

Beate M. Kümmerer

Abstract

Like all the members of the family Flaviviridae the yellow fever virus is an enveloped virus containing a single stranded RNA of positive polarity. After binding of the virion to the cell surface and receptor-mediated endocytosis the nucleocapsid is released into the cytoplasm where translation and RNA replication occurs. Translation results in a single polyprotein which is cleaved co- and posttranslationally by host cell and viral proteases to yield the viral proteins. The N-terminal part encodes the structural proteins C, prM, and E, whereas the remaining C-terminal two-thirds releases the nonstructural proteins NS1, NS2A, NS2B, NS3, NS4A, NS4B, and NS5. Specific functions have been attributed to most of the yellow fever viral proteins. Whereas the structural proteins are involved in virion formation, the nonstructural proteins are mainly involved in RNA replication. RNA replication is regulated leading to an asymmetric ratio of plus strand molecules to minus strand RNA molecules of 10:1. The genome RNAs assemble together with several copies of the viral capsid protein to form nucleocapsids that acquire a lipid bilayer envelope with virion glycoproteins by budding through intracellular membranes. Interestingly, recent studies also indicate a role of nonstructural proteins in infectious particle formation. Those as well as other studies describing functional analyses of yellow fever viral proteins have been performed using reverse genetics. An infectious cDNA clone has been described for the yellow fever virus vaccine strain 17D, which certainly is a great tool to gain further insights into the molecular biology of yellow fever virus.

Classification

The yellow fever virus (YFV) belongs to the Flaviviruses, which represent the prototype genus within the family Flaviviridae (Francki *et al.*, 1991). This virus family also includes the pestiviruses and hepatitis C virus. Most flaviviruses are arthropod-borne and multiply in both vertebrates and arthropods (Burke and Monath, 2001). In case of YFV, transmission occurs mainly by the mosquitoes *Aedes aegypti*. Isolation of the first YFV strain was performed in 1927 from a patient in Africa (Asibi strain) (Stokes *et al.*, 2001). Since then, much progress has been made in understanding the virion and genome structure, the viral life cycle, as well as functions of single viral proteins.

Virion morphology

The YF virions are spherical and enveloped with a diameter of 40 to 60 nm (Murphy, 1980). They possess a spherical nucleocapsid of approximately 30 nm consisting of a single-stranded positive-sense RNA genome and several copies of a small, basic capsid protein. The nucleocapsid is surrounded by a lipid envelope in which two envelope proteins are embedded: the envelope (E) protein and the membrane (M) protein or its precursor prM (Murphy, 1980). Intracellular virions, which only contain prM as well as released extracellular virions, which predominantly contain M have been described. In comparison with intracellular particles, extracellular particles have a greater infectivity than those remaining intracellular. For West Nile virus, the specific infectivity of intracellular virus was demonstrated to be 60-fold lower than the one of extracellular particles (Wengler and Wengler, 1989). Recently, the structure of immature, prM-containing YFV particles was determined to 25 Å resolution by cryoelectron microscopy and image reconstruction techniques (Zhang *et al.*, 2003). The structure showed 60 icosahedrally organized trimeric spikes on the particle surface, each spike consisting of three prM:E heterodimers. Furthermore, a third form of flaviviral particles, so called subviral particles, have been described (Kümmerer and Rice, 2002; Russell *et al.*, 1980). They consist of the envelope and the embedded envelope proteins but lack the nucleocapsid and are therefore not infectious. Expression of prM and E is sufficient for release of subviral particles, which also elicit protective immunity. Mice could be protected from lethal YFV encephalitis by recombinant vaccinia virus producing the prM and E proteins of YFV (Pincus *et al.*, 1992).

Viral life cycle

Binding of YFV to the cell surface is believed to be mediated by the E protein. A specific cellular receptor has not been identified yet for YFV. Most likely, uptake of the virions is mediated by receptor-mediated endocytosis into coated vesicles (Ishak *et al.*, 1988). Virions are later found in uncoated prelysosomal vesicles where the low pH environment results in an acid-catalyzed conformational change of the E protein into a fusogenic form. After fusion of the virion envelope with the endosomal membrane, the viral nucleocapsid is released into the cytoplasm of the cell where translation and RNA replication occurs (Lindenbach and Rice, 2001). In a first step, translation of a single polyprotein occurs, which is cleaved co- and posttranslationally into the single viral structural and nonstructural proteins by host cell and viral proteases. The viral RNA-dependent RNA polymerase assembles with other viral and most likely also host cell proteins to form an RNA replication complex. Starting from the incoming genomic plus strand RNA, synthesis of full-length negative strand RNA intermediates occurs, which then serve as template for the production of additional genomic plus-strand RNA molecules (Westaway, 1987). Plus stranded RNA is packaged into progeny virions and budding occurs through intracellular membranes (Ohyama *et al.*, 1977).

Genome structure

The genomic RNA of YFV is infectious. It consists of a single stranded RNA of positive polarity with a length of approximately 11 kb (Rice, 1985). The genome encodes one large polyprotein which is flanked by a short 5′ untranslated region (UTR) and a 3′ UTR (Figure 1.1) (Chambers *et al.*, 1990a). The 5′ UTR is about 120 nucleotides in length and the 3′ UTR comprises about 500 nucleotides. Similar to eukaryotic RNAs, the YF viral genome contains a 5′ cap structure. Initiation of protein translation presumably occurs by ribosome-scanning. The 3′ terminus is not polyadenylated. Rather, the 3′-terminal 90

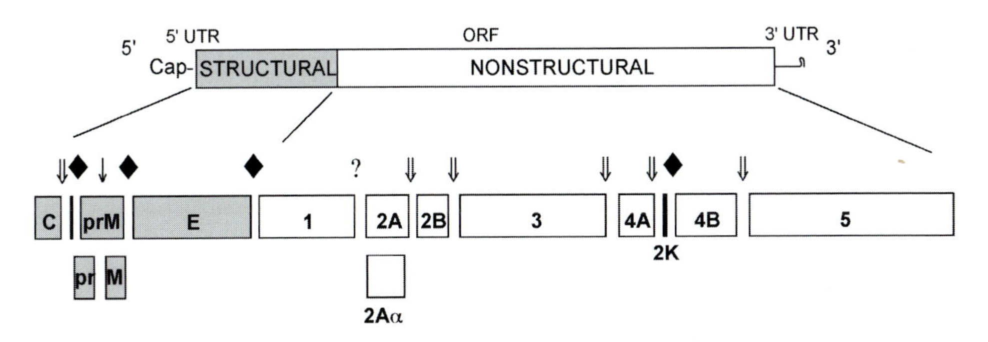

Figure 1.1 Schematic presentation of the yellow fever virus genome structure and polyprotein processing. Top, the YFV genome with the structural and nonstructural protein-coding region, the 5′- UTR with the 5′-cap structure and the 3′- UTR with the potential 3′ secondary structure are shown. Below, the mature YFV proteins generated by proteolytic processing of the polyprotein are demonstrated. Gray boxes represent the structural proteins, white boxes represent nonstructural proteins. In addition, two small hydrophobic fragments are cleaved from the polyprotein (black bars). Cleavage sites for viral serine protease (⇓), the host signalase (♦), furin (↓), or unknown proteases (?) are indicated.

nucleotides of the genome form a conserved secondary structure (Brinton *et al.*, 1986). One copy of a pseudoknotted structure was described for YFV, which is repeated in other mosquito-borne flaviviruses (Olsthoorn and Bol, 2001). In addition, YFV contains a conserved sequence element (5′ CS) in the 5′ region of the genome and two conserved regions (CS1 and CS2) in the 3′ terminal region. In contrast to other mosquito-borne flaviviruses, which contain two CS2 motifs, the YFV possesses only one. The conserved sequences 5′ CS and CS1, as well as the 3′ stem–loop structure, were found to be essential for virus replication, whereas the conserved sequence CS2 and a region containing YFV-specific repeated sequences were dispensable (Bredenbeek *et al.*, 2003). Fine mapping of the *cis*-acting nucleotide sequences within the 5′ region of the YFV genome revealed that 18 nucleotides (residues 146 to 163 of the YFV genome, which encode amino acids 9 to 14 of the capsid protein) are essential for replication of YFV and a slightly longer sequence (residues 146 to 166) is required for full replication (Corver *et al.*, 2003). This region is proposed to be involved in cyclization of YFV RNA and might play an important role in regulating replication, translation, and/or packaging.

Polyprotein processing

The single open reading frame encodes one large polyprotein of about 10 000 bases (Chambers *et al.*, 1990a). It is cleaved co- and posttranslationally by host cell and viral proteases to release the single viral proteins (Figure 1.1, Table 1.1). The structural proteins are encoded from the 5′-terminal quarter of the genome whereas the remaining two-thirds encode the nonstructural proteins. The order of the proteins within the polyprotein is NH_2-C-prM(M)-E-NS1-NS2A-NS2B-NS3-NS4A-NS4B-NS5-COOH. In addition, two small hydrophobic peptides are released from the polyprotein. One is derived from the C-terminus of the anchored capsid protein; after cleavage the mature capsid is released (Amberg *et al.*, 1994). The second one represents a small fragment between the NS4A and the NS4B protein and is called 2K based on its predicted size (Lin *et al.*, 1993).

The structural proteins prM and E as well as the following nonstructural protein NS1 are translocated into the endoplasmic

Table 1.1 Characteristics and functions of yellow fever virus proteins

Protein	MW	Modification	Cleavage site at N-terminus/ protease responsible for cleavage	Function/enzymatic activity
C	11 kD	–	[virC/anchC cleavage site: RR↓S]	Capsid protein Interaction with genomic RNA
prM/ M	26 kD 8 kD	Glycosylation	prM: TGG↓V Signalase M: SRR↓A Furin	Membrane protein (embedded into viral envelope)
E	53–54 kD	Glycosylation[a]	AYS↓A Signalase	Envelope protein Hemagglutination activity Mediates binding to cell surface
NS1	39–41 kD	Glycosylation	VGA↓D Signalase	RNA replication Role in neurovirulence
NS2A	24 kD	–	VTA↓G unknown	Role in production of infectious particles
NS2B	14 kD	–	RR↓S NS2B-3 protease	Cofactor of viral serine protease
NS3	69 kD	–	RR↓S NS2B-3 protease	Trypsine-like serine protease RNA helicase RNA triphosphatase
NS4A	16 kD	–	RR↓G NS2B-3 protease	Presumably role in RNA replication
NS4B	27 kD	Yes—nature of modification unknown	VAA↓N Signalase	Presumably role in RNA replication
NS5	103 kD	Phosphorylation	RR↓G NS2B-3 protease	RNA-dependent RNA polymerase Methyltransferase

[a]Dependent on YFV strain or substrain

reticulum. Cleavages that generate the N-termini of prM and E as well as the C-terminus of E are mediated by the host cell signal peptidase (Lindenbach and Rice, 2001). In contrast, the capsid protein remains in the cytoplasm. Processing to generate the C-terminus of the mature capsid protein is performed by the viral NS2B-3 protease (Amberg *et al.*, 1994). This cleavage is a prerequisite for efficient processing to generate the N-terminus of prM by the signal peptidase (Amberg and Rice, 1999). Therefore, mutations that abolish cleavage to produce the C-terminus of the mature capsid proteins also prevent the production of infectious particles (Amberg and Rice, 1999). Furthermore, mutations that enhance signalase cleavage to generate the N-Terminus of prM are lethal for virus production (Lee *et al.*, 2000).

The protease responsible for processing at the NS1-2A site is assumed to be localized in the endoplasmic reticulum but has not been identified yet. However, for dengue virus it is known that the eight last amino acids of NS1 are required for cleavage at the NS1-2A site (Hori and Lai, 1990; Pethel *et al.*, 1992). Processing in the remaining nonstructural protein region is mainly mediated by the NS2B-3 protease (Lindenbach and Rice, 2001). This viral serine protease is responsible for cleavage at the NS2A/2B, NS2B/3, NS3/4A, NS4A/2K, and NS4B/5 sites. For YFV, an additional cleavage site in the C-terminal region of NS2A has been described (NS2Aα site) (Kümmerer and Rice, 2002; Nestorowicz *et al.*, 1994). Similar to dengue virus, for which a minor cleavage within NS3 has been described (Arias *et al.*, 1993), an additional cleavage within NS3 might also be possible for YFV. The viral serine protease cleavage sites usually consist of two basic amino acids followed by an amino acid with a short side chain. In case of YFV the cleavage sites usually consist of RR↓G/S (Table 1.1) (Chambers *et al.*, 1989; Lindenbach and Rice, 2001; Rice *et al.*, 1986). Little variation is observed for the NS2Aα site (QK↓T) and the NS4A-2K site (QR↓S). In contrast to the majority of cleavage events in the NS region, the N-terminus of NS4B is generated by host cell signal peptidase, but prior cleavage at the NS2A/2K site is required (Lin *et al.*, 1993). Other than the case just mentioned, processing at other sites within the NS region does not take place in an obligate order.

The structural proteins

The capsid (C) protein

The C protein is a small, basic protein of about 11 kDa in size. Owing to the basic character, C binds strongly to RNA. Together with the viral RNA several copies of the C protein form the nucleocapsid. Analysis of purified C protein expressed in *Escherichia coli* revealed that it is largely alpha-helical and forms dimers (Jones *et al.*, 2003).

The membrane (prM/M) protein

The prM protein has a size of about 26 kDa and represents the precursor of the mature M protein (8 kDa). The "pr" segment of YFV contains three potential N-glycosylation sites. However, it is suggested that glycosylation only occurs at two sites (Chambers *et al.*, 1990a). The "pr" portion also contains conserved cysteine residues, which form intramolecular disulfide bridges. The C-terminus of prM/M contains a transmembrane domain, which serves as a endoplasmic reticulum localization signal and plays a role in the formation of YFV envelope (Op De Beeck *et al.*, 2003; Op De Beeck *et al.*, 2004). It is assumed that prM forms heterooligomeric complexes with the E protein in intracellular particles thereby preventing E to undergo an irreversible acid catalyzed conformational change to its fusogenic form (Heinz *et al.*, 1994). This allows the transport of the virions to the cell surface through acidic compartments. During virus release cleavage of prM to M occurs (Guirakhoo *et al.*, 1991). M first remains in a nonfusogenic state that is transformed to a fusion-competent form after binding, entry, and exposure to an acidic environment.

The envelope (E) protein

The E protein represents the major envelope protein of the YFV. It comprises hemagglutination activity and is the primary target for neutralizing antibodies. It has a predicted molecular weight (MW) of 53 to 54 kDa and is a type I membrane protein with a C-terminal membrane anchor (Lindenbach and Rice, 2001). As described for the prM/M protein, the transmembrane domain serves as an endoplasmic reticulum localization signal and is involved in the formation of the envelope (Op De Beeck *et al.*, 2003; Op De Beeck *et al.*, 2004). Glycosylation of the E protein occurs in some but not all YFV strains or substrains. For example, in contrast to the Asibi strain and the vaccine strain 17D-204, 17DD and 17D-213 contain potential N-glycosylation sites at amino acid residues 153 and 151, respectively (Post *et al.*, 1992). Glycosylated E protein is also used to assemble YFV vaccine virions (Post *et al.*, 1992). The ectodomain of E contains 12 conserved cysteine residues. As shown for West Nile virus, they form intramolecular disulfide bridges (Nowak and Wengler, 1987). For the tick-borne encephalitis virus E protein, the X-ray structure of the E-protein ectodomain was determined (Rey *et al.*, 1995). In analogy to the data obtained for the tick-borne encephalitis virus, it can be assumed that the ectodomain of the YFV E-protein also consists of three structural domains in a head-to-tail homodimeric unit. Genetic analyses have also demonstrated that E is an important determinant of YFV virulence and viscerotropism (McArthur *et al.*, 2003; Nickells and Chambers, 2003).

The nonstructural proteins

NS1

The NS1 protein has a predicted MW of 39 to 41 kDa and contains additional modifications by N-glycosylation, which is unique among the flavivirus NS-proteins (Post *et al.*, 1991). NS1 can be found intracellular and on the cell surface. In addition, it is secreted into the extracellular fluid from mammalian but not from insect cells (Post *et al.*, 1991). The intracellular NS1 protein forms homodimers, which are detergent-resistant as well as heat- and acid-labile (Winkler *et al.*, 1988). For other flaviviruses, namely the dengue virus and the tick-borne encephalitis virus, it was shown that the NS1 protein is secreted as a hexamer or other oligomeric forms (Crooks *et al.*, 1994; Flamand *et al.*, 1999). It can be assumed that this is also the case for secreted YFV NS1. In addition, a protein containing NS1 and a portion of NS2A can be found within YFV infected cells (Chambers *et al.*, 1990b).

Analysis of mutants which lack the N-linked glycosylation sites of YFV NS1 protein suggest a role of NS1 in flavivirus RNA replication and pathogenesis (Muylaert *et al.*, 1996). Further genetic analyses involving clustered charged-amino-acid-to-alanine scanning mutations of YFV NS1 resulted in the identification of a temperature-sensitive mutation which blocks RNA accumulation (Muylaert *et al.*, 1997). A role of NS1 in RNA replication was later supported by the analysis of an YFV NS1 deletion mutant. The latter analysis suggests a role for NS1 before or at the initiation of minus-strand RNA synthesis (Lindenbach and Rice, 1997). Further genetic studies demonstrated that YFV NS1 can be complemented in *trans* and that an interaction between NS1 and NS4A (either direct or indirect) is necessary for replicase function (Lindenbach and Rice, 1997; Lindenbach and Rice, 1999).

Furthermore, antibodies to NS1 with complement fixing activity are produced after flaviviral infections (Russell *et al.*, 1980). After active immunization with YFV NS1 or passive immunization with anti-YFV NS1 monoclonal antibodies animals are protected against challenge with YFV (Cane and Gould, 1988; Gould *et al.*, 1986; Schlesinger *et al.*, 1986; Schlesinger *et al.*, 1985). Such protection may occur via antibody-dependent complement-mediated lysis of virus-infected cells expressing NS1 on the surface (Schlesinger *et al.*, 1990). In addition, HLA-B35-restricted CTL epitopes were found on NS1 following YFV vaccination (Co *et al.*, 2002).

NS2A

NS2A is a small hydrophobic protein of about 24 kDa. It contains an additional serine protease-dependent cleavage site, which results in the release of a C-terminal truncated NS2A product of about 22 kDa (NS2Aα) (Nestorowicz *et al.*, 1994). Mutations at this cleavage site region block the production of infectious particles while the release of subviral particles remains unimpaired (Kümmerer and Rice, 2002). Interestingly, not the inhibition of the processing event at the NS2Aα site but the identity of the amino acids at the NS2Aα site seem to be important for this block. For another flavivirus, the Kunjin virus, it was demonstrated that NS2A localizes to presumed sites of RNA replication and that it binds to NS3, NS5 and the 3′ UTR (Mackenzie *et al.*, 1998). This data suggest that NS2A might help to localize viral RNA to the membrane-bound replicase complex.

NS2B

The N-terminus of NS2B, a protein of about 14 kDa, was determined by N-terminal amino acid sequencing (Chambers *et al.*, 1989). It represents the cofactor of the viral trypsine-like serine protease and forms a stable complex with NS3 (Chambers *et al.*, 1993). NS2A contains hydrophobic segments, which mediate association to membranes. To study the role of specific regions of YFV NS2B in processing and NS3 interaction, a series of mutations were introduced in the hydrophobic regions and in a central conserved hydrophilic region (Chambers *et al.*, 1993). Whereas mutations in the hydrophobic regions had subtle effects on proteolytic processing, mutations within the conserved domain dramatically reduced or even abolished cleavage. The conserved hydrophilic domain of NS2B is also important for the complex formation with NS3 (Chambers *et al.*, 1993).

NS3

The NS3 protein has a MW of about 69 kDa. It contains at least three enzymatic activities: a trypsine-like serine protease, an RNA helicase, and an RNA triphosphatase activity (Lindenbach and Rice, 2001). Together with its cofactor NS2B, it forms an active NS2B-3 proteinase complex. The active center of the protease is localized in the N-terminal part of NS3. For YFV the catalytic triad is composed of the NS3 residues histidine-53, aspartic acid-77, and serine-138 (Chambers *et al.*, 1990c). As mentioned earlier, the NS2B-3 protease mediates the majority of cleavages within the NS protein region (2A/2B, 2B/3, 3/4A, 4A/2K, and 4B/5) as well as one cleavage within the structural protein region (virC/anchC) (Lindenbach and Rice, 2001). Furthermore, it mediates additional processing within NS2A and NS3. The cleavage sites recognized by the serine protease usually consist of two basic amino acids followed by an amino acid with a short side chain [(R/Q)-(R/K)↓(G/S/T) in case of YFV]. The determinants of cleavage site specificity of the YFV protease have been studied using site directed-mutagenesis at the NS2B-3 cleavage site (Chambers *et al.*, 1995). At the P1 position, significant levels of cleavage were only retained when the R residue was exchanged to K. At the P1′ and P2 positions, conservative and nonconservative substitutions were tolerated resulting in intermediate levels of cleavage. Exchanges at the P3 and P4 positions had no effects on cleavage efficiencies. Similar results were obtained by mutational analysis of the NS2A/2B cleavage site (Nestorowicz *et al.*, 1994). In addition, substitutions that abolished cleavage did not result in production of infectious virus (Chambers *et al.*, 1995; Nestorowicz *et al.*, 1994).

The C-terminal portion of NS3 contains RNA helicase motifs, suggesting that NS3 plays a direct role in RNA replication (Gorbalenya *et al.*, 1989). The RNA helicase uses the energy from ATP hydrolysis to help to dissociate nascent RNA strands from their template. In addition, the C-terminal part also contains an RNA triphosphatase activity, which most likely is involved in the formation of the 5′-terminal cap structure (Wengler and Wengler, 1993).

NS4A

Little is known about the hydrophobic NS4A protein with a size of approximately 16 kDa. Genetic studies indicated that NS4A interacts with NS1 and that this interaction is important for RNA replication (Lindenbach and Rice, 1997; Lindenbach and Rice, 1999). This suggestion was made based on studies involving an YFV NS1 deletion mutant that requires YFV NS1 in *trans*. In contrast, DEN NS1 was only able to complement in *trans* after a suppressor mutant in YFV NS4A arose (Lindenbach and Rice, 1997). Involvement of NS4A in RNA replication is supported by findings for Kunjin virus, which describe a partial colocalization of NS4A with double-stranded RNA (Mackenzie *et al.*, 1998).

NS4B

NS4B has a predicted MW of 27 kDa. The N-terminus of YFV NS4B was determined by N-terminal amino acid sequencing (Chambers *et al.*, 1989). Owing to its hydrophobic character, it is associated to membranes. Using proteinase K protection analyses of the *in vitro*-transcribed products, a model for the YFV NS4B topology was established (Lin *et al.*, 1993). Studies with other flaviviruses demonstrated, that NS4B initially appears as a 30 kDa protein that decreases to approximately 28 kDa (Preugschat and Strauss, 1991). However, the nature of this posttranslational modification is not known. In addition, NS4B colocalizes with double-stranded RNA at putative sites of viral RNA replication (Westaway *et al.*, 1997).

NS5

With a predicted MW of 103 kDa, NS5 is the largest protein among the flavivirus proteins. It is highly conserved among flaviviruses and contains a glycine-glycine-aspartic acid [(Gly-Gly-Asp (GDD)] motif, which is present in all RNA-dependent RNA polymerases (RDRP). The RNA-dependent RNA polymerase activity was demonstrated for recombinant dengue type 1 virus NS5 expressed in *Escherichia coli* (Tan *et al.*, 1996). It is suggested that the error rate of the YFV RNA-dependent RNA polymerase is as low as $1.9 \times 10(-7)$ to $2.3 \times 10(-7)$ (Pugachev *et al.*, 2004).

The N-terminal part of NS5 (between residues 60 and 145) contains a sequence element that is homologous to methyltransferases (Koonin, 1993). The methyltransferase activity is probably involved in methylation of the 5′ cap structure. Mutations destroying the GDD or methyltransferase motif are lethal for RNA replication (Khromykh *et al.*, 1998). For Kunjin virus it could be shown that those lethal mutations can be complemented by providing NS5 in trans (Khromykh *et al.*, 1998). In addition, YF NS5 is phosphorylated by serine/threonine kinases (Reed *et al.*, 1998). Using monoclonal antibodies,

YF NS5 could also be detected in the nuclei of infected cells (Buckley *et al.*, 1992). This could also be shown for other flaviviruses. In case of dengue virus, two electrophoretically separable phosphorylated species were detected but only the slower migrating form localized within the nucleus (Kapoor *et al.*, 1995). A nuclear localization signal was identified for dengue virus NS5 between the methyltransferase and the RDRP motifs (Forwood *et al.*, 1999). Most likely YFV contains a corresponding signal.

RNA replication

In a first step, the incoming genomic RNA of positive orientation is translated into a polyprotein, which is cleaved to release the single viral NS proteins. The NS proteins assemble to form an RNA replication complex. After synthesis of the complementary RNA strand of negative polarity, further plus stranded RNA molecules are generated (Brinton, 1986; Westaway, 1987). These genomic RNAs serve either as template for the production of further negative strands or they are packaged into progeny virus. The replication of full-length RNA occurs via a semiconservative mechanism. Replicative forms (RFs), which consist of duplex molecules, as well as replicative intermediates (RIs), which contain double-stranded regions and nascent single-stranded RNA, can be detected (Chu and Westaway, 1985; Cleaves *et al.*, 1981). Since YFV RNA replication is mediated by the RNA-dependent RNA polymerase, viral RNA synthesis can be demonstrated by metabolic labeling in the presence of actinomycin D, an inhibitor of DNA-dependent RNA polymerase. Although the mechanism is not know, it is assumed that RNA replication is regulated since an asymmetric ratio of plus stranded molecules to minus stranded molecules of 10:1 can be detected (Cleaves *et al.*, 1981; Lindenbach and Rice, 1997). Replication of subgenomic replicons indicated that the structural proteins are dispensable for replication. However, a small portion of the structural protein coding region has to be retained for autonomous replication (Jones *et al.*, 2005; Molenkamp *et al.*, 2003). This region, which contains the RNA cyclization sequence (Hahn *et al.*, 1987b), is located within the coding sequence of the first 20 codons of the capsid protein. The importance of this region for genome replication was mapped by deletion analysis in YFV (Corver *et al.*, 2003).

Vaccine

The attenuated vaccine strain 17D was obtained by serial passage of the Asibi strain on chicken embryo cells (Theiler and Smith, 1937). Although some fatal adverse events temporally associated with YFV vaccination were reported (Galler *et al.*, 2001; Martin *et al.*, 2001), YFV 17D is still considered to be one of the most successful vaccines developed to date. After a single immunizing dose, the vaccine elicits long-lasting or perhaps even lifelong protective immunity. To date, the molecular basis for the attenuation is not known. Comparison of the nucleotide sequence of the genome from two YFV 17D substrains identified 48 nucleotide sequence differences, which are 17D strain-specific and potentially related to viral attenuation (dos Santos *et al.*, 1995). The 48 changes are scattered throughout the genome; 26 of the changes are silent, whereas the other 22 lead to amino acid substitutions (dos Santos *et al.*, 1995). The importance of those mutations with regard to attenuation still needs to be analyzed. The availability of an infectious cDNA clone of 17D (see below) represents a great tool to introduce single amino acid changes into the YF 17D genome and to study their impact on attenuation.

Reverse genetics

The complete YFV sequences of the virulent Asibi strain as well as the vaccine strain 17D were determined in 1987 (Hahn *et al.*, 1987a). Two years later, the first recombinant YFV was recovered from cDNA for YF 17D (Rice *et al.*, 1989). One main problem in constructing infectious flavivirus cDNA clones is the instability of full-length cDNA clones and their toxic effects on *Escherichia coli*. To circumvent this problem, the first infectious YFV RNA was recovered from a full-length template generated by *in vitro* ligation of two appropriate restriction fragments (Rice *et al.*, 1989). Later, the use of low copy vectors facilitated the generation of stable flavivirus full-length clones. In case of YFV, the complete cDNA was cloned under control of a bacteriophage SP6 promoter in the low copy vector pACYC177 (Figure 1.2) (Bredenbeek *et al.*, 2003). After linearization of the full-length cDNA at the 3′ end of the genome, generation of infectious RNA can be performed by run-off transcription (Figure 1.2) (Bredenbeek *et al.*, 2003). Since the YFV genome is capped, cap analog needs to be included in the *in vitro* transcription reaction to yield 5′ capped transcripts. The *in vitro* transcribed RNA can be used directly for transfection. Alternatively, the template DNA can be removed by treatment with RNase free DNase. This procedure might be required when transfected template DNA might interfere with the detection of low levels of replication. Transfection in eukaryotic cells can be performed using DEAE dextran, cationic liposomes such as Lipofection or electroporation (Ruggli and Rice, 1999). Usually, electroporation yields the highest transfection efficiencies. In case of *in vitro* transcribed YFV RNA, electroporation is mainly performed into BHK cells resulting in a specific infectivity of about 1×10^6 plaque forming units per μg of *in vitro* transcribed RNA.

With the aid of an infectious YFV cDNA clone, requirements for RNA replication could be analyzed. It was demonstrated that the NS3 serine protease is essential for virus replication (Chambers *et al.*, 1990c). Also mutations that inhibit processing at certain NS2B-3-dependent cleavage sites abolish RNA replication (Chambers *et al.*, 1995; Nestorowicz *et al.*, 1994). In addition, functional analyses of YFV proteins can be performed. Mutations are often introduced by site directed mutagenesis and the impact of those mutations on polyprotein processing, RNA replication, viral growth and virulence can be determined allowing conclusions on protein function. Using this technology, it could be demonstrated that mutations of the N-linked glycosylation sites of the YFV NS1 protein effects virus replication and mouse neurovirulence (Muylaert *et al.*, 1996). Mutations in the YFV NS2A protein selectively inhibit production of infectious particles suggesting a role of NS2A in production and/or release of infectious particles (Kümmerer and Rice, 2002).

Besides functional analyses of flaviviral proteins, the availability of an infectious YFV cDNA clone of the 17D vaccine strain also enables the possibility of vaccine production (Rice, 1990). Currently, the vaccine is produced with the use of a seed lot method to minimize batch-to-batch variability. A plasmid containing functional YFV 17D cDNA can now serve as source for vaccine production (Duarte dos Santos *et al.*, 1995; Marchevsky *et al.*, 1995). In addition, the YFV cDNA clone is used as backbone for the production of live-atten-

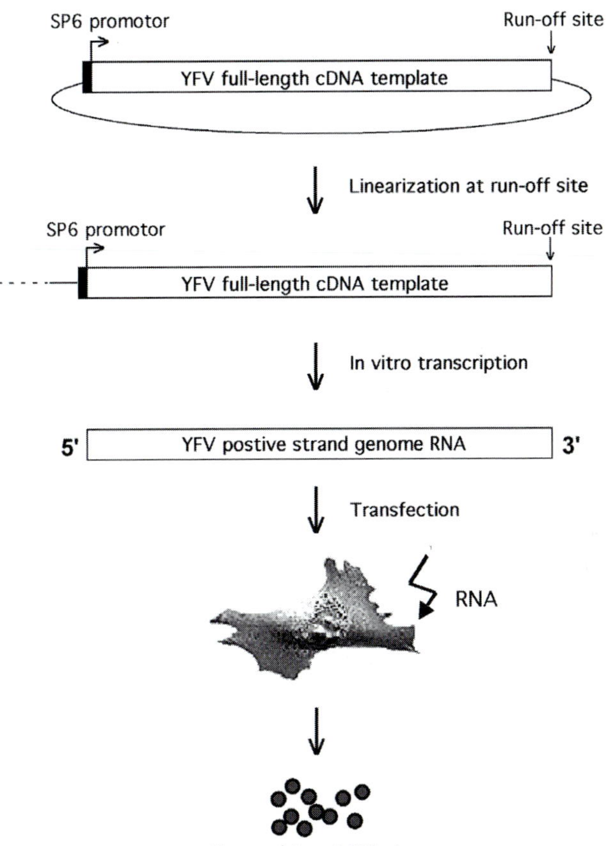

Figure 1.2 Schematic presentation of the recovery of recombinant yellow fever virus from an infectious cDNA clone. See text for further explanation and details.

uated chimeric viruses (ChimeriVax vaccines) (Monath *et al.*, 1999). They are established by replacing the prM and E genes of YFV with the corresponding genes of other flaviviruses. The ChimeriVax vaccine technology was applied for several flaviviruses including dengue virus (DEN)1–4, Japanese encephalitis virus (JE) and West Nile virus (WN) (Arroyo *et al.*, 2004; Guirakhoo *et al.*, 2000; Monath *et al.*, 2000; Monath *et al.*, 2002; Monath *et al.*, 1999). For ChimeriVax-JE, phase II trials have already been performed (Jones, 2003; Monath *et al.*, 2003). ChimeriVax-DEN has been tested so far in nonhuman primates (Guirakhoo *et al.*, 2004a; Guirakhoo *et al.*, 2000; Guirakhoo *et al.*, 2004b; Jones, 2004) and for ChimeriVax-WN, preclinical studies in mice and macaques were performed (Arroyo *et al.*, 2004).

Besides creating candidate vaccines against other flavivirus-mediated diseases, the infectious full-length clone of 17D has also been used to deliver epitopes from unrelated microbial pathogens. The immunodominant protective B cell of the circumsporozoite (CS) protein of the human malaria parasite, *Plasmodium falciparum*, was inserted into the envelope protein of 17D (Bonaldo *et al.*, 2002). The genomic stability of that recombinant virus, named 17D/8, was confirmed by serial *in vitro* passages. Immunization of mice with 17D/8 led to long-lasting production of antibodies to *P. falciparum* sporozoites. In addition, the H-2K(d) restricted CTL epitope of the circumsporozoite protein (CS) of *Plasmodium yoelii*, a rodent malaria parasite, was expressed using the infectious 17D cDNA clone (Tao *et al.*, 2005). In this case, the epitope was inserted between proteins NS2B and NS3. The efficiency of protection could be demonstrated upon sporozoite challenge of immunized mice.

Conclusions

Since the discovery of the YFV much progress has been made on the molecular level of this virus. However, certain steps within the flaviviral life cycle still need to be analyzed in further detail. For example, the cellular receptor for YFV still needs to be determined and little is also known about the mechanism of packaging and assembly. In addition, the role of certain proteins in viral pathogenesis and viral-host cell interaction needs to be elucidated in further detail. The availability of an infectious full-length clone of YFV certainly represents a great tool to gain further insights into this import virus.

References

Amberg, S. M., and Rice, C. M. (1999). Mutagenesis of the NS2B-NS3-mediated cleavage site in the flavivirus capsid protein demonstrates a requirement for coordinated processing. J. Virol. *73*, 8083–8094.

Amberg, S. M., Nestorowicz, A., McCourt, D. W., and Rice, C. M. (1994). NS2B-3 proteinase-mediated processing in the yellow fever virus structural region: *in vitro* and *in vivo* studies. J. Virol. *68*, 3794–3802.

Arias, C. F., Preugschat, F., and Strauss, J. H. (1993). Dengue 2 virus NS2B and NS3 form a stable complex that can cleave NS3 within the helicase domain. Virology *193*, 888–899.

Arroyo, J., Miller, C., Catalan, J., Myers, G. A., Ratterree, M. S., Trent, D. W., and Monath, T. P. (2004). ChimeriVax-West Nile virus live-attenuated vaccine: preclinical evaluation of safety, immunogenicity, and efficacy. J. Virol. *78*, 12497–12507.

Bonaldo, M. C., Garratt, R. C., Caufour, P. S., Freire, M. S., Rodrigues, M. M., Nussenzweig, R. S., and Galler, R. (2002). Surface expression of an immunodominant malaria protein B cell epitope by yellow fever virus. J. Mol. Biol. *315*, 873–885.

Bredenbeek, P. J., Kooi, E. A., Lindenbach, B., Huijkman, N., Rice, C. M., and Spaan, W. J. (2003). A stable full-length yellow fever virus cDNA clone and the role of conserved RNA elements in flavivirus replication. J. Gen. Virol. *84*, 1261–1268.

Brinton, M. A. (1986). Replication of flaviviruses. In The Togaviridae and Flaviviridae, S. Schlesinger, and M. J. Schlesinger, eds. (New York, Plenum Press), pp. 327–365.

Brinton, M. A., Fernandez, A. V., and Amato, J. (1986). The 3′-nucleotides of flavivirus genomic RNA form a conserved secondary structure. Virology *153*, 113–121.

Buckley, A., Gaidamovich, S., Turchinskaya, A., and Gould, E. A. (1992). Monoclonal antibodies identify the NS5 yellow fever virus non-structural protein in the nuclei of infected cells. J. Gen. Virol. *73 (Pt 5)*, 1125–1130.

Burke, D. S., and Monath, T. P. (2001). Flaviviruses. In Fields Virology, D. M. Knipe, and P. M. Howley, eds. (Philadelphia, Lippincott-Raven Publishers), pp. 1043–1125.

Cane, P. A., and Gould, E. A. (1988). Reduction of yellow fever virus mouse neurovirulence by immunization with a bacterially synthesized non-structural protein (NS1) fragment. J. Gen. Virol. *69*, 1241–1246.

Chambers, T. J., McCourt, D. W., and Rice, C. M. (1989). Yellow fever virus proteins NS2A, NS2B, and NS4B: identification and partial N-terminal amino acid sequence analysis. Virology *169*, 100–109.

Chambers, T. J., Hahn, C. S., Galler, R., and Rice, C. M. (1990a). Flavivirus genome organization, expression, and replication. Annu. Rev. Microbiol. *44*, 649–688.

Chambers, T. J., McCourt, D. W., and Rice, C. M. (1990b). Production of yellow fever virus proteins in infected cells: Identification of discrete polyprotein species and analysis of cleavage kinetics using region-specific polyclonal antisera. Virology *177*, 159–174.

Chambers, T. J., Weir, R. C., Grakoui, A., McCourt, D. W., Bazan, J. F., Fletterick, R. J., and Rice, C. M. (1990c). Evidence that the N-terminal domain of nonstructural protein NS3 from yellow fever virus is a serine protease responsible for site-specific cleavages in the viral polyprotein. Proc. Natl. Acad. Sci. USA *87*, 8898–8902.

Chambers, T. J., Nestorowicz, A., Amberg, S. M., and Rice, C. M. (1993). Mutagenesis of the yellow fever virus NS2B protein: Effects on proteolytic processing, NS2B-NS3 complex formation, and viral replication. J. Virol. *67*, 6797–6807.

Chambers, T. J., Nestorowicz, A., and Rice, C. M. (1995). Mutagenesis of the yellow fever virus NS2B/3 cleavage site: Determinants of cleavage site specificity and effects on polyprotein processing and viral replication. J. Virol. *69*, 1600–1605.

Chu, P. W. G., and Westaway, E. G. (1985). Replication strategy of Kunjin virus: Evidence for recycling role of replicative form RNA as template in semiconservative and asymmetric replication. Virology *140*, 68–79.

Cleaves, G. R., Ryan, T. E., and Schlesinger, R. W. (1981). Identification and characterization of type 2 dengue virus replicative intermediate and replicative form RNAs. Virology *111*, 73–83.

Co, M. D., Terajima, M., Cruz, J., Ennis, F. A., and Rothman, A. L. (2002). Human cytotoxic T lymphocyte responses to live attenuated 17D yellow fever vaccine: identification of HLA-B35-restricted CTL epitopes on nonstructural proteins NS1, NS2b, NS3, and the structural protein E. Virology *293*, 151–163.

Corver, J., Lenches, E., Smith, K., Robison, R. A., Sando, T., Strauss, E. G., and Strauss, J. H. (2003). Fine mapping of a cis-acting sequence element in yellow fever virus RNA that is required for RNA replication and cyclization. J. Virol. *77*, 2265–2270.

Crooks, A. J., Lee, J. M., Easterbrook, L. M., Timofeev, A. V., and Stephenson, J. R. (1994). The NS1 protein of tick-borne encephalitis virus forms multimeric species upon secretion from the host cell. J. Gen. Virol. *75*, 3453–3460.

dos Santos, C. N., Post, P. R., Carvalho, R., Ferreira, II, Rice, C. M., and Galler, R. (1995). Complete nucleotide sequence of yellow fever virus vaccine strains 17DD and 17D-213. Virus Res. 35, 35–41.

Duarte dos Santos, C. N., Post, P. R., Carvalho, R., Ferreira, I. I., Rice, C. M., and Galler, R. (1995). Complete nucleotide sequence of yellow fever virus vaccine strains 17DD and 17D-213. Virus Res. 35, 35–41.

Flamand, M., Megret, F., Mathieu, M., Lepault, J., Rey, F. A., and Deubel, V. (1999). Dengue virus type 1 nonstructural glycoprotein NS1 is secreted from mammalian cells as a soluble hexamer in a glycosylation-dependent fashion. J. Virol. *73*, 6104–6110.

Forwood, J. K., Brooks, A., Briggs, L. J., Xiao, C. Y., Jans, D. A., and Vasudevan, S. G. (1999). The 37-amino-acid interdomain of dengue virus NS5 protein contains a functional NLS and inhibitory CK2 site. Biochem. Biophys. Res. Commun. *257*, 731–737.

Francki, R. I. B., Fauquet, C. M., Knudson, D. L., and Brown, F. (1991). Classification and nomenclature of viruses: Fifth report of the international committee on taxonomy of viruses. Arch. Virol. *Suppl. 2*, 223.

Galler, R., Pugachev, K. V., Santos, C. L., Ocran, S. W., Jabor, A. V., Rodrigues, S. G., Marchevsky, R. S., Freire, M. S., Almeida, L. F., Cruz, A. C., *et al.* (2001). Phenotypic and molecular analyses of yellow fever 17DD vaccine viruses associated with serious adverse events in Brazil. Virology *290*, 309–319.

Gorbalenya, A. E., Koonin, E. V., Donchenko, A. P., and Blinov, V. M. (1989). Two related superfamilies of putative helicases involved in replication, recombination, repair and expression of DNA and RNA genomes. Nucleic Acids Res. *17*, 4713–4729.

Gould, E. A., Buckley, A., Barrett, A. D. T., and Cammack, N. (1986). Neutralizing (54K) and non-neutralizing (54K and 48K) monoclonal antibodies against structural and nonstructural yellow fever virus proteins confer immunity in mice. J. Gen. Virol. *67*, 591–595.

Guirakhoo, F., Heinz, F. X., Mandl, C. W., Holzmann, H., and Kunz, C. (1991). Fusion activity of flaviviruses: comparison of mature and immature (prM-containing) tick-borne encephalitis virions. J. Gen. Virol. *72*, 1323–1329.

Guirakhoo, F., Weltzin, R., Chambers, T. J., Zhang, Z. X., Soike, K., Ratterree, M., Arroyo, J., Georgakopoulos, K., Catalan, J., and Monath, T. P. (2000). Recombinant chimeric yellow fever-dengue type 2 virus is immunogenic and protective in nonhuman primates. J. Virol. *74*, 5477–5485.

Guirakhoo, F., Pugachev, K., Zhang, Z., Myers, G., Levenbook, I., Draper, K., Lang, J., Ocran, S., Mitchell, F., Parsons, M., *et al.* (2004a). Safety and efficacy of chimeric yellow fever-dengue virus tetravalent vaccine formulations in nonhuman primates. J. Virol. *78*, 4761–4775.

Guirakhoo, F., Zhang, Z., Myers, G., Johnson, B. W., Pugachev, K., Nichols, R., Brown, N., Levenbook, I., Draper, K., Cyrek, S., *et al.* (2004b). A single amino acid substitution in the envelope protein of chimeric yellow fever-dengue 1 vaccine virus reduces neurovirulence for suckling mice and viremia/viscerotropism for monkeys. J. Virol. *78*, 9998–10008.

Hahn, C. S., Dalrymple, J. M., Strauss, J. H., and Rice, C. M. (1987a). Comparison of the virulent Asibi strain of yellow fever virus with the 17D vaccine strain derived from it. Proc. Natl. Acad. Sci. USA *84*, 2019–2023.

Hahn, C. S., Hahn, Y. S., Rice, C. M., Lee, E., Dalgarno, L., Strauss, E. G., and Strauss, J. H. (1987b). Conserved elements in the 3′ untranslated region of flavivirus RNAs and potential cyclization sequences. J. Mol. Biol. *198*, 33–41.

Heinz, F. X., Stiasny, K., Puschner-Auer, G., Holzmann, H., Allison, S. L., Mandl, C. W., and Kunz, C. (1994). Structural changes and functional control of the tick-borne encephalitis virus glycoprotein E by the heterodimeric association with protein prM. Virology *198*, 109–117.

Hori, H., and Lai, C.-J. (1990). Cleavage of dengue virus NS1-NS2A requires an octapeptide sequence at the C terminus of NS1. J. Virol. *64*, 4573–4577.

Ishak, R., Tovey, D. G., and Howard, C. R. (1988). Morphogenesis of yellow fever virus 17D in infected cell cultures. J. Gen. Virol. *69*, 325–335.

Jones, T. (2003). ChimeriVax-JE. Acambis. Curr. Opin. Investig. Drugs *4*, 1019–1022.

Jones, T. (2004). Technology evaluation: ChimeriVax-DEN, Acambis/Aventis. Curr. Opin. Mol. Ther. *6*, 443–450.

Jones, C. T., Ma, L., Burgner, J. W., Groesch, T. D., Post, C. B., and Kuhn, R. J. (2003). Flavivirus capsid is a dimeric alpha-helical protein. J. Virol. *77*, 7143–7149.

Jones, C. T., Patkar, C. G., and Kuhn, R. J. (2005). Construction and applications of yellow fever virus replicons. Virology *331*, 247–259.

Kapoor, M., Zhang, L., Ramachandra, M., Kusukawa, J., Ebner, K. E., and Padmanabhan, R. (1995). Association between NS3 and NS5 proteins of dengue virus type 2 in the putative RNA replicase is linked to differential phosphorylation of NS5. J. Biol. Chem. *270*, 19100–19106.

Khromykh, A. A., Kenney, M. T., and Westaway, E. G. (1998). trans-Complementation of flavivirus RNA polymerase gene NS5 by using Kunjin virus replicon-expressing BHK cells. J. Virol. *72*, 7270–7279.

Koonin, E. V. (1993). Computer-assisted identification of a putative methyltransferase domain in NS5 protein of flaviviruses and lambda 2 protein of reovirus. J. Gen. Virol. *74*, 733–740.

Kümmerer, B. M., and Rice, C. M. (2002). Mutations in the yellow fever virus nonstructural protein NS2A selectively block production of infectious particles. J. Virol. *76*, 4773–4784.

Lee, E., Stocks, C. E., Amberg, S. M., Rice, C. M., and Lobigs, M. (2000). Mutagenesis of the signal sequence of yellow fever virus prM protein: enhancement of signalase cleavage In vitro is lethal for virus production. J. Virol. *74*, 24–32.

Lin, C., Amberg, S. M., Chambers, T. J., and Rice, C. M. (1993). Cleavage at a novel site in the NS4A region by the yellow fever virus NS2B-3 proteinase is a prerequisite for processing at the downstream 4A/4B signalase site. J. Virol. *67*, 2327–2335.

Lindenbach, B. D., and Rice, C. M. (1997). trans-Complementation of yellow fever virus NS1 reveals a role in early RNA replication. J. Virol. *71*, 9608–9617.

Lindenbach, B. D., and Rice, C. M. (1999). Genetic interaction of flavivirus nonstructural proteins NS1 and NS4A as a determinant of replicase function. J. Virol. *73*, 4611–4621.

Lindenbach, B. D., and Rice, C. M. (2001). *Flaviviridae*: The viruses and their replication. In Fields Virology, D. M. Knipe, and P. M. Howley, eds. (Philadelphia, Lippincott-Raven Publishers), pp. 991–1041.

Mackenzie, J. M., Khromykh, A. A., Jones, M. K., and Westaway, E. G. (1998). Subcellular localization and some biochemical properties of the flavivirus Kunjin nonstructural proteins NS2A and NS4A. Virology *245*, 203–215.

Marchevsky, R. S., Mariano, J., Ferreira, V. S., Almeida, E., Cerqueira, M. J., Carvalho, R., Pissurno, J. W., Travassos da Rosa, A. P. A., Simoes, M. C., Duarte dos Santos, C. N., *et al.* (1995). Phenotypic analysis of yellow fever virus derived from complementary DNA. Am. J. Trop. Med. Hyg. *52*, 75–80.

Martin, M., Tsai, T. F., Cropp, B., Chang, G. J., Holmes, D. A., Tseng, J., Shieh, W., Zaki, S. R., Al-Sanouri, I., Cutrona, A. F., *et al.* (2001). Fever and multisystem organ failure associated with 17D-204 yellow fever vaccination: a report of four cases. Lancet *358*, 98–104.

McArthur, M. A., Suderman, M. T., Mutebi, J. P., Xiao, S. Y., and Barrett, A. D. (2003). Molecular characterization of a hamster viscerotropic strain of yellow fever virus. J. Virol. *77*, 1462–1468.

Molenkamp, R., Kooi, E. A., Lucassen, M. A., Greve, S., Thijssen, J. C., Spaan, W. J., and Bredenbeek, P. J. (2003). Yellow fever virus replicons as an expression system for hepatitis C virus structural proteins. J. Virol. *77*, 1644–1648.

Monath, T. P., Soike, K., Levenbook, I., Zhang, Z. X., Arroyo, J., Delagrave, S., Myers, G., Barrett, A. D., Shope, R. E., Ratterree, M., *et al.* (1999). Recombinant, chimaeric live, attenuated vaccine (ChimeriVax) incorporating the envelope genes of Japanese encephalitis (SA14-14-2) virus and the capsid and nonstructural genes of yellow fever (17D) virus is safe, immunogenic and protective in non-human primates. Vaccine *17*, 1869–1882.

Monath, T. P., Levenbook, I., Soike, K., Zhang, Z. X., Ratterree, M., Draper, K., Barrett, A. D., Nichols, R., Weltzin, R., Arroyo, J., and Guirakhoo, F. (2000). Chimeric yellow fever virus 17D-Japanese encephalitis virus vaccine: dose-response effectiveness and extended safety testing in rhesus monkeys. J. Virol. *74*, 1742–1751.

Monath, T. P., McCarthy, K., Bedford, P., Johnson, C. T., Nichols, R., Yoksan, S., Marchesani, R., Knauber, M., Wells, K. H., Arroyo, J., and Guirakhoo, F. (2002). Clinical proof of principle for ChimeriVax: recombinant live, attenuated vaccines against flavivirus infections. Vaccine *20*, 1004–1018.

Monath, T. P., Guirakhoo, F., Nichols, R., Yoksan, S., Schrader, R., Murphy, C., Blum, P., Woodward, S., McCarthy, K., Mathis, D., *et al.* (2003). Chimeric live, attenuated vaccine against Japanese encephalitis (ChimeriVax-JE): phase 2 clinical trials for safety and immunogenicity, effect of vaccine dose and schedule, and memory response to challenge with inactivated Japanese encephalitis antigen. J. Infect. Dis. *188*, 1213–1230.

Murphy, F. A. (1980). Togavirus morphology and morphogenesis. In The Togaviruses: Biology, Structure, Replication, R. W. Schlesinger, ed. (New York, Academic Press), pp. 241–316.

Muylaert, I. R., Galler, R. G., and Rice, C. M. (1996). Mutagenesis of the N-linked glycosylation sites of the yellow fever virus NS1 protein: effects on virus replication and mouse neurovirulence. Virology *222*, 159–168.

Muylaert, I. R., Galler, R. G., and Rice, C. M. (1997). Genetic analysis of yellow fever virus NS1 protein: Identification of a temperature-sensitive mutation which blocks RNA accumulation. J. Virol. *71*, 291–298.

Nestorowicz, A., Chambers, T. J., and Rice, C. M. (1994). Mutagenesis of the yellow fever virus NS2A/2B cleavage site: Effects on proteolytic processing, viral replication and evidence for alternative processing of the NS2A protein. Virology *199*, 114–123.

Nickells, M., and Chambers, T. J. (2003). Neuroadapted yellow fever virus 17D: determinants in the envelope protein govern neuroinvasiveness for SCID mice. J. Virol. *77*, 12232–12242.

Nowak, T., and Wengler, G. (1987). Analysis of disulfides present in the membrane proteins of the West Nile flavivirus. Virology *156*, 127–137.

Ohyama, A., Ito, T., Tanimura, E., Huang, S.-C., Hsue, J.-Y., and Furu, Y. (1977). Electron microscopic observation of the budding maturation of group B arboviruses. Microbiol. Immunol. *21*, 535–538.

Olsthoorn, R. C., and Bol, J. F. (2001). Sequence comparison and secondary structure analysis of the 3′ noncoding region of flavivirus genomes reveals multiple pseudoknots. RNA *7*, 1370–1377.

Op De Beeck, A., Molenkamp, R., Caron, M., Ben Younes, A., Bredenbeek, P., and Dubuisson, J. (2003). Role of the transmembrane domains of prM and E proteins in the formation of yellow fever virus envelope. J. Virol. *77*, 813–820.

Op De Beeck, A., Rouille, Y., Caron, M., Duvet, S., and Dubuisson, J. (2004). The transmembrane domains of the prM and E proteins of yellow fever virus are endoplasmic reticulum localization signals. J. Virol. *78*, 12591–12602.

Pethel, M., Falgout, B., and Lai, C.-J. (1992). Mutational analysis of the octapeptide sequence motif at the NS1-NS2A cleavage junction of dengue type 4 virus. J. Virol. *66*, 7225–7231.

Pincus, S., Mason, P. W., Konishi, E., Fonseca, B. A. L., Shope, R. E., Rice, C. M., and Paoletti, E. (1992). Recombinant vaccinia virus producing the prM and E proteins of yellow fever virus protects mice from lethal yellow fever encephalitis. Virology *187*, 290–297.

Post, P. R., Carvalho, R., and Galler, R. (1991). Glycosylation and secretion of yellow fever virus nonstructural protein NS1. Virus Res. *18*, 291–302.

Post, P. R., Santos, C. N. D., Carvalho, R., Cruz, A. C. R., Rice, C. M., and Galler, R. (1992). Heterogeneity in envelope protein sequence and N-linked glycosylation among yellow fever virus vaccine strains. Virology *188*, 160–167.

Preugschat, F., and Strauss, J. H. (1991). Processing of nonstructural proteins NS4A and NS4B of dengue 2 virus *in vitro* and *in vivo*. Virology *185*, 689–697.

Pugachev, K. V., Guirakhoo, F., Ocran, S. W., Mitchell, F., Parsons, M., Penal, C., Girakhoo, S., Pougatcheva, S. O., Arroyo, J., Trent, D. W., and Monath, T. P. (2004). High fidelity of

yellow fever virus RNA polymerase. J. Virol. *78*, 1032–1038.
Reed, K. E., Gorbalenya, A. E., and Rice, C. M. (1998). The NS5A/NS5 proteins of viruses from three genera of the family *Flaviviridae* are phosphorylated by associated serine/threonine kinases. J. Virol. *72*, 6199–6206.
Rey, F., Heinz, F. X., Mandl, C., Holzmann, H., Kunz, C., Harris, B., and Harrison, S. C. (1995). Crystal structure of the envelope glycoprotein E from tick borne encephalitis virus. Nature *375*, 291–298.
Rice, C. M. (1990). Overview of flavivirus molecular biology and future vaccine development via recombinant DNA. Southeast Asian J. Trop. Med. Pub. Health *21*, 671–677.
Rice, C. M., Lenches, E.M., Eddy, S.R., Shin, S.J., Sheets, R.L. and Strauss, J.H. (1985). Nucleotide sequence of yellow fever virus: implications for flavivirus gene expression and evolution. Science *229*, 726–733.
Rice, C. M., Aebersold, R., Teplow, D. B., Pata, J., Bell, J. R., Vorndam, A. V., Trent, D. W., Brandriss, M. W., Schlesinger, J. J., and Strauss, J. H. (1986). Partial N-terminal amino acid sequences of three nonstructural proteins of two flaviviruses. Virology *151*, 1–9.
Rice, C. M., Grakoui, A., Galler, R., and Chambers, T. J. (1989). Transcription of infectious yellow fever virus RNA from full-length cDNA templates produced by *in vitro* ligation. New Biol. *1*, 285–296.
Ruggli, N., and Rice, C. M. (1999). Functional cDNA clones of the Flaviviridae: strategies and applications. Adv. Virus. Res. *53*, 183–207.
Russell, P. K., Brandt, W. E., and Dalrymple, J. M. (1980). Chemical and antigenic structure of flaviviruses. In The Togaviruses: Biology, Structure, Replication, R. W. Schlesinger, ed. (New York, Academic Press), pp. 503–529.
Schlesinger, J. J., Brandriss, M. W., and Walsh, E. E. (1985). Protection against 17D yellow fever encephalitis in mice by passive transfer of monoclonal antibodies to the nonstructural glycoprotein gp48 and by active immunization with gp48. J. Immunol. *135*, 2805–2809.
Schlesinger, J. J., Brandriss, M. W., Cropp, C. B., and Monath, T. P. (1986). Protection against yellow fever in monkeys by immunization with yellow fever virus nonstructural protein NS1. J. Virol. *60*, 1153–1155.
Schlesinger, J. J., Brandriss, M. W., Putnak, J. R., and Walsh, E. E. (1990). Cell surface expression of yellow fever virus non-structural glycoprotein NS1: consequences of interaction with antibody. J. Gen. Virol. *71*, 593–599.
Stokes, A., Bauer, J. H., and Hudson, N. P. (2001). The transmission of yellow fever to Macacus rhesus. (1928). Rev. Med. Virol. *11*, 141–148.
Tan, B. H., Fu, J., Sugrue, R. J., Yap, E. H., Chan, Y. C., and Tan, Y. H. (1996). Recombinant dengue type 1 virus NS5 protein expressed in Escherichia coli exhibits RNA-dependent RNA polymerase activity. Virology *216*, 317–325.
Tao, D., Barba-Spaeth, G., Rai, U., Nussenzweig, V., Rice, C. M., and Nussenzweig, R. S. (2005). Yellow fever 17D as a vaccine vector for microbial CTL epitopes: protection in a rodent malaria model. J. Exp. Med. *201*, 201–209.
Theiler, M., and Smith, H. H. (1937). Use of yellow fever virus modified by *in vitro* cultivation for human immunization. J. Exp. Med. *65*, 787–800.
Wengler, G., and Wengler, G. (1989). Cell-associated West Nile flavivirus is covered with E+pre-M protein heterodimers which are destroyed and reorganized by proteolytic cleavage during virus release. J. Virol. *63*, 2521–2526.
Wengler, G., and Wengler, G. (1993). The NS3 nonstructural protein of flaviviruses contains an RNA triphosphatase activity. Virology *197*, 265–273.
Westaway, E. G. (1987). Flavivirus replication strategy. Adv. Virus. Res. *33*, 45–90.
Westaway, E. G., Khromykh, A. A., Kenney, M. T., Mackenzie, J. M., and Jones, M. K. (1997). Proteins C and NS4B of the flavivirus Kunjin translocate independently into the nucleus. Virology *234*, 31–41.
Winkler, G., Randolph, V. B., Cleaves, G. R., Ryan, T. E., and Stollar, V. (1988). Evidence that the mature form of the flavivirus nonstructural protein NS1 is a dimer. Virology *162*, 187–196.
Zhang, Y., Corver, J., Chipman, P. R., Zhang, W., Pletnev, S. V., Sedlak, D., Baker, T. S., Strauss, J. H., Kuhn, R. J., and Rossmann, M. G. (2003). Structures of immature flavivirus particles. EMBO J. *22*, 2604–2613.

Taxonomy and Evolution of the Flaviviridae

2

Ernest A. Gould and Tamara S. Gritsun

Abstract

Currently there are at least 73 characterized flaviviruses which on the basis of their phylogenetic relationships can be broadly divided into four groups. This number is increasing as more viruses are discovered and will undoubtedly continue to increase for some time yet. Some flaviviruses such as yellow fever virus, Japanese encephalitis virus and tick-borne encephalitis virus were first recognized because they caused major human epidemics involving high fatality rates. Others, such as dengue virus appear to have increased their capacity to cause severe disease as they have dispersed more widely throughout the tropics. In contrast others may not have been observed to produce overt disease but have been discovered either using conventional isolation and serological methods during the mid-twentieth century or more recently using improved methods such as RT-PCR sequencing for their recognition and identification. These new molecular methods have also revealed the occurrence of homologous recombination and the presence of flaviviral DNA in infected cells, the significance of which may not yet have been fully recognized.

This review will focus on the taxonomy, evolution, emergence and dispersal of the flaviviruses taking into account recent observations.

Properties of the flaviviruses

The genus *Flavivirus* contains a range of zoonotic viruses that infect both vertebrate and invertebrate species. They are positive-stranded RNA viruses, with a genome length of 10 to 11 kb. The RNA contains a single open reading frame that encodes a polyprotein of about 3400 amino acids that is processed by cellular and viral proteases to produce three structural proteins, the capsid (C), the membrane (M) and the envelope (E) protein and seven nonstructural (NS) proteins, NS1, NS2A, NS2B, NS3, NS4A, NS4B and NS5 proteins (Rice *et al.*, 1985). The E protein hemagglutinates red blood cells, mediates receptor binding and pH-dependent membrane fusion activity as well as inducing neutralizing and/or protective humoral immune responses (Monath and Heinz, 1996) whereas the nonstructural proteins provide the replicative and proteolytic functions, that are essential for virus reproduction (Lindenbach and Rice, 2003). The open reading frame is flanked by 5′ and 3′ untranslated regions (UTRs) that form complex RNA secondary structures which direct virus translation and replication (Brinton *et al.*, 1986; Mandl *et al.*, 1998; Markoff, 2003; Proutski *et al.*, 1999). Elements within the 5′ and 3′ UTR are believed to facilitate replication through enhancer and promoter functions. The lengths of these UTRs are highly variable amongst the different flaviviruses but they possess repeat elements and conserved domains that accurately reflect their evolutionary ancestry (see Chapter 4).

Taxonomy of the flaviviruses

These arthropod-borne viruses (arboviruses), form the genus *Flavivirus* in the family Flaviviridae. They are antigenically and genetically distinct from the other recognized genera in the family, i.e. *Hepacivirus* and *Pestivirus*. Based on the ecological, epidemiological, antigenic and phylogenetic criteria recommended by the International Committee for the Taxonomy of Viruses (ICTV) the flaviviruses can be divided into four groups, namely two mosquito-borne groups, a tick-borne group and a nonvectored group (Gould *et al.*, 2003). They can be further subdivided into 12 subgroups that specifically reflect their phylogenetic and antigenic relationships (Heinz *et al.*, 2000), their ecological characteristics and their disease associations (Gaunt *et al.*, 2001). This scheme will need to be reassessed as more sequence data relating to newly recognized flaviviruses, including insect-only viruses such as cell-fusing agent virus (CFAV), and Kamiti river virus (KRV) (Cook *et al.*, 2006; Cook and Holmes, 2006; Crochu *et al.*, 2004) and nonvectored, bat-borne viruses such as Tamana bat virus (TABV) (de Lamballerie *et al.*, 2002) are included in the analyses. About seventy three flaviviruses are currently recognized (Calisher and Gould, 2003; Gould *et al.*, 2003; Heinz *et al.*, 2000) but this number is constantly being revised with the recognition of new viruses such as Alkhurma virus (Charrel, 2005; Charrel and de Lamballerie, 2003), two new strains of West Nile virus, namely Krasnodar (Lvov *et al.*, 2004) and Rabensburg (Bakonyi *et al.*, 2005), a new flavivirus species, New Mapoon virus, and separately a new subtype, namely TS5273, the latter two of which are in the Kokobera virus (KOKV) complex (Nisbet *et al.*, 2005), and another new species, namely Sitiawan virus (SV) which was isolated from chickens in Malaysia (Kono *et al.*, 2000). This virus is both antigenically and genetically closely related to Israel Turkey meningoencephalomyelitis virus (ITV) and may justifiably be considered to be a subtype of ITV.

The flaviviruses have a worldwide distribution but individually they show characteristic dispersal patterns that largely reflect the particular vector–host relationships with which each is associated (Gaunt *et al.*, 2001; Gould *et al.*, 2003). For example sylvatic yellow fever virus (YFV) is uniquely indigenous to the jungles and savannah regions of central Africa, and South America whereas dengue virus (DENV) is distributed globally throughout the tropics (Gould *et al.*, 2003; Strode, 1951). Tick-borne encephalitis virus (TBEV) circulates between wild animals and ticks on the forest floors of Europe and Asia (Marin *et al.*, 1995; Zanotto *et al.*, 1995). On the other hand, louping ill virus (LIV), Spanish sheep encephalomyelitis virus (SSEV), Turkish sheep encephalomyelitis virus (TSEV) and Greek goat encephalomyelitis virus (GGEV) are found on the sheep-grazing hillsides of Europe presumably having emerged out of the European and Asian forests as sheep and goats were moved onto the hillsides. However, none of these tick-borne flaviviruses is found in the forests of Latin America (Figure 2.1).

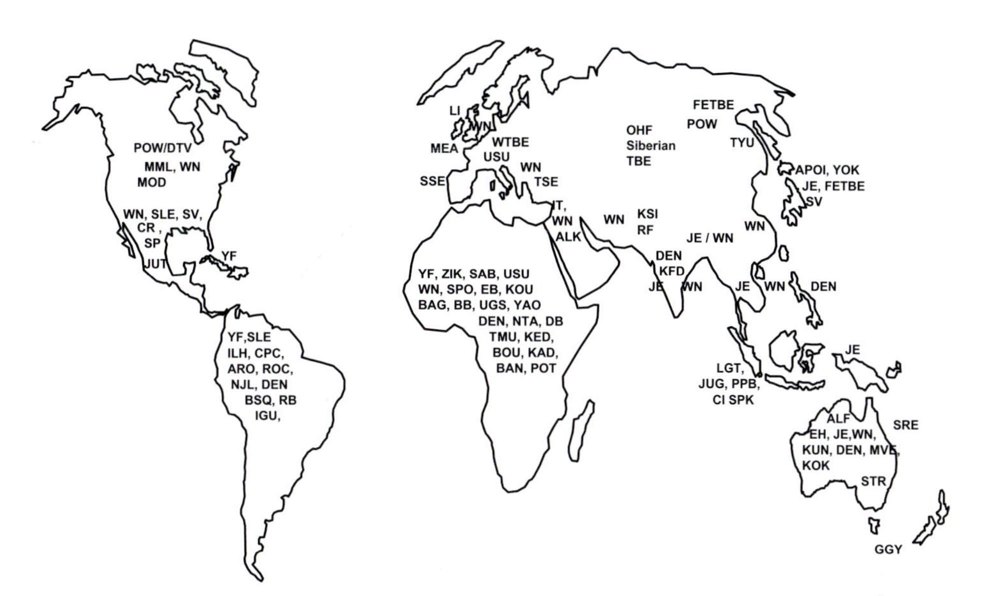

Figure 2.1 Geographic location, by land mass, of the recognized flaviviruses. Some viruses are shown more than once to emphasize their wide geographic distribution. Virus designations are abbreviated for economy of space.

Phylogenetic analysis combined with studies of ecological and epidemiological characteristics and virus dispersion patterns provide the basis for rational explanations of current and future disease patterns. The purpose of this review is to update our understanding of these characteristics for the flaviviruses.

Origins of the flaviviruses

Despite the publication of several detailed phylogenetic trees (Billoir *et al.*, 2000; Cook *et al.*, 2006; Cook and Holmes, 2006; Gould *et al.*, 2001; Gould *et al.*, 2003; Kuno *et al.*, 1998; Marin *et al.*, 1995), it is still not possible to identify precisely the early evolutionary events and the geographic origins of the extant viruses in the genus *Flavivirus*. Indeed, there is still doubt over the origins of some of the more recently emerged flaviviruses such as DENV which is known to have sylvatic cycles in simian species in both Asia and Africa (Gubler, 1997a; Wang *et al.*, 2000). Nevertheless, analysis of the viral genetic relationships within the trees and the branching patterns, taken together with the known biochemical, antigenic, ecological, anthropological and epidemiological characteristics of the viruses provides sufficient information for informed speculation concerning their evolutionary origins. The likelihood that the genus *Flavivirus* emerged in the New World appears to be low compared with the alternative possibility, i.e. that they emerged in the Old World. Until recently the most divergent "insect only" lineages, such as CFAV and KRV, had not been found to circulate as infectious viruses in nature in the Americas. However, this has now changed with the discovery of KRV in Puerto Rico (Cook *et al.*, 2006). The fact that this strain of KRV shares virtually identical sequence with isolates of KRV from Kenya strongly suggests that the Latin American virus was transported from Africa on the ships that crossed the Atlantic Ocean regularly during the slave trading period, together with other viruses such as YFV and DENV. CFAV was isolated from *Aedes aegypti* mosquito cells (Stollar and Thomas, 1975). Two distinct isolates of the genetically related KRV (Crabtree *et al.*, 2003) were obtained from *Aedes macintoshi* larvae and pupae collected in flooded dambos in Central Province, Kenya (Sang *et al.*, 2003). Subsequently, sequences of these RNA viruses were shown to be present and to persist as DNA integrated in the genome of *Aedes* spp. mosquitoes collected from a variety of sources worldwide, inferring long-term evolutionary relationships between these insect-only flaviviruses and the African mosquitoes from which they were isolated (Cook *et al.*, 2006; Crochu *et al.*, 2004). Amongst the most divergent flaviviruses, the only one that appears unique to the New World, i.e. it was isolated only in Trinidad, is TABV, which was isolated from the insectivorous bat *Pteronotus parnellii* (Price, 1978). However, this virus is not known to circulate on mainland South or North America. One can speculate that TABV was introduced into Trinidad during the early trading period between Africa and the Americas. African isolates of this virus or a closely related virus would be needed to support this notion. Although the rodent-associated viruses with no known vector (NKV) (see Figure 2.5) are all found exclusively in the New World all of the related bat-associated NKV, with the exception of Bukalasa bat virus (BBV), are found only in the Old World and APOIV roots the entire NKV clade, when tested using either outgroups or mid-point rooting (Gould *et al.*, 2003). Apoi virus was isolated from rodents in Hokkaido in Japan. Thus, the rodent-associated NKV have their roots in the Old World and we can guess that they were probably introduced into the New World during the early trading period between Africa and the Americas.

Taking into account the reservation that there is no foolproof method of proving these assumptions, the most parsimonious interpretation is that the flaviviruses originated in the Old World and since CFAV and KRV, have similar genome strategies and appear to root the genus, Africa appears to be the most likely candidate for the origin of the flaviviruses.

Evolution of tick-borne and NKV flaviviruses

Early phylogenetic analyses of limited numbers of flavivirus sequences (Mandl *et al.*, 1988; Venugopal *et al.*, 1994) confirmed the previous relationships established on the basis of serology (Calisher, 1988; Porterfield, 1980) and confirmed the differences in selective constraints on mosquito and tick-borne flaviviruses. This was reflected in the observation that the tick-borne flaviruses shared greater sequence identity than that measured between the mosquito-borne viruses (Gritsun *et al.*, 1993; Shiu *et al.*, 1991; Venugopal *et al.*, 1994). Moreover, the genus was shown to be monophyletic (Venugopal *et al.*, 1994) and the tick- and mosquito-borne viruses occupied distinct phylogenetic lineages (Shiu *et al.*, 1991; Venugopal *et al.*, 1994).

Interestingly, West Nile virus (WNV), which is normally vectored by mosquitoes, has also been isolated on several occasions from ticks. However, the nucleotide sequence of tick-borne WNV is virtually identical to that of mosquito-borne WNV; therefore, tick-borne WNV retains its phylogenetic position as a mosquito-borne virus (Gaunt *et al.*, 2001; Lvov *et al.*, 2004), implying that the mosquito is the primary invertebrate host for this virus or, less likely, that the invertebrate host does not significantly influence evolutionary selection. Nevertheless, the significance of the fact that other mosquito-borne viruses such as St. Louis encephalitis virus (SLEV) and Koutango virus (KOUV) are found in ticks in the wild should not be ignored, particularly when considering how these viruses are sustained in specific regions over long periods of time.

Antigenic studies and nucleotide sequence data relating to tick- and mosquito-borne flaviviruses showed that the individual viruses in the tick-borne encephalitis virus (TBEV) complex were more closely related than the individual mosquito-borne viruses are to each other. This was interpreted as indicating that the tick-borne viruses had evolved and dispersed in a continuous manner across Asia and Europe (Gao *et al.*, 1993; Gritsun *et al.*, 1995), contrasting with the more discontinuous evolution of the mosquito-borne viruses. As the phylogenetic trees improved (Gould *et al.*, 2001; Marin *et al.*, 1995; Zanotto *et al.*, 1996) and more sequences became available (Gaunt *et al.*, 2001; Kuno *et al.*, 1998), this distinction in characteristics of evolution and dispersal between the tick- and mosquito-borne viruses became more evident. The relatively low level of genetic variation combined with the continuous branching pattern of viruses in the TBEV complex contrasted with the much more significant genetic changes and then long periods of time before further change was identified, as exhibited by the mosquito-borne viruses (Figure 2.2). As the result of these studies it was proposed that the viruses in the TBEV complex had evolved in a continuous fashion, i.e. clinally across the forests of Asia and Europe (Zanotto *et al.*, 1995) and this is illustrated in Figure 2.3. The genetically related sheep encephalomyelitis viruses, found in Turkey (TSEV), Greece (GGEV), Spain (SSEV), the UK and Norway, i.e. louping ill virus (LIV), were associated with sheep, goats and grouse on the uplands, bordering the forests. The emergence of each of these viruses is presumed to have occurred on the sheep/goat-rearing hillsides bordering the forests. Ticks infected with strains of TBEV on the edges of the forests began to feed on the nearby sheep/goats, thus infecting these animals with viruses that subsequently evolved as sheep encephalomyelitic viruses. A likely temporal and spatial pattern of the emergence and dispersion of these sheep-associated viruses is presented in Figure 2.4.

Previous phylogenetic analyses of the entire genus (Gould *et al.*, 2003) based on sequence data from the NS5 gene (Figure 2.5) presented the viruses as four major groups, namely group I, several viruses of which are associated with *Culex* and other ornithophilic spp., group II some of which are associated with *Aedes* spp., group III, associated with ticks, and group IV, the viruses for which a vector is not known; these are usually referred to as the NKV (Porterfield, 1980). It must be continuously emphasized that whilst groups I and II are clearly distinct on the trees, the specific association of each virus with either *Culex* spp., or *Aedes* spp. is not an absolute requirement. Moreover, the tree topology implies that the group II mosquito-borne viruses emerged earlier than group I mosquito-borne flaviviruses. Interestingly, whilst the tick-borne viruses form a distinct group, some of these viruses can be found in mosquitoes in nature. Nevertheless, the close genetic relationships of the tick-borne viruses assigned to group III indicate the likelihood of similar ecological characteristics and evolutionary constraints. Indeed, WNV has been shown to infect and replicate in more than 60 species of mosquito in North America but this virus has also been isolated from ticks in Russia. Despite the diversity of susceptible mosquito spe-

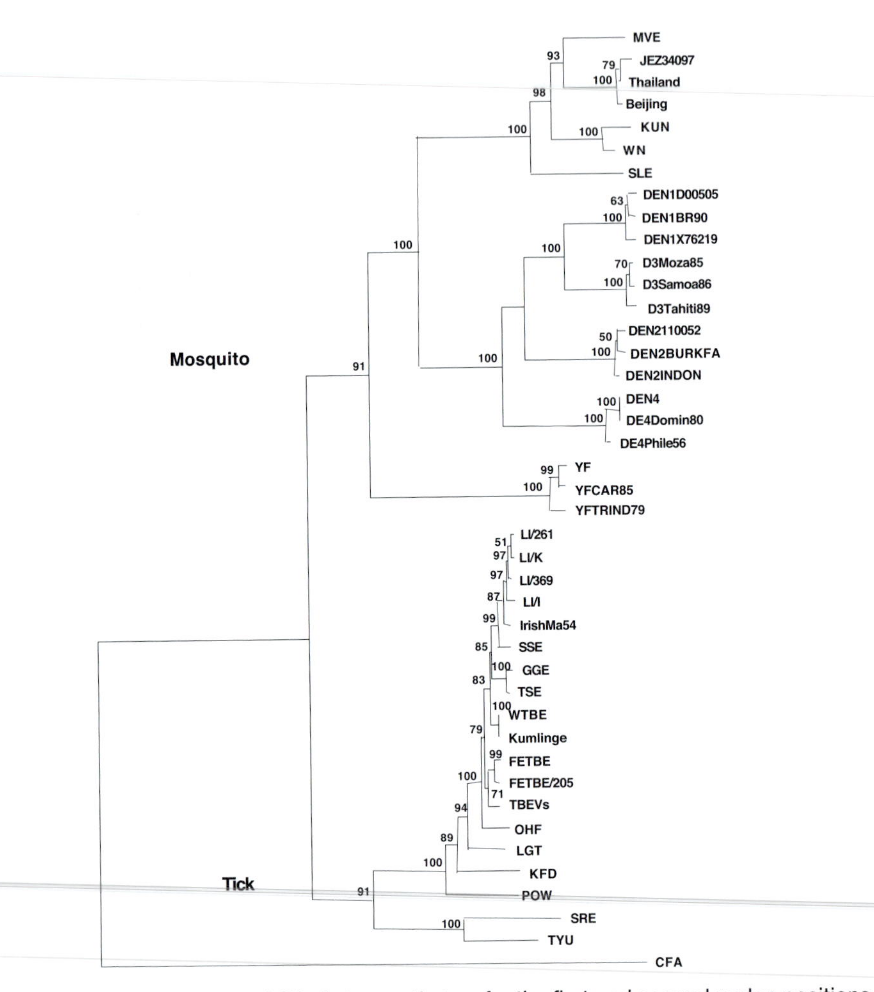

Figure 2.2 Maximum likelihood (ML) phylogenetic tree for the first and second codon positions of the E gene of mosquito- and tick-borne flaviviruses. Percentage bootstrap values are shown above the branches.

cies infected by WNV, the *Culex* spp. are important invertebrate vectors of this virus and the association with this particular invertebrate species is common to several other viruses in the same group. In the most recent flavivirus phylogenetic analyses (Cook and Holmes, 2006), the NS3 gene (Figure 2.6) and whole genome sequences (data not shown) were analyzed using updated and more versatile nucleotide alignment programs. The results support some of the previous data, suggesting that the mosquito-borne flaviviruses represent an outgroup to the remaining flaviviruses, and that tick-borne transmission is probably a derived trait within the genus. Nevertheless, one is still left with the question of whether or not these viruses originated as insect-only viruses and then acquired the capacity to infect vertebrates before (or after) they diverged to produce the mosquito-borne, tick-borne and NKV groups. Further analyses using sequence alignments of the 3′ UTR, may throw some light on this (see Chapter 4).

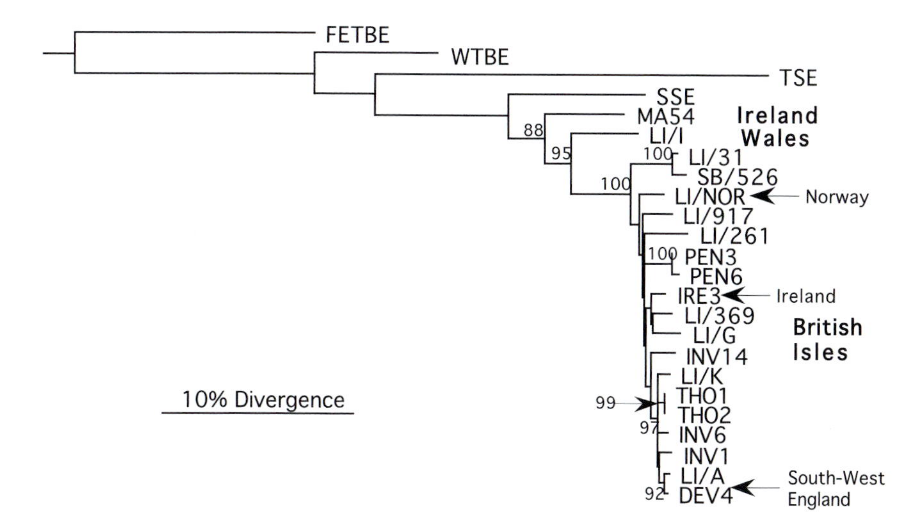

Figure 2.3 ML phylogenetic tree of the E gene of 24 tick-borne flaviviruses. Branch lengths are drawn to scale and all nodes supported by more than 75% bootstraps are indicated. The tree is rooted with the sequence from FETBE virus, strain Sofjin. The three main populations of virus in the British Isles are indicated, along with those viruses secondarily introduced into Ireland and Norway and the viruses found in the south west of England, that appeared following the introduction of red grouse and deer onto the Devon moorlands (McGuire *et al.*, 1998).

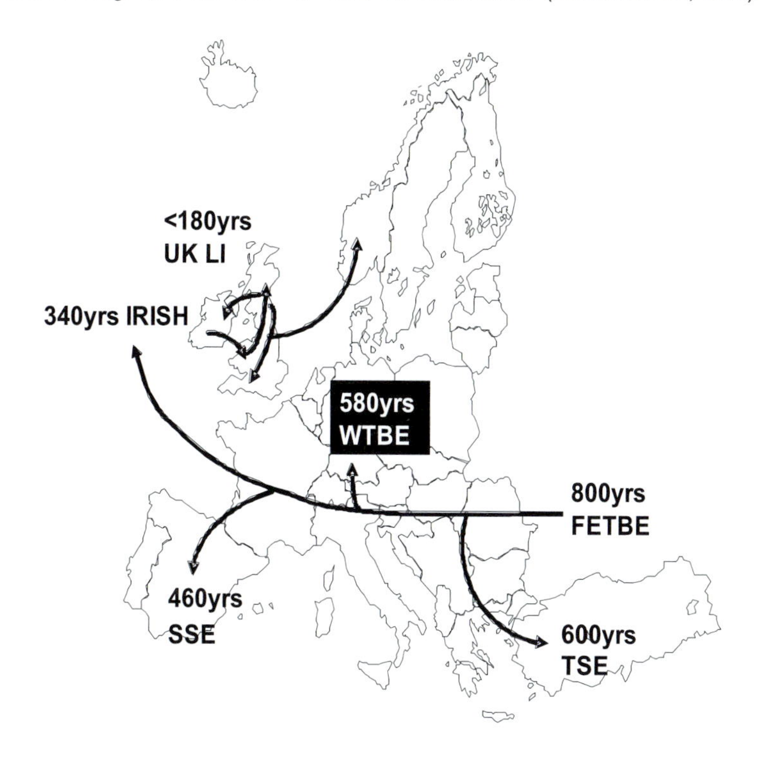

Figure 2.4 Predicted evolution and dispersal pattern of sheep encephalomyelitic viruses across the south western region of Asia, then across Europe reaching the UK less than 200 years ago.

Figure 2.5 ML phylogenetic tree of NS5 nucleotide sequences (990 bp) from 70 flaviviruses. As there are major differences in base composition between the three major groups of viruses (tick-borne, mosquito-borne, no-known vector; Jenkins *et al.*, 2001), and this is known to affect phylogenetic accuracy, trees inferred for these groups were estimated separately and then joined in a final phylogeny in which all branch lengths were reoptimized. A highly variable region of 18 amino acids where the alignment was uncertain was also removed prior to analysis. All trees were constructed using the general time reversible (GTR) model of nucleotide substitution allowing each codon position to have a different rate of change. Bootstrap values (from 1000 replicate neighbor-joining trees estimated under the ML substitution model) are shown where >70%. All analyses were undertaken using the PAUP* package. The tree is mid-point rooted for purposes of clarity only and all horizontal branch lengths are drawn to scale. Modified from Gould *et al.* (2003).

On the basis of comparative nucleotide sequence alignments, estimated substitution rates and knowledge of the date when each virus isolation occurred, it was possible to estimate the approximate number of years in the past when the tick- and mosquito-borne viruses emerged. The estimates showed that this was likely to be since the end of the last major ice-age across the northern hemisphere, i.e. between approximately five and ten-thousand years ago (Zanotto *et al.*, 1996). In the warmer regions, further south, evolution of the most divergent lineages almost

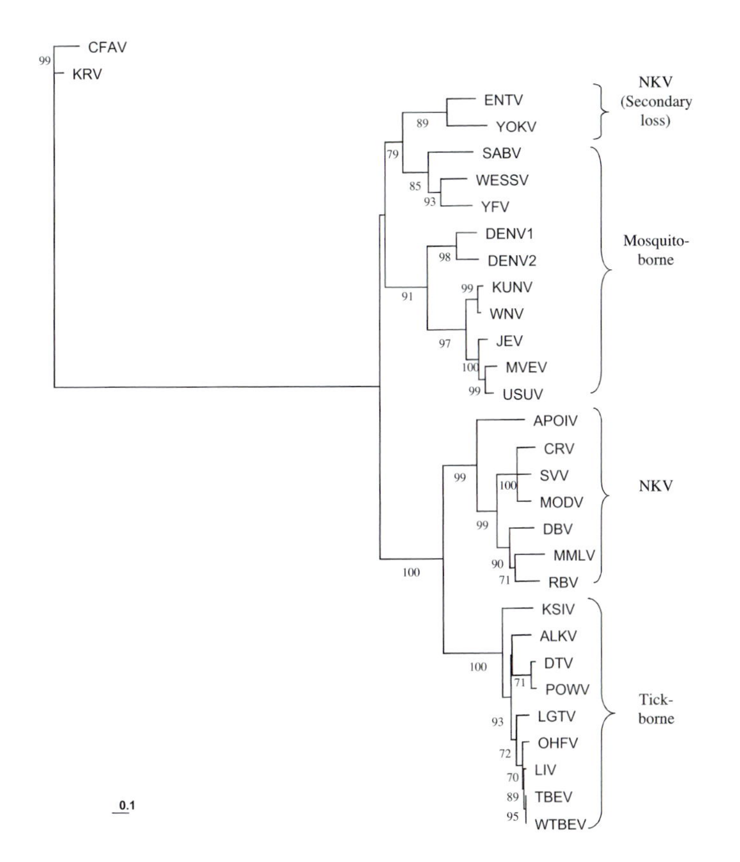

Figure 2.6 Phylogenetic relationships of the genus *Flavivirus* inferred using the NS3 gene. Numbers next to branches depict quartet puzzling support values for clades, which give an indication of the robustness of each node on the current data, with 100 representing maximum support for the branch in question (Cook and Holmes, 2006).

certainly was occurring before the end of the ice-age in the northern regions. Indeed, the sequences of CFAV, KRV and TABV show that these viruses are very distantly related to the recognized viruses in the genus *Flavivirus* (Cook *et al.*, 2006; Crochu *et al.*, 2004; de Lamballerie *et al.*, 2002) and therefore it may be justified to assign them to the family Flaviviridae. New data based on 3′ UTR alignments, currently being prepared for publication, may provide an answer to this question.

Figures 2.5 and 2.6 also show that three NKV which are primarily associated with bats, Yokose virus (YOKV), Entebbe bat virus (EBV) and Sokoluk virus (SOKV), diverged with the mosquito-borne viruses, and subsequently diverged separately from these mosquito-borne viruses, implying that they lost the requirement for vector transmission after diverging from the other NKV.

As indicated earlier, APOIV was isolated from a rodent in Hokkaido. This virus is the most divergent of the NKV. Its geographic separation and significant genetic distance from the other NKV implies that there are probably many more related viruses, in this region of the tree, to be discovered. The same assumption can be made for the insect-only viruses. They are genetically substantially different from the

NKV and the mosquito- and tick-borne viruses. Presumably there are hundreds or maybe even thousands of flaviviruses to be discovered that can fill these evolutionary gaps.

The tick-borne flaviviruses, designated group III (Figure 2.5) separate into two distinct lineages, those associated with seabirds and *Ornithodorus* spp. and those associated with rodents and *Ixodes* spp. The position of Kadam virus (KADV) as basal to the seabird-associated tick-borne viruses is interesting because KADV appears to occupy a different ecological niche in Africa compared with these viruses. Indeed, recent more detailed phylogenetic analyses imply that KADV is sufficiently genetically different from the other tick-borne flaviviruses to justify its inclusion as a separate group (Grard, *et al.*, 2006, manuscript submitted for publication). Moreover, KADV resembles ALKV in being associated with ticks and camels (Charrel, 2005). Whether or not this reflects an African origin for ALKV, and other related viruses such as Kyasanur Forest disease virus (KFDV) is discussed in more detail later.

The mosquito-borne viruses assigned to group II, i.e. those that we have designated as being associated with *Aedes* spp., because they include YFV and DENV, are more divergent than those assigned to group I, i.e. those associated with *Culex* spp. This implies that all the group I mosquito-borne lineages are descended from the group II lineages all of which appear to have their origins in the Old World. Therefore by definition all the group I viruses are rooted in the Old World even though some of them may now be found exclusively in the New World. In conclusion, although we still have much to learn about the evolution of the flaviviruses, phylogenetic analysis has already revealed many interesting features, each of which contributes to our understanding of their evolution.

Nonviremic transmission of tick-borne flaviviruses

One of the most interesting aspects of tick-borne flavivirus evolution and dispersion came to light when the concept of nonviremic transmission (NVT) of tick-borne arboviruses emerged (Jones *et al.*, 1983). Although the original observations were made with the bunyavirus Thogoto virus, and studies of saliva activated transmission, most of the work on transmission in nature dealt with the TBEV complex viruses. Thus, following on from the discovery of NVT with Thogoto virus were the observations that LIV and TBEV were transmissible by ticks co-feeding on wild hares and viremic and nonviremic rodents. Remarkably, NVT was also observed on hares and rodents that had neutralizing antibody to the virus (Jones *et al.*, 1997; Labuda *et al.*, 1997). It is important to emphasize that this mechanism of transmission does not require an animal to become sick or to develop viremia. Subsequently it was shown that co-feeding transmission of TBEV by ticks occurs in specific regions of Europe when particular climatic conditions occur in the forests. Following a rapid fall in ground-level temperatures during the early autumn, large numbers of noninfected larvae pass the winter without feeding. These overwintering noninfected larvae then emerge synchronously with newly emerged infected nymphs in the spring (Randolph *et al.*, 2000; Randolph *et al.*, 1999). These observations provide a rational explanation for the focal distribution of TBEV in the forests of Europe. They also emphasize the importance of the tick in long-term virus survival.

It has been suggested that this co-feeding phenomenon between larvae and nymphs is not representative of the situation in the forests because some observers believe that nymphs and larvae feed on different hosts. However, this is an oversimplification; co-feeding nymphs and larvae are frequently observed feeding together on red deer and rodents in Europe (Dr Annapaola Rizzoli, Centre for Alpine Studies, Trento, Italy, and Dr Milan Labuda, Institute of Virology, Bratislava, Slovakia—independent personal communications).

These results also provide a rational explanation for how LIV and the related sheep encephalomyelitic viruses in Europe and Asia survive long term. For many years it was known that louping ill was a fatal epidemic disease, particularly in sheep on the British moorlands and Irish hillsides and farms. Laboratory experiments with a variety of animals showed that virus transmission to ticks was dependent on critical threshold levels of virus in the blood of the animals (Reid, 1984). It was shown that the fatal epidemics in sheep, and also in grouse, arose when young immunologically naïve animals were introduced onto the moors where infected ticks were present. Professional disease management teams decided that appropriate immunization of the sheep against LIV and treatment with acaricides would eventually eradicate the disease. To some extent this strategy worked. However, at that time, co-feeding transmission by ticks on nonviremic animals was not recognized. In practice the disease returned each time new flocks of nonimmune sheep were introduced during the following years, presumably because the virus had survived long-term in the ticks which were transmitting the virus by co-feeding on mountain hares and other nonsusceptible wildlife species that roamed the moors. Thus, the two modes of virus transmission, NVT and viraemic transmission, provide long-term (TBEV in rodents, or LIV in hares) and short-term (LIV encephalitis in sheep, or human encephalitis) survival strategies for flaviviruses.

Evolution of mosquito-borne flaviviruses

It has been known for many years that the dispersion of arboviruses in general is largely dependent on their adaptation to specific vertebrate and invertebrate hosts in the natural environment (Mattingly, 1960). However, the impact of global human activity (transportation, irrigation, urbanization, agriculture, deforestation, etc.) on virus evolution and dispersal, particularly in the more recent centuries, has been extraordinary and can readily be revealed by combining phylogenetic analysis with ecological, epidemiological and historical data. The mosquito-borne flaviviruses provide an excellent example of how this can be done.

Group II flaviviruses

The yellow fever virus evolved out of Africa. It was one of the first to be studied in detail and many aspects of this fascinating virus have been reviewed extensively (Bloom, 1993; Monath, 1991; Monath and Heinz, 1996; Strode, 1951). Yellow fever first came to prominence following the commencement of the importation of slaves to the New World from Africa, between 400 and 500 years ago. However, with the benefit of hindsight and sequence data it is believed that YFV emerged from a more divergent flavivirus lineage in the African jungles approximately 3000 to 5000 years ago (Zanotto *et al.*, 1996). The virus is transmitted year-round among simian species and their associated *Aedes* spp. mosquitoes. In general YFV is not noticeably virulent for these monkeys implying long term relationships between the monkeys, mosquitoes and virus. The principal YFV vector, *A. africanus* feeds on canopy-dwelling monkeys in the humid equatorial African forests. The surrounding savannah regions support dense monkey populations that are fed upon by a variety of *Aedes* spp. mosquitoes that may also transmit the virus to humans. Human epidemics arise intermittently and are associated most frequently with *Aedes aegypti* which breeds in and around urban dwellings in or near to the savannah area. Interestingly, it is believed that Africans may be more resistant to YFV than Caucasians (Strode, 1951), a characteristic that has also been identified for DENV (Morier *et al.*, 1987). This could be a contributory factor to the relatively limited dispersion of YFV and the evolutionary history of DENV that will be discussed below.

Phylogenetically very closely related, but genetically and biochemically distinguishable YFV also circulates among simian species and mosquitoes (*Haemagogus*) in the humid equatorial forests and surrounding grasslands of Latin America but in contrast with Africa, the monkeys often suffer fatal infections, implying that YFV was introduced into the New World relatively recently. This is further supported by the suggestion that *Aedes* spp. mosquitoes originated in Africa (Strode, 1951 and Tabachnik, 1991; Gould *et al.*, 2003) and were transported together with YFV on the slave ships that crossed the Atlantic Ocean from the fifteenth century onwards. Indeed, urban "pseudoepidemics" of yellow fever were common at ports receiving the slave ships in Europe, North America and Latin America (Bloom, 1993; Monath, 1991; Monath and Heinz, 1996; Strode, 1951). However, of the many countries into which YFV was frequently being introduced together with the slaves, only Latin America provided the jungles, surrounding savannah, mosquitoes and simian species most suited to the establishment of YFV as an endemic virus.

Yellow fever virus and dengue virus dispersion—a paradox?

Yellow fever virus and DENV provide us with an interesting paradox. Both viruses are transmitted to humans by *Aedes aegypti*, both viruses produce hemorrhagic disease in humans and both circulate in sylvatic cycles within the jungles of Africa. However, whilst either DENV or more divergent lineages of this virus have emerged out of Africa, dispersing westwards to Latin America and eastwards to Asia and the Pacific Islands, YFV has only dispersed westwards from Africa to Latin America. This is despite the fact that the climate, the ecology, and the vector species in many parts of Asia appear to be capable of sustaining YFV. What are the reasons for these differences in virus dispersal patterns? The major reason derives from the fact that DENV is a "modern" virus, i.e. in terms of evolution, it is better adapted than YFV to the human environment, perhaps through its higher transmission efficiency between humans, by *Aedes aegypti* and *Aedes albopictus*. As can be seen in the phylogenetic trees (Figures 2.5 and 2.6) and from the estimates of times of emergence (Zanotto *et al.*, 1996), DENV evolved relatively recently from the more divergent lineages that include YFV. The major change that occurred during this evolution was the development of the capacity of DENV to be transmitted from human to human by *Aedes* spp. apparently without requiring the underpinning support of a sylvatic existence that is an essential requirement for YFV. Thus, in contrast with YFV, the human is not a dead-end host for DENV. Dengue virus has adapted to the urban environment and also to a wider variety of relevant *Aedes* spp. than YFV, providing DENV with more effective urban/peri-urban survival strategies.

Another important contributory factor to the greater epidemiological success of DENV is that in contrast with YFV, many human DENV infections do not result in severely incapacitating or fatal infections. Thus viremic humans may continue relatively normal behavioral patterns whilst circulating the virus locally within and between households or villages and towns or they may even transport the virus to other regions of the world whilst incubating it prior to the development of viremia, thereby significantly aiding its worldwide distribution (Gubler, 1997a). In contrast, YFV produces spasmodic outbreaks in urban areas of Africa and South America. However, YFV frequently produces severely incapacitating, clinically very recognizable, and often fatal disease. Therefore few humans, infected by YFV are capable of traveling significant distances. Dengue virus also infects and is transmitted efficiently to humans by *A. albopictus* which has been dispersed globally, mostly in the small pools of water carried in disused car tires that are transported in vast numbers around the world. *A. albopictus* has adapted from the forest to the human environment being found in rural areas as well as peri-urban localities. This mosquito provides an excellent vector for the continuous transmission and survival of DENV through human populations. Another possible factor that could have a major impact on our understanding of mosquito-transmitted arboviruses arises from the novel observation that WNV can be transmitted nonviremically between mosquitoes co-feeding on vertebrate hosts (Higgs *et al.*, 2005a). This phenomenon of nonviremic transmission (NVT) by mosquitoes is important because it means that the virus does not necessarily require the vertebrate host to be susceptible to infection by the virus. If NVT occurs amongst other mosquito-borne arboviruses such as DENV and YFV, then it will be interesting to compare their efficiency of transmission. Other possible contributory factors to the global spread of DENV compared with the relatively limited dispersion of YFV have been discussed in detail previously (Gould *et al.*, 2001; Gould *et al.*, 2003; Gubler, 1997a; Gubler, 1997b; Monath, 1991; Monath, 1994; Monath and Heinz, 1996).

Other group II flaviviruses

With the exception of Jugra virus (JUGV), Sepik virus (SEPV), Edge Hill virus (EHV), YFV and Wesselsbron virus (WESSV), which is closely related to YFV (Gaunt *et al.*, 2001; Kuno *et al.*, 1998) (Gould, unpublished results), all the other viruses that we have assigned to group II, i.e. Saboya (SABV), Potiskum (POTV), Uganda S (UGSV), and Banzi (BANV), have been found only in Africa, presumably existing as sylvatic viruses and thereby avoiding the inadvertent impact of humans moving themselves and animals over long distances. In contrast, JUGV, and EHV have been isolated only outside Africa, in Asia. These viruses occupy different habitats from YFV. It is therefore possible that they, together with others yet to be discovered, emerged from more divergent sylvatic lineages and circulated amongst animals without necessarily causing significant disease. Perhaps they or their more divergent lineages were then unknowingly exported with the animals or by birds, eastwards out of Africa into Asia and Australia where they have adapted to the local ecologies. In other words, viruses such as YFV, JUGV,

WESSV, SEPV and EHV represent the more widely dispersed species of the group II African viruses.

Zika, Spondweni, and Kedougou virus

In addition to DENV which has been discussed above, the other group I *Aedes* spp.-associated viruses include Zika virus (ZIKV) and Spondweni virus (SPOV) which are both found in Africa, are very closely related and have been isolated from humans. Zika virus has also been isolated in Malaysia (Marchette *et al.*, 1969). In Africa the habitat of this virus is similar to that of YFV. It is therefore interesting to speculate why ZIKV was able to disperse out of Africa to Malaysia when YFV has failed to achieve this. There are several possibilities:

- ZIKV, like DENV, emerged more recently than YFV and is therefore more likely to have been suitably adapted to the changing ecology
- ZIKV may infect a wider range of mosquito species than YFV which would facilitate dispersal
- In common with DENV, the lower disease severity of ZIKV in humans might potentiate its mobility.

Whatever the reasons for the eastward dispersal of ZIKV, this virus has not appeared in Latin America where it might reasonably be expected to have been transported together with YFV, during the slave trading period. Presumably this reflects differences in vector–host relationships between YFV and ZIKV which favored YFV when slaves were being transported to the New World.

Kedougou virus (KEDV), which is found in Africa, is the most closely related virus to ZIKV and SPOV but has different ecological associations compared with these viruses although it is associated with *Aedes* spp. and infects humans. Clearly, we need to know a lot more about these viruses before we can explain why they show such different dispersal patterns.

Group I flavivirus dispersion

Group I contains several viruses that circulate amongst *Culex* spp. and/or other ornithophilic mosquitoes and birds although again it must be emphasized that this is not an absolute definition of all group I viruses. Many of these viruses characteristically produce encephalitic infections when they infect humans and in some cases birds. They are found widely dispersed throughout both the New and the Old World. However, currently, only WNV has achieved the status of being present on both sides of the World and its dispersal to the New World occurred very recently. It follows from this observation that whilst birds are clearly important agents of virus dispersion within geographic regions, they do not necessarily directly transport these viruses east or westwards across the major oceans. Indeed, following its introduction to North America, WNV dispersed southwards and northwards from New York reaching Florida within months, but its westwards progress was much slower, reaching California some four years later. This pattern of dispersal correlates closely with the annual north–south migratory flight paths of birds and is relevant to many of the viruses described below.

Japanese encephalitis virus

In terms of the number of fatal human infections due to group I viruses, Japanese encephalitis virus (JEV) is easily the most important pathogen. This virus is very widely distributed throughout Asia, including, Japan, mainland China, Taiwan,

India, Bangladesh, Malaysia, Indonesia and Australasia. Many of these countries have developed extensive pig and poultry industries which provide excellent hosts for amplification of both JEV and its mosquito vectors. Therefore, whilst JEV may circulate naturally between birds and *Culex* spp. in the wild, hardly ever encountering humans, its success as an epidemic virus in humans is enhanced by these commercial activities in rural and urban environments. The fact that JEV is not found in Africa despite its close relationship with many African viruses suggests that the virus may have emerged from a more divergent African lineage that dispersed out of Africa into Asia. This is consistent with the fact that JEV has not been detected on the bird migratory routes that connect Africa with the more westerly parts of Asia and central Europe.

The phylogenetic trees that focus on the genus *Flavivirus* show that JEV is closely related to Usutu virus (USUV) which is believed to have originated in Africa (Bakonyi *et al.*, 2004), and to Murray Valley encephalitis virus (MVEV) and the closely related Alfuy virus (ALFV) both of which are found in Australia. In common with JEV, these viruses are likely to have emerged recently (Uchil and Satchidanandam, 2001). This is shown in Figures 2.5 and 2.6). The ancestral lineage of JEV and these other viruses probably gradually dispersed across Asia into Australasia where MVEV and ALFV presumably emerged. Japanese encephalitis virus has been studied in more detail than many of its close relatives.

The phylogenetic trees that focus on JEV suggest (1) that there are northern and southern Asian genotypes, (2) that the virus has been introduced multiple times from different sources into India and other Asian countries, (3) that strains isolated from the same country are often widely divergent, (4) that strains isolated from geographically distant countries are often virtually identical, and (5) that JEV has recently dispersed eastwards to Australia (Chen and Beaty, 1982; Chen *et al.*, 1992; Chen *et al.*, 1990; Hanna *et al.*, 1996; Paranjpe and Banerjee, 1996; Uchil and Satchidanandam, 2001). Taking all of these observations into account and the suggestion that JEV may have its phylogenetic roots in Malaysia (Solomon *et al.*, 2003) it seems reasonable to conclude that an ancestral lineage of JEV, USUV, MVEV and ALFV dispersed out of Africa and JEV emerged from this lineage in southeast Asia possibly during the past few centuries.

Japanese encephalitis virus then radiated throughout Asia, assisted by migratory birds and human activities such as commercial transportation. This virus has successfully adapted to a very large area of Asia; the possibility that the virus could gradually disperse further westwards beyond India and Pakistan, eventually reaching the southern regions of central and western Europe should not be ignored. Moreover, there is no obvious reason why JEV could not be introduced into the New World in a manner similar to that discussed below for WNV. Whether or not it would be as successful in the New World as WNV, is difficult to predict. On the other hand, in the light of our knowledge of WNV and the available and developing vaccines for these viruses, maybe JEV has left it too late!

Usutu virus

The USUV reference strain SAAR-1776 was originally isolated by Dr. B.M. McIntosh from *Culex neavei* collected in Ndumu, Natal, South Africa, on January 30, 1959. Until the year 2001, USUV had only ever been found in Africa (Henderson *et al.*, 1972; Karabatsos, 1985; Williams *et al.*, 1964), where it was believed to be a relatively widespread but benign virus that produced subclinical infections in birds, other animals and in the case of humans, a febrile illness with a rash (Karabatsos, 1985). However, in the late summer of 2001, investigations of reports of bird deaths in the zoo in Vienna and in neighboring regions revealed that sick and dying birds were infected with USUV. This was the first time the virus had ever been shown to produce fatal encephalitic infections and it was the first report of the presence of USUV outside Africa. Subsequent phylogenetic comparison based on the complete genome sequence of the Austrian USUV isolate with a recognized African strain, showed the viruses to be distinguishable but very similar (Bakonyi *et al.*, 2005). Usutu virus was possibly introduced into Vienna by an infected migratory bird flying in from Africa. If we compare this virus with the dispersal of WNV in North America (see below), following its introduction into Austria it rapidly became established in the indigenous bird species both in and near to Vienna. Nevertheless, during the four years of its presence in Austria, its dispersal as a bird pathogen appears to have been very restricted despite the detection of antibodies in healthy birds several kilometers distant from the initial infection focus.

These observations imply either that USUV is not being regularly reintroduced from Africa by migratory birds or that it is being introduced regularly but causes clinical and fatal infections in only a small proportion of the birds it infects. One reason for this could be the presence of low level immunity in the bird population either due to exposure to nonvirulent or related flaviviruses. There is some support for this latter suggestion based on plaque reduction neutralization tests performed on resident and migratory birds in the United Kingdom (Buckley *et al.*, 2003). Neutralizing antibodies to USUV and WNV were detected in the sera of UK resident birds. A proportion of the surveyed birds contained antibodies that neutralized USUV but not WNV. These results imply that USUV is introduced by migrant birds to the UK and presumably therefore other northern European countries, but the virus does not appear to be virulent for these birds. Until WNV, USUV or another closely related flavivirus has been isolated from healthy birds in the UK, a level of skepticism of this interpretation will remain!

West Nile virus

West Nile virus is antigenically relatively closely related to JEV, MVEV, ALF and USUV, but it diverges distinctly from these viruses (Figures 2.5 and 2.6). West Nile virus was first isolated in 1937 from a febrile patient in the West Nile district of Uganda. Seroepidemiological studies demonstrated that the virus was widespread in birds, horses, humans, and ornithophilic mosquitoes, particularly but not exclusively, *Culex* spp., in Africa (Bernkopf *et al.*, 1953; Melnick *et al.*, 1951; Taylor *et al.*, 1956). However, whilst small epidemic outbreaks of fever and occasional cases of encephalitis were recorded in Africa, in general the virus circulated relatively harmlessly. Infection usually resulted in a self-limiting nonfatal febrile disease rarely associated with manifestations of encephalitis (Taylor *et al.*, 1956). Subsequently, human epidemics and occasional outbreaks of encephalitis in horses and birds in Israel, Romania, Russia, France, Italy, Portugal, and several other southern European countries, were recorded (Bernkopf *et al.*, 1953; Cantile *et al.*, 2000; Filipe and Pinto, 1969; Hayes, 1989; Hubalek and Halouzka, 1999; Juricova *et al.*, 1998; Lvov *et al.*, 2004; Murgue, 2002; Murgue *et al.*, 2001; Platonov *et al.*, 2001; Tsai *et al.*, 1998). However, WNV came to greater prominence as a pathogenic arbovirus when it was inadvertently introduced into North America in 1999. The exact circumstances under which the virus was introduced will never be defined precisely but it was rapidly identified (Briese *et al.*, 1999) as a typical Israel/Egyptian strain of WNV that was known to be circulating in these countries, and in Russia (Lvov *et al.*, 2004) in the late 1990s.

A logical conclusion would therefore be that either an infected bird or mosquito was transported to New York by an incoming flight from, or nearby, one of these countries. This infected bird or mosquito then transmitted the virus to indigenous birds/mosquitoes. It seems unlikely that an infected viremic human introduced the virus as he/she might reasonably be expected to have presented with clinical symptoms and then been traced. An alternative possibility, i.e. that a migrant bird flying from or to Africa/the Middle East or Russia, was blown offwind across the Atlantic Ocean to New York seems highly unlikely. Billions of birds migrate annually, therefore such an event would be expected to occur more than once and there is no evidence that WNV has been introduced into North America on more than one occasion. Once established in and around New York, the introduced virus dispersed rapidly and efficiently throughout North America, the southern regions of Canada, Mexico, Cuba and South America. Detailed descriptions and maps of these dispersal patterns for WNV can be reviewed on the CDC website (http://www.cdc.gov/ncidod/dvbid/westnile/).

On the basis of the phylogenetic evidence (Blitvich *et al.*, 2004) only one strain of WNV has been successfully introduced and then become established, in North America. As discussed above, this supports the argument that the virus was not introduced into North America by migrant birds because, under these circumstances, the likelihood of it being a single introduction is extremely low. This contrasts with Africa, Europe, Asia, India, Australasia and Australia where the virus has been introduced many times from Africa producing a wide range of variants, most of which are closely related to those found in Africa. Early phylogenetic analyses defined at least two distinct lineages (Lanciotti *et al.*, 2002), one containing strains of WNV that have emerged out of Africa, causing encephalitis outbreaks in Europe and Asia and the other containing strains that are mostly sylvatic and have rarely been found outside Africa. However, recent data have identified a strain designated Krasnodar (Lvov *et al.*, 2004) that was isolated from a tick in Russia, and another designated Rabensburg (Bakonyi *et al.*, 2005) that was isolated from a *Culex pipiens* mosquito in the Czech Republic. These isolates are significantly different from previously recognized WNV strains and may need to be reassessed to decide whether or not they are strains of WNV.

It is interesting to note that whilst WNV is considered to be a mosquito-transmitted virus there have been several isolations of WNV from ticks and one can speculate on the possible impact of this? For example, it is now known that WNV can be transmitted nonviremically between co-feeding ticks (Lawrie *et al.*, 2004). As discussed earlier for the TBEV complex viruses, NVT between co-feeding ticks provides a long-term strategy for survival and gradual evolution of the virus in the wild. It also enlarges the range of species that can be involved in virus transmission.

Perhaps even more significantly, it is now known that WNV can be transmitted nonviremically between co-feeding mosquitoes (Higgs *et al.*, 2005b). This important observation might explain the impressive success of WNV as it dispersed across North America, involving a very wide range of vertebrate species, many of which were not necessarily considered to be susceptible to the virus. Nonviremic transmission could explain how these insusceptible species may have contributed significantly to the rapid and widespread dispersal of WNV. The highly successful dispersion of WNV is perhaps even more striking in Asia where WNV is now known to have dispersed across the entire landmass of Russia from the western boundaries with Europe, to the east coast of Russia. In short, NVT could potentially enhance virus survival, transmission and dispersion and obviate the requirement for viremia. For example, all vertebrates, including immune and insusceptible animals might facilitate mosquito infection through NVT. Most significantly these findings question the status of dead-end hosts in the WNV transmission cycle and may partly explain the success with which WNV established and rapidly dispersed throughout North America. If our instincts based on phylogenetic analysis are to be believed, WNV has emerged out of Africa in the recent past and dispersed, probably via migrant birds to every continent except Antarctica, or is it there but not yet recognized?

On the basis of serological evidence in wild birds (Buckley *et al.*, 2003; Juricova *et al.*, 1998) and other species (unpublished data), WNV also circulates in northern Europe, although as yet there is no recognized disease association. Although it has not been possible thus far to isolate WNV from birds in northern Europe, the detection of antibodies in birds is not a surprising observation. Several other arboviruses are known to circulate in northern Europe and Scandinavia. Ockelbo virus, a close relative of the alphavirus Sindbis virus, circulates and causes disease in humans in Scandinavia and is believed to be carried between Africa and Scandinavia by migrating birds (Norder *et al.*, 1996). These alphaviruses share very similar ecological niches with WNV in Africa and like WNV, they do not cause overt disease in the birds or other animals that they infect either in Africa or in Scandinavia.

There is accumulating evidence that WNV can be transmitted between vertebrate species by routes additional to those involving arthropods (CDC, 2002; Iwamoto *et al.*, 2003; Komar *et al.*, 2003; Nir *et al.*, 1965; Pogodina *et al.*, 1983; Ravindra *et al.*, 2004; Sbrana *et al.*, 2005; Tonry *et al.*, 2005) (Gould unpublished observations). Although invertebrates almost certainly contribute significantly to WNV transmission in northern Europe, the relatively lower numbers and therefore density of mosquitoes in these regions might limit their importance in terms of disease transmission efficiency. Furthermore, because we now know that WNV and other mosquito-borne arboviruses can utilize other methods of transmission, such as aerosols, the fecal–oral route, and direct transmission through breast milk, this could at least in part explain how such viruses might circulate between vertebrates in environments previously considered suboptimal for efficient circulation of arboviruses.

St. Louis encephalitis and other related Latin American and African viruses

West Nile virus is the exception, to the rule that the *Culex* spp.-associated viruses, i.e. the group I viruses, are found either in the Old World or the New World but never in both. Prior to the introduction of WNV into North America St. Louis encephalitis virus (SLEV) was the most well recognized

New World group I flavivirus. Its presence in the Americas is almost certainly the result of transportation of either a more divergent lineage, or SLEV itself, from Africa across the Atlantic Ocean during the slave trading period. This conclusion is supported by detailed examination of the phylogenies based on sequences obtained from geographically dispersed SLEV strains. These phylogenies infer that the virus was introduced through South America and it then gradually evolved and dispersed northwards eventually reaching North America (Gould *et al.*, 2004; Kramer and Chandler, 2001). Many phylogenetic trees show that SLEV is closely related to several other New World viruses, namely Rocio virus (ROCV), Ilheus virus (ILHV), Naranjal virus (NJLV), Bussuquara virus (BSQV), Iguape virus (IGUV), and Aroa virus (AROAV), not all of which have been shown to circulate amongst ornithophilic mosquitoes. These viruses are all found in South America but few have ever been isolated in North America supporting the idea that SLEV, and possibly these other viruses, were introduced from Africa through Latin America.

In addition to these New World viruses the SLEV-related clades also include some Old World viruses, e.g. ZIKV, Ntaya virus (NTAV), Tembusu virus (TMUV), Israel turkey meningoencephalomyelitis virus (ITV). This also implies that there have been several introductions of group I flaviviruses from Africa to Latin America during the past few centuries. However, these viruses do not appear to have been as successful as WNV in producing human epidemics in the New World. This is almost certainly due to the fact that WNV was introduced into an immunologically naïve environment, and also because this virus infects a very wide range of arthropod, mammalian, avian and reptilian species in North America as indeed it does in Africa. Such versatility, aided and abetted by the advantages afforded through NVT (Higgs *et al.*, 2005b) virtually guaranteed WNV the uninterrupted dispersion success that has been witnessed over the past five years.

Observations based on recent findings

The clinal evolution, and geographical dispersion of the tick-borne flaviviruses as well as the influence of climate and habitat on virus transmission by co-feeding ticks have been described above and in more detail in other papers and reviews. Here, additional observations, not referred to above, that have recently become apparent will be discussed.

Powassan virus

Many phylogenetic trees show that Powassan virus (POWV) is one of the most divergent viruses within the TBEV complex (Figure 2.5). This virus is found in the far-east Russian forests associated with *Ixodes* spp. ticks and small wild mammals, where far eastern strains of TBEV are also found (Gritsun *et al.*, 2003). In Russia, the virus has also been isolated from mosquitoes (Sobolev *et al.*, 1978) (D Lvov, Moscow, personal communication). Powassan virus is also found in many different forested regions of Canada and the USA where ticks are abundant (Karabatsos, 1985; Telford *et al.*, 1997). Although it has never been associated with mosquitoes in North America, such an association might explain its wide distribution throughout the United States and Canada! An alternative and perhaps more plausible explanation is offered below.

The fact that POWV is present in both the Old and the New World raises the question, where did this virus originate? According to the clinal theory of evolution and dispersal of the TBEV complex viruses (Zanotto *et al.*, 1995), Powassan virus emerged in the Old World. This is also supported by the fact there are no other TBE complex viruses in the Americas, assuming that deer tick virus is a subtype of POWV. One possible explanation is that POWV emerged in Russia, or at least in the Old World and two to three hundred years ago was introduced into Canada by ticks associated with the large numbers of early settlers that moved to Canada with their accompanying animals. Although this concept would not be unprecedented, it can be counter-argued that if POWV was introduced into the Americas by the early settlers, then why was TBEV not also introduced at the same time as POWV?

Two possibilities spring to mind. Firstly, if TBEV emerged only a few centuries ago then it is entirely possible that this virus had not dispersed significantly amongst Russian populations when the early settlers were making their journeys to the New World in large numbers. If this is the case, then it could be argued that a virus equivalent to the Old World TBEV should eventually evolve in North America. Perhaps this is happening but it is too early to be identified! Alternatively, TBEV may have been introduced into the New World at or about the same time as POWV, but it either failed to establish in the local forest habitats or it became established but rarely infects humans and is therefore not yet recognized in North America. It is also worth noting that the widespread distribution of POWV throughout North America is consistent with the fact that many of the early settlers moved southwards from Canada relatively soon after arriving in the New World.

Tick-borne encephalitis virus

Tick-borne encephalitis virus (TBEV) presents a significant threat to human health in the Northern Hemisphere and despite the availability of efficient vaccines the incidence of tick-borne encephalitis continues to grow. Multiple aspects of this and related tick-borne flavivirus infections were described in recent reviews (Charrel *et al.*, 2004; Gritsun *et al.*, 2003; Gritsun, 2003). It has been established on the basis of serological tests that TBEV could be divided into three subtypes, namely Far Eastern (FETBEV), Siberian (SibTBEV) and western European (WE-TBEV). This grouping which was subsequently confirmed by phylogenetic analysis (Zanotto *et al.*, 1995), correlates with their geographical location and pathogenetic characteristics for humans and animals. For example, FE-TBEV is more virulent for laboratory animals and for humans compared with Sib- and WETBEV, with declining estimated fatality rates of 20–60%, 7–8% and 1–2% respectively, moving westwards, for humans. Siberian subtypes of TBEV have also been shown to cause chronic tick-borne encephalitis, a slowly progressive disease that often results in disability and death although this syndrome is not recognized in Western Europe. Thus, as the TBEV complex viruses evolved clinally across the northern hemisphere, the most virulent human strains originated in the forests of far-east Russia and Japan (Gritsun *et al.*, 2003; Hayasaka *et al.*, 2001) and strains of decreasing virulence for humans gradually evolved as the viruses dispersed westwards, reaching the moorlands of Scotland, Wales, England and the Irish hillsides and farmlands where they circulate amongst sheep, red grouse and other animal species (McGuire *et al.*, 1998; Reid, 1984). It is interesting to observe that as they evolved westwards, although their virulence for humans decreased, their virulence for sheep apparently increased.

However, there is now accumulating evidence that the far-eastern strains have gradually moved westwards reaching Latvia

(Lundkvist *et al.*, 2001) and Slovakia (unpublished results), possibly being carried by migratory birds although this mechanism does not appear to be an essential requirement for their gradual dispersal. In addition, new strains of TBEV with altered antigenic properties have been identified in the European part of Russia possibly contributing to the rise of the incidence of TBEV infections in Europe (Pogodina *et al.*, 1992). These observations, together with the appearance of POWV in Canada and ALKV in Saudi Arabia remind us that we must not rule out the possibility of more virulent strains of TBEV emerging in western Europe during the next decades.

Flavivirus infections and their introduction into Latin America

In the context of tick-borne flavivirus dispersion, it is important to note that there are no recognized TBEV complex viruses in Latin America. Although many of the observations presented above can help to explain this, it is also consistent with the discussions relating to the appearance of the group II mosquito-borne viruses in Latin America. We have argued that these group II mosquito-borne viruses were probably introduced to Latin America largely through the trading of slaves from Africa. Many of the TBEV complex viruses are found in Asia but few if any are found in western Africa, from where the ships embarked. Therefore, TBEV would not be expected to be introduced onto the slave ships in Africa and clearly they could not be introduced into Latin America from the slave ships. Interestingly, this is an additional argument for an African origin of the Latin American mosquito-borne flaviviruses, including DENV, rather than an Asian origin of these viruses. It follows from the above argument that if DENV emerged in Africa, rather than Asia, this virus would be expected to be present in South America at the same time as YFV and this is precisely what happened. Indeed, one can argue that if DENV originated in Asia, the American genotype, i.e. the indigenous Latin American DENV (Leitmeyer *et al.*, 1999; Uzcategui *et al.*, 2001) would not exist!

What is the origin of Alkhurma virus?

Another intriguing question arises from the recent discovery of ALKV in Saudi Arabia. The nucleotide sequence of this virus is very similar to that of KFDV (Charrel *et al.*, 2001) which has only ever been reported to have been isolated in India, although there have been unsubstantiated claims of the presence of a closely related virus in China. Alkhurma virus was isolated from hospitalized patients who were probably infected by ticks carried on sheep and camels in Saudi Arabia. Large numbers of sheep and camels are transported annually from different African countries and also from India to Saudi Arabia for the Muslim pilgrimage to Mecca, and the most commonly infected patients were sheep butchers. One plausible explanation for the origin of this virus could therefore be that ALKV-infected ticks were introduced into Saudi Arabia from a trading African country or from India, implying that ALKV or a closely related lineage of this virus circulates amongst ticks and sheep in Africa and India! An alternative possibility is that there is a complex of hemorrhagic disease viruses (including RFV, KSIV, ALKV, etc.) circulating in southern Russia, Saudi Arabia and other nearby regions all of which emerged from an African lineage at some time in the past. Regardless of which possibility is nearest to the truth, we are still left with the problem of how KFDV appeared, apparently in isolation, in the

Kyasanur Forest in the Mysore district of India. If we accept the idea that ALKV originated in Africa, one possibility would be that KFDV was introduced to India during trading across the Indian Ocean between India and Africa. On the other hand, TBEV and KFDV antibodies in humans, horses, donkeys, cattle and camels in Saurashtra, western India, have been recorded (Boshell, 1969; Gubler, 1997a; Smithburn *et al.*, 1954; Varma, 2001), implying that KFDV is widespread throughout western India. The Saurashtra Peninsula, in western India north of Bombay, is on the Arabian sea and this particular part of India has a long history of trade with the Middle East. Thus, KFDV is probably present throughout most of the Western Ghats as cryptofoci in particular ecosystems involving monkeys and ticks. Where such stable ecosystems are disturbed, the silent KFDV focus might become overt. If any of these suggestions proves to be correct, then one is left with the intriguing possibility that a highly pathogenic human hemorrhagic virus circulates amongst sheep, camels and possibly wild animal species in Africa, apparently without currently causing hemorrhagic disease outbreaks in humans.

What is the significance of recombination amongst the flaviviruses?

Recombination is another important mechanism of virus evolution although it is difficult to predict whether or not it will lead to increased or decreased fitness. As methods of comparative sequence analysis have improved, evidence of homologous recombination between closely related strains of dengue virus, Japanese encephalitis virus and St. Louis encephalitis virus has been uncovered (Craig *et al.*, 2003; Gould *et al.*, 2004; Tolou *et al.*, 2001; Twiddy and Holmes, 2002; Uzcategui *et al.*, 2001; Worobey *et al.*, 1999). It has been suggested that evidence of recombination in flaviviruses is controversial (Monath, Kanesa-Thasan *et al.*, 2005) but a similar phenomenon of template switching is also recognized amongst related strains of virus in the genus *Hepacivirus* (Kalinina *et al.*, 2004; Kalinina *et al.*, 2002). Therefore there seems to be little justification for implying that the evidence is controversial. To date, with the exception of one publication based on antigenic evidence (Pogodina, 1977) heterologous flavivirus recombination has not been recognized either in nature or in the laboratory, although heterologous recombination has been detected between a Sindbis-like virus (Aura virus), and a New World alphavirus (Eastern equine encephalitis virus) producing Western equine encephalitis virus (WEEV) which currently circulates in the Americas (Hahn *et al.*, 1988).

The occurrence of such an unlikely event, i.e. between two viruses that might reasonably be expected to circulate in very different habitats, means that we should not ignore the possibility of it happening between related arboviruses in other genera that may share similar habitats, such as JEV and WNV in Asia. The presumed method by which recombination occurs, i.e. the polymerase switching templates in cells infected by two different RNA viruses (Lai, 1992; Simmonds, 2004) seems to present real opportunities for heterologous recombination. Under experimental laboratory conditions, viable heterologous recombinant flaviviruses can be engineered, as shown by the production of chimeric flaviviruses that are being developed as potential live attenuated vaccines (Monath *et al.*, 2000). The likelihood of heterologous recombination resulting in a viable flavivirus in nature is very low (Monath *et al.*, 2005; Seligman and Gould, 2004), and such genetic accidents would probably lead to viruses with reduced rather than increased virulence.

Nevertheless, given the recognized existence of WEEV and the evidence of recombination between two completely unrelated viruses, a coronavirus and influenza C virus (Luytjes *et al.*, 1988) the possibility of heterologous recombination amongst flaviviruses should not be ruled out absolutely. Human, animal and invertebrate densities are on the increase and because of air transportation global virus dispersion can take place within 24–36 hours resulting in larger numbers of epidemic outbreaks due to different viruses and therefore a greater likelihood of mixed infections. Typical examples in nature where flaviviruses might overlap significantly are, YFV and DENV in South America, WNV and SLEV in North America, JEV and WNV in Asia and POWV and WNV in North America.

The phenomenon of nonviremic transmission of TBEV between ticks co-feeding on insusceptible rodent hosts (Labuda *et al.*, 1993) provides ideal conditions for mixed infections. Under these conditions, a healthy infected tick could become infected by a second flavivirus which would then have the opportunity for recombination. Now that we know NVT occurs, at least under laboratory conditions, between co-feeding mosquitoes, there seems no obvious reason why this should not occur in nature. Because it increases the numbers of hosts for transmission, the existence of NVT between mosquitoes therefore increases the perceived likelihood of mixed infections. When mosquitoes become infected they reproduce the virus to high titers and remain infected for life. Thus, the combination of viremic and nonviremic transmission of flaviviruses would present enhanced opportunities for mixed infections to occur in mosquitoes.

How many more flaviviruses are there?

As we have already indicated, several new or currently recognized flaviviruses have emerged in the past decade. What is the likelihood that even more will appear or that currently recognized flaviviruses will continue to spread to new areas? The flavivirus phylogenetic trees all show major areas of divergence between the viruses, for example, we cited APOIV as being widely divergent from the other NKV. This is an obvious area where one might expect to find a number of viruses that would fill this gap. The genetic distance between CFAV, KRV, TABV and the mosquito- and tick-borne viruses is also another major gap that will eventually be filled. Moreover, there is no doubt that with the exception of countries like Cuba and Singapore, mosquito control programs have not been improved sufficiently to impact significantly on mosquito-borne disease incidence. There is also accumulating serological and virological evidence to support the view that vertebrates in countries not normally associated with mosquito-borne flavivirus disease outbreaks are being exposed to these and other arboviruses (Aspock *et al.*, 1972; Buckley *et al.*, 2003; Juricova *et al.*, 1998). It is perhaps a sobering thought that we are only just beginning to discover the myriads of viruses that nature has invented. Virologists should have a great future!

Acknowledgments

We are indebted to Professor Raja Varma for the information and informed opinion concerning KFDV and its epidemiology and dispersal in India. The authors also wish to thank Dr Shelley Cook for her help in supplying Figure 2.6 and for allowing us access to "in press" manuscripts. We are also grateful to Dr Kirsty McGuire for the design of Figure 2.4.

References

Aspock, H., Graefe, G., Kunz, C., and Radda, A. (1972). Antibodies against arboviruses in starlings (*Sturnus vulgaris* L.) in Austria. Zentralbl. Bakteriol. Orig. A *221*, 141–142.

Bakonyi, T., Gould, E. A., Kolodziejek, J., Weissenbock, H., and Nowotny, N. (2004). Complete genome analysis and molecular characterization of Usutu virus that emerged in Austria in 2001: comparison with the South African strain SAAR-1776 and other flaviviruses. Virology *328*, 301–310.

Bakonyi, T., Hubalek, Z., Rudolf, I., and Nowotny, N. (2005). Novel flavivirus or new lineage of West Nile virus, central Europe. Emerging Infect. Dis. *11*, 225–231.

Bernkopf, H., Levine, S., and Nerson, R. (1953). Isolation of West Nile virus in Israel. J. Infect. Dis. *93*, 207–218.

Billoir, F., de Chesse, R., Tolou, H., de Micco, P., Gould, E. A., and de Lamballerie, X. (2000). Phylogeny of the genus flavivirus using complete coding sequences of arthropod-borne viruses and viruses with no known vector. J. Gen. Virol. *81 Pt 3*, 781–790.

Blitvich, B. J., Fernandez-Salas, I., Contreras-Cordero, J. F., Lorono-Pino, M. A., Marlenee, N. L., Diaz, F. J., Gonzalez-Rojas, J. I., Obregon-Martinez, N., Chiu-Garcia, J. A., Black, W. C. t., and Beaty, B. J. (2004). Phylogenetic analysis of West Nile virus, Nuevo Leon State, Mexico. Emerging Infect. Dis. *10*, 1314–1317.

Bloom, K. J. (1993). The Mississippi Valley's great yellow fever epidemic of 1878 (Baton Rouge, Louisiana, Louisiana State University Press).

Boshell, J. (1969). Kyasanur Forest disease: ecologic considerations. Am. J. Trop. Med. Hyg. *18*, 67–80.

Briese, T., Jia, X. Y., Huang, C., Grady, L. J., and Lipkin, W. I. (1999). Identification of a Kunjin/West Nile-like flavivirus in brains of patients with New York encephalitis. Lancet *354*, 1261–1262.

Brinton, M. A., Fernandez, A. V., and Dispoto, J. H. (1986). The 3′-nucleotides of flavivirus genomic RNA form a conserved secondary structure. Virology *153*, 113–121.

Buckley, A., Dawson, A., Moss, S. R., Hinsley, S. A., Bellamy, P. E., and Gould, E. A. (2003). Serological evidence of *West Nile virus*, *Usutu virus* and *Sindbis virus* infection of birds in the UK. J. Gen. Virol. *84*, 2807–2817.

Calisher, C. H. (1988). Antigenic classification and taxonomy of flaviviruses (family Flaviviridae) emphasizing a universal system for the taxonomy of viruses causing tick-borne encephalitis. Acta Virol. 32, 469–478.

Calisher, C. H., and Gould, E. A. (2003). Taxonomy of the Virus Family *Flaviviridae*. Adv. Virus. Res. 59, 1–17.

Cantile, C., Di Guardo, G., Eleni, C., and Arispici, M. (2000). Clinical and neuropathological features of West Nile virus equine encephalomyelitis in Italy. Equine Vet. J. *32*, 31–35.

CDC (2002). Possible West Nile virus transmission to an infant through breast feeding-Michigan. Morbidity Mortality Weekly Rep. *51*, 577–578.

Charrel, M. (2005). Low diversity of alkhurma hemorrhagic fever virus, Saudi Arabia, 1994–1999. Emerging Infect. Dis. *11*, 683–688.

Charrel, M., and de Lamballerie, X. (2003). The Alkhurma virus (family Flaviviridae, genus Flavivirus): an emerging pathogen responsible for hemorrhage fever in the Middle East. Med. Trop. 63, 296–299.

Charrel, R. N., Attoui, H., Butenko, A. M., Clegg, J. C., Deubel, V., Frolova, T. V., Gould, E. A., Gritsun, T. S., Heinz, F. X., Labuda, M., *et al.* (2004). Tick-borne virus diseases of human interest in Europe. Clin. Microbiol. Infect. *10*, 1040–1055.

Charrel, R. N., Zaki, A. M., Attoui, H., Fakeeh, M., Billoir, F., Yousef, A. I., de Chesse, R., De Micco, P., Gould, E. A., and de Lamballerie, X. (2001). Complete coding sequence of the Alkhurma virus, a tick-borne flavivirus causing severe hemorrhagic fever in humans in Saudi Arabia. Biochem. Biophys. Res. Commun. *287*, 455–461.

Chen, B. H., and Beaty, B. J. (1982). Japanese encephalitis vaccine (2–8 strain) and parent (SA 14 strain) viruses in *Culex tritaeniorhynchus* mosquitoes. Am. J. Trop. Med. Hyg. *31*, 403–407.

Chen, W. R., Rico Hesse, R., and Tesh, R. B. (1992). A new genotype of Japanese encephalitis virus from Indonesia. Am. J. Trop. Med. Hyg. *47*, 61–69.

Chen, W. R., Tesh, R. B., and Rico-Hesse, R. (1990). Genetic variation of Japanese encephalitis virus in nature. J. Gen. Virol. *71*, 2915–2922.

Cook, S., Bennett, S. N., Holmes, E. C., De Chesse, R., Moureau, G., and de Lamballerie, X. (2006). Isolation of a new strain of the flavivirus cell fusing agent virus in a natural mosquito population from Puerto Rico. J. Gen. Virol. *87*, in press.

Cook, S., and Holmes, E. C. (2006). A multigene analysis of the phylogenetic relationships among the flaviviruses (Family: Flaviviridae) and the evolution of vector transmission. Arch. Virol. *151*, 309–325.

Crabtree, M. B., Sang, R. C., Stollar, V., Dunster, L. M., and Miller, B. R. (2003). Genetic and phenotypic characterization of the newly described insect flavivirus, Kamiti River virus. Arch. Virol. *148*, 1095–1118.

Craig, S., Thu, H. M., Lowry, K., Wang, X. F., Holmes, E. C., and Aaskov, J. G. (2003). Diverse dengue type 2 virus populations contain recombinant and both parental viruses in a single mosquito host. J. Virol. *77*, 4463–4467.

Crochu, S., Cook, S., Attoui, H., Charrel, R. N., De Chesse, R., Belhouchet, M., Lemasson, J. J., de Micco, P., and de Lamballerie, X. (2004). Sequences of flavivirus-related RNA viruses persist in DNA form integrated in the genome of *Aedes* spp. mosquitoes. J. Gen. Virol. *85*, 1971–1980.

de Lamballerie, X., Crouchu, S., Billoir, F., Neyts, J., de Micco, P., Holmes, E. C., and Gould, E. A. (2002). Genome sequence analysis of Tamana bat virus and its relationship with the genus Flavivirus. J. Gen. Virol. *83*, 2443–2454.

Filipe, A. R., and Pinto, M. R. (1969). Survey for antibodies to arboviruses in serum of animals from southern Portugal. Am. J. Trop. Med. Hyg. *18*, 423–426.

Gao, G. F., Hussain, M. H., Reid, H. W., and Gould, E. A. (1993). Classification of a new member of the TBE flavivirus subgroup by its immunological, pathogenetic and molecular characteristics: identification of subgroup-specific pentapeptides. Virus Res. *30*, 129–144.

Gaunt, M. W., Sall, A. A., Lamballerie, X., Falconar, A. K., Dzhivanian, T. I., and Gould, E. A. (2001). Phylogenetic relationships of flaviviruses correlate with their epidemiology, disease association and biogeography. J. Gen. Virol. *82*, 1867–1876.

Gould, E. A., de Lamballerie, X., Zanotto, P. M. A., and Holmes, E. C. (2001). Evolution, epidemiology and dispersal of flaviviruses revealed by molecular phylogenies. Adv. Virus. Res. *57*, 71–103.

Gould, E. A., de Lamballerie, X., Zanotto, P. M. d. A., and Holmes, E. C., eds. (2003). Origins, evolution and vector/host co-adaptations within the genus *Flavivirus* (San Diego, Academic Press).

Gould, E. A., Moss, S. R., and Turner, S. L. (2004). Evolution and dispersal of encephalitic flaviviruses. Arch. Virol. Suppl. *(18)*, 65–84.

Gritsun, T. S., Holmes, E. C., and Gould, E. A. (1995). Analysis of flavivirus envelope proteins reveals variable domains that reflect their antigenicity and may determine their pathogenesis. Virus Res. 35, 307–321.

Gritsun, T. S., Lashkevich, V. A., and Gould, E. A. (1993). Nucleotide and deduced amino acid sequence of the envelope glycoprotein of Omsk haemorrhagic fever virus; comparison with other flaviviruses. J. Gen. Virol. *74*, 287–291.

Gritsun, T. S., Lashkevich, V. A., and Gould, E. A. (2003). Tick-borne encephalitis. Antiviral Res. 57, 129–146.

Gritsun, T. S., Nuttall, P. A., and Gould, E. A. Advances in Virus Research 60, 318–360. (2003). Virus epidemiology, ecology and emergence. Tick-borne flaviviruses. In The Flaviviruses: Current Molecular Aspects of Evolution, Biology, and Disease Prevention, C. T. J., and T. P. Monath, eds.

Gubler, D. J. (1997a). Dengue and dengue haemorrhagic fever: its history and resurgence as a global public health problem. (New York, CAB International).

Gubler, D. J. (1997b). The emergence of dengue/dengue haemorrhagic fever as a global public health problem. In Factors in the Emergence of Arbovirus Diseases, J. F. Saluzzo, and B. Dodet, eds. (Paris, Elsevier), pp. 83–92.

Hahn, C. S., Lustig, S., Strauss, E. G., and Strauss, J. H. (1988). Western equine encephalitis virus is a recombinant virus. Proc. Natl. Acad. Sci. USA 85, 5997–6001.

Hanna, J., Ritchie, S. A., Phillips, D. A., Shield, J., Bailey, M. C., Mackenzie, J. S., Poidinger, M., McCall, B. J., and Mills, P. J. (1996). An outbreak of Japanese encephalitis in the Torres Strait, Australia. Med. J. Aust. *165*, 256–260.

Hayasaka, D., Ivanov, L., Leonova, G. N., Goto, A., Yoshii, K., Mizutani, T., Kariwa, H., and Takashima, I. (2001). Distribution and characterization of tick-borne encephalitis viruses from Siberia and far-eastern Asia. J. Gen. Virol. *82*, 1319–1328.

Hayes, C. G. (1989). West Nile fever. In The arboviruses, epidemiology and ecology, M. T. P., ed. (CRC Press), pp. 59–88.

Heinz, F. X., Collett, M. S., Purcell, R. H., Gould, E. A., Howard, C. R., Houghton, M., Moormann, R. J. M., Rice, C. M., and Thiel, H. J. (2000). Family *Flaviviridae*. In *Virus Taxonomy. 7th Report of the International Committee for the Taxonomy of Viruses.*, M.

H. V. Regenmortel, C. M. Fauquet, D. H. L. Bishop, E. Carstens, M. K. Estes, S. Lemon, J. Maniloff, M. A. Mayo, D. McGeogch, C. R. Pringle, and R. B. Wickner, eds. (San Diego, Academic Press.), pp. 859–878.

Henderson, B. E., McCrae, A. W., Kirya, B. G., Ssenkubuge, Y., and Sempala, S. D. (1972). Arbovirus epizootics involving man, mosquitoes and vertebrates at Lunyo, Uganda (1968). Ann. Trop. Med. Parasitol. *66*, 343–355.

Higgs, S., Schneider, B. S., Vanlandingham, D. L., Klingler, K. A., and Gould, E. A. (2005a). Non-viraemic transmission of West Nile virus. Proc. Natl. Acad. Sci. USA *In press*.

Higgs, S., Schneider, B. S., Vanlandingham, D. L., Klingler, K. A., and Gould, E. A. (2005b). Non-viremic transmission of West Nile virus. Proc. Natl. Acad. Sci. USA *102*, 8871–8874.

Hubalek, Z., and Halouzka, J. (1999). West Nile fever—a reemerging mosquito-borne viral disease in Europe. Emerging Infect. Dis. *5*, 643–650.

Iwamoto, M., Jerrigan, D. B., Guasch, A., Trepka, M. J., Blackmore, C. G., Hellinger, W. C., Pham, S. M., Zaki, S., Lanciotti, R. S., Lance-Parker, S. E., *et al.* (2003). Transmission of West Nile virus from an organ donor to four transplant recipients. N. Engl. J. Med. *348*, 2196–2203.

Jenkins, G. M., Pagel, M., Gould, E. A., de A Zanotto, P. M., and Holmes, E. C. (2001). Evolution of base composition and codon usage bias in the genus Flavivirus. J. Mol. Evol. *52*, 383–390.

Jones, L. D., Davies, C. R., Steele, G. M., and Nuttall, P. A. (1983). A novel mode of arbovirus transmission involving a nonviraemic host. Science *237*, 775–777.

Jones, L. D., Gaunt, M., Hails, R. S., Laurenson, K., Hudson, P. J., Reid, H., Henbest, P., and Gould, E. A. (1997). Transmission of louping ill virus between infected and uninfected ticks co-feeding on mountain hares. Med. Vet. Entomol. *11*, 172–176.

Juricova, Z., Pinowski, J., Literak, I., Hahm, K. H., and Romanowski, J. (1998). Antibodies to alphavirus, flavivirus, and bunyavirus arboviruses in house sparrows (*Passer domesticus*) and tree sparrows (*P. montanus*) in Poland. Avian Dis. *42*, 182–185.

Kalinina, O., Norder, H., and Magnius, L. O. (2004). Full-length open reading frame of a recombinant Hepatitis C virus from St Petersburg: proposed mechanism for its formation. J. Gen. Virol. *85*, 1853–1857.

Kalinina, O., Norder, H., Mukomolov, S., and Magnius, L. O. (2002). A natural intergenotypic recombinant of hepatitis C virus identified in St Petersburg. J. Virol. *76*, 4034–4043.

Karabatsos, N. (1985). International Catalogue of Arthropod-borne viruses. 3rd ed, San Antonio, Texas: American Society for Tropical Medicine and Hygiene, [Suppl 1]: 137–152.

Komar, N., Langevin, S., Hinten, S., Neneth, N., Edwards, E., Hettler, D., Davis, B., Bowen, R., and Bunning, M. (2003). Experimental infection of North American birds with the New York strain of West Nile virus. Emerging Infect. Dis. *9*, 311–327.

Kono, Y., Tsukamoto, K., Abd Hamid, M., Darus, A., Lian, T. C., Sam, L. S., Yok, C. N., Di, K. B., Lim, K. T., Yamaguchi, S., and Narita, M. (2000). Encephalitis and retarded growth of chicks caused by Sitiawan virus, a new isolate belonging to the genus Flavivirus. Am. J. Trop. Med. Hyg. *63*, 94–101.

Kramer, L. D., and Chandler, L. (2001). Phylogenetic analysis of the envelope gene of St. Louis encephalitis virus. Arch. Virol. *146*, 2341–2355.

Kuno, G., Chang, G. J., Tsuchiya, K. R., Karabatsos, N., and Cropp, C. B. (1998). Phylogeny of the genus Flavivirus. J. Virol. *72*, 73–83.

Labuda, M., Kozuch, O., Zuffova, E., Eleskova, E., Hails, R. S., and Nutall, P. (1997). Tick-borne encephalitis virus transmission between ticks cofeeding on specific immune natural hosts. Virology *235*, 138–143.

Labuda, M., Nuttall, P. A., Kozuch, O., Eleckova, E., Williams, T., Zuffova, E., and Sabo, A. (1993). Non-viraemic transmission of tick-borne encephalitis virus: a mechanism for arbovirus survival in nature. Experientia *49*, 802–805.

Lai, M. M. (1992). RNA recombination in animal and plant viruses. Microbiol. Rev. *56*, 61–79.

Lanciotti, R., Ebel, G. D., Deubel, V., Kerst, A. J., Murri, S., Meyer, B., Bowen, M. D., McKinney, N., Morril, W. E., Crabtree, M. B., *et al.* (2002). Complete genome sequences and phylogenetic analysis of West Nile virus strains isolated from the United States, Europe and the Middle East. Virology *298*, 96–105.

Lawrie, C. H., Uzcategui, N. Y., Gould, E. A., and Nuttall, P. A. (2004). Ixodid and Argasid Tick species and West Nile virus. Emerging Infect. Dis. *10*, 653–657.

Leitmeyer, K. C., Vaughn, D. W., Watts, D. M., Salas, R., Villalobos, I., de, C., Ramos, C., and Rico Hesse, R. (1999). Dengue virus structural differences that correlate with pathogenesis. J. Virol. *73*, 4738–4747.

Lindenbach, B. D., and Rice, C. M. (2003). Molecular Biology of Flaviviruses. Adv. Virus. Res. *59*, 23–47.

Lundkvist, K., Vene, S., Golovljova, I., Mavtchoutko, V., Forsgren, M., Kalnina, V., and Plysusnin, A. (2001). Characterisation of tick-borne encephalitis virus from Latvia: evidence for co-circulation of three distinct subtypes. J. Med. Virol. *65*, 730–735.

Luytjes, W., Bredenbeek, P. J., Noten, A. F., Horzinek, M. C., and Spaan, W. J. (1988). Sequence of mouse hepatitis virus A59 mRNA 2: indications for RNA recombination between coronaviruses and influenza C virus. Virology *166*, 415–422.

Lvov, D. K., Butenko, A. M., Gromashevsky, V. L., Kovtunov, A. I., Prilipov, A. G., Kinney, R., Aristova, V. A., Dzharkenov, A. F., Samokhvalov, E. I., Savage, H. M., *et al.* (2004). West Nile virus and other zoonotic viruses in Russia: examples of emerging-reemerging situations. Arch. Virol. (Suppl) *18*, 85–96.

Mandl, C. W., Heinz, F. X., and Kunz, C. (1988). Sequence of the structural proteins of tick-borne encephalitis virus (western subtype) and comparative analysis with other flaviviruses. Virology *166*, 197–205.

Mandl, C. W., Holzmann, H., Meixner, T., Rauscher, S., Stadler, P. F., Allison, S. L., and Heinz, F. X. (1998). Spontaneous and engineered deletions in the 3′ noncoding region of tick-borne encephalitis virus: construction of highly attenuated mutants of a flavivirus. J. Virol. *72*, 2132–2140.

Marchette, N. J., Garcia, R., and Rudnick, A. S. O. (1969). Isolation of Zika virus from *Aedes aegypti* mosquitoes in Malaysia. The Am. J. Trop. Med. Hyg. *18*, 411–415.

Marin, M. S., Zanotto, P. M., Gritsun, T. S., and Gould, E. A. (1995). Phylogeny of TYU, SRE, and CFA virus: different evolutionary rates in the genus Flavivirus. Virology *206*, 1133–1139.

Markoff, L. (2003). 5′- and 3′-Noncoding Regions in Flavivirus RNA. Adv. Virus. Res. *59*, 177–223.

Mattingly, P. F. (1960). Ecological aspects of the evolution of mosquito-borne virus diseases. Trans. R. Soc. Trop. Med. Hyg. *54*, 97–112.

McGuire, K., Holmes, E. C., Gao, G. F., Reid, H. W., and Gould, E. A. (1998). Tracing the origins of louping ill virus by molecular phylogenetic analysis. J. Gen. Virol. *79*, 981–988.

Melnick, J. L., Paul, J. R., Riordan, J. T., Barnett, V. H., Golblum, M., and Zabin, E. (1951). Isolation from human sera in Egypt of a virus apparently identical to West Nile virus. Proc. Soc. Exp. Biol. Med. *77*, 661–665.

Monath, T. P. (1991). Yellow fever, Victor, Victoria? Conqueror, conquest? Epidemics and research in the last forty years and prospects for the future. Am. J. Trop. Med. Hyg. *45*, 1–43.

Monath, T. P. (1994). Dengue: the risk to developed and developing countries. Proc. Natl. Acad. Sci. USA *91*, 2395–2400.

Monath, T. P., and Heinz, F. X. (1996). Flaviviruses. In Virology, B. N. Fields, ed. (Philadelphia-New York, Lippincott-Raven), pp. 961–1034.

Monath, T. P., Kanesa-Thasan, N., Guirakhoo, F., Pugachev, K., Almond, J., Lang, J., Quentin-Millet, M. J., Barrett, A. D., Brinton, M. A., Cetron, M. S., *et al.* (2005). Recombination and flavivirus vaccines: a commentary. Vaccine *23*, 2956–2958.

Monath, T. P., Levenbook, I., Soike, K., Zhang, Z. X., Ratterree, M., Draper, K., Barrett, A. D., Nichols, R., Weltzin, R., Arroyo, J., and Guirakhoo, F. (2000). Chimeric yellow fever virus 17D-Japanese encephalitis virus vaccine: dose-response effectiveness and extended safety testing in rhesus monkeys. J. Virol. *74*, 1742–1751.

Morier, L., Kouri, G., Guzman, G., and Soler, M. (1987). Antibody-dependent enhancement of dengue 2 virus in people of white descent in Cuba. Lancet *1*, 1028–1029.

Murgue, B. (2002). The Ecology and Epidemiology of West Nile Virus in Africa, Europe and Asia In: Mackenzie, J. M., Barrett, A. D., Deubel, V. (eds) Japanese Encephalitis and West Nile viruses. *267*.

Murgue, B., Murri, S., Zientara, S., Durand, B., Durand, J. P., and Zeller, H. (2001). West Nile outbreak in horses in southern France, 2000: The return after 35 years. Emerging Infect. Dis. *7*, 792–796.

Nir, Y., Beemer, A., and Goldwasser, R. A. (1965). West Nile virus infection in mice following exposure to a viral aerosol. Br. J. Exp. Pathol. *46*, 443–449.

Nisbet, D. J., Lee, K. J., van den Hurk, A. F., Johansen, C. A., Kuno, G., Chang, G. J., Mackenzie, J. S., Ritchie, S. A., and Hall, R. A. (2005). Identification of new flaviviruses in the Kokobera virus complex. J. Gen. Virol. *86*, 121–124.

Norder, H., Lundstrom, J. O., Kozuch, O., and Magnius, L. O. (1996). Genetic relatedness of Sindbis virus strains from Europe, Middle East and Africa. Virology 222, 440–445.

Paranjpe, S., and Banerjee, K. (1996). Phylogenetic analysis of the envelope gene of Japanese encephalitis virus. Virus Res. *42*, 107–117.

Platonov, A. E., Shipulin, G. A., Shipulina, O. Y., Tyutyunnik, E. N., Frolochkina, T. I., Lanciotti, R. S., Yazyshina, S., Platonova, O. V., Obukhov, I. L., Zhukov, A. N., *et al.* (2001). Outbreak of West Nile virus infection, Volgograd Region, Russia (1999). Emerg. Infect. Dis. *7*, 128–132.

Pogodina, V. V. (1977). Interspecies interactions of arboviruses. II. Participation of the genomes of two flaviviruses, West Nile and Japanese encephalitis, in formation of a virus clone with dual antigenic determinants. Acta Virol. *21*, 8–14.

Pogodina, V. V., Bochkova, N. G., Dzhivanian, T. I., Levina, L. S., Karganova, G. G., Riasova, R. A., Sergeeva, V. A., and Lashkevich, V. A. (1992). The phenomenon of antigenic defectiveness in naturally circulating strains of the tick-borne encephalitis virus and its possible connection to seronegative forms of the disease. Voprosy Virusologii *37*, 103–107.

Pogodina, V. V., Frolova, M. P., Malenko, G. V., Fokina, G. I., Koreshkova, G. V., Kiseleva, L. L., Bochkova, N. G., and Ralph, N. M. (1983). Study on West Nile virus persistence in monkeys. Arch. Virol. *75*, 71–86.

Porterfield, J. S. (1980). Antigenic characteristics and classification of Togaviridae. In *The Togaviruses*, R. W. Schlesinger, ed. (New York, Academic Press), pp. 13–46.

Price, J. L. (1978). Isolation of Rio Bravo and a hitherto undescribed agent, Tamana bat virus, from insectivorous bats in Trinidad, with serological evidence of infection in bats and man. Am. J. Trop. Med. Hyg. *27*, 153–161.

Proutski, V., Gritsun, T. S., Gould, E. A., and Holmes, E. C. (1999). Biological consequences of deletions within the 3′-untranslated region of flaviviruses may be due to rearrangements of RNA secondary structure. Virus Res. *64*, 107–123.

Randolph, S. E., Green, R. M., Peacey, M. F., and Rogers, D. J. (2000). Seasonal synchrony: the key to tick-borne encephalitis foci identified by satellite data. Parasitology *121*, 15–23.

Randolph, S. E., Miklisova, D., Lysy, J., Rogers, D. J., and Labuda, M. (1999). Incidence from coincidence: patterns of tick infestations on rodents facilitate transmission of tick-borne encephalitis virus. Parasitology *118*, 177–186.

Ravindra, K. V., Friefeld, A. G., Kalil, A. C., Mercer, D. F., Grant, W. J., Botha, J. F., Wrenshall, L. E., and Stevens, R. B. (2004). West Nile virus-associated encephalitis in recipients of renal and pancreas transplants: case series and literature review. Clin. Infect. Dis. *38*, 1257–1260.

Reid, H. W. (1984). Epidemiology of louping ill. In Vectors in virus biology, M. A. Mayo, and K. A. Harrap, eds. (London, Academic Press), pp. 161 -178.

Rice, C. M., Lenches, E. M., Eddy, S. R., Shin, S. J., Sheets, R. L., and Strauss, J. H. (1985). Nucleotide sequence of yellow fever virus: implications for flavivirus gene expression and evolution. Science *229*, 726–733.

Sang, R. C., Gichogo, A., Gachoya, J., Dunster, M. D., Ofula, V., Hunt, A. R., Crabtree, M. B., Miller, B. R., and Dunster, L. M. (2003). Isolation of a new flavivirus related to cell fusing agent virus (CFAV) from field-collected flood-water *Aedes* mosquitoes sampled from a dambo in central Kenya. Arch. Virol. *148*, 1085–1093.

Sbrana, E., Tonry, J. H., Xiao, S. Y., Travassos da Rosa, A. P. A., Higgs, S., and Tesh, R. B. (2005). Oral transmission of West Nile virus in a hamster model. Am. J. Trop. Med. Hyg. 72.

Seligman, S. J., and Gould, E. A. (2004). Live flavivirus vaccines: reasons for caution. Lancet *363*, 2073–2075.

Shiu, S. Y., Ayres, M. D., and Gould, E. A. (1991). Genomic sequence of the structural proteins of louping ill virus: comparative analysis with tick-borne encephalitis virus. Virology *180*, 411–415.

Simmonds, P. (2004). Genetic diversity and evolution of hepatitis C virus—15 years on. J. Gen. Virol. *85*, 3173–3188.

Smithburn, K. C., Kerr, J. A., and Gatne, P. B. (1954). Neutralizing antibodies against certain viruses in the sera of residents of India. J. Immunol. *72*, 248–257.

Sobolev, S. G., Shestopalova, N. M., Linev, M. B., and Rubin, S. G. (1978). [Electron microscopic study of the An-750 strain of Powassan virus isolated in the Soviet Union]. Vopr. Virusol. 3, 359–366.

Solomon, T., Ni, H., Beasley, D. W., Ekkelenkamp, M., Cardosa, M. J., and Barrett, A. D. (2003). Origin and evolution of Japanese encephalitis virus in southeast Asia. J. Virol. *77*, 3091–3098.

Stollar, V., and Thomas, V. L. (1975). An agent in the *Aedes aegypti* cell line (Peleg) which causes fusion of *Aedes albopictus* cells. Virology *64*, 367–377.

Strode, G. K. (1951). Yellow fever (New York, McGraw-Hill).

Taylor, R. M., Work, T. H., Hurlbut, H. S., and Rizk, F. (1956). A study of the ecology of West Nile virus in Egypt. Am. J. Trop. Med. 5, 579–620.

Telford, S. R., 3rd, Armstrong, P. M., Katavolos, P., Foppa, I., Garcia, A. S., Wilson, M. L., and Spielman, A. (1997). A new tick-borne encephalitis-like virus infecting New England deer ticks, *Ixodes dammini*. Emerging Infect. Dis. 3, 165–170.

Tolou, H. J. G., Couissinier-Paris, P., Durand, J.-P., Mercier, V., de Pina, J.-J., de Micco, P., Billoir, F., Charrel, R. N., and de Lamballerie, X. (2001). Evidence for recombination in natural populations of dengue virus type 1 based on the analysis of complete genome sequences. J. Gen. Virol. *82*, 1283–1290.

Tonry, J. H., Xiao, C. Y., Siirin, M., Chen, H., Travassos da Rosa, A. P. A., and Tesh, R. B. (2005). Persistent shedding of West Nile virus in urine of experimentally infected hamsters. Am. J. Trop. Med. Hyg. *72*, 320–324.

Tsai, T. F., Popovici, F., Cernescu, C., Campbell, G. L., and Nedelcu, N. I. (1998). West Nile encephalitis epidemic in southeastern Romania. Lancet 352, 767–771.

Twiddy, S. S., and Holmes, E. C. (2002). The extent of homologous recombination in the genus Flavivirus. J. Gen. Virol. *84*, 429–440.

Uchil, P. D., and Satchidanandam, V. (2001). Phylogenetic analysis of Japanese encephalitis virus: envelope gene based analysis reveals a fifth genotype, geographic clustering and multiple introductions of the virus into the Indian subcontinent. Am. J. Trop. Med. Hyg. 65, 242–251.

Uzcategui, N. Y., Camacho, D., Comach, G., Cuello de Uzcategui, R., Holmes, E. C., and Gould, E. A. (2001). Molecular epidemiology of dengue type 2 virus in Venezuela: evidence for in situ virus evolution and recombination. J. Gen. Virol. *82*, 2945–2953.

Varma, M. G. R. (2001). Kyasanur Forest disease. In Encyclopedia of arthropod-transmitted infections of man and domesticated animals, M. W. Service, ed. (Wallingford UK, CAB International), pp. 254–260.

Venugopal, K., Gritsun, T., Lashkevich, V. A., and Gould, E. A. (1994). Analysis of the structural protein gene sequence shows Kyasanur Forest disease virus as a distinct member in the tick-borne encephalitis virus serocomplex. J. Gen. Virol. *75*, 227–232.

Wang, E., Ni, H., Xu, R., Barrett, A. D., Watowich, S. J., Gubler, D. J., and Weaver, S. C. (2000). Evolutionary relationships of endemic/epidemic and sylvatic dengue viruses. J. Virol. *74*, 3227–3234.

Williams, M. C., Simpson, D. I., Haddow, A. J., and Knight, E. M. (1964). The isolation of West Nile virus from man and of Usutu virus from the bird-biting mosquito *Mansonia aurites* (Theobald) in the Entebbe area of Uganda. Ann. Trop. Med. Parasitol. *58*, 367–374.

Worobey, M., Rambaut, A., and Holmes, E. C. (1999). Widespread intra-serotype recombination in natural populations of dengue virus. Proc. Natl. Acad. Sci. USA *96*, 7352–7357.

Zanotto, P. M., Gao, G. F., Gritsun, T., Marin, M. S., Jiang, W. R., Venugopal, K., Reid, H. W., and Gould, E. A. (1995). An arbovirus cline across the northern hemisphere. Virology *210*, 152–159.

Zanotto, P. M., Gould, E. A., Gao, G. F., Harvey, P. H., and Holmes, E. C. (1996). Population dynamics of flaviviruses revealed by molecular phylogenies. Proc. Natl. Acad. Sci. USA 93, 548–553.

3 Origin, Evolution, and Function of Flavivirus RNA in Untranslated and Coding Regions: Implications for Virus Transmission

T. S. Gritsun, A. Tuplin, and E. A. Gould

Abstract

In this review we analyze the research that specifically targets flavivirus RNA secondary structures. We focus mainly on data related to the 5′ and 3′ untranslated regions (5′UTR and 3′UTR) but the limited data relating to stable conserved secondary structures in the coding region of the flavivirus genome are also discussed. We propose that the 3′UTR, and possibly the open-reading frame, evolved through multiple duplication events of an RNA domain approximately 200 nucleotides in length, the remnants of which will be demonstrated in tick-borne flaviviruses. Subsequently, these repeat sequences and the associated RNA secondary structures may have evolved into stem–loop conformations with promoter and enhancer functions that impact on virus replication efficiency. The viral promoter probably folds as a complex transitional flexible RNA structure consisting of a number of transient stems and loops conserved between all flaviviruses. One of the transient forms of the promoter is formed due to the physical interaction between multiple complementary sequences in the 3′UTR, the 5′UTR and the coding region resulting in genome circularization. The folding of the 3′UTR, independently from the 5′UTR, revealed other transient promoter elements that might occur before or after circularization. These include a terminal 3′ stable long hairpin (3′LSH) with an adjacent dumbbell-like structure DB1. The folding of the 5′UTR predicts the formation of a conserved terminal Y-shaped structure that is essential for virus infectivity and might contribute to the promoter function. The replication enhancer is located in the 3′UTR between the stop codon and the promoter. It contains repeated conserved sequences and secondary structures but it is more variable between different flaviviruses than the promoter. Although the enhancer function may not be essential for virus viability under experimental conditions in the laboratory, it might play a significant role in nature where the rate of virus replication could be critical for virus transmission and dissemination between vertebrates and invertebrates. The conserved RNA elements predicted in the coding region of the flavivirus genome might also function to accelerate virus replication in the environment thereby enhancing the likelihood of virus survival.

Footnote: Figures 3.2, 3.3, 3.4, 3.10, 3.11, 3.14, 3.17, and 3.19 can be viewed only on the web: http://www.horizonpress.com/hsp/supplementary/flavi/

Introduction

The flaviviruses comprise a globally distributed group of arboviruses, transmitted mainly by tick or mosquito vectors. They have been the focus of extensive scientific research for about 70 years due to the severe illnesses they produce in humans and animals. The most significant flaviviruses are the mosquito-transmitted dengue viruses that cause hemorrhagic fever, and Japanese encephalitis virus (JEV) that causes encephalitis in the tropical and subtropical regions of the world. A group of tick-transmitted flaviviruses previously known collectively as the tick-borne encephalitis virus (TBEV) complex is found in the northern hemisphere and is particularly prevalent in Europe and Russia where they show clinal evolutionary characteristics (Charrel *et al.*, 2004; Gritsun *et al.*, 2003a; Gritsun *et al.*, 2003b; Zanotto *et al.*, 1995).

The flaviviruses are included in the genus *Flavivirus* within the family Flaviviridae (Heinz *et al.*, 2000) and ecologically are subdivided into three groups—mosquito-borne, tick-borne and nonvectored flaviviruses (MBFV, TBFV and NKV respectively). The recognized phylogenetic relationships between the main virus groups show a high degree of concordance with the previous relationships that were based on serological analysis (Gaunt *et al.*, 2001).

The flaviviruses are small (50 nm in diameter) enveloped viruses with single-stranded RNA of positive polarity (~11 kb in length). The genome contains a single open reading frame (ORF) that encodes a polyprotein of about 3400 amino acids. Co-translationally the polyprotein is processed into the individual polypeptides by cellular and viral proteases. The virion envelope contains the structural glycoproteins E and prM (in immature virions) or M (in mature virions) and the capsid is formed by the structural protein C. The nonstructural proteins NS1, NS2A, NS2B, NS3, NS4A, NS4B, and NS5 comprise the virus replication complex (Lindenbach and Rice, 2001).

The open reading frame is flanked by untranslated regions (UTRs), 5′UTR and 3′UTR. Both UTRs form complex RNA structures containing functional domains that are believed to play a role in virus translation, replication and/or assembly.

The mode of replication of flavivirus RNA is described as semiconservative and asymmetric, with an average of only one nascent positive-sense strand per negative-sense template and with an excess of positive-strand relative to negative-strand RNA synthesis (Chu and Westaway, 1985; Cleaves *et al.*, 1981; Westaway *et al.*, 1999). No free minus-strand RNA has been revealed in analysis of infected cells (Khromykh and Westaway, 1997).

The UTRs have attracted a lot of scientific interest because genetic modifications within these regions are known to attenuate flaviviruses without altering their antigenic specificity making them potential candidates for live attenuated vaccines (Cahour *et al.*, 1995; Durbin *et al.*, 2001; Mandl *et al.*, 1998; Pletnev, 2001; Proutski *et al.*, 1997a; Proutski *et al.*, 1999; Troyer *et al.*, 2001). They are also good targets for antisense oligonucleotides that may act as antiviral by preventing virus replication (Deas *et al.*, 2005; Kinney *et al.*, 2005). Previously there have been two reviews of flavivirus UTRs and their secondary structures (Markoff, 2003; Thurner *et al.*, 2004). In this review we trace the origin and development of genetic elements within flavivirus UTRs and the coding region in relation to their secondary RNA structures and potential role in the virus life cycle. It is hoped that this will further our conceptual understanding of the molecular organization of flavivirus RNA secondary structures within the UTRs and the ORF and in particular their evolution and relevance to virus replication and transmission between vertebrate and invertebrate hosts.

The 3′ untranslated region of flaviviruses

As with other positive-stranded RNA viruses the 3′UTR of flaviviruses plays an important role in virus replication through the initiation of both negative-strand synthesis and translation of the polyprotein. Studies into the structure and function of flavivirus 3′ UTRs may be subdivided into experimental and theoretical approaches. Both will be reviewed here in the context of RNA secondary structure and used to explain how flavivirus 3′ UTRs evolved, through the conservation of functionally important RNA structures.

The mean pairwise sequence identity between tick- and mosquito-borne flaviviruses at the nucleotide level (less than 50% for the 3′UTR) is too low to enable the automated construction of a robust alignment. Even a nucleotide sequence alignment between groups within the mosquito-borne viruses has to be limited to certain regions of reduced heterogeneity. Such regions are often associated with the conservation of particular RNA folding structures or signal sequences within the 3′UTR. The rate of nucleotide substitutions within the 3′ UTRs of flaviviruses is obviously different from that of the open reading frames (ORFs) where the requirement of coding for the polyprotein places additional constraints on sequence variation at nonsynonymous sites. The conservation of functional RNA structure and potential signal sequences within the 3′UTR results in constraints on sequence variation that may be different from those observed along unstructured regions of the ORF.

Because flaviviruses are arboviruses, they need to adapt to two evolutionarily distinct types of host species, namely vertebrates (mammalian/avian) and invertebrates (ticks/mosquitoes). Consequently, at the molecular level, virus proteins are under pressure to interact with two different host molecular pathways and their evolution is subjected to double constraints. A comparison of mosquito- and tick-borne flavivirus genomes suggests that the evolution of virus proteins has generally occurred through the gradual accumulation of single nucleotide substitutions across the genome, with only a low number of small deletions or insertions during a period of divergence of at least 10 000 years (Zanotto *et al.*, 1996). This correlates with the fact that flavivirus proteins exhibit a high degree of antigenic cross-reactivity and amino acid sequence identity.

Although viral proteins are quite sensitive to change, and point mutations could have a distinctive effect on their function, some regions of RNA in the 3′UTR appear to be less important for virus reproduction and have been shown experimentally to be nonessential for virus infectivity. For example, a characteristic feature of both mosquito- and tick-borne flavivirus 3′ UTRs is a variable domain (approximately 300 nucleotides) located between the stop codon and a more conserved downstream domain, termed the core or conserved 3′UTR (C3′ UTR) region (also approximately 300 nucleotides). This "redundant" region was, for example, described for TBEV (Gritsun *et al.*, 1997; Hayasaka *et al.*, 2001; Wallner *et al.*, 1995) and its redundancy for virus infectivity has been confirmed experimentally (Mandl *et al.*, 1998). Also within the C3′ UTR region two distinct types of sequence elements have been identified; those that are essential for virus replication and those that are not essential but modulate the efficiency of virus replication (Mandl *et al.*, 1998). Moreover, a number of structural RNA domains within the C3′ UTR are highly conserved, even between mosquito and tick-borne flaviviruses although considerable variability was exhibited outside the C3′ UTR. Such a high level of conservation reflects the functional significance of these domains, suggesting a major role in virus replication (and possibly translation) through interactions with the host or cellular proteins involved in the complex virus replication machinery.

Comparison between even closely related flavivirus species demonstrated that in contrast to the ORF, sequences within the UTRs have undergone such significant molecular changes (large deletions, insertions and duplications) that it is now difficult to trace their gradual evolution. Moreover, the failure of anyone, to date, to produce a robust alignment between the 3′ UTRs of the tick- and mosquito-borne flaviviruses prompted us to review these two groups individually. We have also included the NKV in the comparative alignment.

The 3′UTR of mosquito-borne flaviviruses (MBFV)

Evolution of the mosquito-borne flavivirus 3′UTR

Phylogenetic analysis of the three mosquito-borne flavivirus (MBFV) groups (see Chapter 2) demonstrates a higher degree of relatedness between the dengue virus (DENV) serotypes 1–4 and JEV than between either of these groups and yellow fever virus (YFV). However, comparative alignment of 3′UTR sequences between these three groups of viruses, using such tools as ClustalW and MEGA, have proven to be unable to provide adequate alignments, mostly due to the wide divergence between the sequences. Nevertheless, within the 3′UTR, there are short (~25 nucleotides) but highly conserved sequence motifs (CSs) designated CS1, CS2 and CS3 and their repeats, designated RCS2 and RCS3 respectively (Figure 3.1A) (Hahn *et al.*, 1987). Assuming that the MBFV diverged from a common ancestor, we aligned sequences manually through the insertion of gaps between these "classical" CSs that we used as anchors for the alignment (Figure 3.2, http://www.horizonpress.com/hsp/supplementary/flavi/). The resulting alignments revealed "sequence remnants" that were presumably preserved between CSs despite the extensive deletions and possibly even recombination events that have occurred during evolution of the flavivirus 3′ UTRs. A further guide in constructing the alignment was comparative analysis of previously recognized conserved RNA secondary structures, such as stem–loops and tertiary structures such as pseudoknots and genome circularization sequences. We superimposed these structures onto the alignment to understand the manner by which the 3′ UTRs have evolved and the "evolutionary guide" they used to preserve conserved elements. Figure 3.2 (http://www.horizonpress.com/hsp/supplementary/flavi/) illustrates an abridged version of this alignment, which includes representative examples of each MBFV group.

It is already known (see also Figures 3.1A and 3.2, http://www.horizonpress.com/hsp/supplementary/flavi/), that only CS1 (25 nucleotides) and CS2 (23 nucleotides) are shared between all three groups of MBFV (Hahn *et al.*, 1987). Moreover, the RCS2 (repeat of CS2) was shared only between the DENV and JEV groups, thus

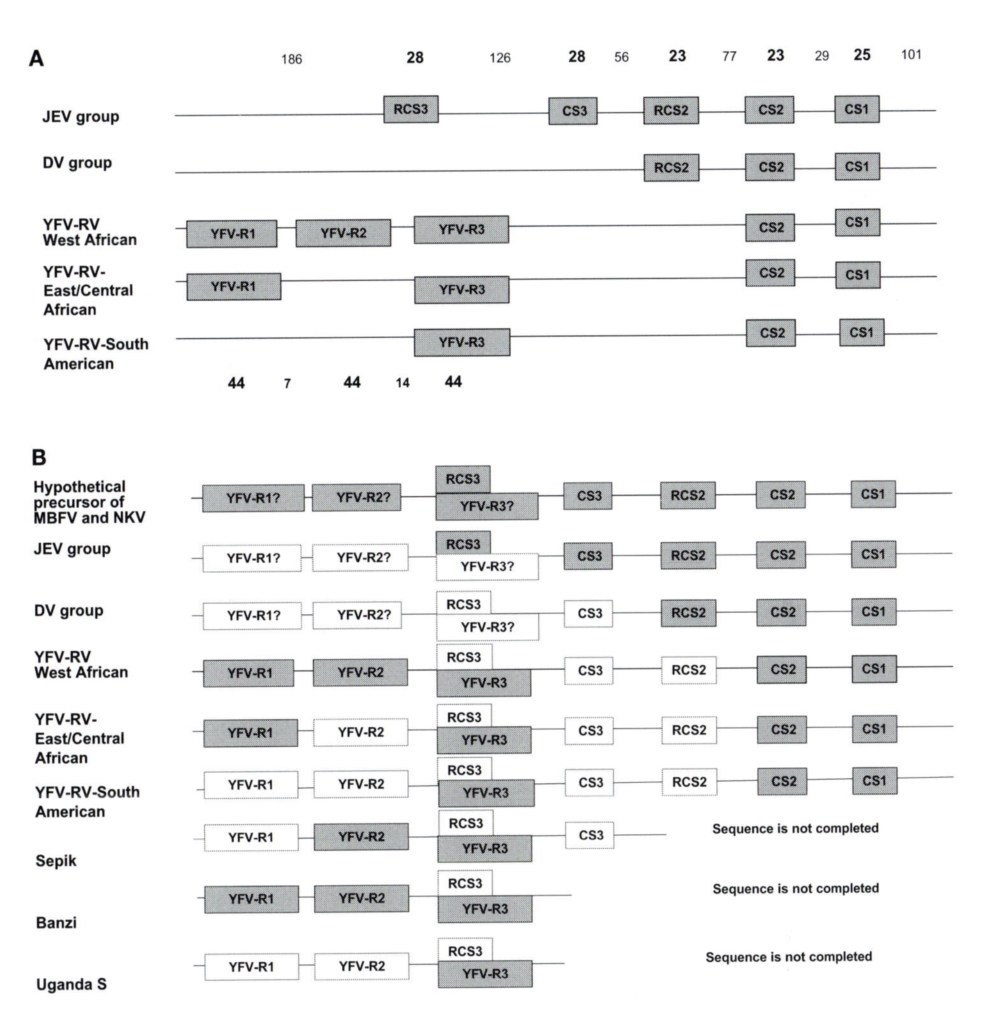

Figure 3.1 Schematic organization of conserved sequences (CSs) within the MBFV 3′UTR (A) adapted from (Hahn *et al.*, 1987) with modifications (Mutebi *et al.*, 2004) and (B) deduced from the alignment in Figure 3.2 (http://www.horizonpress.com/hsp/supplementary/flavi/). The regions of high nucleotide conservation (conserved sequences, CS and their repeats, RCS) are boxed, and regions of nucleotide variability between them are depicted as solid lines. Numbers on the top (for JEV group) and bottom (for YFV group) show the number of nucleotides for each CS (big letters) and variable regions (small letters) between them. Shadowed boxes specify intact CSs/RCSs whereas partial conservation of the appropriate CSs/RCSs is depicted as striped boxes.

excluding the YFV group. CS3 and its repeat RCS3 (28 nucleotides) were described only within the JEV group.

Independent of this, YFV has three longer tandem repeats YF-R1, YF-R2 and YF-R3, each about 45 nucleotides long and located almost immediately after the stop-codon (Hahn *et al.*, 1987). However, it was subsequently demonstrated that these three repeats were present only in the West African genotype of YFV; two other YFV genotypes, i.e. East/Central African and South American have two and one copy of this repeat respectively (Mutebi *et al.*, 2004; Wang *et al.*, 1996). Additionally, the viruses that are distantly related to

YFV, namely Banzi, Sepik and Uganda S virus, preserved three, two and one intact repeat respectively (Mutebi *et al.*, 2004) although these repeats have diverged further than those within the YFV group (Figures 3.1A and 3.3, http://www.horizonpress.com/hsp/supplementary/flavi/). These intact repeats were also used initially as "anchors" to align the quite divergent 3′UTR sequences of the YFV-related viruses, Banzi, Sepik and Uganda S, with the YFV genotypes (Figure 3.3, http://www.horizonpress.com/hsp/supplementary/flavi/) and subsequently these sequences were incorporated into the alignment that was constructed for the 3′UTR of the DENV, JEV and YFV groups (Figure 3.4, http://www.horizonpress.com/hsp/supplementary/flavi/).

In summary, all the repetitive elements previously noted are present in the JEV group; CS3 and RCS3 are not detected in the DENV group and the YFV group shares only CS1 and CS2 with the DENV and JEV group. In addition some strains of YFV possess unique repeats (YF-R1, YF-R2 and YF-R3) that are not shared with any other flaviviruses (Hahn *et al.*, 1987). The occurrence of the CSs and their repeats has been a puzzle since their discovery in 1987 (Hahn *et al.*, 1987). Why are they conserved between such divergent viruses and why have they remained unchanged during the evolution of the genus? What is the biological significance of these CSs and repeated sequences, do they have independent functions or do they need to work together to effect genome function?

A number of RNA viruses are known to possess repetitive sequences within their genomes (Bryan *et al.*, 1992; Faragher and Dalgarno, 1986; Peerenboom *et al.*, 1997; Shi *et al.*, 1997; Warren and Murphy, 2003; Santagati *et al.*, 1994). Such repeats may be associated with the ability of RNA-polymerase to detach from its template during RNA synthesis and to reattach to the same template downstream (deletion) or upstream (generation of repeat) of the point of initial detachment (copy-choice mechanism or strand slippage). New insertions appear when RNA polymerase attaches to and copies a second template RNA (trans) before returning to copy the original RNA template (Copper *et al.*, 1974; Hajjou *et al.*, 1996; Lai, 1992; Pilipenko *et al.*, 1995). The recombinant insertions or stretches of repeated sequence represent sites of new point mutations, deletions and insertions. These molecular events underpin a mechanism of generating new genetic elements that display novel structure and/or function. In this context it is reasonable to ask whether the duplicated sequences in the 3′UTR of MBFV (CS2-RCS2, CS3-RCS3 and YFV-R1-R3-R2) confer an advantageous phenotype or have they remained functionless and essentially neutral throughout the evolutionary development of the virus genus? A number of duplicated repeats have remained unchanged within widely divergent flavivirus groups (for example, CS3-RCS3 within the JEV group or YFV tandem repeats) whilst they have been lost within other similarly divergent groups. This suggests that initially the duplicated repeats conferred a level of selective advantage which lessened in some of the virus groups over time, perhaps due to a change in host or virus replication strategy, resulting in the loss of the repetitive motifs from some but not all flavivirus groups.

A more detailed analysis of the UTR alignments may provide some of the answers to these puzzling questions. This has been achieved by rigorous manual editing of the alignment originally produced using ClustalW for three major MBFV groups. To align sequences between CSs, with a very low homology, large gaps were introduced to imitate the natural deletions. Close inspection of the improved alignment shows that in some stretches of nucleotides, outside the CSs, the YFV sequence is quite similar either to the DENV or the JEV group but not necessarily to both within the same nucleotide region, and the frequency of occurrence of these regions of homology decreases moving upstream towards the stop codon (see Figure 3.2, http://www.horizonpress.com/hsp/supplementary/flavi/). Moreover, the conventional image of specific conservation of "classical" CSs only for certain virus groups (Figure 3.1A) needs reassessment because remnants of similar sequences within RCS2, CS3 and RCS3 could be identified for all groups of MBFV including YFV (Figure 3.2, http://www.horizonpress.com/hsp/supplementary/flavi/).

Details that might reveal the evolutionary development of the YFV-specific repeat sequences YFV-R1, YFV-R2 and YFV-R3 were also observed in the improved alignment. Previously (Mutebi *et al.*, 2004; Wang *et al.*, 1996), it had been reported that strains of YFV from different geographic locations contain different combinations of the intact YFV-specific repeats, originally identified in (Hahn *et al.*, 1987). West African strains of YFV were shown to contain intact YFV-R1, YFV-R2 and YFV-R3, whereas east and central African strains apparently contained only YFV-R1 and YFV-R3. On the other hand, south American strains of YFV were considered to have only YFV-R3. It was suggested that deletions of the repeat sequences could have occurred as the viruses dispersed from West Africa, to east Africa, losing one repeat sequence, or westwards to South America, losing two repeat sequences. However, the alignment presented in Figure 3.3 (http://www.horizonpress.com/hsp/supplementary/flavi/) shows that this process of deletion of the repeat sequences was not an "all or nothing" phenomenon. It seems more likely that a series of mutations and deletions has taken place as the viruses have adapted to their new environments. In fact, it is very important to note that traces or remnants of each repeat sequence remain in the geographically distinct YFV groups. As can be seen in Figure 3.3 (http://www.horizonpress.com/hsp/supplementary/flavi/), West African strains of YFV contain all three intact YFV-specific repeat sequences (YFV-R1, YFV-R2 and YFV-R3), whilst the East and Central African YFV strains contain intact YFV-R1, YFV-R3 and partial YFV-R2. The South American YFV strains contain intact YFV-R3, and partial YFV-R2 and YFV-R1 sequences. This is further supported by the evidence that the most closely YFV-related viruses, Banzi (BANV), Uganda S (UGSV) and Sepik (SEPV) also contained both intact and partial fragments of the supposedly YFV-specific repeats (Mutebi *et al.*, 2004). Moreover, in contrast to previous observations, the YFV-R1, YFV-R2 and YFV-R3 may not be not unique to the YFV group viruses. The remnants of repeats YFV-R2 and YFV-R3 were also traced in the JEV and DENV-group viruses although it was not possible to find even "remnants" of YFV-R1 in these two virus groups (Figures 3.2–3.4, http://www.horizonpress.com/hsp/supplementary/flavi/).

From these observations it appears that before the YFV and the JEV/DENV groups emerged, a founder lineage contained all of the repeat sequences described above. As the viruses emerged and evolved, the individual repeat sequences were then either totally or partially conserved according to the particular selective constraints that were placed on them. A summary of the relationships of these repeat sequences within the virus groups is presented in Figure 3.1B and the specific details can be seen in the alignment in Figures 3.2–3.4 (http://www.horizonpress.com/hsp/supplementary/flavi/).

Mutebi and co-workers (Mutebi *et al.*, 2004) noted that a partial sequence of the YFV-repeat sequences, 5′-AACCGGGATACAAC-3′, is also present in the *Drosophila melanogaster* genome. Therefore, one could speculate that YFV-R was acquired early in the evolution of the genus by recombination between a divergent/ancestral flavivirus and an infected host such as *Drosophila*. If such recombination occurred, these sequence elements may then have extended and triplicated followed by the subsequent reduction of individual nucleotides within YFV-R1 and YFV-R2.

In summary, our alignment of the 3′ UTRs of three major MBFV groups has revealed (1) residual homology of regions located between "classical" CSs within each MBFV group and (2) the presence of all three repeated CSs (CS2-RCS2, CS3-RCS3 and YFV-R1, YFV-R2 and YFV-R3), either intact or as remnants in each of the three major flavivirus groups. We have therefore extended the original concept of the previously defined repeat sequences (Figure 3.1A) (Hahn *et al.*, 1987) and proposed the existence of a new one (Figure 3.1B). This provides the basis for an explanation of the origin and subsequent development of flavivirus repeat sequences within the 3′UTR. Initially an ancient precursor of the MBFV group contained all the repeat sequences that are currently described for each individual MBFV group. Subsequently during diversification of the viruses, as a result of their dispersal and adaptation to different hosts, different elements of the repeat sequences were preserved or deleted providing each virus group with specific advantages that ensure their successful circulation and dissemination in nature.

Thus, within each virus group the short repeat sequences, of the 3′UTR are being preserved. This might indicate that they have a signaling function but does not explain the duplication of the conserved sequences. It could be that these repeats initiate the rapid and efficient assembly of proteins of the replication machinery (around the 3′UTR), the function of which is enhanced by their oligomerization, again contributing to the rate of virus replication.

A further question to resolve is what were the molecular mechanisms that resulted in the appearance of short (~25 nucleotide) repeats separated by longer (~50–130 nucleotides) nonrepeated sequences. The commonly accepted concept of complex rearrangements of the virus genome could be appropriate (Dominguez *et al.*, 1990). Alternatively, it is possible that the initial repeats within the 3′UTR of the MBFV were longer than the currently described 23–25 nucleotide CSs. These current repeated CSs probably represent remnants of longer tandem repeats, possibly more than 100 nucleotides long, which have gradually been deleted and now contain only remnants of the original tandem repeat sequences. This prompted a search for longer repeat sequences in the TBFV group that has diverged less during the past 10 000 years compared with the MBFV (Zanotto *et al.*, 1996). Indeed, longer repeat sequences (about 200 nucleotides in length) were found (see later).

The secondary RNA structures of the 3′UTR of mosquito-borne flaviviruses

The application of computer programs using different algorithms to predict RNA secondary structures on the basis of their minimum free energy, resulted in three models of 3′UTR folding of MBFV (Gritsun *et al.*, 1997; Hahn *et al.*, 1987; Khromykh *et al.*, 2001; Olsthoorn and Bol, 2001; Proutski *et al.*, 1997a; Proutski *et al.*, 1997b; Rauscher *et al.*, 1997). In each set of data the confirmation of predicted stems was supported by compensatory (co-variant) substitutions between closely related strains and virus species; the thermodynamic justification of the structures has been published in detail and will not be discussed here. The purpose of this review is to present the current understanding of the nature of flavivirus 3′ UTRs, and to define the evolutionary pressures driving the variability and conservation within the 3′UTR.

Initially a long (~90 nucleotides) stable hairpin (3′LSH) was observed at the terminus of the 3′UTR. The structure of this 3′LSH was very similar in tick- and mosquito-borne flaviviruses, in spite of the fact that there was a high level of sequence divergence within this region (Figure 3.5) (Brinton *et al.*, 1986; Rice *et al.*, 1985; Takegami *et al.*, 1986; Wengler and Castle, 1986; Zhao *et al.*, 1986). A further stable hairpin, termed stem–loop 2 (SL2), was identified upstream of, but adjacent to the 3′LSH (Shi *et al.*, 1996a) (Figure 3.5). A pseudoknot tertiary interaction has been shown between the SL2 and 3′ SLH that is highly conserved between divergent MBFV providing further evidence for a functional role for SL2 and 3′LSH (Shi *et al.*, 1996a).

The stable folding of the 3′LSH and SL2 as independent structures was confirmed when other stem–loop structures were discovered in the 3′UTR of TBFV

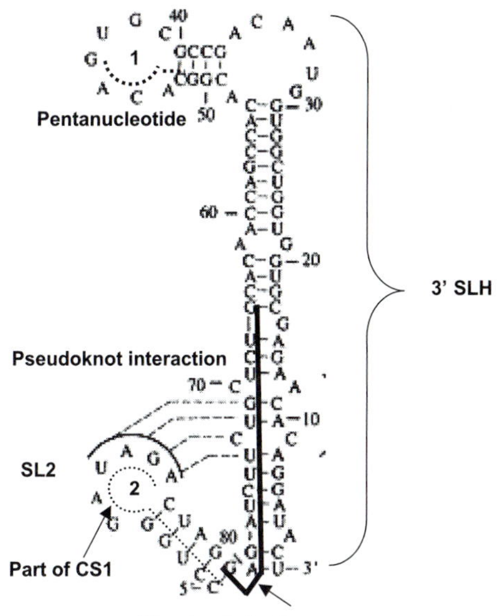

Figure 3.5 Detailed structure of the 3′LSH and SL2. Formation of the pseudoknot between nucleotides within the SL2 (highlighted by the line above letters) and single-stranded region of the stem within 3′LSH (joined by dashed lines) as adapted from (Shi *et al.*, 1996a). The conserved pentanucleotide CACAG within loop 1 is underlined using a thick dashed line. Circularization sequences within the CS1 of the SL2 are underlined by a thin dotted line. The circularization sequence outside CS1 is underlined by a thick line.

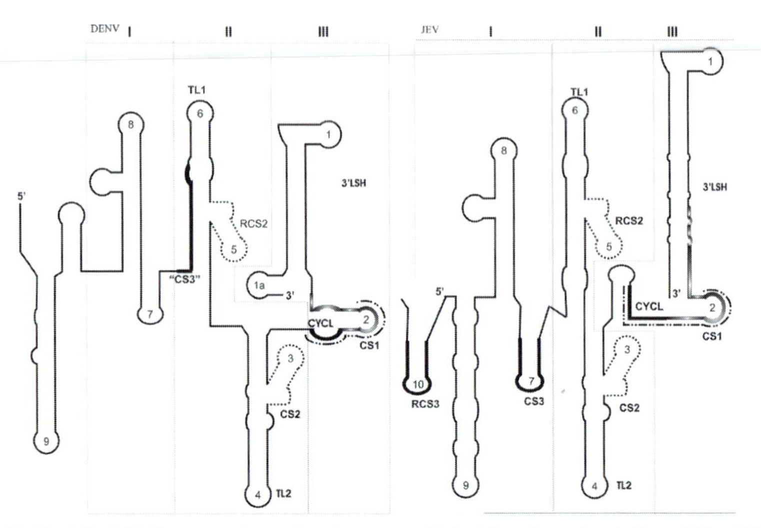

Figure 3.6 Model of RNA secondary structures predicted for domains I, II and III of DENV (A) and JEV (B) using the GA algorithm (Proutski *et al.*, 1997b). The repeated sequences CS2 and RCS2 are shown by dotted lines. The repeated sequences CS3 and RCS3 conserved within JEV group are specified by thick lines. The remnants of CS3 found in DENV are specified as "CS3." The CYCL sequence is depicted as a thick bar and other circularization sequences are specified by grey-shadowed bars extended from CYCL. The dashed-dotted line around CYCL and loop 2 specifies the position of the CS1. The enumeration of loops corresponds to those introduced into Figure 3.4 (http://www.horizonpress.com/hsp/supplementary/flavi/), in addition to terminal loops 1 and 2 (TL1 and TL2). Adapted from (Proutski *et al.*, 1997b).

(Gritsun *et al.*, 1997; Proutski *et al.*, 1997a; Proutski *et al.*, 1997b; Rauscher *et al.*, 1997). Figure 3.6 illustrates the RNA secondary structures that were predicted for almost the full-length of the 3′UTR of DENV and JEV. Initially, an algorithm was used that simulates the natural folding pathway (genetic algorithm, GA) that occurs during RNA elongation (Proutski *et al.*, 1997a; Proutski *et al.*, 1997b). This GA model predicts the transient existence off metastable structures that appear during RNA synthesis. A group of stem–loop structures was conditionally subdivided into three domains each of which was folded independently from the others. These data showed a close similarity of stem–loop patterns between the DENV and JEV groups (Figure 3.6) but a different conformation for the YFV group outside the 3′LSH (not shown here). These results are consistent with the phylogenetic analysis of flaviviruses based on ORF data. For the DENV the 3′LSH appeared shorter but the authors argued that a longer structure, more similar to the rest of the genus might be formed at higher energy levels. According to this research, the 3′LSH was the only similar structure between the three major MBFV groups. The entire 3′UTR of TBEV folded by this GA algorithm also produced a different set of stem–loop structures except for the 3′LSH, which will be discussed later.

A second set of data was produced as the result of the search for pseudoknots upstream of the 3′LSH, coinciding with domain II (Olsthoorn and Bol, 2001). This work predicted more extensive structural similarity between the three major MBFV groups (Figure 3.7) and even with the TBFV (see below). This region was predicted to form two dumbbell-like structures (DB1 and DB2) that are sup-

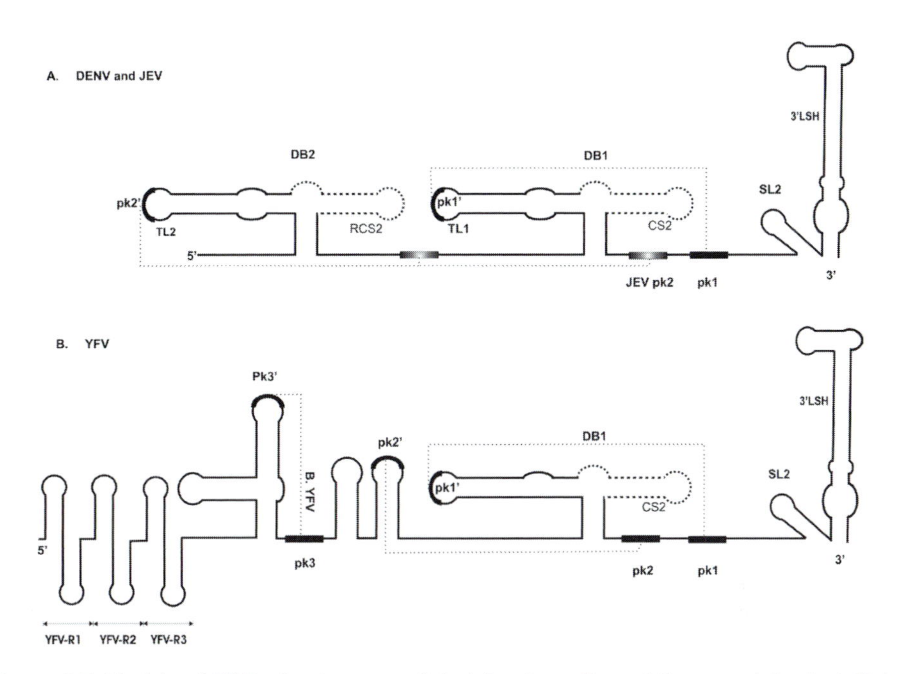

Figure 3.7 Models of RNA structures predicted the formation of the pseudoknots (pk) for domain II of the MBFV and repeated sequences of the YFV. The positions of the 3′LSH and SL2, two dumbbell-like structures (DB1 and DB2) and YFV repeats (YFV-R1, YFV-R2 and YFV-R3) are specified. Conserved sequences and their repeats (CS2 and RCS2) are shown as dashed lines. JEV and DENV (A) are similar in shape and shown as one model whereas YFV has some differences (B). Pseudoknots common between DENV and JEV are shown as black boxes and individual pseudoknots (for JEV or DENV) are shown in grey-shadowed boxes. Terminal loops 1 and 2 (TL1 and TL2) that were identified in the model of (Proutski *et al.*, 1997b) are indicated. Dotted lines join the interacting pseudoknots. Adapted from (A) (Olsthoorn and Bol, 2001) and (B) (Mutebi *et al.*, 2004).

ported by distant tertiary interactions between pseudoknots. Similar dumbbell-like structures were also described using the other RNA-folding algorithm (Rauscher *et al.*, 1997). The DB1 structures were similar in shape between the three MBFV groups and even (although partially) with TBFV. Therefore, this type of analysis reveals more extensive structural similarity between the three MBFV groups and the TBFV group. These studies also demonstrated what type of secondary structures the flavivirus 3′UTR was tending to preserve during its evolution and illustrated how the conserved elements CS2 and RCS2 are exposed on the surface implying a possible signaling function for them. As would be expected, the DENV and JEV shared structural similarity. The YFV and TBFV groups demonstrated different folding patterns in the DB2 region although the same principle of distant pseudoknot interactions was demonstrated.

The third set of data considered RNA at the stage that probably directly precedes the initiation of negative RNA strand synthesis from the positive-strand—the circularization between 5′ and 3′ ends of flavivirus genomes (Khromykh *et al.*, 2001; Thurner *et al.*, 2004). The idea of circularization of flavivirus genomes originated from the discovery of the "classical" absolutely conserved (between MBFV) circularization motif (CYCL) CATATTGA,

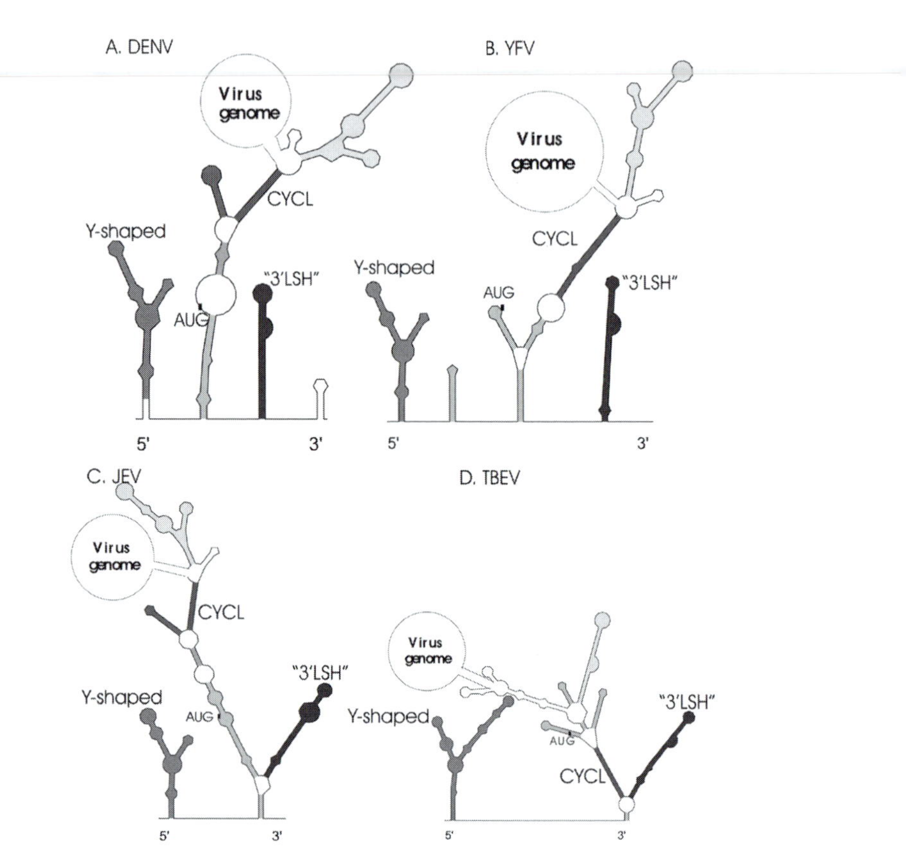

Figure 3.8 Model of RNA secondary structures that predicts the circularization of the genome of flaviviruses. The positions of the AUG initiation codon, CYCL sequence and elements of circularization stem have been indicated. The Y-shaped structure at the 5′UTR that is folded independently from the circularization stem is shown. The part of the side stem–loop that overlapped with the 3′LSH is specified as "3′LSH." Adapted from (Thurner *et al.*, 2004)

that is a part of CS1 and as an inverted repeat (complement) is also present in the capsid protein gene, in close proximity to the 5′UTR (see below) (Hahn *et al.*, 1987). Figure 3.8 illustrates the latest circularization model of the secondary structures produced for MBFV and TBFV (Thurner *et al.*, 2004). As can be observed, in addition to the "classical" CYCL, several other short inverted (complementary) repeats within the 5′ and 3′ UTRs and also coding regions have formed long "circularization stem." However, some sequences within the 3′UTR still fold independently of the "circularization stem" branching as a side stem–loop within the 5′UTR and 3′UTR (discussed later).

The question arises, which structures in the 3′UTR are real? RNA is a flexible molecule and thermodynamic computer-simulations only predict minimum energy conformations that are likely to be the most stable. None of the programs considers that within a cell environment the RNA interacts with virus proteins (see later) which can alter the energy of folding and therefore allow the formation of structures that are thermodynamically less preferable. An evolutionary approach in this situation might complement the structural predictions by revealing compensatory mutations in stem structures between closely related viruses and thus could provide evidence for the stability and functional significance of these structures. In addition the absolute conservation of sequences that form loops or stems could indicate a signaling function, i.e. whether or not these structures could interact with proteins or with other single-stranded RNA regions.

It is quite possible that all the predictions for the 3′UTR of flaviviruses (i.e. GA of 3′UTR during the synthesis of positive-strand RNA from the double-stranded template, formation of pseudoknots due to a distal RNA–RNA interaction, circularization of virus RNA to initiate synthesis of negative RNA strand from the positive-strand) reflect different stages of the life cycle of the positive-strand RNA in infected cells and each of the described structures could have an independent function during each stage. From this point of view the 3′UTR of flaviviruses may represent a multifunctional domain, which could change its conformation during transition from one stage to another and thus make RNA available for multiple interactions that are required during different stages of virus replication.

Taking these considerations into account, what is the purpose of the CSs for the MBFV? Do they form stems and loops or do they form double helices to circularize the flavivirus genome? Do they have any relation to pseudoknots or structures that are supported by the pseudoknots?

Links between phylogenetic and structural analysis of the 3′UTR of mosquito-borne flaviviruses

In search of the answers to some of the questions posed above, the CSs were superimposed onto the secondary structures depicted in Figures 3.5–3.8 and then the three groups of secondary structures described above were superimposed on the comparative alignment of the MBFV groups. The annotated alignment (Figure 3.2, http://www.horizonpress.com/hsp/supplementary/flavi/) specifies the individual elements of the secondary structures according to the predictions of all three structural models, i.e. GA-model of the 3′UTR, the formation of the pseudoknots and circularization. For convenience the analysis will commence with domain III, and will then follow with domain II and finally domain I.

Domain III

The 3′LSH (~90 nucleotides long) comprises the main structure of domain III and is located at the 3′ end of the 3′UTR and although its sequences are not highly conserved, they still show significant homology within the MBFV group. As the alignment shows this region contains features associated with circularization. The distal end of the 3′LSH (Figure 3.5) is also folded as a stem–loop structure that branches independently from the long "circularization stem" (Figure 3.8). These structures share terminal loop 1 with the pentapeptide CACAG which is conserved between the MBFV (Brinton *et al.*, 1986). The regions involved in the formation of this side stem structure (underlined in Figure 3.2, http://www.horizonpress.com/hsp/supplementary/flavi/) were still formed by similar but variant sequences of MBFV. The other feature of this domain is the absolute conservation between all flaviviruses of two terminal nucleotides CU_{OH} (see below).

The proximal part of the 3′LSH from its 5′ end contains sequences that are directly involved in genome RNA circularization (Figures 3.2, http://www.horizonpress.com/hsp/supplementary/flavi/, and

3.8). Thus two possible functions could be assigned to this region—the formation of 3′LSH and direct involvement in circularization. Absolute conservation of the pentapeptide CACAG as a single-stranded region implies a signaling function, possibly recognition by some proteins of the polymerase complex. Both structures, i.e. 3′LSH and the "circularization stem" are considered to be thermodynamically stable and supported by the co-variant analysis. It is possible that both exist, at different stages of the RNA life cycle and are equally functionally significant. Therefore, the conservation of sequences in this region may be under two forms of selective pressure, firstly to preserve 3′LSH and, secondly to circularize the virus genome. The function of the 3′LSH could be firstly to expose loop 1, containing the conservative pentapeptide, which would interact with protein(s) of the replication complex and secondly to preserve the stem and circularization sequences for long range interactions with the 5′ end of the genome.

Domain III also contains another stem-structure, SL2, that is involved in the formation of a pseudoknot with the stem of 3′LSH (Figure 3.5). The alignment shows that SL2 is partially formed by CS1 that is located downstream of CYCL (Figure 3.2, http://www.horizonpress.com/hsp/supplementary/flavi/). On the other hand, although CS1 encodes the circularization sequences they are formed by different parts of CS1 for the different MBFV (Figure 3.2, http://www.horizonpress.com/hsp/supplementary/flavi/). The sequence of the 3′LSH involved in forming a pseudoknot with SL2 is also highly conserved (Figure 3.2, http://www.horizonpress.com/hsp/supplementary/flavi/). Thus CS1 is not covered completely by any secondary structures compatible with either the GA or circularization models. Therefore, in addition to CYCL and the SL2, described within CS1, it might also function as a single-stranded region that is essential for protein(s) recognition.

The other unanswered question is, what is the reason for the absolute conservation of the CYCL motif in comparison with other more variable "nonclassical" circularization motifs? The double-stranded regions of the "circularization stem" do not need to be formed by the perfectly conserved sequences; the stems could be supported through less well conserved sequences by compensatory mutations. The absolute conservation of CYCL motifs at both 5′ and 3′ ends almost certainly indicates that these sequences are directly involved in interaction with proteins, either individually as single-stranded stretches or as a part of the "circularization" stem or both.

In summary, based on only theoretical predictions it is very difficult to assign any specific function to the CS1 region and more research is required to understand the reasons for its high degree of conservation.

Domain II

In the GA model, domain II of DENV and JEV are similar to each other but different from YFV; domain II is formed from two symmetrical hairpins with terminal loops TL1 and TL2 (Figure 3.6). The positions of the stems and TLs are specified on the alignment (Figure 3.2, http://www.horizonpress.com/hsp/supplementary/flavi/). Two symmetrical bulges that branch from each of the two hairpins (numbered 3 and 5) were formed by the highly conserved CS2 and RCS2 (Figure 3.6) indicating that they could be recognition signals for other molecular interactions. Similarly the sequences within loop TL1 were perfectly conserved between DENV and JEV and therefore could have an exposed position suitable for interactions with proteins or pseudoknot formation. However, the sequences that form the TL2 in DENV and JEV are different. They are also different within the JEV group although within the DENV group the TL2 sequences are conserved.

According to the GA model YFV has a completely different folding pattern within this region; the CS2 region is involved in the formation of a short hairpin that is not similar to the structures formed by the equivalent CS2 of DENV or JEV (not shown here) (Proutski *et al.*, 1997b). By superimposing these structures onto the alignment it can be seen that for YFV this region was formed by different sequences from those used for the formation of the domain II for the DENV and JEV groups (Figure 3.2, http://www.horizonpress.com/hsp/supplementary/flavi/).

However, an alternative folding pattern—based on the formation of pseudoknots—revealed more similarity between the three major MBFV groups (Figure 3.7) (Olsthoorn and Bol, 2001). Although the sequences and positions of the pseudoknots were different between DENV/JEV, compared with YFV, the RNA secondary structures were very similar in conformation, comprising two major dumbbell-like structures supported by pseudoknots. The pseudoknot model of folding also illustrates the possible significance of highly conserved "classical" CS2 and RCS2; these sequences are exposed on the top of the dumbbell-like structures DB1 and DV2, supported by the pseudoknots. These two structures may have been preserved because of the functional necessity to act together, whether or not they are similar, as for the DENV/JEV group (DB2), or different, as for the YFV group (D1). In other words the function of DB1 might be significant and associated with the exposure of the CS2. However, it could depend on the function of the second dumbbell-like domain DB2, that evolved into differently shaped structures for YFV (Figure 3.7) and TBFV (see Figure 3.13). CS2 and RCS2 may be signals for the protein, the functional activity of which depends upon dimerization/multimerization and the presence of the second signal could speed up the process of assembly required for virus replication.

In summary there are two possible models of folding for this region. Both include TL1 and TL2 as single-stranded loops supported by double-stranded stems (compare Figures 3.6 and 3.7); the CS2 and RCS2 for DENV and JEV fold in a similar manner in both models. Nevertheless, the pseudoknot model reveals greater structural similarity between the three major groups of MBFV and adequately correlates with phylogenetic studies. The most important discovery is the similar folding of the CS2 regions between all three groups of MBFV which makes sense in evolutionary terms, i.e. the high level of structural conservation with the exposed CS2 as a signaling sequence suggests a similar function within all three virus groups.

In this context should we then "discard" the discovery of the domain II as a gradually folded structure? Probably not, if we again picture RNA not as a static molecule but a floating structure that changes conformation during transition from one stage of the virus cycle to the other. The structures discovered in the GA model for domain II may be involved in the stage preceding the formation of more stable pseudoknot-supported dumbbell-like structures; this would explain why these models partially overlap. The formation of metastable structures as intermediates probably has less evolutionary significance than the formation of dumbbell-like structures and explains the structural differences between domains II for DENV/JEV and YFV predicted by the GA model. This might explain why metastable structures are less conserved between divergent viruses than those supported by pseudoknots. Thus, with the CS2 exposed on the surface of these structures, they could be directly involved in interactions within the complex replication machinery.

Domain I

The boundary between domains I and II for the DENV and JEV groups is located in different regions of the alignment (Figure 3.2, http://www.horizonpress.com/hsp/supplementary/flavi/). For the JEV group it starts from the CS3 that has been defined as a conserved element specific to the JEV group although "remnants" of this sequence might be traced in the other MBFV groups. The position of the DENV and JEV group hairpins (enumerated, 7, 8 and 9 on Figure 3.6) demonstrates that they are formed within regions of poor-alignment. The absence of strict conservation in both sequences and structures suggests that the GA model predicts intermediate metastable structures which present the initial stage in the folding of the entire 3′UTR as an RNA domain. Therefore, these hairpin structures might provide a supportive role for the subsequent more significant folding of domains II and III. As intermediate structures they probably do not provide a signaling function and their conservation is not essential although it is preserved structurally between the more closely related DENV and JEV groups. Domain I is the most highly variable region in the MBFV and only remnants of sequence similarity remain. Its close vicinity to the nonessential domain of the flavivirus 3′UTR also suggests that domain I might merely enhance the functioning of the whole 3′UTR rather than being essential for virus infectivity.

Based on theoretical research it is difficult to understand the significance of the CS3 and RCS3 for the JEV group. The GA model established that CS3 accommodates hairpin 7 within the JEV group; later this folding was extended to the RCS3 showing similar presentation of the RCS3 at the distal part of the hairpin (Figure 3.6) (Proutski *et al.*, 1999). Whether or not these two regions are preserved in the JEV group due to supportive activities essential for the 3′UTR is not clear. They could represent the remains of earlier sequences that have effectively been lost during evolution. However, the absolute conservation of sequences within this large divergent subgroup suggests that they, like CS2/RCS2, might provide a signaling function, for example facilitating protein dimerization or interacting specifically with other regions of the RNA.

Additionally predictions have been made for domain I of YFV and the YFV-related viruses (Figure 3.7). Using the same GA algorithm for RNA folding, it was established that each YFV repeat (YFV-R1, YFV-R2 and YFV-R3) forms symmetrical short stem–loop structures; predictions were based on both thermodynamic characteristics and co-variant analysis of substitutions between repeat sequences of each YFV and YFV-related flavivirus (Mutebi *et al.*, 2004). Whether or not these structures play a significant role in functioning of the virus genome or merely support the folding of a downstream domain (acting as a spacer) remains to be determined.

In summary the theoretical predictions can be interpreted as follows:

1. Domain III is at least bi-functional, it is involved in formation the 3′LSH with a highly conserved pentapeptide exposed as a loop and it also circular-

izes viral RNA through a complementary interaction with the 5′ end. The function of CS1 and CYCL as a 5′ part of CS1 is still not absolutely clear although the conservation suggests a signaling function, probably through direct interaction with protein(s) of the polymerase complex.

2. Domain II is formed by the two dumbbell-like structures DB1 and DB2 supported by pseudoknots. Their possible function is to expose conserved regions CS2/RCS2 for interaction with proteins and/or to support folding before/during the initiation of minus strand RNA synthesis.
3. Domain I is probably required only to ensure efficient folding of domains II and III, for example by separating the folding of domains II and III from the coding part of the genome. Therefore the role of secondary structures in domains I might be restricted by the principle only of "noninteraction" with domains II and III. However, in the JEV group the conserved elements CS3 and RCS3 might have a signaling function. Whether or not YFV-R1, YFV-R2 and YFV-R3 in domain I have a significant function in virus replication remains to be determined; their role might be only supportive for efficient folding of other downstream domains and their conservation only in some virus strains might be explained by slow evolutionary development of these particular strains in their particular environment. Nevertheless the fact of their conservation indicates the presence of signaling activity in this region for this particular group of strains.

Do experimental data support these suggestions? Most experiments were performed before the complex RNA-folding algorithms were applied to analyze virus UTRs and therefore they were not designed according to these concepts. Nevertheless, we can still learn much by analyzing the results of such experimental data.

Experimental evidence of RNA structures in mosquito-borne flaviviruses

Initial evidence for RNA structure in 3′LSH came from the partial digestion of RNA when the regions resistant to single-strand specific ribonucleases were identified (Brinton *et al.*, 1986; Mohan and Padmanabhan, 1991). Further evidence came from the discovery of self-priming 3′UTR during cDNA synthesis *in vitro* (Hahn *et al.*, 1987). Circular dichroism spectral analysis and ribonuclease probing provided evidence not only for the 3′LSH and SL2 but also for the formation of a pseudoknot structure between the SL2 stem-part of the 3′LSH (Shi *et al.*, 1996a). Nevertheless, the possibility that the formation of this structure could be an artifact of the deproteinization of RNA during phenol extraction could not be excluded.

Evidence of the structure for the 3′SLH was found in work in which mutations that destroy the formation of the distal part of the stem were lethal for the virus (Zeng *et al.*, 1998) and significantly inhibited the synthesis of RNA *in vitro* (You *et al.*, 2001). However, other explanations have also been proposed (see below). The formation of a pseudoknot between SL2 and the bulging region of the long stem at 3′LSH (Figure 3.5) was confirmed by experiments *in vitro* and *in vivo*: a single nucleotide substitution that destroyed the formation of the pseudoknot significantly reduced the synthesis of RNA from the isolated 3′UTR and was also lethal for the virus (You *et al.*, 2001).

Functional significance of the 3′UTR and the individual elements of the mosquito-borne flaviviruses

Since its discovery, the 3′UTR was believed to be a domain that interacts with the viral RNA polymerase; it was assumed that it acts as a promoter of virus replication. This idea arose from comparison with the 3′ UTRs of plant viruses and bacteriophages with similar RNA genome strategies. Therefore, experimental research was designed to (1) look for a polymerase function, (2) understand the significance of the conserved elements in their interaction with viral and cellular proteins, and (3) understand their role in virus replication.

Experiments on the entire 3′UTR and 3′LSH

In many experiments the properties of the whole 3′UTR domain and 3′LSH were studied simultaneously and therefore to avoid repetition they will be considered together.

The 3′UTR and 3′LSH bind NS5 and NS3 proteins of the viral polymerase complex

The first evidence of the functional significance of the 3′UTR as a domain that initiate viral RNA synthesis came from experiments on the isolated 3′UTR of DENV that bound the bacterially expressed recombinant NS5 protein (viral RdRp) (Tan *et al.*, 1996). Subsequent supporting evidence came from studies of WNV and JEV: the isolated positive-stranded 3′UTR was shown to bind to the NS5 protein (110kDa) and also to the viral helicase protein (NS3 protein, 71 kDa) in infected BHK cells (Chen *et al.*, 1997). Subsequently it was demonstrated that the binding of NS5 protein to the 3′UTR is carried out through the 3′LSH whilst the NS3 protein binds nonspecifically to different parts of the 3′UTR (Chen *et al.*, 1997; Cui *et al.*, 1998; You and Padmanabhan, 1999), as well as to other RNAs. Whether the NS5 protein binds to the 3′LSH specifically or through interactions with cellular proteins, or other RNA domains remains to be determined.

The 3′UTR directs RNA synthesis in an in vitro *RdRp assay*

The 3′UTR was shown to direct RNA synthesis in an *in vitro* RdRp assay using an expressed recombinant NS5 polymerase, in the absence of other viral proteins (Tan *et al.*, 1996). However, the product was characterized as double-stranded RNA corresponding in size to the recombinant 3′UTR and was believed to have been synthesized not by initiation but by the copy-back mechanism. In contrast, in other experiments it was demonstrated that cytoplasmic extracts from infected mammalian or mosquito cells, or the purified NS5 protein of DENV failed to initiate RNA synthesis *in vitro* on the 3′UTR unless the 5′end of the genome (with the CYCL sequence) was supplied in *cis* (on the same molecule as the 3′UTR) or in *trans* (from a different RNA molecule) (Ackermann and Padmanabhan, 2001; You *et al.*, 2001; You and Padmanabhan, 1999). This implied that circularization of the genome was essential for the initiation of viral RNA transcription (see below).

3′UTR and 3′LSH interacts with cellular proteins

The interaction of cellular proteins with the 3′UTR was established in several studies. In some of them the experiments were carried out for the whole 3′UTR domain whilst for others the research was performed only on the 3′LSH. Firstly, using WNV, an interaction between the 3′LSH and cytoplasmic cellular proteins from uninfected BHK cells with molecular masses of 105, 84 and 56 kDa was observed (Blackwell and Brinton, 1995; Blackwell and Brinton, 1997).

The protein p56 was identified as a 50 kDa translation elongation factor alpha (EF1-α). A protein with identical molecular mass which interacted with the isolated 3′UTR of JEV was identified in the brains of neonatal mice (Ta and Vrati, 2000) and in a mosquito cell line (De Nova Ocampo *et al.*, 2002). The normal function of this protein in cells is the formation of a ternary complex EF1-α-GTP-aa-tRNA to facilitate the binding of ribosomes to mRNA before the initiation of translation. The activity of the ternary complex depends on the phosphorylation of the EF1-α and the phosphorylation of EF1-α was required for binding with 3′LSH. This supports similar data for other positive-stranded RNA viruses for which different cellular translation factors were shown to be involved in viral RNA synthesis. In addition to the initiation of translation, the EF1-α might also be involved in targeting mRNA to the cell-specific sites of translation. Therefore, for WNV the function of the EF1-α was assumed to be the assembly of a polymerase complex around the 3′ SLH and also targeting it to the internal cell membranes, that provide the optimal environment for viral RNA synthesis (Blackwell and Brinton, 1997).

Two other unidentified proteins p105 and p84 in uninfected BHK cells bind to different sites of the 3′ SLH. It is believed that these two proteins specifically recognize the 3′LSH prior to their interaction with viral NS5 and NS3 proteins (Blackwell and Brinton, 1995). In mosquito cells the 3′LSH bound the same set of proteins as the whole 3′UTR region although the amount of individual proteins varied between these two binding complexes (De Nova Ocampo *et al.*, 2002). Whether or not these complexes have any functional significance for the virus life cycle remains to be determined.

Similar binding experiments using mouse brains established that in addition to p50, the 36-kDa protein was probably involved in an interaction with the whole 3′UTR. The protein was assumed to be murine Mov34, a protein that is required for normal progression of the cell cycle. For example it has been shown that interaction of Mov34 with one of the HIV proteins arrested the cell cycle and similar observations have been made in cells infected with JEV (Ta and Vrati, 2000). The Mov34 protein is a component of the mouse cellular 26S proteosome, a multifunctional protein complex that is involved in, degradation of ubiquitinated proteins, cell control, early steps of the immune response, and regulation of transcription. It is believed that in relation to JEV-3′UTR this protein like EF-1α directs transport and localization of the virus RNA to the replication site (Ta and Vrati, 2000).

In addition recombinant human La autoantigen (52 kDa) and PTB protein (57/60 kDa) have been shown to bind to the 3′UTR of DENV and other RNA. The normal cellular function of La has not been defined precisely but it might act as a chaperone protein supporting RNA in a conformation that either favors translation, or protects RNA from degradation. The PTB normal cellular function is splicing and it remains to be determined how these two proteins are involved in flavivirus replication (De Nova Ocampo *et al.*, 2002).

Eight proteins with molecular masses 34, 39, 51, 52, 56, 62, 72 and 84 kDa from both infected and uninfected mosquito cells have been shown to react with the entire 3′UTR and its truncated forms (De Nova Ocampo *et al.*, 2002); they have not yet been identified.

The involvement of the 3′LSH in host protein interactions is also evident from experiments involving mutagenesis on an infectious clone of DENV2: substitution of the proximal double-stranded region of the 3′LSH of DENV2 for the equivalent region of WNV resulted in more severe growth restriction of the mutant virus in the mosquito cells in comparison with the primate cells indicating that this domain of the 3′UTR is recognized by host factors (Zeng *et al.*, 1998).

The 3′UTR and 3′LSH might influence the initiation of virus translation

As with many other cellular and viral mRNAs, the 3′UTR of flaviviruses may function to enhance virus translation in a manner analogous to the 3′ poly(A) tail (Holden and Harris, 2004). The efficiency of translation *in vivo* was evaluated using synthetic RNA transcribed from a plasmid containing a reporter luciferase gene sited between the DENV2 5′UTR and 3′UTR. The 3′UTR increased translation of the DENV2 5′UTR and also other recombinant capped 5′ UTRs and even the IRES. Approximately half of the translation efficiency was due to the terminal 3′LSH domain. However, mutations located upstream of the 3′LSH of DENV2 also contributed to the efficiency of translation (Edgil *et al.*, 2003). Neither viral proteins nor direct RNA–RNA interactions were required to mediate this effect. It was suggested that the 3′UTR/3 LSH probably binds directly to a protein(s) that participates in the assembly of the ribosomal initiation complex (Holden and Harris, 2004).

However, this contradicts other data (Li and Brinton, 2001) where it has been shown both *in vitro* and *in vivo* that the WNV 3′LSH inhibits translation of the reporter CAT protein in chimeric RNA with the CAT gene placed between the 5′UTR and 3′LSH. The 3′LSH also reduced the translation of capped and uncapped mRNAs with either viral or nonviral 5′UTR in both *cis* and *trans* suggesting that direct contact between 3′UTR and 5′UTR of flaviviruses was not involved. It was proposed that competition between the 5′UTR and 3′UTR for access to cellular proteins involved in translation initiation could explain the translation inhibition observed.

Similar conclusions were drawn from experiments using a WNV replicon, where it was shown that removal of the whole 3′UTR or CS2- CS1–3′LSH slightly increased (by about 20%) translation (Tilgner *et al.*, 2005). In contrast to (Edgil *et al.*, 2003) the deletion of the region upstream of the 3′LSH, between the stop codon and RCS2, did not affect translation (Tilgner *et al.*, 2005). The removal of either CS1 or CS2 also did not alter virus translation (Lo *et al.*, 2003). These data are in good agreement with studies of the inhibitory effects of antisense oligonucleotides designed to bind to specific regions of the WNV 3′UTR (Deas *et al.*, 2005). Oligonucleotides (20-mers) that bind the CS1, loop1 which includes the conserved pentanucleotide CACAG, and stem of the 3′LSH, inhibit replication but not translation of virus-based replicons.

In summary, the preceding description implies that the regulation of DENV and WNV translation utilizes different strategies. Alternatively, of course, this could merely reflect the differences in design and interpretation of the experiments.

The 3′LSH is absolutely required for virus viability

In experiments with DENV and YFV infectious clones, deletion of either the entire or a part of the 3′LSH totally abolished virus replication (Bredenbeek *et al.*, 2003; Men *et al.*, 1996) whereas the substitution of the DENV 3′LSH with the equivalent WNV 3′LSH resulted in severe growth restriction of the virus (Zeng *et al.*, 1998). However, sequences involved in the formation of the 3′LSH and in the circularization are overlapped (Figures 3.2, http://www.horizonpress.com/hsp/supplementary/flavi/ and 3.8 and see Figure 3.10, http://www.horizonpress.com/hsp/supplementary/flavi/), therefore, these experiments probably indicate that the abrogation of circularization rather than perturbation of 3′LSH folding *per se* could explain the loss of virus reproduction (see below). There are no experiments that differentiate between the two folding models, 3′LSH or circularization, but the linear sequences involved in either model, are indispensable for virus viability.

The 3′LSH is involved in replication and probably encapsidation of the virus genome

Later we shall demonstrate how the 3′LSH sequence is involved in replication of the flavivirus genome through its circularization, but here we show that it is probably also involved in other stages of the virus life cycle. The influence on viral RNA synthesis of the 3′LSH on its own, and/or as a part of circularization sequences is evident from the experiments in which elements of the DENV 3′LSH were substituted with those of WNV (You *et al.*, 2001; Zeng *et al.*, 1998). The *in vitro* replication rate of the chimeric 3′LSH was compared with the level of virus infectivity in cells. Two groups of mutants were identified. In the first group the infectivity of chimeric DENV/WNV-3′LSH mutants (Zeng *et al.*, 1998) correlated directly with efficiency of transcription *in vitro* (You *et al.*, 2001). This implies that the integrity of the 3′LSH sequence is essential for virus replication. However, in the second group, whilst the replication rate *in vitro* was high virus infectivity was either very low or completely absent; it was suggested that the 3′LSH region on its own or as a part of the circularization stem could be involved other stages of the virus life cycle, such as encapsidation.

Alterations in the 3′LSH sequence reduce virus infectivity in mosquito cells more than in mammalian cells

Host-range phenotypic mutants were produced by substitution of the proximal part of the 3′LSH of DENV2 with the equivalent portion of the WNV genome; this mutant replicated efficiently in primate cells whereas in mosquito cells its growth was severely restricted (Zeng *et al.*, 1998). However, an *in vivo* RdRp assay showed little difference between either infected primate or mosquito cells (You *et al.*, 2001). Therefore in addition to replication, the 3′LSH is probably involved in other functions that have different requirements in different hosts.

The conserved pentanucleotide CACAG in the 3′LSH is essential for virus replication

Initially the significance of the conserved pentanucleotide CACAG was demonstrated by mutating the first three nucleotides of the Kunjin virus replicon (Khromykh *et al.*, 2003). Subsequently each individual nucleotide was substituted (Tilgner *et al.*, 2005) and it was shown that each is critical for RNA replication. The conserved nucleotides at the 1st, 2nd, 3rd, and 5th positions were also essential for RNA synthesis whereas substitutions in the 4th position of the pentanucleotide allowed virus replication albeit at a lower rate. This explains why substitution in the 4th position for Murray Valley encephalitis virus (MVEV) and cell fusing agent virus (CFAV) are not lethal (Charlier *et al.*, 2002). In addition, substitution of the nucleotide U (which is partially conserved in the genus *Flavivirus*) immediately downstream of the pentanucleotide did not totally inhibit replication of the replicon (Khromykh *et al.*, 2003) although its rate was reduced. This supports the observation that U→C substitutions at the same position in the infectious clone of TBEV caused minor attenuation of virus reproduction in cell culture and mice (Gritsun *et al.*, 2001).

After the sequences of some NKV viruses became available (Charlier *et al.*, 2002), it became clear that conservation of the pentanucleotide sequence occurs only amongst vectored flaviviruses. All nonvectored flaviviruses have C or U in the 2nd position and additionally APOIV virus has a U in the 3rd position. Therefore it is possible that this pentanucleotide reflects the adaptation of viruses to different hosts (through interaction with host-specific proteins) or to mutated NS5 protein.

The next section reviews experiments performed to understand the significance of the individual elements of the 3′UTR other than 3′LSH and how they make this domain function as a virus promoter and enhancer.

Experiments on the individual elements of the 3′UTR

Terminal CU_{OH} nucleotides

The two terminal nucleotides CU_{OH} of the 3′LSH are highly conserved among all flaviviruses. They are also present at the 5′UTR terminus as the complementary inverted dinucleotide 5′pppAG, thus comprising the CU_{OH} of the 3′end of the negative RNA strand. The significance of this conservation was demonstrated in experiments on the replicon where it was shown that mutations of these nucleotides at both 5′ and 3′ ends completely blocked Kunjin virus replication (Khromykh *et al.*, 2003). Conservation of the two terminal nucleotides is recognized for many RNA viruses and imposes stringent selective constraints from the active site of RdRp that accommodates these two nucleotides in a specific molecular tunnel of viral RNA polymerase (van Dijk *et al.*, 2004).

Conserved sequences (CSs)

CS1 is essential for virus survival.

The infectious clone of DENV4 (Men *et al.*, 1996) and of YFV (Bredenbeek *et al.*, 2003) were each used to introduce deletions into the 3′UTR to establish the significance of conserved linear elements. The removal of CS1 alone or in combination with CS2 completely abolished infectivity of both viruses indicating that CS1 is essential for virus replication; this is reflected in its virtually complete conservation amongst widely divergent MBFV.

CS2, RCS2, CS3, RSC3, YF-R1, 2, and 3 are nonessential for virus viability but essential for efficient replication

The significance of the region between the stop-codon and CS1 has been analyzed using DENV and YFV infectious clones (Bredenbeek *et al.*, 2003; Men *et al.*, 1996). The removal of CS2 was not lethal for DENV or YFV although both viruses had reduced replication rates. Additional deletions from CS2 progressively upstream towards the stop codon that remove CS2 in combination with RSC3 or with other upstream sequences, were introduced leaving the region downstream of CS2 intact (Men *et al.*, 1996). All these viruses remained infectious although they exhibited a spectrum of growth restrictions in cell culture. The boundary between viable and lethal deletions within the region CAAAAA (between nucleotides 107 and 113) lies immediately upstream of the CS1; deletion mutants within this sequence yielded infectious virus whereas the equivalent virus with this sequence removed was not infectious (Figure 3.4 and see Figure 3.11, http://www.horizonpress.com/hsp/supplementary/flavi/).

Notably, even the entire removal of the region between and including CS2 and the stop codon produced virus that replicated in mosquito but not in simian cells. In general, the impact of 3′UTR deletion mutants in this region was more profound in simian than in mosquito cells implying that this region could determine host cell susceptibility to the virus (Men *et al.*, 1996).

Subsequently these deletions were transposed into the RNA secondary structures predicted by the GA model (Proutski *et al.*, 1999). This analysis was undertaken in an attempt to understand how sequences of virus deletion mutants could be correlated with changes to the secondary RNA structures that such deletion might produce. Figure 3.9 illustrates the type of analysis used in this research.

Most of the deletions removed only one of the symmetrical structures specified in the GA model, either TL1 or TL2, with or without CS2 and RCS2 respectively (Proutski *et al.*, 1999). Removal of one structure, CS2 or RCS2, resulted in relatively mild phenotypic consequences. As these structures were similar in shape and they exposed similar sequence motifs, it was suggested that their functions might be similar and compensatory. This was supported by the observation that removal of both structures resulted in more significant consequences, i.e. the mutant described above with both TL1/ CS2 and TL2/RCS2 deleted can replicate only in mosquito cells. Interestingly, removal of TL2/RCS2 alone reduced virus growth less significantly than the removal of TL2/RCS2 together with upstream sequences corresponding to the nonessential region of the 3′UTR. This could indicate that the term nonessential in this respect relates only to virus infectivity, i.e. not to efficiency of replication.

Analysis of RNA secondary structures also explains why shorter deletions sometimes have more drastic effects on virus reproduction than longer deletions: they could produce more significant perturbations in 3′UTR folding than longer deletions and they could interfere with the correct folding of distant domains. For example, the sequence CAAAAA which has been reported to influence "infectivity or lethality" may be a spacer sequence that provides independent and efficient folding of the 3′LSH region and its removal could promote misfolding of the 3′LSH disabling its function. One might further speculate that the deletion of CAAAAA destroys one of the pseudoknots (P1) that is perfectly conserved between the three MBFV groups (Figure 3.2, http://www.horizonpress.com/hsp/supplementary/flavi/). Another example is the observation that point mutations in the 3′UTR of YFV might be partly responsible for the attenuated characteristics of the YFV-17D vaccine as these mutations are conformational for native folding structures (Proutski *et al.*, 1997a).

Similar conclusions resulted from experiments on the infectious clone of YFV (Bredenbeek *et al.*, 2003). Deletion of the CS1 element abolished virus replication whereas deletion of CS2, RSC2, or YFV-

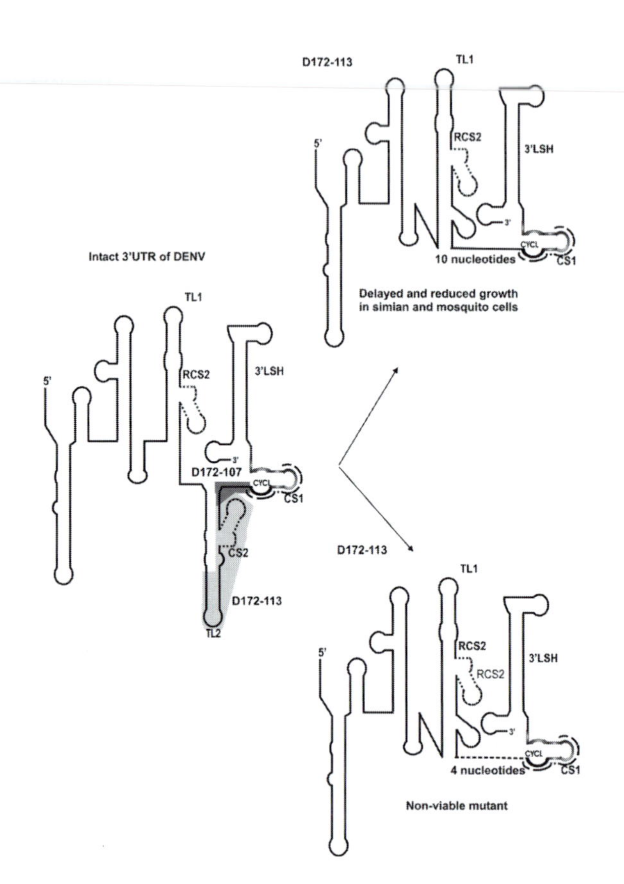

Figure 3.9 This example illustrates the effect of two overlapping mutations on folding of the 3′UTR of DENV and establishes the nucleotide boundary between viable and nonviable viruses. Shorter deletion D172–113 (the region between nucleotides 172–113) is shown in light grey shading and the deletion of an additional nucleotide in the mutant D172–107 is shaded dark grey. The other features of the 3′UTR are the same as on Figure 3.8. Adapted from (Proutski *et al.*, 1999).

R1, YFV-R2 and YFV-R3 still yielded viable virus. Nevertheless mutants with deleted CS2/RSC2 or YFV-R1–YFV-R3 regions formed smaller plaques in comparison with parent virus. These experiments clearly indicate that although they are non-essential for virus replication, CS2 and the three YF-specific tandem repeat sequences accelerate virus replication which might explain their preservation during the evolution of flaviviruses.

Similar results have also been observed with the CS3 and RCS3 of viruses in the JEV group. Instead of infectious clones, these studies utilized replicons with which it is possible to distinguish between replication and other stages of virus reproduction. For Kunjin virus, deletion of 352 nucleotides including the two conserved sequences RCS3 and CS3, significantly inhibited RNA replication (Khromykh and Westaway, 1997) once again implying that these JEV-group-specific elements facilitated rapid replication but were not essential for virus infectivity. Subsequent analysis of the RNA secondary structures of these modified replicons revealed that in addition to CS3-RCS3, the construct was also effectively deficient in TL1 (Proutski *et al.*, 1999). Therefore the reduction of replication could result from the cumulative effect of the simultaneous removal of several elements.

Additionally the individual deletion of CS2, RCS2, CS3 or RCS3 in a WNV replicon (JEV group) did not totally abolish virus RNA synthesis but it was significantly reduced (Lo *et al.*, 2003). Nevertheless, complete deletion between the RCS2 and the stop codon, with retention of the region downstream of the CS2, completely abolished virus replication (Lo *et al.*, 2003; Tilgner *et al.*, 2005) probably through structural misfolding of the mutated 3′UTR.

In separate experiments the region of the 3′UTR that includes CS2 and RCS2 was shown to bind to mosquito cellular proteins that are believed to be involved in virus replication, suggesting that CSs might have signaling activity (De Nova Ocampo *et al.*, 2002). No protein-binding activity was detected with the region upstream of the RCS2 of DENV (De Nova Ocampo *et al.*, 2002). However, the fragment of 3′UTR that contains the CS1–3′LSH did bind to the same mosquito proteins (although at a lower level) that had been observed for the whole 3′UTR or the region between RCS2–3′LSH. This probably indicates a "supportive role" of the RCS2- CS2 fragment for the more efficient binding of proteins to CS1–3′LSH.

Altogether experiments indicate that:

1. CS1 and 3′LSH are essential for virus replication.
2. MBFV can tolerate considerable structural perturbations within the 3′UTR if they do not interfere with CS1 and 3′LSH folding. The degree of structural alterations rather than the length of a deletion might be the most important factor that impacts on virus reproduction.
3. The different elements of the 3′UTR between the stop-codon and CS1 including CS2, RCS2, CS3, RCS3 and tandem repeats of YFV 3′UTR, have a cumulative effect on virus replication rate although none of these elements is essential for virus viability.
4. The part of the 3′UTR that contains the 3′LSH and SL2 (this includes CS1) is sufficient for virus infectivity in mosquito cells. However, the replication of the MBFV in mammalian cells additionally requires the CS2 or RCS2 elements.
5. Viruses tolerate structural perturbation of the 3′UTR in mosquito cells better than in mammalian cells.

Experimental evidence for the circularization of flavivirus genomes

As computer predictions demonstrated, the 5′ and 3′ UTRs of different flaviviruses interact with each other through inverted complementary sequences to circularize the virus genome (Figure 3.8). These sequences are identified on the alignment shown in Figure 3.2 (http://www.horizonpress.com/hsp/supplementary/flavi/).

Previously published results (Zeng *et al.*, 1998) in which different parts of the DENV2 3′LSH were substituted with equivalent parts of WNV could now be explained by circularization of the virus genome. The distal part of the 3′ SLH was substituted for WNV sequences without the loss of virus infectivity whereas sequence exchanges within the proximal part of the 3′LSH stem were lethal for the virus. These sequences overlapped with circularization sequences (Figures 3.2, http://www.horizonpress.com/hsp/supplementary/flavi/, and 3.8) and because in the case of WNV and DENV they are significantly different, the circularization of the virus genome RNA was not an efficient process.

In contrast the distal part of the 3′LSH coincides with the side stem–loop structures that branch from the "circularization stem" and is not involved in circularization (Figures 3.2 and 3.10, http://www.horizonpress.com/hsp/supplementary/flavi/). In terms of sequences and structures this genetic exchange between DENV2 and WNV might be acceptable and therefore explains why these chimeric viruses were viable, although they had reduced infectivity possibly due to imperfect circularization. It is worth noting that complete substitution of the 3′LSH of DENV with that of WNV resulted in considerable inhibition of virus replication (Zeng *et al.*, 1998) indicating that stable folding of the 3′LSH *per se* is insufficient to ensure virus infectivity. Interaction with other parts of the genome, probably at the level of circularization is also required although there is an alternative possibility: the stability of the 3′LSH is supported by interactions with the upstream domains of the 3′UTR.

Initially, direct evidence for circularization was provided only for CYCL, the "classical" sequence that is absolutely conserved within the MBFV group and is also present in the C gene as an inverted repeat sequence. Recombinant RdRp or RdRp from cytoplasmic extracts of infected mosquito cells was used to study the RNA synthesis of DENV *in vitro*. RNA synthesis from the 3′UTR required the simultaneous presence of both 5′ and 3′ ends each containing the CYCL sequence; the mutations that disrupt pairing between 5′ and 3′ CYCL abolished or reduced RNA transcription (Ackermann and Padmanabhan, 2001; You and Padmanabhan, 1999). The physical interaction between 5′and 3′ CYCL also *in vitro* was demonstrated by Psoralen-UV cross-linking (You *et al.*, 2001).

Similar conclusions were drawn from experiments on replicons constructed for viruses from the JEV group; several nucleotide substitutions that disrupt RNA circularization were introduced into CYCL in both the 5′ and the 3′ UTRs. These mutant replicons did not replicate whereas the introduction of compensatory mutations that restore the nucleotide pairing between the UTRs, reinstated virus RNA synthesis although not to the level of wild-type replicons (Khromykh *et al.*, 2001; Lo *et al.*, 2003).

The significance in circularization of other sequences that are encoded outside CYCL, was demonstrated firstly *in vitro*; the presence of a 130-nucleotide fragment of C protein gene that contains CYCL was not enough to initiate replication from the isolated 3′UTR; in addition, a 96-nucleotide 5′UTR was required that on its own did not initiate transcription (Charrel *et al.*, 2001). Similarly, in experiments with replicons of YFV it was demonstrated that CYCL on its own is not sufficient for replication. Longer neighboring regions of C protein, predicted to form the circularization stem, were also required (Corver *et al.*, 2003).

Although quite conclusive these experiments did not provide an explanation for the absolute conservation of CYCL: the stems do not need to preserve sequences as compensatory mutations would still support their folding. The absolute conservation strongly suggests a signaling function and it is possible that beside circularization this domain also binds protein(s), as a single or double-stranded region in a manner similar to that demonstrated for bluetongue virus (Markotter *et al.*, 2004). It was suggested that CYCL could be involved in RNA packaging into the virion rather then in replication (Proutski *et al.*, 1997b). This might simply imply that CYCL is involved in both processes of replication and packaging and its absolute conservation is probably more important for packaging. Therefore, as a bifunctional domain or as a signaling domain, CYCL might be under strong evolutionary pressure to be conserved. Indeed, it was shown that a group of chimeric 3′LSH mutants (between DENV and WNV) that were lethal for the virus or severely reduced its growth, were nevertheless quite efficient in the initiation of RNA synthesis *in vitro* implying that other processes could affect virus virulence (You *et al.*, 2001). Moreover, less efficient replication was demonstrated with mutated CYCL (Khromykh *et al.*, 2001; Lo *et al.*, 2003). Therefore, it is possible that absolute conservation of this domain is important not for replication *per se* but for efficient replication leading to the production of high virus titers.

It was also proposed that circularization could be involved in translation, since bringing two ends together could influence the interaction of RNA with the translation machinery (Corver *et al.*, 2003). However, silent mutations in the CYCL of the replicons did not interrupt the translation that was detected early after RNA transfection, whereas the replication observed later, i.e. after RNA transfection, was completely abolished (Lo *et al.*, 2003).

General concepts from theoretical and experimental research on mosquito-borne flaviviruses

Based on theoretical predictions and the experimental evidence, it was proposed that the 3′UTR of the MBFV could be subdivided into two structurally and functionally well-defined domains (Proutski *et al.*, 1999). This concept is developed further below. One domain refers to the virus promoter which physically coincides with domain III as predicted in the GA model. Folded independently from the other RNA regions, this domain forms two stem–loop structures, i.e. 3′LSH and SL2 that are highly conserved in shape between all flaviviruses. They interact with each other through the highly conserved pseudoknot. The 3′LSH exposes loop 1 with the highly conserved pentanucleotide carrying an essential signaling function. Experiments demonstrated that conservation of the shape of the 3′LSH and sequence of loop 1 are essential for virus infectivity. Domain III binds some cellular proteins and also the viral proteins NS2A, NS5 and NS3 that are involved in the replication complex. However, this domain failed to initiate synthesis of the viral RNA *in vitro* if the 5′UTR was not supplied *in cis* or *in trans*. RNA folding predicts an interaction between the 3′UTR and 5′UTR with the formation of a long circularization stem. The mutagenesis experiments proved that complementary nucleotide pairing between 5′UTR and 3′UTR circularization sequences is vital for initiation of virus replication. Therefore, the 3′LSH- SL2 structures might present an intermediate stage in the formation of a virus promoter, before or after circularization has occurred.

However, the experiments and structural predictions do not explain the high conservation of the CS1 sequence. Linear CS1 could be subdivided into an absolutely conserved CYCL sequence (also present as an inverted repeat in the capsid gene) and another region with relatively high conservation between the MBFV sequences, both of which are involved in genome circularization. These sequences also participate partially in the formation of SL2. It is possible that the circularization stem not only provides interaction between 5′ and 3′UTR but also acts as a double-stranded region presenting CS1 as a signal for recognition by cellular or viral proteins that might be required for replication and/or other stages of the virus life cycle, such as encapsidation. However, an independent signaling function of CS1 as a single-stranded RNA region cannot be excluded.

It is also possible that the multifunctionality ensures the high conservation of CS1 and also other sequence elements within domain III. Deletion of the region 3′LSH- SL2 together with the adjacent CS1 is lethal for the virus whereas on its own without the other upstream sequences, it retained infectivity in mosquito cells. Moreover, nonlethal mutations within the promoter (in the 3′LSH) had a more profound effect on virus replication in mosquitoes than in mammalian cells implying that the flavivirus promoter has been subjected to significant selection pressures in invertebrates (Zeng *et al.*, 1998).

However, for replication in mammalian cells the 3′LSH- SL2 region of the virus promoter is extended and must include one of the similar dumbbell-like structures DB1 or DB2, formation of which depends on the pseudoknots. DB1 that is located upstream of the 3′LSH- SL2 region is highly conserved in shape between different MBFV and exposes CS2 that is also highly conserved. DB2 is presented only in the DENV and JEV groups and exposes a repeat of CS2 (RCS2). In YFV this structure is different in shape but also like DB1, is supported by pseudoknots.

On the other hand, laboratory experiments on the 3′UTR clearly demonstrated that the region upstream of the promoter that starts with DB2, with an exposed RCS2 and other conserved repeats (domains II and I) is not essential for virus viability but still contributes to virus replication rate. In this respect this region could be interpreted as an enhancer of the virus promoter that accelerates the folding of the 3′LSH and/or prepares it for circularization. Domains I and II enhance the signaling function of domain III—cellular and viral proteins bind to domain III better in the presence of domains I and II. How this enhancing function takes place is not yet clear. Possibly efficient folding creates intermediate signals (such as RNA stem–loops or proteins) that facilitate folding of domain III. It is also not clear when the formation of domain III occurs—before or after the formation of the circularization stem.

Based on experimental evidence the enhancer that includes domains I and II appears not to be essential for virus viability but it appears to increase virus replication rate. The MBFV group evolved the ability to be transmitted from the blood of an infected vertebrate host to noninfected mosquitoes, when they take a blood meal. In nature mosquitoes feed for seconds or at the most a few minutes. Consequently, the frequency of successful viral transmission from the vertebrate to the mosquito is greatly enhanced if the virus is present in the blood at a high titer. Moreover, if the rate of virus replication in the infected vertebrate host is also high this increases the rate of virus turnover and therefore the opportunity for other noninfected mosquitoes to become infected during periods of high mosquito density. Laboratory experiments appear to support this concept; deletions in the enhancer part of the 3′UTR reduce the replication rate of the mutant viruses in both simian and mosquito cells, although to a greater extent in simian cells. This may indicate that the evolution of enhancer elements was of greater significance for replication in vertebrate cells where it is important for the virus to reach a high titer in a short period of time.

The existence of an enhancer element, essential for rapid virus replication but not essential for replication *per se*, could explain two apparently contradictory observations—firstly the conservation of RCS2, CS3, RCS3 or the three YFV tandem repeats and secondly the fact that they are not essential for virus viability during laboratory passage. The reasons for preservation of the repeat sequences as enhancer elements is not clear, it is possible that repeat sequences multiply initial signals (for RNA or protein interactions) that speed up folding of the promoter and/or assembly of the polymerase complex and this provides an advantage to the virus. The effect of each individual element of the replication enhancer is relatively moderate, but their cumulative action may accelerate virus replication and this may be an essential requirement for efficient virus transmission and dissemination in the more fastidious natural environment. The relatively lower importance of the enhancer for virus viability also explains why its formation was probably under less selection pressure than that of the promoter and why domains II and I evolved faster than domain III. This also explains why under laboratory conditions the enhancer could accept more variation than domain III.

The 3′UTR of tick-borne flaviviruses

The TBFV are subdivided into two ecological groups, the Mammalian and Seabird groups (Heinz *et al.*, 2000) that probably diverged in different tick species less than 10 000 years ago (Zanotto *et al.*, 1996). Amongst the TBFV, the Mammalian group contains the viruses for which there is most information concerning the 3′UTR.

Evolution of the tick-borne flavivirus 3′UTR

Figure 3.10 (http://www.horizonpress.com/hsp/supplementary/flavi/) presents a comparative alignment of the 3′UTR of all known TBFV sequences. The length of the longest 3′UTR (730 nucleotides) of the TBFV is longer by 100 nucleotides than that of the longest 3′UTR of the MBFV (WNV, 634 nucleotides). The same principles that were established for the MBFV have been applied to the organization of the 3′UTR of the TBFV. Initially two functionally different linear regions were identified, one conserved region that is located at the 3′ end and a variable region located between the conserved region and the stop-codon (Wallner *et al.*, 1995). It is known that TBFV isolated from ticks that were not cultured multiple times in cell lines or mice have the longest UTRs. In contrast, viruses cultured in the CNS of mice or passaged in mammalian cell cultures have deletions or insertions in the variable or redundant region of the 3′UTR that is located between the stop codon and the more conserved (C3′ UTR), or core region (Gritsun *et al.*, 1997; Hayasaka *et al.*, 2001; Wallner *et al.*, 1995). The precise boundary between the variable and the core region has not been identified. However, the longest spontaneous deletions observed following serial culture of these viruses is about 400 nucleotides, thus the core region is probably about 325 nucleotides long.

The 3′ UTRs of the TBFV share greater homology than the MBFV and this corresponds with their antigenic and phylogenetic relationships based on the analysis of ORF sequences. Analysis of the comparative alignment shows that in contrast with the MBFV, the 3′ UTRs of the TBFV group have, to a large extent evolved gradually by point mutations. Clearly, deletions and insertions have occurred but they are relatively small and can only be seen between more distantly related viruses (Figure 3.10, http://www.horizonpress.com/hsp/supplementary/flavi/). Thus, there is a high level of sequence conservation throughout the 3′UTR of the TBFV. The putative circularization domain of 11 nucleotides that also occurs as an inverted repeat sequence in the 5′UTR (Mandl *et al.*, 1993) is not the only absolutely conserved sequence; other quite long conserved regions can readily be identified along the entire 3′UTR.

This high level of conservation enables identification of traces of early development of the 3′UTR (Figures 3.11, http://www.horizonpress.com/hsp/supplementary/flavi/, and 3.12). Three repeat sequences R1/R1′, R2 and R3, separated by long stretches of nucleotides were initially described within the 3′UTR (Wallner *et al.*, 1995). The R2 and R1 sequences were also found at the boundary between the NS5 protein and the 3′UTR (Figure 3.11, http://www.horizonpress.com/hsp/supplementary/flavi/). Two direct repeats DR1 and DR2, each about 30 nucleotides long and separated by 100 nucleotides were also described for the Vasilchenko strain of TBEV (Siberian subtype) and also LIV, with the DR2 located in the redundant region where other TBFV have a large deletion (Gritsun *et al.*, 1997). However, with more virus sequences now available for the analysis, longer repeated sequences have been identified (Figures 3.11, http://www.horizonpress.com/hsp/supplementary/flavi/, and 3.12). Each repeat sequence includes continuous sequences R3+R2+R1 in this order. Additional sequences following R1 have been designated R4. The long repeat sequence (LRS) is therefore designated R3+R2+R1+R4; within the alignment it occupies a region of about 200 nucleotides. Four of these LRSs have now been identified in the 3′UTR and two in the upstream region of the NS5 gene.

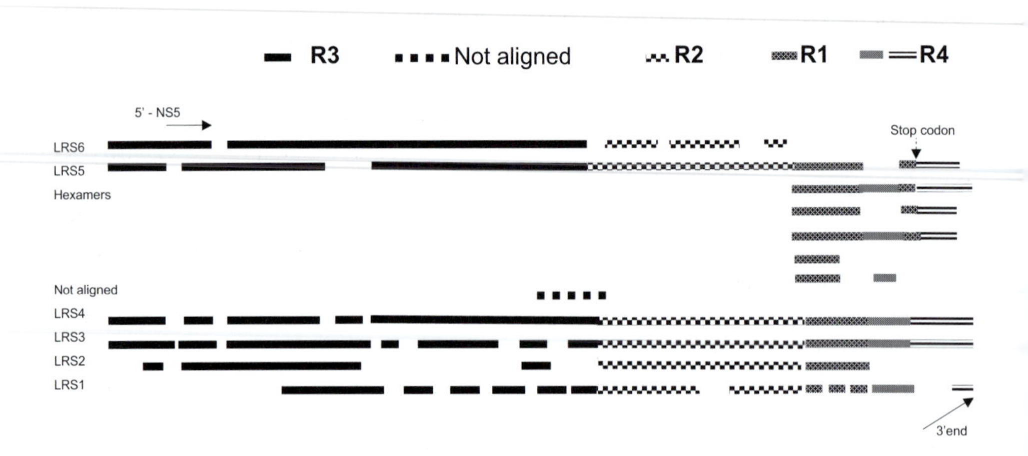

Figure 3.12 Hypothetical development of the 3′UTR of the TBFV group from six LRSs as deduced from Figure 3.11 (http://www.horizonpress.com/hsp/supplementary/flavi/). The continuous TBFV 3′UTR sequence was subdivided into six LRSs that are placed under each other in descending order. Individual elements of each LRS, i.e. R3, R2, R1 and R4 are shown in bars shadowed as specified.

Different LRSs were not an exact copy of each other and the strict alignment was not obvious along the entire 3′UTR. However, the appearance of many similar nucleotide regions interrupted by less similar regions (remnants of repeat sequences?) was reproduced in each copy of the LRS (Figures 3.11, http://www.horizonpress.com/hsp/supplementary/flavi/, and 3.12). Double repeat sequences of similar or even larger size were also revealed in the UTRs of other viruses (Bodaghi *et al.*, 2000; Edwards, 1995; Peerenboom *et al.*, 1997; Shi *et al.*, 1997). Tandem four repeats of the same sequence have also been reported (Faragher and Dalgarno, 1986).

One possible interpretation of the role of these LRSs is that they represent early evolutionary stages of the formation of the TBFV. One can imagine repeated duplication of the LRS mediated by the RNA polymerase with its propensity to pause, detach from the template and translocate downstream before continuing to elongate the RNA along the template. Subsequently, these duplicated LRSs evolved independently through point mutations, deletions or insertions. These evolutionary processes would have led to the formation of stems and loops acting as promoters or enhancers in the 3′UTR (see later).

Based on analysis of ORF sequences, it is known that in general the TBFV have evolved at a slower rate than the MBFV (Gould *et al.*, 2003; Gritsun *et al.*, 1995) and this is also reflected in the 3′UTR. The slow rate of TBFV evolution is largely attributable to the slow life cycle of the tick (3–5 years). In many vertebrate hosts on which ticks feed, TBFV replication occurs only in the feeding ticks and perhaps to a lower extent, simultaneously, in the Langerhans and other cells of the local feeding site in the skin of the host (mammals/birds). Dissemination of TBFV occurs during nonviremic transmission, when infected and noninfected ticks co-feed at the same site on the animal (Gould *et al.*, 2004; Jones *et al.*, 1997; Labuda *et al.*, 1996). Ticks rarely feed more than once or twice per year, resulting in a relatively low turnover rate of TBFV which may explain their slow rate of evolution in comparison with the MBFV.

One of the most obvious consequences of this relatively low turnover of TBFV compared with the MBFV is that some of the early features of their evolution, such as the 6 LRSs described herein, are quite literally "frozen" in time. However, the long conserved regions within the TBFV 3′UTR tend to mask the signaling sequences that are preserved perfectly as CSs and RCSs in the MBFV, between the highly variable regions. Therefore in the TBFV, the signaling domains can only be assumed by comparison with equivalent domains in the MBFV.

The predicted secondary RNA structures in the 3′UTR of tick-borne flaviviruses

Three models for the folding of the TBFV 3′UTR were initially proposed (Gritsun *et al.*, 1997; Proutski *et al.*, 1997b; Rauscher *et al.*, 1997). Although different algorithms were used for these predictions, they all produced a similar folding pattern for the TBFV, and particularly for the loops which were absolutely compatible between all three models. However, the structure of some stems was different in the three models (not shown). It is likely that the similarity in secondary structures within the TBFV species results primarily from the similarity of their sequences rather than from the accuracy of the secondary structure predictions. As an example, Figure 3.13a demonstrates the folding pattern of the distal 333 nucleotides similar for

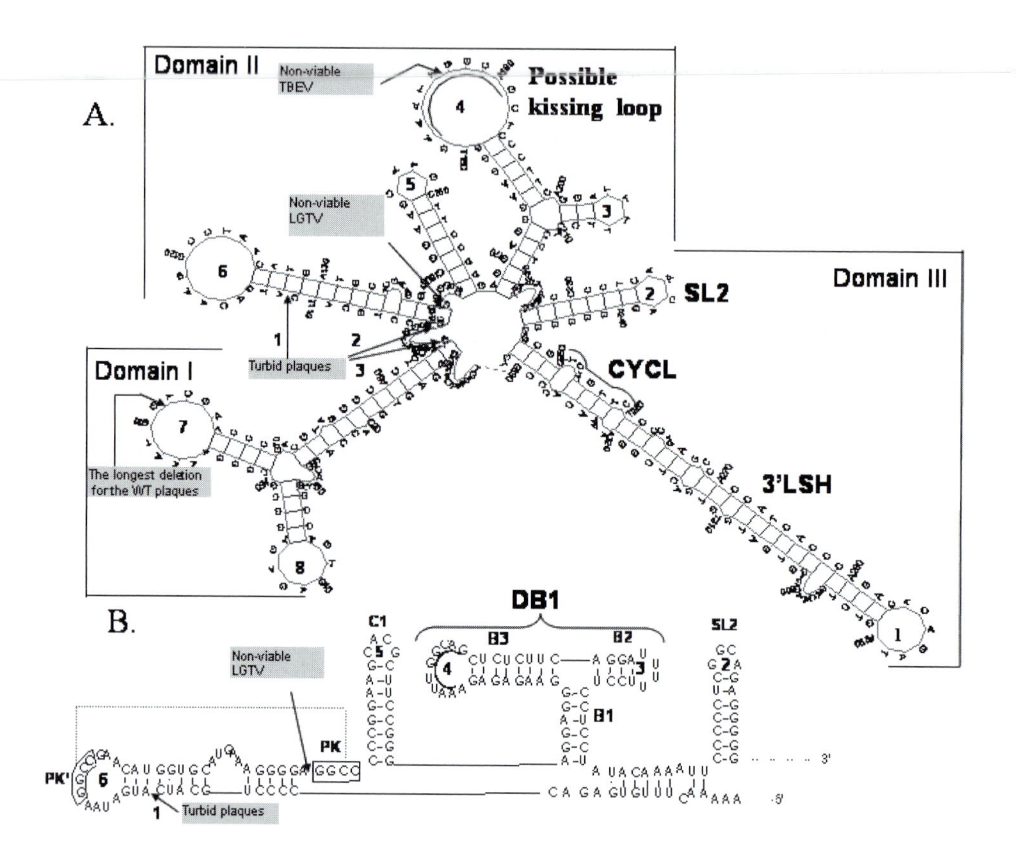

Figure 3.13 Secondary RNA structures predicted for the terminal 333 nucleotides of TBFV as adapted from A) (Gritsun *et al.*, 1997) and B) domain II (Olsthoorn and Bol, 2001). The annotated features include: circularization sequences (CYCL) as part of the 3′LSH, SL2, and dumbbell-like structure (DB1). Numbers specify loops on both, A and B. Domains I, II and III were adapted from (Proutski *et al.*, 1997b). Sequences interacting in pseudoknots (PK and PK' on B) are shown in open boxes and joined by a dotted line. The sequence elements C1, B1, B2 and B3 are specified on the alignment of the 3′UTR for TBFV (Figure 3.10, http://www.horizonpress.com/hsp/supplementary/flavi/). The sequences of loop 4 that are presented as an inverted repeat in loop 6 in the capsid gene (possible kissing loops) are underlined (see also Figure 3.10). Text in shaded boxes specifies the results of the engineered deletions on the infection clone as in Figure 3.10 (http://www.horizonpress.com/hsp/supplementary/flavi/).

different virus species within the TBFV group (Gritsun *et al.*, 1997). Stem–loops were grouped into three domains equivalent with those observed in the MBFV 3′UTR folding (Proutski *et al.*, 1997b). Domain III in the TBFV that contains the 3′LSH- SL2 is similar to domain III for the MBFV. However, there was no indication of the pseudoknot-like interactions between the 3′LSH and SL2 as described for the MBFV.

Domain II of the TBFV contains stem–loops 3, 4, 5 and 6, which are compatible between different folding models. Domain I contains a Y-like structure with loops 7 and 8 varying in conformation between different TBFV species. As the alignment demonstrates (see below), these conformational differences might be due to the presence of large deletions generated in this region during virus passage in laboratory culture systems (see below). Ideally, the folding of this region needs further re-evaluation within the entire wild-type 3′UTR sequence rather than within truncated regions of laboratory virus isolates. In all the models studied thus far, domains II and I were different in shape from the MBFV. However, in the pseudoknot-prediction model, domain II of the TBFV folds similarly to the MBFV and a dumbbell-like structure, DB1, has been identified (Figure 3.13B).

The circularization model as applied to the TBFV predicted the formation of a long circularization stem as described for the MBFV (Figure 3.8). The first circularization domain includes CYCL (Figure 3.10, http://www.horizonpress.com/hsp/supplementary/flavi/), although other sequences in the 3′UTR and coding region from the 5′end also contribute to this interaction (Figure 3.8) (Thurner *et al.*, 2004).

Links between phylogenetic and structural analysis of the 3′UTR of tick-borne flaviviruses

In Figure 3.10 (http://www.horizonpress.com/hsp/supplementary/flavi/) the relative positions of secondary RNA structures predicted by the different algorithms (Figures 3.8 and 3.13) have been superimposed onto the comparative TBFV alignments. In domain III the 3′LSH with the conserved pentanucleotide CACAG in loop 1 partially overlaps with the side of the SL that forms as the result of circularization between the 5′UTR and 3′UTR. The sequences that include the circularization motif form the proximal part of the 3′LSH. The length of sequence involved in circularization varies between different TBEV (not shown). Domain III also includes the short SL2 which has a similar shape to those predicted for the MBFV. Loop 2 is formed from variable sequences in the different viruses whereas the short stem exhibits a high degree of conservation.

Domain II contains four loops, enumerated 3, 4, 5 and 6. The sequences within loops 3 and 4 are more conserved than those of the stems that carry them and are more likely to possess a signaling function. It is worth noting that a conserved sequence within loop 4 (Figures 3.10, http://www.horizonpress.com/hsp/supplementary/flavi/, and 3.13) is an inverted complementary repeat of a sequence in the loop of the SL6 of the capsid gene (see later) suggesting the possibility of "kissing loops" being involved in "long range" RNA–RNA interactions. The sequence of loop 5 is nonconserved whereas some stretches within loop 6 are conserved. Additionally, SL3 and SL4 within domain II for the TBFV, overlap with the dumbbell-like structure (Figure 3.13) that was interpreted as the equivalent of DB1 for the MBFV (Olsthoorn and Bol, 2001).

Domain I is formed by the Y-shaped structures carrying SL7 and SL8. Although within the different TBFV the lengths of the stems vary, the lengths of the loops are similar. There is some variation within loop 7 but it is mostly formed by conserved stretches of nucleotides. Loop 8 is formed by quite conserved sequences with some exceptions and only POWV and Alkhurma virus (ALKV) have deletions in this region. However, POWV and ALKV still form Y-shaped structures that do not interfere with the formation of the other SLs.

It is possible that as with the MBFV, the different models for the 3′UTR of the TBFV represent different secondary RNA structures in different periods of the virus life cycle rather than representing an optimal RNA structure. The partial overlap of predicted structures derived using the GA model and the pseudoknot-predicting models seem to support this—the folding of the stem and loops in the GA model could precede the secondary and tertiary structures involved in the formation of pseudoknots.

Experimental research on the 3′UTR of tick-borne flaviviruses

In attempts to understand the role of the 3′UTR, a series of deletions was progressively introduced into the 3′UTR of a TBEV infectious clone. The effects of these deletions were evaluated in a variety of assays (Mandl *et al.*, 1998). The predicted consequence on RNA secondary structure, of each deletion, was also analyzed. The results are summarized in Figures 3.10 (http://www.horizonpress.com/hsp/supplementary/flavi/) and 3.13. Deletion of the entire variable region containing SL7-SL8 did not affect virus plaque morphology or growth curve characteristics. Further progressive deletions in the core region including the SL7 also did not affect the virus properties in cell culture. However, the partial deletion of SL6 produced a virus with turbid plaques and reduced growth characteristics. Additionally, the deletion of the variable region that preserves SL6 and the downstream regions of the Siberian TBFV, still produced a virus with turbid plaque morphology (unpublished). Similar results were produced for Langat virus (LGTV, see Figure 3.13) (Pletnev, 2001). However, nonviable TBEV was produced by deletion of the region between the stop codon and SL4 (with destruction of SL4) (Mandl *et al.*, 1998) and for LGTV by deletion of the entire region between the stop codon and SL5, including the complete deletion of SL6 (Pletnev, 2001). Therefore, the boundary between infectious and noninfectious virus appears to be within the SL6 sequence (Figures 3.10, http://www.horizonpress.com/hsp/supplementary/flavi/, and 3.13): the deletion that retained the loop sequence of SL6 but destroyed its predicted structure produced a virus with turbid plaques whereas the complete elimination of SL6 resulted in a noninfectious virus. However, there is another possible explanation: a region of SL6 overlaps with the circularization sequence (not shown) and the deletions, for both TBEV and LGTV, removed this vital RNA region. Moreover this region also accommodates a pseudoknot which if removed might also destroy virus infectivity (Figures 3.10, http://www.horizonpress.com/hsp/supplementary/flavi/, and 3.13).

Additional tests demonstrated that all the deletion mutants, including those with altered SL7- SL8, were more attenuated in mice compared with wild-type virus or mutants that preserved SL7- SL8. All these experiments were carried out in mammalian cells since, to date, no tick cell lines have been available to identify the minimum required region for virus replication in the arthropod vector as has been established for the MBFV.

On the basis of these experiments it is proposed that the promoter region absolutely vital for virus infectivity in mammals includes the 3′LSH- SL2, pseudoknot-supported DB1 (equivalent of SL3- SL4), SL5 and the distal part of the SL6. The upstream region might function as an enhancer that is vital for virus survival in the environment. Although the TBFV have a tendency to lose the variable region during serial passage in mice and cell culture, freshly isolated strains have a longer 3′UTR. Therefore, the entire region between the stop-codon and the virus promoter is required to sustain virus infectivity in the environment. In common with the MBFV it possibly ensures efficient virus transmission through mediation of rapid virus replication. Additional research is needed to reveal the secondary structures in this region.

The 3′UTR of flaviviruses with no known vector (NKV)

Linear 3′UTR sequences of the no known vector flaviviruses

Flaviviruses with no known arthropod vector (NKV) have been isolated from rodents or bats and they represent a third ecological group within the genus *Flavivirus* (see Chapter 2). Although there is partial sequence data for most NKVs in the NS5 protein, currently the UTRs of only four NKV have been sequenced (Charlier *et al.*, 2002; Leyssen *et al.*, 2002). Three, Montana myotis leukoencephalitis virus (MMLV), Rio Bravo virus (RBV) and Modoc virus (MODV), were isolated from rodents in the New World (the Americas) and Apoi virus (APOIV) from the Old World. Figure 3.14 (http://www.horizonpress.com/hsp/supplementary/flavi/) presents an alignment based on the 3′UTR which is consistent with alignments and trees based on NS5 data showing the NKV to be quite distantly related to each other.

The APOIV 3′UTR is the longest of those sequenced in the NKV and contains 576 nucleotides. This is still shorter than the longest 3′UTR of the vectored flaviviruses. MODV contains the shortest 3′UTR (366 nucleotides); this is equivalent to the length of the promoter region in the vector-transmitted flaviviruses. It is possible that the UTRs of the NKV are shorter due to passage of these viruses in laboratory culture systems. Alternatively, their short 3′ UTRs might be the result of natural deletions within virus genomes that may replicate only in mammalian cells. In either case the "laboratory acquired" or "natural" deletions should occur in a variable region of the UTR, located between the stop codon and the promoter region. However, comparative alignments between NKV show that deletions are present throughout the whole 3′UTR rather than being concentrated beyond the stop codon. Thus, these deletions are probably not the result of laboratory passage and are more likely to reflect a simplified requirement for the capacity to multiply in a single host species, i.e. rodents.

Despite the wide divergence between the four NKV for which UTR sequence data are available, stretches of conserved nucleotide sequence could be identified through the introduction of large gaps in the comparative alignments with vectored viruses. The CS2 element was identified for all four NKV and was similar to the CS2 of the MBFV. APOIV also has a recognizable RCS2 and the three other NKV have sequence "remnants" of this region. The preservation of CS2 and RCS2 sequences in the NKV and MBFV, suggests a common evolutionary origin and possibly reflects a closer relationship between these viruses than that with the TBFV, which lack the corresponding motifs.

Other short conserved nucleotide sequences, in addition to CS2, were identified between the four NKV along the 3′UTR, but their significance was only revealed after the secondary structures were modeled for this domain (next section).

Secondary RNA structures in the 3′UTR of flaviviruses with no known vector

Detailed analyses of the secondary structures of the NKV 3′UTR have been published (Charlier *et al.*, 2002; Leyssen *et al.*, 2002). Comparisons were made between the NKV and also with vectored viruses. The predicted secondary structures of the NKV are presented in Figure 3.15. Four domains shared between all NKV were identified. Domain IV contains a 3′LSH and a short SL2, similar to those in the MBFV and TBFV. A pentanucleotide CACAG within loop 1, conserved between the MBFV and TBFV, was also present in loop 1 of the NKV although with substitutions in the second or third positions: CCTAG in APOIV, CCCAG in Rio Bravo virus (RBV) and CTCAG in MMLV and MODV. Whether or not these substitutions are associated with the abrogation of vector transmissibility remains to be determined.

The Y-shaped structure upstream of the 3′LSH of the NKV is similar to SL3 and SL4 described for the TBFV, including identical sequences in loop 3 and loop 4 (domain III on Figure 3.15). This structure was also interpreted as the TBFV-equivalent of DB1 described for MBFV (Olsthoorn and Bol, 2001). On the other hand, a dumbbell-like structure with an exposed CS2, similar in shape to DB1 of the MBFV, was observed upstream of the TBFV-like Y-shaped structure. Therefore, both of these structures may have existed in ancestral lineages. One can envisage a situation in the early period of evolution of these viruses into the TBFV, MBFV and NKV when both DB1 and the Y-shaped structure, were preserved only in the NKV, the TBFV having lost DB1 and the MBFV having lost the Y-shaped structure. Perhaps the functions of DB1 and the Y-shaped structure are sufficiently similar for them to be able to compensate each other as was described for CS2 and RCS2 (see above).

In addition APOIV formed a second similar dumbbell-like structure, in which the corresponding RCS2 was exposed. The pseudoknots were also identified although their positions were not conserved within the NKV group (Charlier *et al.*, 2002). Thus, the NKV share sequence elements

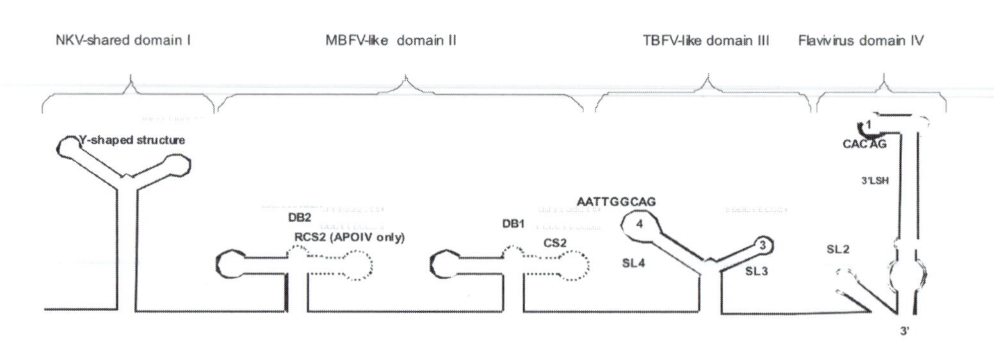

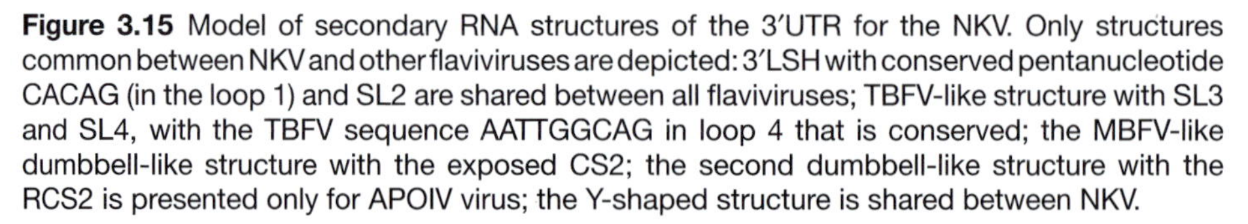
Figure 3.15 Model of secondary RNA structures of the 3′UTR for the NKV. Only structures common between NKV and other flaviviruses are depicted: 3′LSH with conserved pentanucleotide CACAG (in the loop 1) and SL2 are shared between all flaviviruses; TBFV-like structure with SL3 and SL4, with the TBFV sequence AATTGGCAG in loop 4 that is conserved; the MBFV-like dumbbell-like structure with the exposed CS2; the second dumbbell-like structure with the RCS2 is presented only for APOIV virus; the Y-shaped structure is shared between NKV.

and secondary structures with both the MBFV and TBFV (Charlier *et al.*, 2002). This might indicate that both "mosquito-specific" and "tick-specific" structures were present in an ancient flavivirus lineage before it diverged to form the TBFV, MBFV and NKV groups. Whether or not the NKV secondary structures that resemble those of the MBFV and TBFV play a similar role in virus replication remains to be determined.

The circularization model was also applied to fold the 3′UTR of the NKV. However, unlike the MBFV and TBFV, the NKV 3′LSH sequences do not form a side stem–loop structure, instead they interact completely with the matching sequence at the 5′ end of the UTR (Charlier *et al.*, 2002). Moreover, the four NKV are so divergent that when examined on the comparative alignment (Figure 3.14), the region for circularization at the 3′UTR of MODV did not contain the absolute sequence conservation that was assigned to the MBFV and TBFV as classical CYCL. Therefore, in contrast with the MBFV it is difficult at this stage to trace the sequence conservation in the NKV that provides the simultaneous functioning of sequence elements in independent folding of the 3′UTR and circularization. Presumably as more virus sequences and more experimental results become available it will become possible to identify the significance of the structural elements in virus replication, and their adaptation to mammalian hosts.

The 5′UTR of flaviviruses

The most obvious function that has been assigned to the flavivirus 5′UTR is translation, since it is capped and 90–130 nucleotides long, features that are compatible with the organization of typical untranslated regions of cellular mRNA.

However, as will be demonstrated below, the 5′UTR of flaviviruses has another virus function, namely replication; it might also contain signals for virion assembly although there is no current evidence for this. There are two essential differences in the organization of the 5′UTR between the TBFV on the one hand and the MBFV and NKV on the other. The 5′UTR of the TBFV is slightly longer in comparison with the MBFV and NKV, containing about 130 nucleotides. As indicated earlier, in the TBFV this region contains a conserved sequence element of 11 nucleotides that is involved in circularization. In contrast, the 5′UTR of the MBFV and NKV is shorter (90–96 nucleotides for the JEV and DENV groups, ~120 nucleotides for the YFV group, and ~110 for the NKV) and sequences involved in circularization of the MBFV and NKV are located downstream of the 5′UTR, in the capsid gene. Although strictly speaking the circularization region does not belong to the 5′UTR, it will be referred to as though it is a part of the same functional domain as the 5′UTR of TBFV. In the following sections, the format used for the description of the 3′UTR will be followed; the secondary structures of the 5′UTR will be compared in terms of their linear sequences and then these data will be correlated with the experimental research.

Structural predictions for the flavivirus 5′UTR

Initially the secondary structures within the 5′UTR were described for the MBFV (Brinton and Dispoto, 1988). These foldings and later computer-simulated predictions demonstrated that the MBFV, TBFV and NKV have a similar Y-shape structure (~80 nucleotides in length for the MBFV and NKV and about 100 nucleotides for

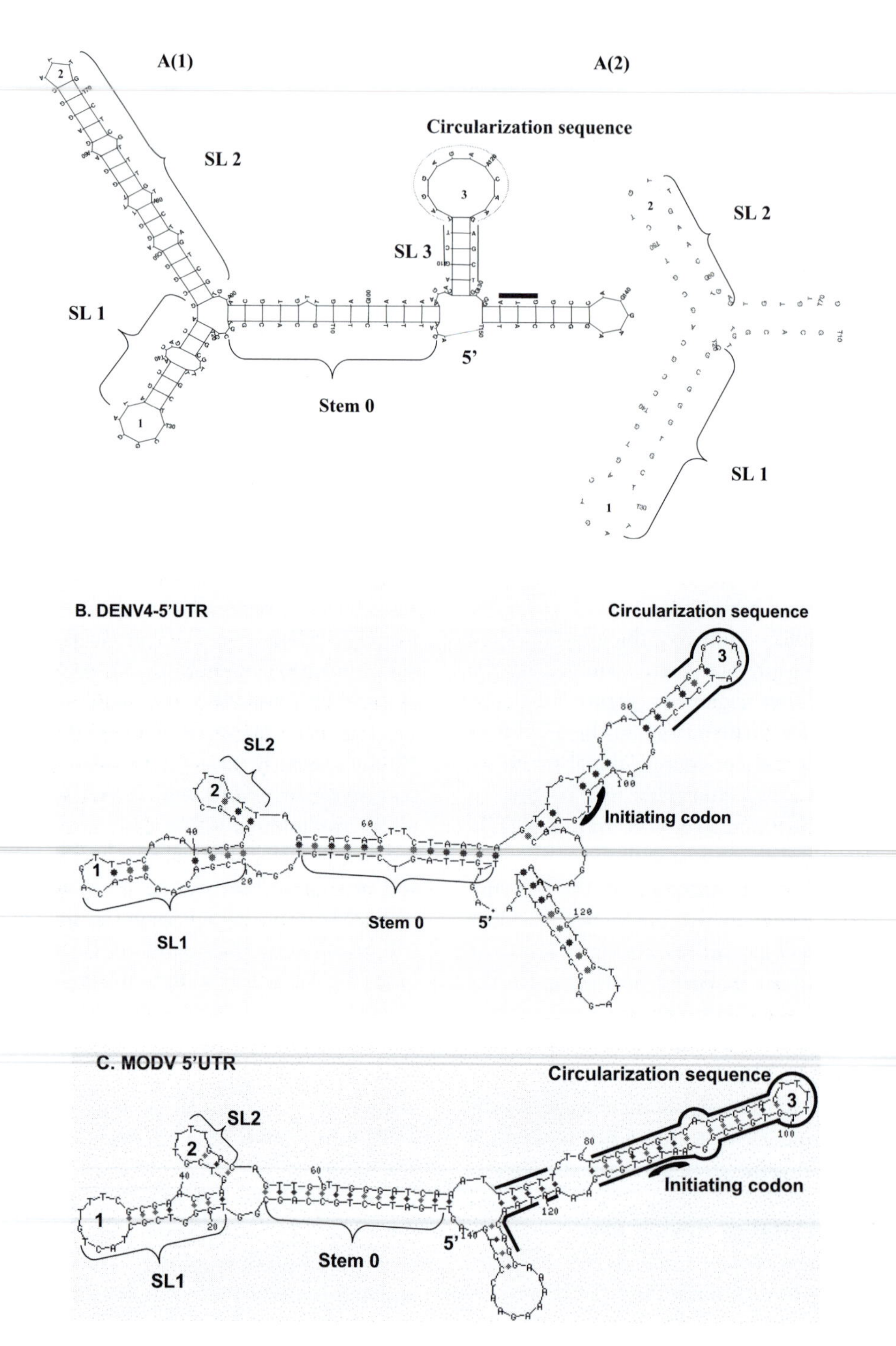

Figure 3.16 Computer-simulated RNA secondary structures for the 5′UTR of flaviviruses. The elements of the Y-shaped structure, i.e. stem 0, SL1 and SL2 are specified. The CYCL and other circularization sequences are highlighted by a solid line. A. A model for two TBFV, i.e. TBEV (A1) and POWV (A2) is adapted from (Gritsun *et al.*, 1997). Models for B) DENV4 (MBFV group) and C) MODV (NKV group) are adapted from (Leyssen *et al.*, 2002).

the TBFV) from the proximal end of the 5′UTR (Gritsun *et al.*, 1997; Leyssen *et al.*, 2002; Thurner *et al.*, 2004) (Figure 3.16). Three main structural elements were described for all three major flavivirus groups—stem 0 and two branching stem–loop structures 1 and 2 (SL1 and SL2 respectively). Although generally preserved in shape, the length of the stems and the size of the loops varied slightly between different virus groups. Stem 0 is formed from 10–14 paired nucleotides. The unpaired 5′-terminal nucleotides, AG, were highly conserved and are not involved in formation of the stem. The length of the first branching SL1 was also similar between different flavivirus groups and formed by 6–8 paired nucleotides. The SL2 was the most variable in length between different flavivirus groups and even between viruses that belong to the same group. For example for POWV the SL2 was much shorter than for the rest of the TBFV (see Figure 3.21) but in general most of TBFV group have longer SL2 than viruses in the MBFV and NKV groups.

All flaviviruses form a third SL3 that accommodates the initiation codon AUG (Figure 3.16). However, it is not clear whether or not SL3 performs any functionally significant role other than circularization of the virus genome as described above. It is notable that the folding of the 5′UTR is not affected by circularization of the genome (Figure 3.8) and the 5′UTR domain retains it original shape independently of the downstream region involved in circularization (Thurner *et al.*, 2004).

Evolution of the flavivirus 5′UTR

5′UTR alignment of mosquito-borne flaviviruses

An initial alignment of the 5′ UTRs between different MBFV revealed significant homology among members of the same flavivirus subgroup, and almost complete conservation between different strains of the same virus. However, only short regions within the 5′UTR were described as conserved among the three different MBFV groups, i.e. JEV, DENV and YFV (Brinton and Dispoto, 1988). As more sequences became available, more comprehensive alignments became possible, as shown in Figure 3.17 (http://www.horizonpress.com/hsp/supplementary/flavi/). Since the sequence involved in circularization of the virus genome for MBFV extends into the coding region, the capsid protein gene is also included in the alignment.

The original alignment produced using ClustalW was subsequently refined manually (Figure 3.17, http://www.horizonpress.com/hsp/supplementary/flavi/) in order to reveal more information than was hitherto possible. In contrast with the 3′UTR, no extensive regions of conservation were identified between the three major MBFV groups. Only short stretches of identical nucleotides were observed and generally similarity was confined to individual stretches between two rather than three of the MBFV groups, i.e. YFV-DENV, YFV-JEV and JEV-DENV.

It is worth noting that the alignment revealed an identical position for the initiation AUG codon for JEV and YFV groups, whereas the position of the initiation codon for the DENV group was shifted. The sequences that form secondary structures for each MBFV group in the 5′UTR are highlighted in the alignment (Figure 3.17, http://www.horizonpress.com/hsp/supplementary/flavi/). Sequence conservation was observed in the region that forms stem 0. Although not highly conserved between the three major MBFV groups, the sequences of stem 0 were more conserved within individual virus groups and were reliably aligned. This indicates that stem 0 could function as a double-stranded region with a signaling function limited to group-specific virus factors such as viral proteins. The 5′ distal region of sequence that is involved in circularization of the virus genome was not conserved between the three MBFV groups but they are all located next to the AUG codon (Figure 3.17, http://www.horizonpress.com/hsp/supplementary/flavi/). The second region of nucleotides involved in circularization is CYCL; it located in the capsid gene and contains eight absolutely conserved nucleotides (TCAATATG) that appear as an inverted repeat sequence in the 3′UTR and as discussed earlier, in addition to circularization might also have a signaling function.

5′UTR alignment of the tick-borne flaviviruses

Fourteen TBFV 5′UTR sequences are now available for analysis (Figure 3.18). Not unexpectedly, the 5′UTR of the TBFV exhibited greater conservation than the MBFV. The highest variability was in SL2; POWV had a deletion that shortens the stem of SL2 and lengthens the stem of SL1 indicating that the shape of both SL1 and SL2 is not essential for virus infectivity. Sequences of the SL1 were quite variable and include compensatory mutations that preserve the shape of this structure. The sequence of loop 1 was not conserved and therefore no signaling function could be assigned either to SL1 or SL2. The most conserved region was in SL3 which overlaps completely with the circularization sequence and includes the 11 nucleotide inverted complementary repeat sequence (CYCL) that occurs in the 3′UTR (Figure 3.18). In common with the MBFV, sequences within stem 0 were the second most conserved region and only a single compensatory substitution was revealed between some of the TBFV. Perhaps the function of SL1 and SL2 is "supportive," i.e. to arrange the formation of stem 0 and/or other regions of the RNA for circularization. Owing to the higher level of conservation of stem 0 in comparison with the more divergent SL1 and SL2, a signaling function may be assigned to this double-stranded region.

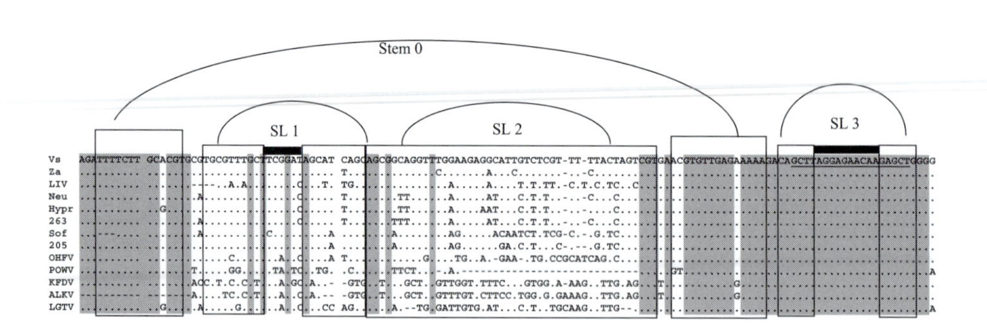

Figure 3.18 Alignment between 5′UTR of TBFV. The sequence of the Vs virus is depicted on the top as a consensus and only differences in the sequences of other viruses are shown; identical nucleotides are presented as dots. Annotated features include the elements of secondary RNA structures that are depicted in boxes in Figure 3.16A: stem 0, SL1 and SL2. The sequences of the loops are depicted as thick black bars. Sequences involved in stems are joined by oval lines. Circularization sequences are shown by underlined letters. Shadow background depicts absolutely conserved sequences.

5′UTR alignment of the no known vector flaviviruses

Only two sequences are available to compare the 5′UTR of the NKV flaviviruses (Figure 3.19, http://www.horizonpress.com/hsp/supplementary/flavi/). In common with the MBFV and TBFV, a region of stem 0 exhibits greater conservation than either SL1 or SL2. However, as more sequences become available this conclusion will be tested more thoroughly. Conservation of the region involved in circularization was not detected in these two viruses although traces of conservation were detectable in the nucleotide region that is complementary to the equivalent of the CYCL domain in the 3′UTR. Taking into account the data on the 5′UTR for the TBFV, MBFV and NKV it is possible that stem 0, being more conserved than SL1 and SL2, is involved in signaling activity important for virus replication. The next section will show how experimental data correlate with these theoretical predictions.

The role of the flavivirus 5′UTR in virus translation and replication

A role for the flavivirus 5′UTR in virus translation and replication is evident from experiments on mutagenesis of this region using the DENV4 infectious clone. Progressive deletions of 5 or 6 nucleotides were introduced into the 5′UTR (Figure 3.20) and their effect was evaluated by *in vitro* translation experiments and virus replication studies in cell culture (Cahour *et al.*, 1995). These deletions have been su-

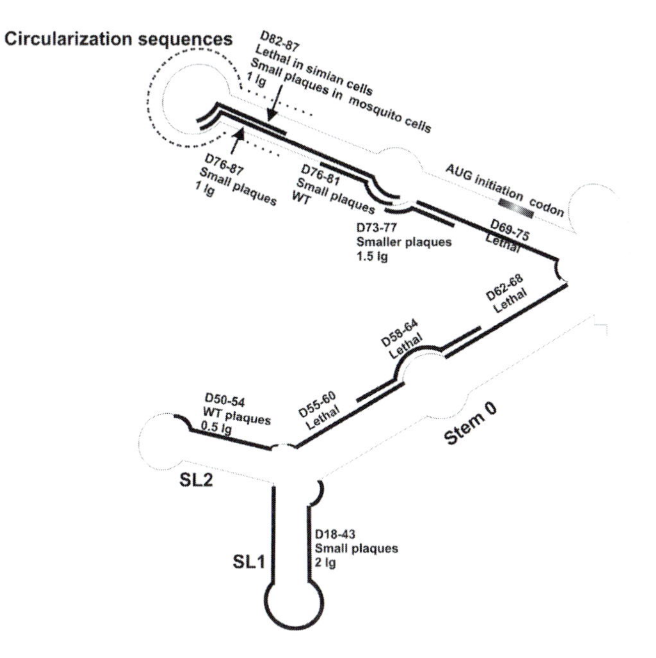

Figure 3.20 Mutagenesis in the 5′UTR of DENV4. The deleted regions are specified as thick black bars; enumeration after D (deletion) shows the boundaries of nucleotide deletion. The first line after the enumerated deletion shows alteration of plaque phenotype in comparison with the wild-type (WT) virus in simian and mosquito cells. The second line shows the approximate reduction of virus titer in $\log_{10}$ (lg) in comparison with WT in simian and mosquito cells. Circularization sequences are shown by a dashed line. The elements of the Y-shaped structure, i.e. stem 0, SL1 and SL2 are denominated and the initiation AUG codon is shown in the shaded box. Adapted from (Cahour *et al.*, 1995).

perimposed onto the computer-simulated secondary structures of the 5′UTR of DENV4 (not shown) and the results correlate with protein translation *in vitro* and virus infectivity in mammals and mosquito cell lines (summarized in Figure 3.20).

The reduction of *in vitro* translation efficiency in some mutants could be explained by the formation of thermodynamically rigid structures that probably slow-down the movement of ribosomes along the 5′UTR. Indeed, secondary structure predictions of such mutants revealed the formation of long stable hairpin structures, energetically less favorable for translation than the original (nonmutated) RNA structure (not shown). In contrast, deletion mutants that demonstrate higher translation rates compared with WT RNA, produced less stable RNA structures, such as short hairpins that might facilitate ribosome scanning along the RNA template. In general it seems that the preservation of the Y-shaped structure was not required to direct efficient protein translation from the 5′UTR of DENV4; variations in translation efficiency *in vitro* were in good agreement with conventional thermodynamic rules deduced for the translation of mRNA (Kozak, 1991; Kozak, 2002) rather than with the stability of the Y-shaped structure.

Moreover, it was subsequently demonstrated that efficiency of virus translation depends on the presence of 3′UTR or 3′LSH sequences; as described above there are two conflicting results demonstrating, enhancement (Holden and Harris, 2004), or inhibition (Li and Brinton, 2001; Tilgner *et al.*, 2005) of translation in the presence of the 3′LSH. The common finding between these two results is that the effect of the 3′UTR on translation from the 5′UTR does not depend on physical interaction between 5′UTR and 3′UTR since the 3′UTR or 3′LSH has the same effect on translation of other nonflavivirus UTRs. This implies that the influence of the 3′UTR on translation directed by the flavivirus 5′UTR is mediated by the cellular protein(s) that may bring the 5′ and 3 ends to close proximity forming a closed-loop complex that facilitates binding of ribosomes to the initiation complex in a manner similar to that of cellular mRNA (Gale *et al.*, 2000; Michel *et al.*, 2000) or viral RNA (Diez *et al.*, 2000; Vende *et al.*, 2000). Such a 3′UTR-dependent mode of translation does not depend on direct contact between the 3′UTR and 5′UTR. Moreover, the 3′UTR can also direct translation from a variety of nonviral templates. Thus, if flavivirus translation is dependent on the 3′UTR, then stem 0 could support 5′UTR–3′UTR interaction, mediated by cellular proteins. This suggestion would explain why the destruction of stem 0 abolishes virus reproduction.

Sequences downstream of stem 0 are involved in interactions with translated regions of the virus genome (SL3 on Figure 3.16). The distal domain of SL3 contains sequences that mediate circularization of the virus genome (highlighted). The partial deletion of these sequences (mutants D76–87 and D82–87) did not render DENV4 noninfectious although its infectivity was considerably reduced. It is interesting to note that the smaller deletion D82–87 produced more severe restrictions on virus growth then the larger deletion D76–87—this virus retained infectivity only in mammalian cells and was completely noninfectious in insect cells.

From the data (Cahour *et al.*, 1995) it is not clear if virus infectivity is directly dependent upon the efficiency of *in vitro* protein translation. Lethal mutations in the 5′UTR were accompanied by either lower or higher levels of *in vitro* translation. On the other hand, four nonlethal mutants D18–43, D73–77 D76–87 and D82–87 produced lower levels of *in vitro* translation than the original RNA; all four mutant viruses had reduced infectivity in cell culture. Moreover, none of the lethal mutations completely abolished *in vitro* translation. One lethal mutant D58–64 had nearly the same level of *in vitro* translation as the original RNA suggesting that translation is not a critical determinant of virus infectivity. It appears from these experiments that alteration of the 5′UTR affects virus reproduction through replication rather than translation.

Experiments by (Cahour *et al.*, 1995) indicate that stem 0 might play a major role in virus replication and that SL1 and SL2 are nonessential for virus infectivity. All deletions introduced into stem 0 (D55–60, D58–64, D62–68) were lethal for the virus (Figure 3.20). In contrast, the total deletion of SL1 that resulted in the elongation of stem 0 and reduced translation *in vitro*, still produced infectious virus albeit with a reduced replication rate. Similarly, deletion in SL2 (D50–54) had almost no effect on virus translation or virus infectivity. This is also supported by the observation that the conformations of the POWV SL1 and SL2 are different from those of other TBFV (Figure 3.16A(2)) due to a deletion in SL2, indicating the absence of constraints on these two structures.

Therefore, it could be that the Y-shaped structure in the 5′UTR of flaviviruses might evolve in a manner that is essential for virus RNA replication rather than protein translation. As a translation-driving domain, the 5′UTR can tolerate far higher levels of sequence variation which would explain why lethal mutations for DENV4 (Cahour *et al.*, 1995) did not entirely destroy *in vitro* translation. The high level of divergence for SL1 and SL2 suggests that their function is to maintain optimal folding of stem 0. This suggestion is supported by the high level of compensatory mutations seen in SL1 and SL2 (Figure 3.16) and also by the observation that they do not provide essential functions for virus infectivity (Figure 3.20). On the other hand, as compensatory mutations are selected for SL1 and SL2, these two structures could provide functions that are only recognizable under the more extreme conditions encountered in the wild.

However, all mutations that affect stem 0 or destroy the Y-shaped structure were lethal for the virus (Figure 3.20) and the Y-shaped structure is quite conserved among all three major virus groups, despite their sequence divergence. Thus, as a double-stranded region, stem 0 could provide a signaling function by binding virus or host proteins required for subsequent interaction with the 3′UTR. For some RNA viruses such as picornaviruses (Agol *et al.*, 1999; Barton *et al.*, 2001) and influenza virus (Crow *et al.*, 2004) the initial binding and activation of RdRp occurs at the 5′ end or internal parts of the virus genome, followed by binding of RdRp to the 3′ end and the initiation of RNA synthesis. For poliovirus the interaction between 5′UTR and 3′UTR not only initiates minus RNA synthesis but also regulates the termination of translation (Barton *et al.*, 2001). For Brome Mosaic virus it was suggested that the secondary structure at the 5′UTR might participate in partition of the plus and minus strands before initiation of the synthesis of plus strand on the minus template (Pogue *et al.*, 1994). For alphaviruses the 5′UTR also plays a key role in the synthesis of both negative- and positive-strand RNAs (Gorchakov *et al.*, 2004).

Extending this analogy to flaviviruses, the possibility cannot be excluded that stem 0 provides the binding and initial activation of viral or host factors that subsequently assemble into the RdRp to initiate negative-strand RNA synthesis from the 3′ end. If stem 0 is adapted to virus proteins, it may have co-evolved simultaneously with the evolution of viral protein(s) which might explain why this region is better conserved within individual virus groups rather than between major virus groups.

Additionally, stem 0 might specifically bind viral/cellular proteins that have a second domain for interaction with sequences in the 3′UTR and this could require the initiation of translation implying that the 5′UTR might have evolved as a bi-functional RNA domain.

Does the 5′UTR function as 3′UTR in the minus-sense RNA?

For many viruses sequences complementary to the 5′UTR are effectively the 3′UTR in the minus RNA strand and they initiate plus-strand RNA synthesis (Pogue *et al.*, 1994). This region binds viral and host elements (Lai, 1998) which regulate asymmetric synthesis of plus-strand RNA that is proportionately in excess compared with minus-strand RNA, a feature that is common for many RNA viruses.

An obvious question arises, does the 3′UTR of a flavivirus minus RNA strand fold as a stable structure and bind viral or host elements that initiate RNA synthesis of positive strand? Currently there has been no theoretical research to investigate the folding of the negative-strand flavivirus 3′UTR and adjacent regions within the capsid gene, or to identify stable structures involved in switching over RNA synthesis from the positive to the negative template. Ribonuclease probing demonstrated the existence of a stable hairpin in the 3′UTR of the minus-strand for the WNV (Shi *et al.*, 1996b) that is different from the predicted Y-shaped structure in the positive-sense RNA (Thurner *et al.*, 2004). However, a Y-shaped structure similar to that in the positive-strand was proposed in DENV (Yocupicio-Monroy *et al.*, 2003), although no evidence was presented. It is possible that stem 0 forms a stable duplex in the negative-strand as a part of the Y-shaped structure of the plus-strand 5′UTR. Owing to its conservation a signaling function may be assigned to it. For example it might interact with host proteins that have been shown to bind WNV and DENV negative-strand 3′ UTRs (Shi *et al.*, 1996b; Yocupicio-Monroy *et al.*, 2003). Therefore, more research at both the theoretical and experimental level is required to investigate details of the RNA structures in the negative orientation of the flavivirus genome and mechanisms of initiation of plus-strand RNA synthesis from the negative template.

Secondary RNA structures within the translational regions of flaviviruses

Conserved secondary structures within the capsid gene of TBFV have been described (Gritsun *et al.*, 1997). For the predicted secondary structure of the 5′ terminal 333 nucleotides, using the Mfold program, see Figure 3.21. In addition to the conserved pattern of stem–loop structures that have been identified in the 5′UTR, computer-assisted analysis of the secondary RNA structures of the capsid gene region (following the 5′UTR) revealed two other stem–loops, designated SL5 and SL6 (Gritsun *et al.*, 1997). These structures were perfectly conserved in shape between all members of the TBFV group. Surprisingly the sequence within loop 6 that is an inverted repeat was present in loop 4 of the

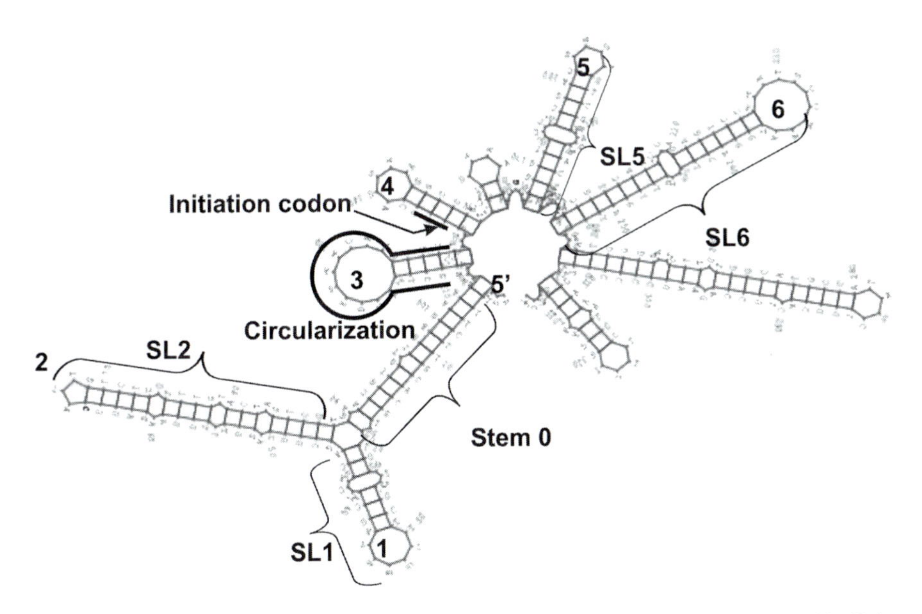

Figure 3.21 The folding of the 5′ terminal 333 nucleotides for the TBFV as described in (Gritsun *et al.*, 1997). The conserved (among the TBFV) elements of the 5′UTR (SL1, SL2 and SL3) and capsid gene (SL5 and SL6) are included in brackets. The initiation codon and the circularization sequences are highlighted by the thick lines.

3′UTR suggesting that kissing loops might contribute to the circularization process (Gritsun *et al.*, 1997). These sequences also appear in loop 4 in the model of the dumbbell-like structures (Olsthoorn and Bol, 2001). Moreover SL5 and SL6 also showed a stable conformation in the folding of the minus-strand RNA (Gritsun *et al.*, 1997)

The use of more advanced phylogenetic algorithms has provided global RNA folding of long molecules and has located the position of conserved RNA secondary structures along entire flavivirus virus genomes (Thurner *et al.*, 2004). This folding confirmed the formation of SL5 and SL6 previously observed in the TBFV group (Gritsun *et al.*, 1997). Many other stem–loop structures were predicted, their numbers and locations being widely different between the divergent flavivirus groups (Figure 3.22).

There has been no targeted research specifically designed to investigate the role of coding region RNA secondary structures in the replication of flaviviruses. The conservation of SL6 implies it has a specific function which in this case could be packaging, replication or interaction with the 3′UTR. Nevertheless, the deletion of this structure (Kofler *et al.*, 2002) demonstrated that it is not essential for virus infectivity although this deletion did reduce virus replication rate, probably due to the corresponding deletion within the C protein.

Thus, initial experimental evidence appears to contradict the concept of a significant role for stable conserved RNA secondary structures in the coding region. However, the existence of similar structures in the coding region of other RNA viruses that are not essential for virus replication but improve its efficiency

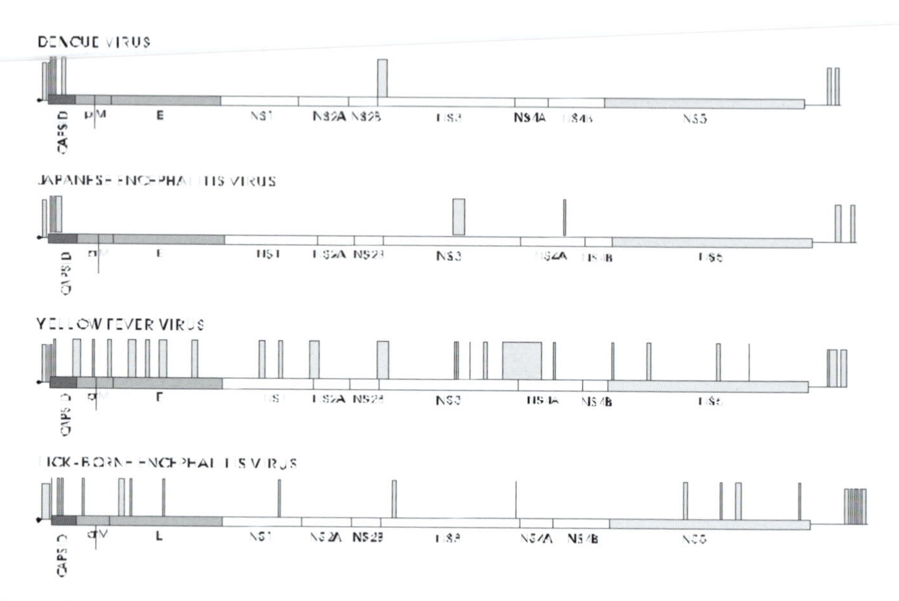

Figure 3.22 Genome map of location of conserved predicted RNA secondary structures in the coding part of the genome for the DENV, YFV and TBEV. Putative conserved secondary structures are indicated by the boxes above the RNA sequence. Kindly provided by Dr. Thurner (Thurner *et al.*, 2004).

(French and Ahlquist, 1987; Goodfellow *et al.*, 2000; Klovins *et al.*, 1998; Lin and Lai, 1993; McKnight and Lemon, 1996; Schuppli *et al.*, 1998) supports the notion of a role for secondary structures in the flavivirus open-reading frame. Perhaps each flavivirus group developed individual secondary structures in the coding region, that are not essential, but their cumulative action might provide the virus genome with greater efficiency of replication. They could act as additional virus enhancers that ensure rapid and efficient onset of virus replication, critical in real (not laboratory) conditions, possibly influencing virus transmission efficiency.

Conclusions

We have reviewed the published studies of flavivirus UTRs and also applied a detailed analysis of new data that have become available through our improved comparative alignments of flavivirus sequences. The following conclusions have been drawn as the result of this work.

The flavivirus 3′UTR, and perhaps also the ORF, might originate from multiple duplication events

Alignment of sequences of the 3′UTR of TBFV revealed four long imperfect repeat sequences each of which contained long repeat sequence (LRS) (~200 nucleotides on the alignment), that were also detected in the distal part of the ORF (as two long repeats in the NS5 protein). From this observation, one can speculate that the 3′UTR originated as the result of multiple duplication of these LRSs. In addition, duplication of smaller repeat sequences was detected within the 3′UTR and may represent other evolutionary remnants of early duplication events. Since the TBFV group evolved more slowly than the MBFV, they have preserved more remnants of the original repeated sequences. The short (20–25 nucleotides long) conserved sequences and their repeat sequences (CSs and RCSs) that were found in the MBFV and in some of the NKV could be conceived as remnants, or "fossils" that represent earlier longer repeated sequences similar to those that can still be traced in the 3′UTR of the TBFV.

Flavivirus RNA has evolved an enhancing function for virus replication that may be essential for virus transmission

Analysis of the different models of RNA secondary structures and correlation with experimental data revealed conserved structural and functional organization of the flavivirus 3′UTR throughout the genus. The active virus promoter that initiates virus RNA synthesis is formed by direct interaction between the 5′- and 3′ termini of the virus genome. The RNA folding predicts the formation of a long circularization stem that contains stretches of highly conserved sequences (CSs). One of these CSs called CYCL is absolutely conserved within each MBFV (eight nucleotides) and TBFV (11 nucleotides); its absolute conservation indicates that in addition to circularization, this sequence might also provide a signaling function for events such as replication, encapsidation or both. However, the distinct folding of the 3′UTR and 5′UTR predicts the formation of stable structures other than the circularization stem. The secondary structure of the most distal (about 100 nucleotides) region of the 3′UTR is perfectly conserved between the three major flavivirus groups and absolutely essential for virus infectivity. This region contains a long stable hairpin 3′LSH and a short SL2. The 3′LSH exposes a highly conserved pentapeptide as a part of loop 1 that provides an essential signaling function for replication. The 3′LSH- SL2 domain could represent the minimal requirement of a promoter since it binds viral RdRp and in the presence of the 5′UTR directs the synthesis of flavivirus RNA *in vitro*. For the MBFV, this part of the 3′UTR on its own supports residual virus infectivity in mosquito cells whereas for replication in mammalian cells the extended promoter is required that in addition to the 3′LSH- SL2 contains the dumbbell-like structure DB1 (about 70 nucleotides long) that lies upstream of the 3′LSH- SL2 and is supported by pseudoknots.

The sequences that form the 3′LSH-SL2 overlap with sequences that enable circularization of the virus genome through interaction with the 5′UTR, to initiate viral RNA synthesis. It is not clear if circularization of the virus genome excludes the independent folding of the 3′UTR and 5′UTR. It is possible that a virus promoter, i.e. the part of the virus RNA that interacts with the RdRp and provides initiation of the RNA synthesis, presents a flexible complex machinery that alters conformation during transition from one stage to another. In this respect the predicted circularization stem and DB1-SL2–3′LSH that share the same sequences might simulate different essential intermediate stages of highly ordered RNA–RNA or RNA-protein interactions that selectively channel the viral RNAs into the initiation of replication and/or virion assembly. The transitions between these different stages are largely unexplored, and much remains to be learned concerning the recognition of cis-acting RNA signals and how they promulgate RNA replication or virion assembly.

The region of the 3′UTR of flaviviruses that lies between the promoter and stop codon could be experimentally deleted without the complete loss of virus infectivity but it hampers virus replication in both arthropod and mammalian cells. For some MBFV and NKV the computational folding of this region predicts the formation of a second dumbbell-like structure DB2 that duplicates DB1 in structure and in exposing RCS2. Additionally, other duplicated secondary structures, with exposed conserved repeated sequences were revealed in this part of the 3′UTR. The presence of conserved repeated sequences might imply that they have a signaling function for proteins whose activity depends on polymerization. The individual deletion of these structures did not abolish virus infectivity but did reduce its replication rate. This region was therefore interpreted as having an enhancer function for virus replication; the experimental data indicate that the rate of virus replication depends on the degree of perturbation of the enhancer's secondary structure. The individual structural elements of the enhancer, acting cooperatively, comprise the entire complex machinery designed to provide rapid and efficient virus replication that could be absolutely critical for virus transmission between the vector and vertebrate host and provide their successful dissemination in nature.

The flavivirus 5′UTR might contribute to the promoter function

The 5′UTR of flaviviruses is about 100 nucleotides long and its most important function for positive RNA viruses is to initiate virus translation after the uncoating of viral RNA in the cell cytoplasm. However, some features of this region predict that the 5′UTR might also be an essential element of the flavivirus promoter. One of these features is the presence of the circularization sequences, including the conserved CYCL. Additionally, the proximal part of the 5′UTR (about 70–100 nucleotides) is not involved in the circularization of the virus genome and is predicted to fold as a Y-shaped structure that is formed by an approximately 10–14-nucleotide long double-helix region (stem 0) with two branching stem–loops (SL1 and SL2). Stem 0 is formed by sequences more conserved than two branching SLs suggesting its functional significance. Experimental data on deletion within this region suggests that stem 0, both as a structure and as a sequence, might be essential for virus infectivity, not through interference with the translation of viral polyprotein but at the level of virus replication. Stem 0 may contribute to the interaction of the 5′UTR with the 3′UTR, although not directly through RNA–RNA contacts but by binding to different domains of the same cellular or virus protein. In this sense the flavivirus replication process may not be unique: the involvement of the 5′UTR in the initiation of replication from the 3′UTR has been demonstrated for other positive- and negative-stranded RNA viruses. However, the 5′UTR acting as the 3′UTR of a minus RNA strand could also contribute to the initiation of plus-strand RNA synthesis from the minus template but this process has not been elucidated.

Conserved stem–loop structures were discovered not only in the UTRs but also in the coding region of the flavivirus genome

Stem–loop structures in the capsid coding region of the genome may act as multiple enhancers of viral RNA replication, by analogy with the strategy of the replication for other RNA viruses. However, their precise role in the functioning of the virus genome has not yet been adequately investigated. Like the enhancers in the UTRs, they might act cooperatively providing a cumulative enhancing effect on viral RNA synthesis that is critical for virus transmission between vertebrates and invertebrates.

References

Ackermann, M., and Padmanabhan, R. (2001). De novo synthesis of RNA by the dengue virus RNA-dependent RNA polymerase exhibits temperature dependence at the initiation but not elongation phase. J. Biol. Chem. *276*, 39926–39937.

Agol, V. I., Paul, A. V., and Wimmer, E. (1999). Paradoxes of the replication of picornaviral genomes. Virus Res. *62*, 129–147.

Barton, D. J., O'Donnell, B. J., and Flanegan, J. B. (2001). 5′ cloverleaf in poliovirus RNA is a cis-acting replication element required for negative-strand synthesis. EMBO J. *20*, 1439–1448.

Blackwell, J. L., and Brinton, M. A. (1995). BHK cell proteins that bind to the 3′ stem–loop structure of the West Nile virus genome RNA. J. Virol. *69*, 5650–5658.

Blackwell, J. L., and Brinton, M. A. (1997). Translation elongation factor-1 alpha interacts with the 3′ stem–loop region of West Nile virus genomic RNA. J. Virol. *71*, 6433–6444.

Bodaghi, S., Ngon, A. Y. M., and Dodds, J. A. (2000). Heterogeneity in the 3′-terminal untranslated region of tobacco mild green mosaic tobamoviruses from *Nicotiana glauca* resulting in variants with three or six pseudoknots. J. Gen. Virol. *81*, 577–586.

Bredenbeek, P. J., Kooi, E. A., Lindenbach, B., Huijkman, N., Rice, C. M., and Spaan, W. J. (2003). A stable full-length yellow fever virus cDNA clone and the role of conserved RNA elements in flavivirus replication. J. Gen. Virol. *84*, 1261–1268.

Brinton, M. A., and Dispoto, J. H. (1988). Sequence and secondary structure analysis of the 5′-terminal region of flavivirus genome RNA. Virology *162*, 290–299.

Brinton, M. A., Fernandez, A. V., and Dispoto, J. H. (1986). The 3′-nucleotides of flavivirus genomic RNA form a conserved secondary structure. Virology *153*, 113–121.

Bryan, G. T., Gardner, R. C., and Forster, R. L. (1992). Nucleotide sequence of the coat protein gene of a strain of clover yellow vein virus from New Zealand: conservation of a stem–loop structure in the 3′ region of potyviruses. Arch. Virol. *124*, 133–146.

Cahour, A., Pletnev, A., Vazielle Falcoz, M., Rosen, L., and Lai, C. J. (1995). Growth-restricted dengue virus mutants containing deletions in the 5′ noncoding region of the RNA genome. Virology *207*, 68–76.

Charlier, N., Leyssen, P., Pleij, C. W. A., Lemey, P., Billoir, F., Van Laethem, K., Vandamme, A. M., De Clercq, E., de Lamballerie, X., and Neyts, J. (2002). Complete genome sequence of Montana Myotis leukoencephalitis virus, phylogenetic analysis and comparative study of the 3′ untranslated region of flaviviruses with no known vector. J. Gen. Virol. *83*, 1875–1885.

Charrel, R. N., Attoui, H., Butenko, A. M., Clegg, J. C., Deubel, V., Frolova, T. V., Gould, E. A., Gritsun, T. S., Heinz, F. X., Labuda, M., *et al.* (2004). Tick-borne virus diseases of human interest in Europe. Clin. Microbiol. Infect. *10*, 1040–1055.

Charrel, R. N., Zaki, A. M., Attoui, H., Fakeeh, M., Billoir, F., Yousef, A. I., de Chesse, R., De Micco, P., Gould, E. A., and de Lamballerie, X. (2001). Complete coding sequence of the Alkhurma virus, a tick-borne flavivirus causing severe hemorrhagic fever in humans in Saudi Arabia. Biochem. Biophys. Res. Commun. *287*, 455–461.

Chen, C. J., Kuo, M. D., Chien, L. J., Hsu, S. L., Wang, Y. M., and Lin, J. H. (1997). RNA-protein interactions: involvement of NS3, NS5, and 3′ noncoding regions of Japanese encephalitis virus genomic RNA. J. Virol. *71*, 3466–3473.

Chu, P. W., and Westaway, E. G. (1985). Replication strategy of Kunjin virus: evidence for recycling role of replicative form RNA as template in semiconservative and asymmetric replication. Virology *140*, 68–79.

Cleaves, G. R., Ryan, T. E., and Schlesinger, R. W. (1981). Identification and characterization of type 2 dengue virus replicative intermediate and replicative form RNAs. Virology *111*, 73–83.

Copper, P. D., Steiner-Pryor, A., Scotti, P. D., and Delong, D. (1974). On the nature of poliovirus genetic recombinants. J. Gen. Virol. *23*, 41–49.

Corver, J., Lenches, E., Smith, K., Robison, R. A., Sando, T., Strauss, E. G., and Strauss, J. H. (2003). Fine mapping of a cis-acting sequence element in yellow fever virus RNA that is required for RNA replication and cyclization. J. Virol. *77*, 2265–2270.

Crow, M., Deng, T., Addley, M., and Brownlee, G. G. (2004). Mutational analysis of the influenza virus cRNA promoter and identification of nucleotides critical for replication. J. Virol. *78*, 6263–6270.

Cui, T., Sugrue, R. J., Xu, Q., Lee, A. K., Chan, Y. C., and Fu, J. (1998). Recombinant dengue virus type 1 NS3 protein exhibits specific viral RNA binding and NTPase activity regulated by the NS5 protein. Virology *246*, 409–417.

De Nova Ocampo, M., Villegas Sepulveda, N., and del Angel, R. M. (2002). Translation elongation factor-1alpha, La, and PTB interact with the 3′ untranslated region of dengue 4 virus RNA. Virology *295*, 337–347.

Deas, T. S., Binduga-Gajewska, I., Tilgner, M., Ren, P., Stein, D. A., Moulton, H. M., Iversen, P. L., Kauffman, E. B., Kramer, L. D., and Shi, P. Y. (2005). Inhibition of flavivirus infections by antisense oligomers specifically suppressing viral translation and RNA replication. J. Virol. *79*, 4599–4609.

Diez, J., Ishikawa, M., Kaido, M., and Ahlquist, P. (2000). Identification and characterization of a host protein required for efficient template selection in viral RNA replication. Proc. Natl. Acad. Sci. USA *97*, 3913–3918.

Dominguez, G., Wang, C. Y., and Frey, T. K. (1990). Sequence of the genome RNA of rubella virus: evidence for genetic rearrangement during togavirus evolution. Virology *177*, 225–238.

Durbin, A. P., Karron, R. A., Sun, W., Vaughn, D. W., Reynolds, M. J., Perreault, J. R., Thumar, B., Men, R., Lai, C. J., Elkins, W. R., *et al.* (2001). Attenuation and immunogenicity in humans of a live dengue virus type-4 vaccine candidate with a 30 nucleotide deletion in its 3′-untranslated region. Am. J. Trop. Med. Hyg. *65*, 405–413.

Edgil, D., Diamond, M. S., Holden, K. L., Paranjape, S. M., and Harris, E. (2003). Translation efficiency determines differences in cellular infection among dengue virus type 2 strains. Virology *317*, 275–290.

Edwards, M. C. (1995). Mapping of the seed transmission determinants of barley stripe mosaic virus. Mol. Plant Microbe Interact. *8*, 906–915.

Faragher, S. G., and Dalgarno, L. (1986). Regions of conservation and divergence in the 3′ untranslated sequences of genomic RNA from Ross River virus isolates. J. Mol. Biol. *190*, 141–148.

French, R., and Ahlquist, P. (1987). Intercistronic as well as terminal sequences are required for efficient amplification of brome mosaic virus RNA3. J. Virol. *61*, 1457–1465.

Gale, M., Jr., Tan, S. L., and Katze, M. G. (2000). Translational control of viral gene expression in eukaryotes. Microbiol. Mol. Biol. Rev. *64*, 239–280.

Gaunt, M. W., Sall, A. A., Lamballerie, X., Falconar, A. K., Dzhivanian, T. I., and Gould, E. A. (2001). Phylogenetic relationships of flaviviruses correlate with their epidemiology, disease association and biogeography. J. Gen. Virol. *82*, 1867–1876.

Goodfellow, I., Chaudhry, Y., Richardson, A., Meredith, J., Almond, J. W., Barclay, W., and Evans, D. J. (2000). Identification of a cis-acting replication element within the poliovirus coding region. J. Virol. *74*, 4590–4600.

Gorchakov, R., Hardy, R., Rice, C. M., and Frolov, I. (2004). Selection of functional 5′ cis-acting elements promoting efficient sindbis virus genome replication. J. Virol. *78*, 61–75.

Gould, E. A., de Lamballerie, X., Zanotto, P. M., and Holmes, E. C. (2003). Origins, evolution, and vector/host coadaptations within the genus Flavivirus. Adv. Virus. Res. *59*, 277–314.

Gould, E. A., Moss, S. R., and Turner, S. L. (2004). Evolution and dispersal of encephalitic flaviviruses. Arch. Virol. Suppl, 65–84.

Gritsun, T. S., Desai, A., and Gould, E. A. (2001). The degree of attenuation of tick-borne encephalitis virus depends on the cumulative effects of point mutations. J. Gen. Virol. *82*, 1667–1675.

Gritsun, T. S., Holmes, E. C., and Gould, E. A. (1995). Analysis of flavivirus envelope proteins reveals variable domains that reflect their antigenicity and may determine their pathogenesis. Virus Res. *35*, 307–321.

Gritsun, T. S., Lashkevich, V. A., and Gould, E. A. (2003a). Tick-borne encephalitis. Antiviral Res. *57*, 129–146.

Gritsun, T. S., Nuttall, P. A., and Gould, E. A. (2003b). Tick-borne flaviviruses. Adv. Virus. Res. *61*, 317–371.

Gritsun, T. S., Venugopal, K., Zanotto, P. M., Mikhailov, M. V., Sall, A. A., Holmes, E. C., Polkinghorne, I., Frolova, T. V., Pogodina, V. V., Lashkevich, V. A., and Gould, E. A. (1997). Complete sequence of two tick-borne flaviviruses isolated from Siberia and the UK: analysis and significance of the 5′ and 3′- UTRs. Virus Res. *49*, 27–39.

Hahn, C. S., Hahn, Y. S., Rice, C. M., Lee, E., Dalgarno, L., Strauss, E. G., and Strauss, J. H. (1987). Conserved elements in the 3′ untranslated region of flavivirus RNAs and potential cyclization sequences. J. Mol. Biol. *198*, 33–41.

Hajjou, M., Hill, K. R., Subramaniam, S. V., Hu, J. Y., and Raju, R. (1996). Nonhomologous RNA–RNA recombination events at the 3′ nontranslated region of the Sindbis virus ge-

nome: hot spots and utilization of nonviral sequences. J. Virol. *70*, 5153–5164.
Hayasaka, D., Ivanov, L., Leonova, G. N., Goto, A., Yoshii, K., Mizutani, T., Kariwa, H., and Takashima, I. (2001). Distribution and characterization of tick-borne encephalitis viruses from Siberia and far-eastern Asia. J. Gen. Virol. *82*, 1319–1328.
Heinz, F. X., Collett, M. S., Purcell, R. H., Gould, E. A., Howard, C. R., Houghton, M., Moormann, R. J. M., Rice, C. M., and Thiel, H. J. (2000). Family *Flaviviridae*. In *Virus Taxonomy. 7th International committee for the Taxonomy of Viruses.*, R. M.H.V., C. M. Fauquet, D. H. L. Bishop, E. Carstens, M. K. Estes, S. Lemon, J. Maniloff, M. A. Mayo, D. McGeogch, C. R. Pringle, and R. B. Wickner, eds. (San Diego, Academic Press.), pp. 859–878.
Holden, K. L., and Harris, E. (2004). Enhancement of dengue virus translation: role of the 3′ untranslated region and the terminal 3′ stem–loop domain. Virology *329*, 119–133.
Jones, L. D., Gaunt, M., Hails, R. S., Laurenson, K., Hudson, P. J., Reid, H., Henbest, P., and Gould, E. A. (1997). Transmission of louping ill virus between infected and uninfected ticks co-feeding on mountain hares. Med. Vet. Entomol. *11*, 172–176.
Khromykh, A. A., Kondratieva, N., Sgro, J. Y., Palmenberg, A., and Westaway, E. G. (2003). Significance in replication of the terminal nucleotides of the flavivirus genome. J. Virol. *77*, 10623–10629.
Khromykh, A. A., Meka, H., Guyatt, K. J., and Westaway, E. G. (2001). Essential role of cyclization sequences in flavivirus RNA replication. J. Virol. *75*, 6719–6728.
Khromykh, A. A., and Westaway, E. G. (1997). Subgenomic replicons of the flavivirus Kunjin: construction and applications. J. Virol. *71*, 1497–1505.
Kinney, R. M., Huang, C. Y., Rose, B. C., Kroeker, A. D., Dreher, T. W., Iversen, P. L., and Stein, D. A. (2005). Inhibition of dengue virus serotypes 1 to 4 in vero cell cultures with morpholino oligomers. J. Virol. *79*, 5116–5128.
Klovins, J., Berzins, V., and van Duin, J. (1998). A long-range interaction in Qbeta RNA that bridges the thousand nucleotides between the M-site and the 3′ end is required for replication. RNA *4*, 948–957.
Kofler, R. M., Heinz, F. X., and Mandl, C. W. (2002). Capsid protein C of tick-borne encephalitis virus tolerates large internal deletions and is a favorable target for attenuation of virulence. J. Virol. *76*, 3534–3543.
Kozak, M. (1991). Structural features in eukaryotic mRNAs that modulate the initiation of translation. J. Biol. Chem. *266*, 19867–19870.
Kozak, M. (2002). Pushing the limits of the scanning mechanism for initiation of translation. Gene *299*, 1–34.
Labuda, M., Austyn, J. M., Zuffova, E., Kozuch, O., Fuchsberger, N., Lysy, J., and Nuttall, P. A. (1996). Importance of localized skin infection in tick-borne encephalitis virus transmission. Virology *219*, 357–366.
Lai, M. M. (1992). RNA recombination in animal and plant viruses. Microbiol. Rev. *56*, 61–79.
Lai, M. M. (1998). Cellular factors in the transcription and replication of viral RNA genomes: a parallel to DNA-dependent RNA transcription. Virology *244*, 1–12.
Leyssen, P., Charlier, N., Lemey, P., Billoir, F., Vandamme, A. M., De Clercq, E., de Lamballerie, X., and Neyts, J. (2002). Complete genome sequence, taxonomic assignment, and comparative analysis of the untranslated regions of the Modoc virus, a flavivirus with no known vector. Virology *293*, 125–140.
Li, W., and Brinton, M. A. (2001). The 3′ stem–loop of the West Nile virus genomic RNA can suppress translation of chimeric mRNAs. Virology *287*, 49–61.
Lin, Y. J., and Lai, M. M. (1993). Deletion mapping of a mouse hepatitis virus defective interfering RNA reveals the requirement of an internal and discontiguous sequence for replication. J. Virol. *67*, 6110–6118.
Lindenbach, B. D., and Rice, C. M. (2001). *Flaviviridae*: The Viruses and Their Replication. In Fields Virology, D. M. Knippe, and P. M. Howley, eds. (London-New York-Tokyo, Lippincott Williams & Wilkins), pp. 991–1042.
Lo, M. K., Tilgner, M., Bernard, K. A., and Shi, P. Y. (2003). Functional analysis of mosquito-borne flavivirus conserved sequence elements within 3′ untranslated region of West Nile virus by use of a reporting replicon that differentiates between viral translation and RNA replication. J. Virol. *77*, 10004–10014.
Mandl, C. W., Holzmann, H., Kunz, C., and Heinz, F. X. (1993). Complete genomic sequence of Powassan virus: evaluation of genetic elements in tick-borne versus mosquito-borne flaviviruses. Virology *194*, 173–184.
Mandl, C. W., Holzmann, H., Meixner, T., Rauscher, S., Stadler, P. F., Allison, S. L., and Heinz, F. X. (1998). Spontaneous and engineered deletions in the 3′ noncoding region of tick-borne encephalitis virus: construction

of highly attenuated mutants of a flavivirus. J. Virol. 72, 2132–2140.

Markoff, L. (2003). 5′- and 3′-noncoding regions in flavivirus RNA. Adv. Virus. Res. 59, 177–228.

Markotter, W., Theron, J., and Nel, L. H. (2004). Segment specific inverted repeat sequences in bluetongue virus mRNA are required for interaction with the virus non structural protein NS2. Virus Res. 105, 1–9.

McKnight, K. L., and Lemon, S. M. (1996). Capsid coding sequence is required for efficient replication of human rhinovirus 14 RNA. J. Virol. 70, 1941–1952.

Men, R., Bray, M., Clark, D., Chanock, R. M., and Lai, C. J. (1996). Dengue type 4 virus mutants containing deletions in the 3′ noncoding region of the RNA genome: analysis of growth restriction in cell culture and altered viremia pattern and immunogenicity in rhesus monkeys. J. Virol. 70, 3930–3937.

Michel, Y. M., Poncet, D., Piron, M., Kean, K. M., and Borman, A. M. (2000). Cap-Poly(A) synergy in mammalian cell-free extracts. Investigation of the requirements for poly(A)-mediated stimulation of translation initiation. J. Biol. Chem. 275, 32268–32276.

Mohan, P. M., and Padmanabhan, R. (1991). Detection of stable secondary structure at the 3′ terminus of dengue virus type 2 RNA. Gene 108, 185–191.

Mutebi, J. P., Rijnbrand, R. C., Wang, H., Ryman, K. D., Wang, E., Fulop, L. D., Titball, R., and Barrett, A. D. (2004). Genetic relationships and evolution of genotypes of yellow fever virus and other members of the yellow fever virus group within the Flavivirus genus based on the 3′ noncoding region. J. Virol. 78, 9652–9665.

Olsthoorn, R. C., and Bol, J. F. (2001). Sequence comparison and secondary structure analysis of the 3′ noncoding region of flavivirus genomes reveals multiple pseudoknots. RNA 7, 1370–1377.

Peerenboom, E., Jacobi, V., Cartwright, E. J., Adams, M. J., Steinbiss, H. H., and Antoniw, J. F. (1997). A large duplication in the 3′-untranslated region of a subpopulation of RNA2 of the UK-M isolate of barley mild mosaic bymovirus. Virus Res. 47, 1–6.

Pilipenko, E. V., Gmyl, A. P., and Agol, V. I. (1995). A model for rearrangements in RNA genomes. Nucleic Acids Res. 23, 1870–1875.

Pletnev, A. G. (2001). Infectious cDNA clone of attenuated Langat tick-borne flavivirus (strain E5) and a 3′ deletion mutant constructed from it exhibits decreased neuroinvasiveness in immunodeficient mice. Virology 282, 288–300.

Pogue, G. P., Huntley, C. C., and Hall, T. C. (1994). Common replication strategies emerging from the study of diverse groups of positive-strand RNA viruses. Arch. Virol. Suppl 9, 181–194.

Proutski, V., Gaunt, M. W., Gould, E. A., and Holmes, E. C. (1997a). Secondary structure of the 3′-untranslated region of yellow fever virus: implications for virulence, attenuation and vaccine development. J. Gen. Virol. 78, 1543–1549.

Proutski, V., Gould, E. A., and Holmes, E. C. (1997b). Secondary structure of the 3′ untranslated region of flaviviruses: similarities and differences. Nucleic Acids Res. 25, 1194–1202.

Proutski, V., Gritsun, T. S., Gould, E. A., and Holmes, E. C. (1999). Biological consequences of deletions within the 3′-untranslated region of flaviviruses may be due to rearrangements of RNA secondary structure. Virus Res. 64, 107–123.

Rauscher, S., Flamm, C., Mandl, C. W., Heinz, F. X., and Stadler, P. F. (1997). Secondary structure of the 3′-noncoding region of flavivirus genomes: comparative analysis of base pairing probabilities. RNA 3, 779–791.

Rice, C. M., Lenches, E. M., Eddy, S. R., Shin, S. J., Sheets, R. L., and Strauss, J. H. (1985). Nucleotide sequence of yellow fever virus: implications for flavivirus gene expression and evolution. Science 229, 726–733.

Santagati, M. G., Itaranta, P. V., Koskimies, P. R., Maatta, J. A., Salmi, A. A., and Hinkkanen, A. E. (1994). Multiple repeating motifs are found in the 3′-terminal non-translated region of Semliki Forest virus A7 variant genome. J. Gen. Virol. 75 (Pt 6), 1499–1504.

Schuppli, D., Miranda, G., Qiu, S., and Weber, H. (1998). A branched stem–loop structure in the M-site of bacteriophage Qbeta RNA is important for template recognition by Qbeta replicase holoenzyme. J. Mol. Biol. 283, 585–593.

Shi, B., Ding, S., and Symons, R. H. (1997). Two novel subgenomic RNAs derived from RNA 3 of tomato aspermy cucumovirus. J. Gen. Virol. 78 (Pt 3), 505–510.

Shi, P. Y., Brinton, M. A., Veal, J. M., Zhong, Y. Y., and Wilson, W. D. (1996a). Evidence for the existence of a pseudoknot structure at the 3′ terminus of the flavivirus genomic RNA. Biochemistry 35, 4222–4230.

Shi, P. Y., Li, W., and Brinton, M. A. (1996b). Cell proteins bind specifically to West Nile virus minus-strand 3′ stem–loop RNA. J. Virol. *70*, 6278–6287.

Ta, M., and Vrati, S. (2000). Mov34 protein from mouse brain interacts with the 3′ noncoding region of Japanese encephalitis virus. J. Virol. *74*, 5108–5115.

Takegami, T., Washizu, M., and Yasui, K. (1986). Nucleotide sequence at the 3′ end of Japanese encephalitis virus genomic RNA. Virology *152*, 483–486.

Tan, B. H., Fu, J., Sugrue, R. J., Yap, E. H., Chan, Y. C., and Tan, Y. H. (1996). Recombinant dengue type 1 virus NS5 protein expressed in Escherichia coli exhibits RNA-dependent RNA polymerase activity. Virology *216*, 317–325.

Thurner, C., Witwer, C., Hofacker, I. L., and Stadler, P. F. (2004). Conserved RNA secondary structures in Flaviviridae genomes. J. Gen. Virol. *85*, 1113–1124.

Tilgner, M., Deas, T. S., and Shi, P. Y. (2005). The flavivirus-conserved penta-nucleotide in the 3′ stem–loop of the West Nile virus genome requires a specific sequence and structure for RNA synthesis, but not for viral translation. Virology *331*, 375–386.

Troyer, J. M., Hanley, K. A., Whitehead, S. S., Strickman, D., Karron, R. A., Durbin, A. P., and Murphy, B. R. (2001). A live attenuated recombinant dengue-4 virus vaccine candidate with restricted capacity for dissemination in mosquitoes and lack of transmission from vaccinees to mosquitoes. Am. J. Trop. Med. Hyg. *65*, 414–419.

van Dijk, A. A., Makeyev, E. V., and Bamford, D. H. (2004). Initiation of viral RNA-dependent RNA polymerization. J. Gen. Virol. *85*, 1077–1093.

Vende, P., Piron, M., Castagne, N., and Poncet, D. (2000). Efficient translation of rotavirus mRNA requires simultaneous interaction of NSP3 with the eukaryotic translation initiation factor eIF4G and the mRNA 3′ end. J. Virol. *74*, 7064–7071.

Wallner, G., Mandl, C. W., Kunz, C., and Heinz, F. X. (1995). The flavivirus 3′-noncoding region: extensive size heterogeneity independent of evolutionary relationships among strains of tick-borne encephalitis virus. Virology *213*, 169–178.

Wang, E., Weaver, S. C., Shope, R. E., Tesh, R. B., Watts, D. M., and Barrett, A. D. (1996). Genetic variation in yellow fever virus: duplication in the 3′ noncoding region of strains from Africa. Virology *225*, 274–281.

Warren, C. E., and Murphy, J. F. (2003). The complete nucleotide sequence of Pepper mottle virus-Florida RNA. Arch. Virol. *148*, 189–197.

Wengler, G., and Castle, E. (1986). Analysis of structural properties which possibly are characteristic for the 3′-terminal sequence of the genome RNA of flaviviruses. J. Gen. Virol. *67*, 1183–1188.

Westaway, E. G., Khromykh, A. A., and Mackenzie, J. M. (1999). Nascent flavivirus RNA colocalized in situ with double-stranded RNA in stable replication complexes. Virology *258*, 108–117.

Yocupicio-Monroy, R. M., Medina, F., Reyes-del Valle, J., and del Angel, R. M. (2003). Cellular proteins from human monocytes bind to dengue 4 virus minus-strand 3′ untranslated region RNA. J. Virol. *77*, 3067–3076.

You, S., Falgout, B., Markoff, L., and Padmanabhan, R. (2001). In vitro RNA synthesis from exogenous dengue viral RNA templates requires long range interactions between 5′- and 3′-terminal regions that influence RNA structure. J. Biol. Chem. *276*, 15581–15591.

You, S., and Padmanabhan, R. (1999). A novel *in vitro* replication system for Dengue virus. Initiation of RNA synthesis at the 3′-end of exogenous viral RNA templates requires 5′- and 3′-terminal complementary sequence motifs of the viral RNA. J. Biol. Chem. *274*, 33714–33722.

Zanotto, P. M., Gao, G. F., Gritsun, T., Marin, M. S., Jiang, W. R., Venugopal, K., Reid, H. W., and Gould, E. A. (1995). An arbovirus cline across the northern hemisphere. Virology *210*, 152–159.

Zanotto, P. M., Gould, E. A., Gao, G. F., Harvey, P. H., and Holmes, E. C. (1996). Population dynamics of flaviviruses revealed by molecular phylogenies. Proc. Natl. Acad. Sci. USA *93*, 548–553.

Zeng, L., Falgout, B., and Markoff, L. (1998). Identification of specific nucleotide sequences within the conserved 3′- SL in the dengue type 2 virus genome required for replication. J. Virol. *72*, 7510–7522.

Zhao, B., Mackow, E., Buckler White, A., Markoff, L., Chanock, R. M., Lai, C. J., and Makino, Y. (1986). Cloning full-length dengue type 4 viral DNA sequences: analysis of genes coding for structural proteins. Virology *155*, 77–88.

Cis- and Trans-acting Determinants of Flaviviridae Replication

4

Sven-Erik Behrens and Olaf Isken

Abstract

The past decade has seen a considerable boost in knowledge on the molecular determinants and mechanisms that orchestrate the intracellular replication of the genomic RNA of Flaviviridae members such as the flaviviruses Kunjin/West Nile and yellow fever, the pestiviruses bovine viral diarrhea virus (BVDV) and classical swine fever virus (CSFV), and hepatitis C virus (HCV). This progress was particularly triggered by the availability of cDNA constructs that permit the instant *in vitro* transcription of full-length infectious viral RNA molecules and of subgenomic replication competent RNAs, the so called "RNA replicons." Moreover, novel experimental strategies combined genetics studies of the viral RNAs with biochemical methods and assay systems and allowed (1) to characterize the functions of the viral proteins and to determine whether the proteins operate in *cis* or *trans* throughout the assembly of the viral replication complex, (2) to define the role of RNA elements in the translation and replication process of the viral genome, and (3) to identify and characterize the function of host-factors participating in the viral life cycle. Along these lines, a particularly important question that is yet unsolved with all positive-strand RNA viruses concerns how viral protein and RNA synthesis is coordinated. This review provides a synopsis on the current state of the art and proposes a model on the initial steps of the viral RNA replication pathway.

Introduction

The Flaviviridae genome is a single stranded, nonsegmented RNA molecule of positive polarity that consists of a long open reading frame (ORF) and nontranslated regions (NTRs) at the 5′ and 3′ ends. As with all positive-strand RNA viruses, the viral genome holds a double-function in the host cell's cytoplasm. Following infection and uncoating, the viral RNA first operates as a messenger RNA. Translation of the ORF is mediated by the cellular translation machinery and initiates either in a type I cap-dependent manner (flaviviruses) or through a type IV internal ribosomal entry site (IRES), a complex RNA element located in the 5′ NTR (pestiviruses and HCV). The resultant polyprotein is co- and post-translationally processed by viral and cellular proteases to yield the viral structural (virion forming) and nonstructural (NS) proteins (Figure 4.1). At a yet incompletely defined stage, the viral genome swaps roles and locks into a membrane-associated replication complex, RC (Egger *et al.*, 2002; El-Hage *et al.*, 2003).

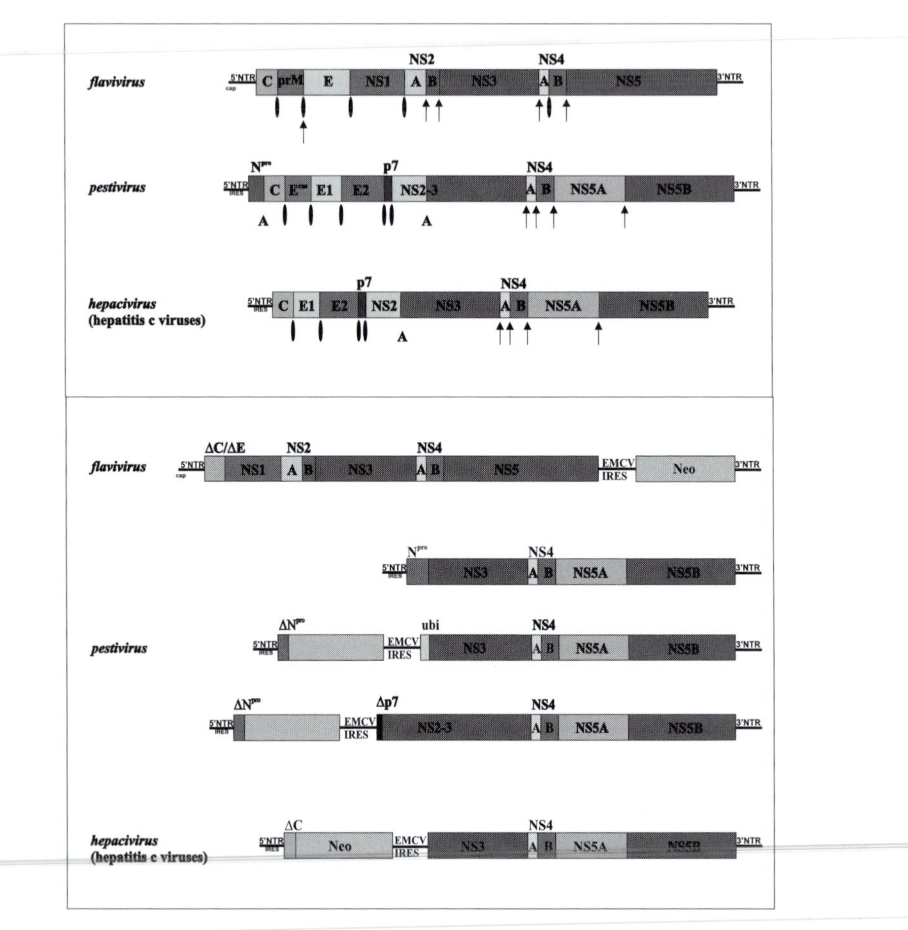

Figure 4.1 Organization of flaviviral, pestiviral and hepaciviral (HCV) full-length genomes (upper panel) and RNA replicons (lower panel). The NTRs are schematized as lines; genetic units are represented as differently shaded boxes. C, prM, E1, E^{rns}, E2 and p7 are the considered viral structural proteins (for a review, see Lindenbach and Rice 2001); the diverse NS proteins are described in the text. Proteolytic cleavage sites in the polyprotein are marked as follows: arrow-cleavage by the NS3 protease complex; circle-cleavage by signal peptidases; A-autoprotease activity; ubi-cleavage by ubiquitine carboxy hydrolase. The shown flaviviral replicon represents a bi-cistronic construct described by Khromykh and Westaway (Khromykh and Westaway, 1997), which, besides the viral proteins that are essential for the replication process expresses a G418/neomycin-resistance gene serving as a selectable marker. Three pestiviral (BVDV) replicons are depicted. The mono-cistronic BVDV replicon designated as "BVDV DI9c," was assembled in accordance to a natural isolate of a defective interfering particle (Meyers *et al.*, 1996). The authentic N-terminus of NS3 is generated by the N^{pro} autoprotease, which has no further function during RNA replication (Behrens *et al.*, 1998). The bi-cistronic BVDV replicon constructs, which, besides the viral proteins, are capable of expressing foreign gene products (gene cassettes downstream of the genuine BVDV 5′ NTR), have been established by Tautz *et al.* In analogy to naturally occurring BVDV biotypes, replicons that express the fully processed NS3 protein are cytopathic, whereas replicons that predominantly express NS2–3 are noncytopathic (Tautz *et al.*, 1999). Functional G418/neomycin-selectable HCV replicons were first described by Lohmann *et al.* (Lohmann *et al.*, 1999). Note that with each of the replicon constructs, short portions of the ORF coding region were fused to the 5′ NTR in order to enable efficient IRES activity (see text). With some constructs, an ubiquitine gene allowed the generation of the authentic N-terminus of the NS3 protein by cellular ubiquitin C-terminal hydrolases.

As outlined in detail in the subsequent sections, the initial RC is believed to include the nascent viral NS proteins along with the viral RNA-dependent RNA polymerase (RdRp), cellular proteins and both termini of the viral genome. The RdRp transcribes the genomic RNA in 3′ to 5′ direction to generate negative-strand RNA copies. The resulting RNA–RNA duplex most probably forms a modified RC, which uses the negative-strand RNA intermediate as a template to synthesize an excess of progeny positive-strand RNA molecules. Novel positive-strand RNA molecules enter further rounds of translation/replication or assemble with the viral structural proteins to yield new virus particles.

With several members of each of the three Flaviviridae genera, it has been possible to assemble cDNA constructs that permit the straightforward transcription of functional viral RNA molecules as well as their mutagenesis (Rice *et al.*, 1985; Kapoor *et al.*, 1995a; Meyers *et al*, 1996a; Meyers *et al.*, 1996b; Moormann *et al.*, 1996; Ruggli *et al.*, 1996; Khromykh and Westaway, 1997; Vassilev *et al.*, 1997; Yamshchikov *et al.*, 2001). Along this line, studies aimed at identifying and characterizing the molecular determinants of the intracellular replication of the viral RNA were particularly facilitated by the discovery that subgenomic RNAs replicate autonomously and independently of the formation of virus particles in transfected host cells (Khromykh and Westaway 1997; Behrens *et al.*, 1998; Moser *et al.*, 1999; Lohmann *et al.*, 1999). Thus, "RNA replicons" of the flaviviruses Kunjin/West Nile and yellow fever consist of the NTRs and the coding units of seven NS proteins, i.e. the glycoprotein NS1, NS2a, the protease cofactor NS2b, the protease and helicase NS3, NS4a, NS4b, and the methyltransferase (MT) and RdRp NS5 (see Figure 4.1). RNA replicons of pestiviruses and HCV essentially consist of the NTRs and the coding region of five NS proteins including the protease and helicase NS3, the protease co-factor NS4A, NS4B, NS5A and the RdRp NS5B (Figure 4.1).

Referring to experimental data that derived to a significant part from studies that applied cDNA constructs of infectious viral genomes or RNA replicons, we discuss the documented functions of viral proteins, genomic RNA motifs and cellular proteins in the viral RNA replication pathway. The homologies and differences between flaviviruses, pestiviruses and HCVs are particularly emphasized.

Functions of viral proteins in Flaviviridae replication

Viral RNA replication involves a defined set of nonstructural proteins

Subgenomic replicon RNAs were either recovered from full-length cDNA constructs by stepwise deletion mutagenesis (Khromykh and Westaway, 1997) or derived from naturally occurring defective interfering particles (Behrens *et al.*, 1998). In any case, the fact that these RNAs turned out to be replication competent in transfected host cells already illustrated that only a restricted set of viral proteins is inevitably involved in the catalysis of both replication steps (Figure 4.1). To ultimately confirm this perception, several laboratories performed systematic mutagenesis studies of the NS protein-coding units of flaviviral, pestiviral or HCV replicon RNAs. The difficulty that genetic alterations of the ORF coding region not only affected the encoded viral proteins but simultaneously modified the coding viral RNA was addressed by control experiments which evaluated the individual ORF mutations also in terms of their impact on translation and on RNA structure and RNA stability. Along the same line, the ORF mutations were assessed with respect to their complementability (see below) and concerning their effects on the viral protease, helicase, methyltransferase (MT) and RdRp activities. In sum, these data supported the concept that on the side of the viral genomes, the proper expression, maturation and structural integrity of the proteins expressed by seven (flaviviruses; NS1, NS2A, NS2B, NS3, NS4A, NS4B, NS5) or five (pestiviruses and HCV; NS3, NS4A, NS4B, NS5A, NS5B) genetic units is necessary and sufficient for the catalysis of the RNA replication process (Khromykh *et al.*, 2000; Grassmann *et al.*, 2001, Kolykhalov *et al.*, 2000; Appel *et al.*, 2005). Most importantly, studies, which specifically investigated the effect of ORF mutations on the first replication step strongly suggested that all mature nonstructural proteins and associated enzymatic activities are already involved in the formation of the initial RC (Grassmann *et al.*, 2001; Khromykh *et al.*, 2001, reviewed by Westaway *et al.*, 2003; for a more thorough discussion of this aspect, see below).

Properties and functions of viral proteins involved in viral RNA replication

The viral enzymes

Studies with the BVDV replicon revealed that even marginal changes in the activity of the NS3/4A protease, in the kinetics of proteolysis, or in the molar ratio of proteolytic cleavage products, severely interfered with the activity of the catalytic replication complex. Proteolytic maturation of the viral nonstructural polyprotein by NS3/4A thus turned out to be a finely balanced process that is closely linked to the catalysis of the RNA replication pathway (Grassmann *et al.*, 1999; Grassmann *et al.*, 2001). As expected, similar evident correlations were found between the activity of the HCV RdRp and the level of RNA replication (Cheney *et al.*, 2002). Further experimental data demonstrated that the NS3 triphosphatase/RNA helicase activity is an essential determinant of the RNA replication process (Grassmann *et al.*, 1999; Khromykh *et al.*, 2000), although the exact function of this interesting virus-encoded enzyme remains so far unknown. A crucial role in replication was also demonstrated for the MT activity that is encoded by the N-terminus of the flaviviral NS5. It is believed to participate in the formation of the flaviviral genomic 5′ cap structure (reviewed by Westaway *et al.*, 2003).

The importance of membrane association

In accord with the observation that assembly of active HCV RCs involves the formation of a so-called membranous web in the cell (Gosert *et al.*, 2003), most of the HCV NS proteins were shown to associate with lipid rafts (Gao *et al.*, 2004; Svitkin *et al.*, 2005). Thus, several lines of evidence suggest that the HCV NS4B acts as a membrane anchor of the RC (Lundin *et al.*, 2003; Elazar *et al.*, 2004), and an analogous function was postulated for the hydrophobic flaviviral NS2A and NS4A proteins (Westaway *et al.*, 2003). Likewise, the flavivirus-specific NS1 glycoprotein, which is assumed to locate in the ER lumen, was defined as a crucial component of the flaviviral RC (Westaway *et al.*, 1997, Muylaert *et al.*, 1997; Lindenbach and Rice, 1997; Brinton, 2002). Finally, it is important to note that correct membrane-anchoring of the HCV NS5A and NS5B proteins was found to be an indispensable prerequisite for replication (Penin *et al.*, 2004; Moradpour *et al.*, 2004).

The role of protein phosphorylation

The HCV and pestiviral NS5A proteins and the N-terminus of the flaviviral NS5 are highly phosphorylated by serine kinases (Reed *et al.*, 1998). The phosphorylation state of these proteins was indicated having an important impact on the modulation of the RNA replication process by affecting protein–protein interactions in the RC (Kapoor *et al.*, 1995b; Blight *et al.*, 2000; Guo *et al.*, 2001; Lohmann *et al.*, 2001; Neddermann *et al.*, 2004; Evans *et al.*, 2004). In fact, the HCV NS5A was found to interact with several host factors and has gained much attention as a potential key player in viral pathogenesis (reviewed by Macdonald and Harris 2004; for a further discussion, see end of this chapter).

Protein interactions in the viral replication complex

Besides the well-studied interactions of the NS3 serine protease domains with NS2B or NS4A acting as co-factors in the flaviviral and pestiviral/HCV protease complex (for a review, see Lindenbach and Rice, 2001), there is little information on functional interactions of the NS proteins with each other and other components of the replication complex. Genetic studies indicated a functional interaction of the yellow fever NS1 and NS4A (Lindenbach and Rice, 1999). *In vitro* studies with HCV suggested the association of NS4B, NS3 and NS5B (Piccininni *et al.*, 2002), NS4A, NS4B and NS5A (Lin *et al.*, 1997), and of NS5A with NS5B (Shirota *et al.*, 2002). While it is currently uncertain whether the findings of Piccininni *et al.* and Lin *et al.* are applicable to the situation *in vivo*, the interaction between NS5A and NS5B was shown to be relevant for RNA replication (Shimakami *et al.*, 2004). Moreover, several lines of experimental evidence that will be described in the following sections of this review indicate that replicative proteins such as NS3, NS5 and NS5B specifically interact with the termini of the viral genomes (Chen *et al*, 1997; Bartholomeuz and Thompson, 1999; Ackermann and Padmanabhan, 2001; Banerjee and Dasgupta, 2001; Xiao *et al.*, 2004) and/or with cellular host-factors (Tu *et al.*, 1999; Wang *et al.*, 2005; Isken *et al.*, submitted for publication; see also Host-factors involved in Flaviviridae replication).

Nonstructural proteins participate in the assembly of virus particles

Some of the nonstructural proteins that are essential for genome replication turned out to possess also critical roles in the assembly and release of virus particles. Thus, a "linker function" between RNA replication and packaging was indicated for the flaviviral NS2A and NS4B proteins (Kummerer and Rice, 2002; Liu *et al.*, 2003, Pugachev *et al.*, 2004), and for the pestiviral NS2–3 (the precursor of NS3) and NS5B proteins (Ansari *et al.*, 2004; Agapov *et al.*, 2004).

In conclusion of this section, it is important to point out that except for the viral protease and RdRp, the precise circumstances of how the diverse NS proteins of flaviviruses, pestiviruses and HCVs participate in the catalysis of RNA replication process still need to be established. Most strikingly, in contrast with other positive-strand RNA viruses such as Picornaviruses and Togaviruses, where intermediates of the polyprotein precursor were shown having important regulatory functions (reviewed by Agol *et al.*, 1999; Kaariainen and Ahola, 2002), it is yet unclear whether intermediates of the Flaviviridae polyprotein (such as for example NS3-NS4A-NS4B) hold an activity during viral replication.

Cis and trans activity of viral proteins

The flaviviral and pestiviral/HCV ORF coding regions show different complementation patterns

Most of the experimental work that intended to define the functions of the viral nonstructural proteins also addressed the question whether, during the assembly of the catalytic RC, the proteins were acting in *cis* (on the parental translation template) or in *trans* (on different translation templates). This aspect was particularly important as it enabled insights into the physical organization of the viral replication machinery. Interestingly, these studies revealed notable differences between flaviruses on the one and pestiviruses/HCVs on the other hand. This was in evident correlation with the fact that within the Flaviviridae family, the pestiviruses and HCVs show nearly identical genomic organizations and modes of gene expression, while both genera reveal a considerably lower degree of similarity with flaviviruses (Figure 4.1). Thus, during complementation assays that applied mutant full-length RNAs and complementing replicons, the flaviviral proteins NS1 and NS5 and, to a lower degree, NS3 were indicated to act in *trans*. In opposition, flaviviral RNAs with mutations in any of the other four nonstructural genes NS2A, NS2B, NS4A, and NS4B were not complemented by the helper systems (Jones *et al.*, 2005; Khromykh *et al.*, 2000; Khromykh *et al.*, 1999; Khromykh *et al.*, 1998; Lindenbach and Rice, 1997). With the pestivirus BVDV, where complementation experiments were performed with genetically distinguishable mutant and helper replicons, only mutations in the NS5A protein-coding unit could be complemented in *trans* (Grassmann *et al.*, 2001). Using the same approach as with BVDV or, alternatively, a transient complementation assay that applied proteins expressed from a nonreplicating helper RNA, the same scenario was recently reported for HCV (Appel *et al.*, 2005).

The significance of cis and trans activity of viral proteins

So far, it can only be speculated on the rationale of the dissimilarities between flaviviruses and pestiviruses/HCV. Technical issues such as the different applied complementation systems could explain them; otherwise, these observations may indicate elementary differences in the functional organization of the flaviviral and pestiviral/HCV replication complexes. For several reasons, a constellation where the majority of the nonstructural proteins and/or their respective precursors are operating preferentially in *cis* appears reasonable. For example, the inability of the BVDV and HCV NS4A, NS4B and NS5B to act in *trans* may be explained such that the presumed interactions of these proteins with membranes can only occur when they are nascent or newly synthesized, which would ensure a correct topology of the assembling replication complex. Such a scenario may also be postulated for the noncomplementable hydrophobic flaviviral proteins NS2A, NS2B, NS4A or NS4B. The *cis* activity of the BVDV and HCV NS3, NS4A, and NS5B, which implies that correctly folded proteins lack the ability to associate with other RNA molecules than the original translation template, could ensure a privileged replication of functional viral genomes and thus represent some kind of quality control of the assembling RC. The strict coupling of translation, proteolysis and activity of the nonstructural proteins "in statu nascendi" would thus be a way to increase the fidelity of the viral replication machinery, which, further speculated, may be particularly important for positive-strand RNA viruses that replicate at lower levels such as pestiviruses and HCV. In support of this idea, it is important to note that with mutant BVDV or HCV replicons (containing mutations in the ORF or in the NTRs) that replicated less efficient than the wild-type replicon, the ratio of negative-strand versus progeny positive-strand RNA was generally unchanged with respect to the wild-type situation. That is, even with sensitive detection procedures as RNase protection or RT-PCR, the synthesis of negative-strand intermediate was only detectable with the simultaneous synthesis of progeny positive-strand RNA (Grassmann *et al.*, 1999; Yu *et al.*, 1999; Yu *et al.*, 2000; Grassmann *et al.*, 2001; Isken *et al.*, 2003; Grassmann *et al.*, 2005). This scenario may indicate a fundamental difference between the multiplication strategy of pestiviruses and HCV and that of other positive-strand RNA viruses such as picorna- and togaviruses, where both replication steps can be uncoupled (Andino *et al.*, 1990: Lemm *et al.*, 1994). Thus, in conjunction with the observation that NS3, NS4A, NS4B and NS5B mainly act in *cis*, these data substantiate the notion that the pestiviral and HCV replication complexes are finely adjusted and rather closed up functional entities. Additional studies will be required to further explain this point and to understand the reasons for the "less tight" organization of the flaviviral RC.

The pestiviral and HCV NS5A protein

The fact that the BVDV and HCV NS5A protein, an essential replication factor of uncertain function, turned out to be the only nonstructural protein that acts in *trans* during replication bears several interesting aspects that are worth discussing. On the one hand, the *trans*-activity of NS5A may be simply caused by a looser association of the protein with membranes that provide the scaffold of the RC. This may come from the fact that NS5A is anchored to endoplasmic reticulum membranes by an amphipathic α-helix (Penin *et al* 2004) and not by a transmembrane domain, as it is the case with NS4B, NS5B, and, via NS4A, with the NS3/NS4A protease complex (for a review on this issue, see Salonen *et al.*, 2005). Alternatively, the ability of the protein to shuttle between different RCs may suggest a particular role of NS5A during RNA replication. Thus, an attractive hypothesis involves that NS5A mediates functionally important interactions between viral proteins. Such a scenario may be implied by the aforementioned *in vitro* co-precipitation experiments, where the HCV NS5A was shown to interact with NS5B and with NS4A and NS4B (Shirota *et al.*, 2002; Lin *et al.*, 1997). Other support for this notion comes from the observation that with BVDV as well as with HCV the replication of NS5A mutant RNAs was only rescued by helper replicon-mediated expression of NS3 to NS5B (BVDV, HCV) or by transient co-expression of the NS3 to NS5A polyprotein (HCV), while solitary expressed NS5A did not *trans*-complement (Grassmann *et al.*, 2001; Appel *et al.*, 2005). These data moreover suggest that the other NS proteins somehow affect the production of a functional NS5A protein, which, in turn, is in line with the observation that phosphorylation of NS5A requires the expression of the protein in the context of the same polyprotein (Neddermann *et al.*, 1999; Koch and Bartenschlager 1999). The second attractive setting that may explain the *trans*-activity of NS5A involves that the protein mediates interactions of cellular components with the RC. Numerous reports indicating contacts of NS5A with various cellular proteins support this idea

(Macdonald and Harris 2004). The most striking data derived from studies, which demonstrated an interaction of NS5A and the human vesicle-associated membrane protein-associated protein A, hVAP-A (Tu *et al.*, 1999; Zhang *et al.*, 2004; Gao *et al.*, 2004). The binding of NS5A to hVAP-A was shown to be required for efficient RNA replication and an inverse correlation was found to exist between NS5A phosphorylation and hVAP-A interaction (Evans *et al.*, 2004). The latter point is particularly interesting considering that independent studies showed that hyperphosphorylation of NS5A inhibits HCV RNA replication (Blight *et al.*, 2000; Guo *et al.*, 2001; Lohmann *et al.*, 2001; Neddermann *et al.*, 2004). Along the same line, NS5A was shown to form a complex with the cellular protein FBL2, and complex formation was shown to be essential for HCV replication (Wang *et al.*, 2005). Moreover, it is worth mentioning that there is evidence for an indirect interaction of NS5A and NF90/NFAR-1, a member of the so-called NF/NFAR proteins. The NF/NFAR proteins represent a group of cellular factors that associate specifically with the termini of the BVDV and HCV RNA and appear to be involved in the regulation of translation and replication of the viral genome (Isken *et al.*, 2003; Isken *et al.*, 2004: Isken *et al.*, submitted for publication; for further details, see Host-factors involved in Flaviviridae replication). This fuels the speculation that *trans*-active NS5A may take on the role as a translational feed back inhibitor; similarly as this has been proposed for the poliovirus 3CD protein (Garmanik and Andino, 1998; see The coordination of viral translation and RNA replication). Thus, an important task of future work will be to examine NS5A as a potential regulator of the HCV translation and replication process, which may also help to clarify the suspected activity of this protein in HCV pathogenesis.

The function of RNA elements in Flaviviridae replication

RNA elements involved in the replication of flaviviruses

The organization of the nontranslated regions and of other functional RNA elements significantly varies between different flaviviral genomes depending on the antigenic subgroup and transmitting vector (i.e. whether the viruses are mosquito-borne or tick-borne). Here, we will focus on a portrayal of the most conserved RNA features; for information on functional RNA motifs that are characteristic for specific members of the flavivirus genus, the reader is referred to articles with emphasis on this subject.

The flaviviral 5′ NTR

As delineated in the Introduction, in contrast with pestiviruses and HCV the flaviviral 5′ NTRs contain no IRES, and translation of the RNA genome is believed to initiate by ribosome scanning from the type I 5′ cap structure. While only small segments in the short (i.e. ca. 100 nucleotide long) 5′ NTRs are conserved among different flavivirus subgroups, considerable sequence homologies exist among members and strains of the same subgroup. Regardless of the limited conservation of the 5′ NTR primary sequence, a consensus RNA secondary structure was proposed (Brinton and Dispoto, 1988; Figure 4.2). This motif, which consists of a long stem–loop structure (5′ SL), is essential for replication (Cahour *et al.*, 1995), and it was indicated to have a modulating role in translation initiation and to determine virus host range and virulence (reviewed by

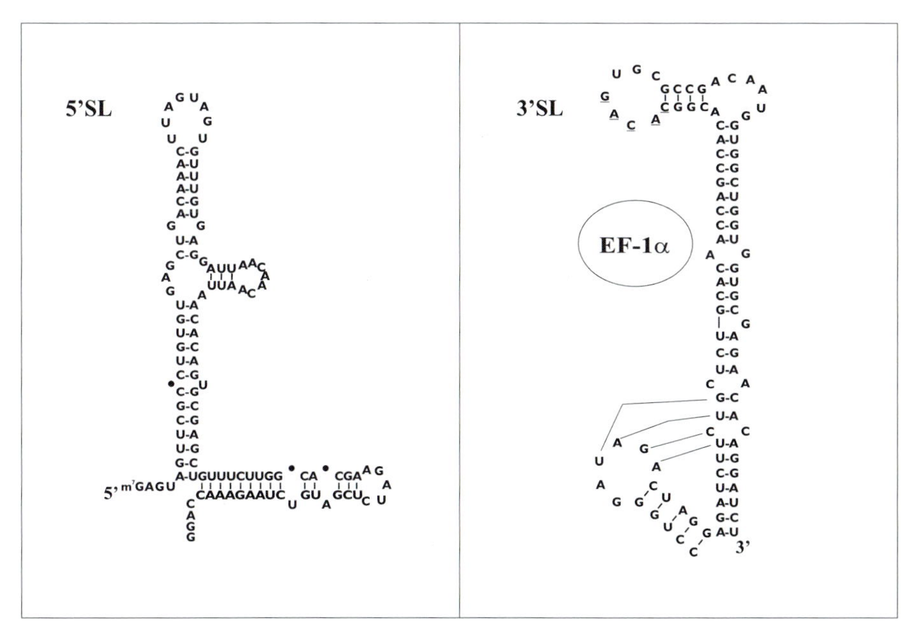

Figure 4.2 Structures of the 5′ and 3′ termini of the flaviviral genome. The figure shows the secondary structures of the 5′ SL and 3′ SL motifs of the West Nile virus genome as proposed by Brinton and Dispoto 1988 and Brinton *et al.*, 1986. Underlining emphasizes the conserved pentanucleotide in the 3′ SL motif, and the EF-1α binding site is indicated (for further information, see RNA elements involved in the replication of flaviviruses).

Markoff, 2003). Moreover, in conjunction with the so called "cyclization sequences" (these elements are described in more detail below), several short sequence elements in the flaviviral 5′ NTR were hypothesized to interact via hydrogen bonding with complementary sequence stretches in the 3′ NTR (Khromyk *et al.*, 2001; Thurner *et al.*, 2004; Alvarez *et al.*, 2005).

The most apparent of these sequence elements is a di-nucleotide, AG, at the 5′-terminus of all flavivirus genomes that is assumed to base pair with complementary UC residues at the genomic 3′-terminus. Another important task of the flaviviral 5′ NTR resides in the negative-strand copy, which forms the complementary of the consensus stem–loop structure 5′ SL near the 3′-terminus. This motif, termed (−)3′ SL, was proposed to be part of the RC that synthesizes progeny positive-strand RNA molecules in the second replication step (Li *et al.*, 2002). Interestingly, four host proteins associate with this region of the negative-strand intermediate of West Nile virus (Shi *et al.*, 1996), one of which was identified as to be TIAR, an RNA binding protein that was demonstrated to be essential for West Nile virus replication (Li *et al.*, 2002; for more details, see Host-factors involved in Flaviviridae replication).

The flaviviral 3′ NTR

The most conserved feature of the flaviviral 3′ NTR is an approximately 100-nucleotide long sequence stretch at the immediate 3′-terminus of the genome. This region folds into a stable stem–loop structure designated as 3′ SL (Rice *et al.*, 1985; Grange *et al.*, 1985; Brinton *et al.*, 1986; Wengler and Castle, 1986; Zhao *et al.*, 1986; Takegami *et al.*, 1986; Hahn *et al.*, 1987; Sumiyoshi *et al.*, 1987; Mandl *et al.*, 1989; Mandl *et al.*, 1991; Proutski *et al.*, 1997a; Proutski *et al.*, 1997b; Rauscher *et al.*, 1997; Figure 4.2). In analogy with the situation with the flaviriral 5′ NTR, the primary sequence of 3′ SL is most preserved among viruses of the same subgroup but significantly less conserved between different viral subgroups. Correct formation of 3′ SL was shown to be essential for replication (Zeng *et al.*, 1998; Yu *et al.*, 2005), and the viral replicase proteins NS3 and NS5 were indicated to interact directly with this motif during formation of the initial RC (Chen *et al.*, 1997; Cui *et al.*, 1998). 3′ SL was moreover suggested to determine the host-specificity of viral replication (Zeng *et al.*, 1998), and in line with this finding, three cellular proteins with molecular weights of 52, 84 and 105 kDa were found to bind to 3′ SL (Blackwell and Brinton 1995).

Interestingly, the 52 kDa protein was identified as the major translation factor EF-1α that appears to be also involved in the replication of rather different types of positive-strand RNA viruses such as the bacteriophage Qβ and poliovirus (Blumenthal and Carmichael, 1979; Harris *et al.*, 1994; Zeenko *et al.*, 2002; De Nova-Ocampo *et al.*, 2002). While EF-1α was shown to associate with the stem of 3′ SL (Blackwell and Brinton, 1997), the top-loop of 3′ SL contains a so-called pentanucleotide 5′-CACAG-3′, which is nearly 100% conserved among all flaviviruses (Figure 4.2). Genetic studies with an infectious clone of West Nile virus recently demonstrated that the pentanucleotide represents an essential *cis*-acting replication signal (Elghonemy *et al.*, 2005).

While the 3′ NTRs of mosquito-borne flaviviruses exhibit unique organizations, which strongly vary between different members of this group, the 3′ NTRs of tick-borne flaviviruses are composed of distinguishable variable and conserved ("core") portions and thus display some analogies with the 3′ NTRs of pestiviruses and HCV (see following chapters). The upstream variable region of the tick-borne flaviviral 3′ NTR, which often contains poly-A stretches, was found to be mainly dispensable for viral replication in cell culture. The downstream conserved region is essential for replication; it contains 3′ SL and most sequence elements (see below) that are assumed to mediate a closed loop conformation of the viral RNA (Wallner *et al.*, 1995; Mandl *et al.*, 1998).

The flaviviral "cyclization sequences"

Besides the 3′ SL motif at the genomic 3′ end, the 3′ NTRs of mosquito-borne and tick-borne flaviviruses enclose a specific pattern of conserved sequence and structure motifs that are not shared between the two virus groups. Thus, a highly conserved ca. 25 nucleotide long region, CS1, is located immediately upstream of 3′ SL in the 3′ NTR of several mosquito-borne viruses such as dengue 2, West Nile and yellow fever virus (Shurtleff *et al.*, 2001). CS1 was predicted to base pair with the most conserved sequence feature in the 5′-terminus of the flavivirus genome, the so-called cyclization sequence, 5′ CS (Hahn *et al.*, 1987), which, in the case of mosquito-borne flaviviruses, resides ca. 30–40 nucleotides downstream of the translational start-codon in the capsid-coding region of the ORF. In accord with this notion, genetic studies with different virus systems demonstrated that the complementarity of CS1 and 5′ CS (which does not necessarily involve preservation of the original genomic sequence composition) represents an important prerequisite for a template-specific activity of the RdRp and for viral RNA replication (You and Padmanabhan 1999; You *et al.*, 2001; Khromykh *et al.*, 2001; Corver *et al.*, 2003). Interestingly, a portion of CS1 was found capable to also form a pseudoknot structure with an internal loop of 3′ SL. This fueled the speculation that alternative interactions of CS1, i.e. with the genomic 5′ end to form a loop-like structure of the entire genome or with 3′ SL to form an internal pseudoknot, are crucial determinants of the regulation of translation and replication of the viral RNA (Shi *et al.*, 1996; Khromykh *et al.*, 2001; see below for further discussion). Tick-borne flaviviruses also contain highly conserved sequence elements in both termini of the genome, which are supposed to base pair and to support in this way a loop-like conformation of the viral RNA genome. These concern the so-called R3 and PR elements in the conserved region of the 3′ NTR and a 5′ CS element, which, in contrast to the situation with mosquito-borne flaviviruses, is positioned in the 5′ NTR (Khromykh *et al.*, 2001; Mandl *et al.*, 1993). Finally, it is worth noting that the 3′ NTRs of different flavivirus subgroups often include additional conserved sequence elements as well as several copies of the cyclization sequences, the functions of which remain to be defined (reviewed by Lindenbach and Rice, 2003).

In sum, it is important to emphasize that all flaviviral genomes harbor conserved RNA elements in the genomic 5′ and 3′-terminus, which modulate translation and/or that are essential for the RNA replication process. Moreover, the available experimental data clearly indicate that both termini of the flaviviral RNA are functional parts of a common, complex RNA structure that involves the formation of a loop-like arrangement of the viral genome. In fact, recent studies performed with atomic force microscopy revealed that the dengue virus genome adopts a circular structure, the formation of which was shown to essentially depend on the presence of complementary sequence-elements in both ends of the viral RNA (Alvarez *et al.*, 2005). The circular RNA conformation was suggested to be an import precondition for the regulation of viral protein and RNA synthesis (for a further discussion of this aspect, see below).

RNA elements involved in the replication of pestiviruses and HCV

RNA motifs that were characterized as crucial determinants of the translation and/or replication process of the pestiviral and HCV genomes show a surprisingly high degree of conservation between subtypes of the same genus and even between both genera. For that reason, we discuss the current state of knowledge in the pestivirus and HCV systems side-by-side.

The pestiviral and HCV 5′ NTR

Computer predictions and structure probing experiments performed with the 5′ NTRs of pestiviral (BVDV-bovine viral diarrhea virus types 1 and 2, CSFV-classical swine fever virus, BDV-border disease virus) and HCV (HCV subtypes 1–5) genomes revealed a significant extent of sequence conservation and evident similarities of the RNAs secondary and tertiary structures (Figure 4.3).

The IRES The most obvious common features of the pestiviral and HCV 5′ NTRs are the downstream portions, which, in each case, fold into extensive stem–loop structures designated as domains II, III and IV. Domains II, III and IV (domains III and IV, which form a distinctive pseudoknot structure near the translational start codon) were demonstrated to include the major functional elements of the type IV internal ribosomal entry site (IRES) (Honda *et al.*, 1996; Poole *et al.*, 1995; Rijnbrand *et al.*, 1995; Rijnbrand

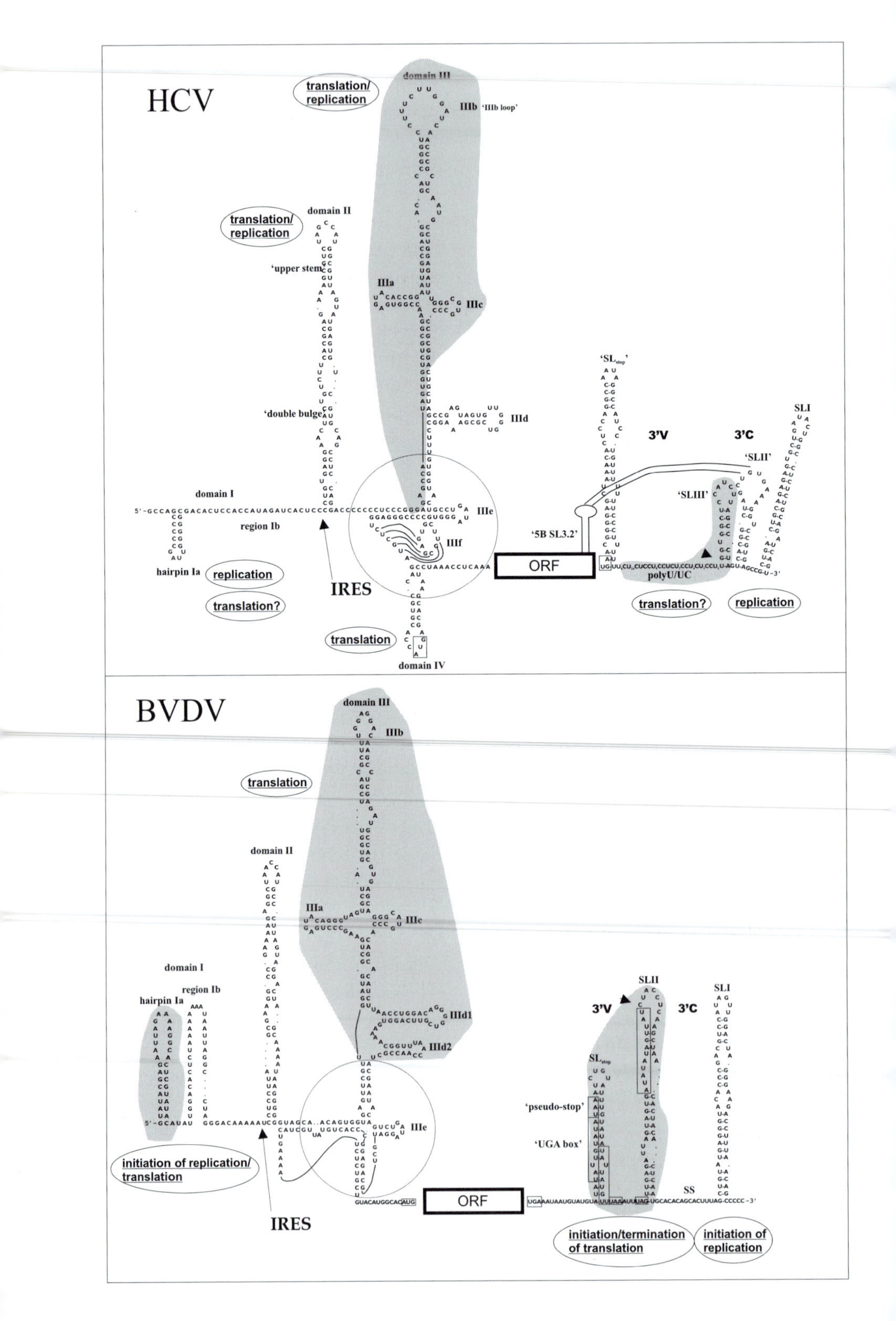
HCV
translation/ replication
domain III
IIIb 'IIIb loop'
domain II
'upper stem'
IIIa
IIIc
'double bulge'
IIId
domain I
IIIe
region Ib
IIIf
hairpin Ia
replication
translation?
IRES
translation
domain IV
ORF
'SLstop'
'5B SL3.2'
3'V
3'C
SLI
'SLII'
'SLIII'
polyU/UC
replication
BVDV
IIIb
IIIa
IIIc
IIId1
IIId2
region Ib
hairpin Ia
initiation of replication/ translation
SLII
SS
'pseudo-stop'
'UGA box'
initiation/termination of translation
initiation of replication

et al., 1997; Rijnbrand and Lemon, 2000; Tsukiyama-Kohara *et al.*, 1992; Wang *et al.*, 1995). Importantly, with both viruses, the presence of the adjoining ORF was shown to be an important determinant of efficient IRES activity (Lu and Wimmer, 1996; Behrens *et al.*, 1998; Tautz *et al.*, 1999; Myers *et al.*, 2001; Rijnbrand *et al.*, 2001; Kim *et al.*, 2003).

The pestiviral and HCV IRESes exhibit a unique mode of activity such that translation initiation is mediated by direct positioning of the 40S ribosome subunit at the translation initiation codon in the absence of additional translation factors (Pestova *et al.*, 1998; Hellen and Sarnow, 2001; Pestova and Hellen, 1999). Structural analysis of the HCV IRES revealed that all composing RNA elements adopt a tertiary structure that binds the translation initiation complex at high affinity (Kieft *et al.*, 2002; Kieft *et al.*, 1999; Lukavsky *et al.*, 2003; Spahn *et al.*, 2001; Collier *et al.*, 2002; Lukavsky *et al.*, 2000; Lytle *et al.*, 2002). The sites for the assembly of the 43S translation preinitiation complex and of the 80S ribosome have been mapped (Lytle *et al.*, 2001; Sizova *et al.*, 1998; Kolupaeva *et al.*, 2000a; Kolupaeva *et al.*, 2000b). They involve the most 3′-terminal region of the IRES, also termed as "core-domain" (Figure 4.3), the structure of which was proposed to be conserved among all viral IRES elements (Le *et al.*, 1996).

The role of domain I during IRES-mediated translation Compared with domains II and III, the upstream domain I regions of the pestiviral and HCV 5′ NTRs reveal a significantly lower degree of similarity. With both viral RNAs, domain I forms a small stem–loop structure at or near the genomic 5′-terminus, which was uniformly termed as hairpin Ia, although the composition of the pestiviral (e.g. BVDV) hairpin Ia is more complex than that of the HCV Ia. Hairpin Ia and domain II are separated by the spacer region Ib, which appears to exist predominantly in a single-stranded conformation (Honda *et al.*, 1999; Honda *et al.*, 1996; Grassmann *et al.*, 2005; see Figure 4.3). The minimum IRES RNA sequences that are capable of initiating translation do not necessarily include domain I, i.e. hairpin Ia and region Ib (Honda *et al.*, 1996; Rijnbrand *et al.*, 1995; Rijnbrand *et al.*, 1997; Rijnbrand and Lemon, 2000). Nevertheless, several lines of evidence (though some of the available data are inconsistent) indicate that the presence of domain I has a marked impact on the efficiency of the IRES. Along this line, some work suggested that the presence and correct fold of Ia supports IRES function ([with HCV] Fukushi *et al.*, 1994; Luo *et al.*, 2003; [with BVDV] Yu *et al.*, 2000). Other studies indicated that a deletion of Ia had little consequences ([with BVDV] Chon *et al.*, 1998; [with HCV] Friebe *et al.*, 2001) or stimulated

---

Figure 4.3 (Opposite) Structures and suggested functions of the HCV and BVDV 5′ and 3′ NTRs. The depicted structures (HCV strain 1B; BVDV strain CP7) were either experimentally proven or predicted (experimentally unverified motifs are marked by quotations). The translational start and stop-codons are boxed; the suggested "core region" of the HCV and BVDV IRES is marked by a circle. Arrows specify the considered 5′ boundaries of the IRES domains in the 5′ NTRs and the suggested borders between the 3′ V and 3′ C regions in the 3′ NTRs. Functional RNA motifs and their assignment during translation and RNA replication are specifically indicated (see text). Shadowed regions designate the defined NF/NFAR protein binding sites (Isken *et al.*, submitted for publication; see also RNA elements involved in the replication of pestiviruses and HCV).

IRES function ([with HCV] Honda *et al.*, 1996; Kamoshita *et al.*, 1997; Rijnbrand *et al.*, 1995; [with CSFV] Rijnbrand *et al.*, 1997).

Recently, we performed a comparative study of the BVDV and HCV 5′ NTR, which demonstrated that the BVDV hairpin Ia and region Ib support IRES activity while translation experiments with HCV RNA revealed an inhibitory effect of Ia and basically no effect of Ib on IRES function (Grassmann *et al.*, 2005). We explained these findings, which were in agreement with most other reports, by a generally different functional organization of the HCV and BVDV 5′ NTR (see below).

Replication signals in the pestiviral and HCV 5′ NTR The identification of RNA determinants of the viral replication process was particularly facilitated by genetic studies of so-called "bi-cistronic" RNA replicon constructs where the authentic viral 5′ NTR precedes a reporter gene while a second IRES element enables expression of the viral nonstructural genes independently of mutations in the NTRs (Tautz *et al.*, 1999; Lohmann *et al.*, 2001).

Thus, with the BVDV system, studies with full-length viral genomes or replicons with mutant 5′ NTRs identified three short sequence stretches in the 5′-terminal stem of hairpin Ia and in the single-stranded Ib region that are crucial for the catalysis of the first replication step (Frolov *et al.*, 1998; Yu *et al.*, 2000; Grassmann *et al.*, 2005). Other experimental data defined the authentic fold of hairpin Ia as a prerequisite for efficient BVDV RNA replication (Becher *et al.*, 2000; Yu *et al.*, 2000, see Figure 4.3). In contrast, domains II and III of the BVDV 5′ NTR were shown to be redundant for viral replication (Frolov *et al.*, 1998; Becher *et al.*, 2000; Grassmann *et al.*, 2005). Accordingly, as a main characteristic of the BVDV 5′ NTR, the signals that are important for the initiation of RNA replication are separated from the functional IRES (Figure 4.3). The only exception from this rule is the BVDV domain I, which turned out to be important for efficient translation of the viral proteins as well as for the initiation of the replication cycle. Thus, due to this dual role, the BVDV domain I is believed to be a crucial regulator of viral protein and RNA synthesis (Isken *et al.*, 2003; Yu *et al.*, 2000; this issue will be further discussed below).

In remarkable contrast with the pestiviral 5′ NTRs, genetic studies of several laboratories implied the presence of replication signals in domains II and III, i.e. in the IRES forming region of the HCV 5′ NTR (Friebe *et al.*, 2001; Kim *et al.*, 2002; Reusken *et al.*, 2003). Recently, we identified some of these elements on the nucleotide level. Thus, sequence and structure motifs in domain II, namely the upstream portion of the HCV-specific "double-bulge II region" and the sequence composition of the "upper stem" of domain II, turned out to be pure replication signals, because mutations compromised the multiplication of the viral RNA but did not interfere with translation (see Figure 4.3). Other replication signals in the HCV domains II and III definitely overlapped with functional parts of the IRES. This was most obvious with the structure of the "upper stem region" of domain II, the formation of which was found to be important for replication but also for proper translation. Other replication signals that overlapped with the IRES were identified in the downstream portion of domain II and in the "loop-IIIb" region of domain III (Grassmann *et al.*, 2005; Figure 4.3).

Similar to their BVDV counterparts, the HCV hairpin Ia and Ib regions were shown to be crucial replication signals (Luo *et al.*, 2003; Kim *et al.*, 2002; Reusken *et al.*, 2003; Friebe *et al.*, 2001). However, in contrast to the situation with BVDV, the function of HCV Ia was found to be nearly exclusively dependent on the formation of the RNA structure. Interestingly, besides domain I, only one sequence element in domain II (in the HCV-specific "double-bulge II"; see Figure 4.3) was found to be essential for replication, while all other defined replication signals solely contributed to the efficiency of the replication process. This suggests that the string of replication elements in the HCV domains I, II and III are components of a higher-order RNA motif. Thus, in view of the proposed tertiary structure of the HCV IRES it is tempting to assume that a major part, if not the entire HCV 5′ NTR, participates in protein as well as RNA synthesis and, most probably, in the regulation of both processes (Grassmann *et al.*, 2005).

In sum, it is important to note that despite the striking resemblance of the HCV and BVDV 5′ NTRs, several RNA elements apply different modes of activity. In the BVDV 5′ NTR, replication signals are restricted to the 5′-terminal domain I. In contrast, with HCV, there is a significant overlap of replication signals with the functional IRES. While the BVDV domain I supports IRES activity, the HCV domain I appears to down-regulate IRES function.

It may be speculated that the overlay of translation and replication signals in the HCV 5′ NTR and the HCV-specific downregulation of viral protein synthesis by hairpin Ia serve the same purpose, namely to adjust HCV replication to a lower level. Further speculated, low levels of RNA replication, low levels of viral gene expression and a noncytopathic phenotype of RNA replication, may be reasons that cause a lower immunogenicity and thus contribute to the persistence of HCV infections (for a review, see Hahn, 2003). Along these lines, the different functional organization of the BVDV and HCV 5′ NTR was indicated contributing to differences in the life cycle of both related viruses.

The pestiviral and HCV 3′ NTR

In contrast to the situation with the 5′ NTRs, there exist no sequence homologies between the pestiviral and HCV 3′ NTRs. However, in analogy to the 3′ NTRs of tick-borne flaviviruses, the pestiviral and HCV 3′ NTRs are each composed of a less conserved and a highly conserved portion. The functional relevance of this organization was most thoroughly studied with the BVDV 3′ NTR.

The pestiviral 3′ NTR Sequence alignments of the genomes of representative pestivirus strains revealed that the 3′ NTR consists of a conserved (3′ C) and a variable (3′ V) region (Deng and Brock, 1993). Computer modeling and experimental structure probing demonstrated that the 3′ C portion, which includes the immediate 3′-terminus of the viral genome, exhibits a pronounced secondary structure consisting of two extensive stem–loop structures, SLI and SLII that are separated by a highly conserved single-stranded intervening region termed as SS (Deng and Brock, 1993; Yu *et al.*, 1999; see also Figure 4.3). Applying a systematic mutagenesis approach to BVDV replicons, we could differentiate individual sequence and structural elements within SLI and SS, which are essential for the initiation of the replication cycle (Yu *et al.*, 1999). Since none of the 3′ C mutations affected translation initiation or RNA stability (Isken *et al.*, 2004; H. Yu, unpublished data), we hence considered SS and SLI as functional components of the genomic "promoter" for the viral RC to initiate negative-strand RNA synthesis. The variable 3′ V region of the 3′ NTR, which is located immediately downstream of the ORF (Figure 4.3), exhibits a remarkable heterogeneity in terms of size and sequence composition between different virus strains (Deng and Brock, 1993; Becher *et al.*, 1998). However, despite its genetic drift, 3′ V comprises structure as well as sequence motifs that are conserved between all pestiviral strains. Thus, with each pestiviral genome, 3′ V forms the upstream portion of SLII and an unstable hairpin structure named "SL_{stop}" downstream of the translational stop codon. 3′ V was moreover shown to harbor either single or multiple copies of a conserved 12-nucleotide sequence element designated as "UGA box" (Figure 4.3).

Another conspicuous feature of the BVDV 3′ V region that is conserved among all pestiviruses concerns the distance between the UGA box motifs and the viral ORF, which, interestingly, can be measured in triplet units. This peculiarity, which prompted the speculation that some kind of translational reading frame was maintained in the 3′ NTR (see below), also includes triplets that resemble translational stop codons and were termed "pseudo-stop-codons" (Isken *et al.*, 2004). Using a combination of genetic and structure-probing experiments, we could show that the different conserved features of the BVDV 3′ V region are functional parts of a common, complex RNA signal. This motif is necessary for efficient translation initiation (see below) and, most interestingly, for accurate termination of translation at the ORF stop codon of the viral RNA. Efficient termination of translation, in turn, was shown to be an essential prerequisite for efficient replication of the viral RNA (Isken *et al.*, 2004). The pestiviral 3′ V region was thus indicated to act as a kind of safe-lock system that avoids interferences of the translation machinery with the assembling replication complex at the conserved 3′-terminus of the viral RNA (for further discussion, see below and Figure 4.4).

In summary, these data revealed that the bipartite composition of the pestiviral 3′ NTR has an evident functional impact: the conserved and stably structured 3′ C region was shown to be part of the nascent replication complex; the variable 5′ portion (3′ V) was indicated to take on the role as a coordinator of the viral translation and replication cycle. Along this line, it is important to emphasize that the presence of an intact 3′ V region in the 3′ NTR significantly stimulated the 5′ NTR/IRES-mediated translation of the BVDV RNA. 3′ V was moreover found to act as a specific binding site of the NF/NFAR group of cellular proteins, and several lines of evidence support the notion that formation of this viral/cellular RNP complex is an essential requirement for the indicated functions of 3′ V during translation initiation and translation termination (Isken *et al.*, 2004). Since the NFAR proteins were shown to associate also with the BVDV 5′ NTR and to support interactions of the 5′ and 3′-termini of the viral RNA (Isken *et al.*, 2003, see Host-factors involved in Flaviviridae replication), the 3′ V/NFAR ribonucleoprotein complex is suspected to be part of a closed-loop conformation of the viral RNA, which may be necessary for the regulation of translation and RNA replication (Isken *et al.*, 2003; see The coordination of viral translation and RNA replication).

The HCV 3′ NTR The HCV 3′ NTR is composed of three structurally distinct regions: a highly conserved 3′-terminal segment (3′ X) that was indicated to

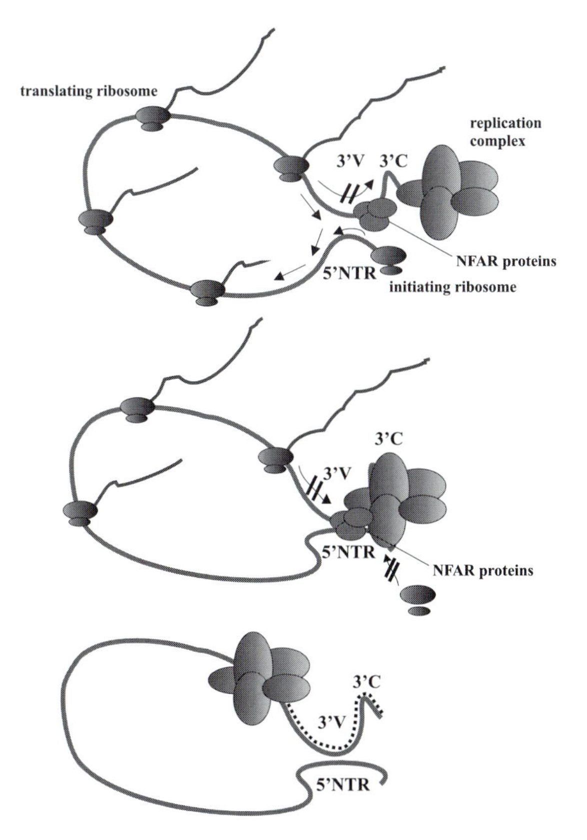

Figure 4.4 Speculative model on the initial steps of the BVDV replication pathway. This model is in detail discussed in The coordination of viral translation and RNA replication.

form three stem–loop structures (termed as SL1, SL2, and SL3, in 3–5′ direction; Blight and Rice, 1997), a unique, extended poly(U)-poly(UC) [poly(U/UC)] tract, and an upstream variable region (Ito and Lai., 1997; Kolykhalov *et al.*, 1996; Tanaka *et al.*, 1995; Figure 4.3). The poly(U/UC) tract and 3′ X segment were each shown to be required for infectivity of genome-length RNA transcripts inoculated into chimpanzees (Kolikhalov *et al.*, 2000; Yanagi *et al.*, 1999). Systematic genetic studies with RNA replicons confirmed that the conserved 3′-terminal portion and the 3′-portion of the poly(U/UC) tract contain RNA signals that are essential for replication, while the upstream variable region was shown to contribute to efficient RNA replication (Friebe and Bartenschlager, 2002; Yi and Lemon 2003a, b).

Other studies suggested that, in addition to playing a major role in RNA replication, the 3′ NTR, in particular 3′X, stabilizes the RNA and enhances IRES-mediated translation (Ito *et al.*, 1998; Spangberg *et al.*, 2001; Michel *et al.*, 2001; contradictory findings were published by Kong and Sarnow, 2002). These effects were indicated to require the binding of cellular factors such as PTB, the autoantigen La (see Host-factors involved in Flaviviridae replication), or the binding of certain ribosomal proteins to the 3′ NTR (Ito and Lai, 1997; Ito and Lai, 1999; Spangberg *et al.*, 1999; Wood *et al.*, 2001). In analogy with the situation with BVDV, we recently found that members of the cellular NF/NFAR proteins specifically interact with both the termini of the HCV RNA. Moreover, as with BVDV, we observed that formation of the HCV 3′ NTR/NFAR protein complex is an essential prerequisite for HCV RNA replication, and translation experiments suggested that the HCV 3′ NTR/NFAR protein complex (as the BVDV 3′ NTR/NFAR protein complex) supports efficient translation termination (Isken *et al.*, unpublished data).

Thus, despite of significant differences in the sequence composition of the BVDV and HCV 3′ NTR, these data pointed on similar roles of the respective 3′C and 3′V regions in the regulation of viral protein and RNA synthesis.

Cis-replicating elements

Following predictions of Tuplin and coworkers (Tuplin *et al.*, 2002), three different laboratories recently identified *cis*-replicating elements (CREs) in the coding sequence of the HCV polyprotein. A CRE-like function was specifically demonstrated for two highly conserved stem–loop structures in the C-terminal NS5B coding region. One of these stem–loop structures termed as 5B SL3.2 or SLV was analyzed in depth, and sequence and structural features of this motif defined as essential replication signals (You *et al.*, 2004; Lee *et al.*, 2004; Friebe *et al.*, 2005; Figure 4.3). Most interestingly, in analogy with picornavirus CREs (see below), the function of 5B SL3.2 was shown to be context independent, i.e. a blockage of RNA replication caused by mutational disruption of the 5B SL3.2 structure could be restored when an intact copy of this element was inserted into the 3′ NTR. Moreover, 5B SL3.2 was indicated to form a pseudo-knot with SL2 in the HCV 3′ NTR (indicated in Figure 4.3), and this structure was demonstrated to be essential for replication (Friebe *et al.*, 2005). The discovery of a CRE in the HCV genome may be of striking importance considering that picornavirus CRE were implied to play a direct role in the initiation of RNA replication (Goodfellow *et al.*, 2000; Gerber *et al.*, 2001; Lobert *et al.*, 1999; McKnight *et al.*, 1998; Morasco *et al.*, 2003; Yin *et al.*, 2003). In support of this notion, Lee and colleagues indeed showed an interaction of the HCV RdRp NS5B with SLV/5B SL3.2 (Lee *et al.*, 2004). Thus, an important subject of future investigations will be to understand how this interaction of NS5B with SLV/5B SL3.2 contributes to the recognition of the HCV genome to initiate the first replication step.

Taken together, in analogy to the flaviviruses, conserved RNA elements in the 5′ and 3′-termini of the pestiviral and HCV genomes modulate translation, and both termini were shown to be essentially involved in the initiation of the RNA replication cycle. Though obvious cyclization sequences do not exist in the pestiviral and HCV genomes, the available experimental information clearly suggests that the viral genomes adopt a closed loop conformation that involves RNA–RNA interactions. In fact, Thurner and colleagues reported short, complementary sequence stretches in the 5′ and 3′ ends of the HCV genome (Thurner *et al.*, 2004), which were not further explored. However, besides RNA/RNA interactions, several lines of evidence support the notion that circularization of the viral genome is supported by the formation of an unusual viral RNA/cellular RNP complex that includes the NFAR proteins and perhaps other host factors such as PTB and La (see below). As with the flaviviruses, a circular RNA conformation is suggested to be an import determinant of the switch from viral protein to viral RNA synthesis (see The coordination of viral translation and RNA replication).

Host factors involved in Flaviviridae replication

Several lines of evidence suggest that the multiplication of Flaviviridae is not self reliant and that the viruses subvert cellular proteins to become part of their replication strategy (reviewed by Lai, 1998; Tellinghuisen and Rice, 2003). While ample reports exist on physical contacts of cellular factors with the viral NTRs or with viral proteins, for most of these interactions a functional relevance was not determined. Accordingly, we focus here on the description of cellular factors that were shown having an evident impact on the replication process of Flaviviridae members. For information on interactions of cellular proteins with viral components that were not further analyzed or on cell–virus interactions that are believed to play a role in the innate and cellular immune response, the reader is referred to reviews that specifically deal with these subjects.

TIA-1/TIAR are crucial determinants of West Nile virus replication

As one of several proteins that were found to specifically interact with the (−)3′ SL motif of the West Nile virus negative-strand RNA intermediate (Shi *et al.*, 1996), Li and colleagues identified a 42 kDa protein as to be TIAR (Li *et al.*, 2002; see also The flaviviral 5′ NTR). TIAR and the closely related protein TIA-1 are members of the RNA recognition motif (RRM) family of RNA binding proteins, and both proteins bind at high affinity to the (−)3′ SL motif *via* this RRM motif. Most importantly, and in significant contrast with five other types of viruses, West Nile virus was observed to replicate clearly less efficient in murine TIAR knockout cell lines than in control cells (Li *et al.*, 2002). Both, TIAR and TIA-1, are postulated to be major players in regulating the translational arrest that occurs during cell stress (Kedersha *et al.*, 1999). TIA-1/TIAR were also suggested to function as specific translational silencers (Gueydan *et al.*, 1999, Piecyk *et al.*, 2000), as alternative splicing regulators (Del Gatto-Konczak *et al.*, 2000; Le Guiner *et al.*, 2001) and to have yet undefined functions during apoptosis (Tian *et al.*, 1995) and development (Beck *et al.*, 1998). The data of Li *et al.* proved that TIAR and, to a lesser extent, TIA-1, provide a necessary function for West Nile virus in the course of its intracellular replication process. In view of the scenario that the 3′-terminus of the negative-strand intermediate is thought to be an essential determinant of the second replication step (see The flaviviral 5′ NTR), the function of the TIAR/TIA-1 interaction may include assisting the formation or stabilization of (−)3′ SL in replication complexes and/or in the recognition of the negative-strand template by the viral replicase (Li *et al.*, 2002).

The role of PTB and La in the replication of HCV

PTB belongs to the hnRNP family of RNA binding proteins (Ghetti *et al.*, 1992). The exact function of PTB, which is mainly in the nucleus but also shuttles to the cell's cytoplasm, is unknown; initially the protein has been suggested to be involved in the regulation of mRNA splicing (Gil *et al.*, 1991). PTB has been shown to facilitate translation directed by picornavirus IRES elements (Gosert *et al.*, 2000) and to bind to multiple pyrimidine tracts in the HCV genome. These sequence elements reside in the 5′ NTR, the core protein-coding unit, and in the 3′ NTR (Chung and Kaplan, 1999; Ali and Siddiqui, 1995). Though some of the experimental data are contradictious, they indicate that the binding of PTB to the HCV NTRs modulates the activity of the IRES (Anwar *et al.*, 2000, Ito and Lai, 1999, Ito *et al.*, 1998; Tardif *et al.*, 2002; Tischendorff *et al.*, 2004). Most interestingly, PTB was found to interact with the 3′ end of the HCV RNA at two conserved stem–loop structures, SL3 and SL2, both which were shown to be critical for RNA replication (Tsuchihara *et al.*, 1997; Ito and Lai, 1997; Friebe *et al.*, 2001; see also The HCV 3′ NTR).

The autoimmune antigen La is a conserved RNA-binding phosphoprotein that associates with pre-tRNAs and other nascent pol III transcripts. It is believed that La transiently stabilizes and accompanies these RNA during processing and maturation (Intine *et al.*, 2000; reviewed by Maraia 2001). Most (ca. 80%) of La resides in the nucleus; the cytoplasmic fraction may vary depending on the environmental conditions (Maraia, 2001). La can interact with the HCV 5′ NTR (near the initiator AUG) and with the 3′ NTRs (within the poly-U/UC region) (Ali and Siddiqui, 1997; Spangberg *et al.*, 1999), and the association of La with the 5′ NTR was indicated to be important for efficient IRES-mediated translation of the HCV RNA (Ali and Siddiqui, 1997; Izumi *et al.*, 2004).

Although the functional roles of PTB and La are far from being understood, their importance for the viral replication process was recently demonstrated by RNAi knockdown experiments. That is, even modest reductions of the expression of La or PTB resulted in a substantial inhibition of HCV replication (Domitrovich *et al.*, 2005; Zhang *et al.*, 2004).

NSAP1/hnRNP Q/SYNCRIP supports the translation of HCV RNA

NS1-associated protein 1, NSAP1, was originally identified as a human cellular protein that interacts with the NS1 nonstructural protein of mouse minute virus (Harris *et al.*, 1999). NSAP1 is identical with the human hnRNP Q protein (Mourelatos *et al.*, 2001) and with the mouse SYNCRIP (synaptotagmin binding, cytoplasmic RNA-interacting protein; Mizutani *et al.*, 2000). NSAP1/hnRNP Q/SYNCRIP (here further designated as hnRNP Q) reveals a high homology with hnRNP R. However, in contrast to hnRNP R and other splicing variants (hnRNP Q2 and Q3), hnRNP Q is distributed throughout the cytosol instead of being localized in the nucleus (Mitzutani *et al.*, 2000). hnRNP Q binds to RNA *in vitro*, preferentially to poly(A) or poly(U), in a phosphorylation-dependent manner (Hresko and Mueckler, 2000). Although hnRNP Q was reported to modulate *c-fos* mRNA stabilization by forming a protein complex with other translation factors such as Unr, PABP, and PAIP1 (Grosset *et al.*, 2000), the physiological role of the protein in the cytoplasm is not yet understood.

hnRNP Q was recently found to considerably augment HCV mRNA translation through a specific interaction with an adenosine-rich region in the core-coding region, i.e. that region, which was previously shown to be necessary for efficient IRES-mediated translation of HCV RNA (Lu and Wimmer, 1996; see also 'RNA elements involved in the replication of pestiviruses and HCV'). Thus, over-expression of hnRNP Q specifically enhanced HCV IRES-dependent translation, and siRNA-mediated knockdown specifically inhibited the translation of HCV mRNA (Kim *et al.*, 2004). The data of Kim *et al.* moreover suggested that hnRNP Q mediates the stimulating effect on the HCV IRES in conjunction with other RNA binding factors such as hnRNP L, which also was found to interact with the HCV core-coding region and perhaps PTB, which, in turn, was observed to interact with hnRNP L (Hahm *et al*, 1998).

hVAP-33 is a functional component of HCV replication.

Human vesicle-associated membrane protein-associated protein A (hVAP-33) is a commonly expressed, endoplasmic reticulum/Golgi-localized protein, which is suggested to play a fundamental role in intracellular vesicle biogenesis and trafficking. hVAP-33 was originally found to stably bind to and to co-localize with the HCV NS5A in the cell, but further characterization revealed also interactions with the NS5B RdRp: NS5A binds to the C-terminus, while NS5B binds to the N-terminus of hVAP-33 (Tu *et al.*, 1999). Accordingly, hVAP-33 has been hypothesized to function as a tie up site for assembly or localization of the HCV replication complex (Tu *et al.*, 1999). In support of this notion, inhibition of hVAP-A through either RNA interference or expression of truncated, dominant negative fragments of the protein inhibited HCV RNA replication (Gao *et al.*, 2004; Zhang *et al.*, 2004). Most interestingly, Evans and colleagues recently demonstrated a striking correlation between the ability of NS5A to interact with hVAP-33 and to participate in HCV RNA replication (Evans *et al.*, 2004). Further experimental data revealed an inverse correlation between NS5A phosphorylation and hVAP-33 interaction and suggested a model in which the hyperphosphorylation of the HCV NS5A disrupts interaction with hVAP-33 and negatively regulates viral RNA replication (Evans *et al.*, 2004; see also The role of the pestiviral and HCV NS5A protein).

The role of FBL2 in HCV replication

Recently, Wang and colleagues demonstrated that the geranylgeranylated cellular protein FBL2 forms a stable (i.e. immunoprecipitable) complex with the HCV NS5A protein (Wang *et al.*, 2005). By over expression of a dominant-negative FBL2 and by siRNA-mediated knockdown of FBL2, the authors revealed an essential role of this protein for HCV replication. Since the interaction of FBL2 and NS5A was found to depend on the geranylgeranylation of FBL2, it was speculated that the cellular protein enters the membrane associated RC of HCV via this interaction (Wang *et al.*, 2005; Ye *et al.*, 2003).

The role of NF/NFAR proteins in BVDV and HCV replication

Recently, we found that a cellular protein complex comprising the proteins NF45, NF90/NFAR-1 (also known as DRBP76), and RNA helicase A (RHA; also known as NDHII) (Liao *et al.*, 1998) associates at high specificity with the 5′ as well as with the 3′ NTR of the BVDV and HCV genomes while interacting poorly with other regions of the viral genomes. Most interestingly, the interaction sites of this complex were found to comprise exactly those RNA elements of the NTRs that were previously suggested being involved in the regulation of viral translation and RNA replication. That is, with BVDV, the cellular proteins were found to bind to the "3′ V" region of the 3′ NTR and to domains I and III of the 5′ NTR (Isken *et al.*, 2003). With HCV, the binding sites were shown to involve the polyUC/polyU tract and SL3 motifs of the 3′ NTR and the domain III region of the 5′ NTR (Isken *et al* 2003; Isken *et al.*, 2004; see also Chapter 'RNA elements involved in the replication of pestiviruses and HCV').

NF90/NFAR-1 and RHA encode double-strand RNA-binding motifs (dsRBM) and hence belong to a protein family with widespread functional diversity including, for instance, the Vaccinia E3L, the RNase III-like proteins Dicer and Drosher, and the double strand RNA regulated protein kinase PKR. Though their exact function(s) in the naïve cell remain to be determined, NF45, NF90/NFAR-1 and RHA (summarized as "NF/NFAR proteins") were indicated to be involved in the regulation of gene expression on the transcriptional and post-transcriptional level (for a review, see Saunders and Barber 2003).

In HCV and BVDV infected cells, significant fractions of the NF/NFAR proteins which, in naïve cells are predominantly nuclear, accumulate by a yet undefined mechanism at the sites of viral replication in the cellular cytoplasm. In the cytoplasm, NF90/NFAR-1 was shown to interact with the viral RNA and, interestingly, with the viral NS5A protein. However, the association with the viral protein was found to depend on the formation of the functional replication complex rather than on direct protein–protein contacts (Isken *et al.*, submitted for publication). Using precipitation experiments and electron microscopy, we could show that the NF45/NF90/RHA complex not only favors 5′/3′ interactions but even triggers the arrangement of a stable loop-like structure of the HCV RNA. Genetic studies confirmed that formation of the HCV loop-structure is crucial for viral replication, and cells were incapable to support HCV replication if the cytoplasm was depleted of RHA by RNAi (Isken *et al.*, 2003; Isken *et al.*, submitted for publication). The observed loop-structure explained experimental data, which showed that during the mRNA phase of the viral genome, formation of the viral 3′ NTR/NFAR ribonucleoprotein complex is an important prerequisite for efficient 5′ NTR/IRES-mediated translation as well as for efficient termination of translation at the translational stop-codon of the viral ORF. Thus, at the initial stage of the viral infection, the 3′ NTR/NFAR complex apparently supports translation and prohibits interferences of the translation machinery with the nascent viral replication complex (Isken *et al.*, 2004). In sum, the available experimental data provide solid evidence that NF/NFAR proteins are recruited as functional components of the HCV and BVDV replication machinery and that the proteins are involved in the coordination of translation and replication of the viral RNAs (see below).

There is no shortage of reports on contacts of host and virus encoded components, and recent developments of RNAi technologies as well as the availability of knockout mice models (at least for flaviviruses) are expected to further increase the number of putative host-factors (i.e. cellular proteins that are somewhat involved in viral replication). Nevertheless, the most challenging tasks of future research remain to elucidate the actual functions of host components in the viral life cycle and to define how host–virus interactions affect viral pathogenesis.

The coordination of viral translation and RNA replication

As a common feature of all positive-strand RNA viruses, the viral genome holds two major functions in the cytoplasm of the infected host-cell. On the one hand, the RNA is translated in 5′–3′ direction, on the other hand, the genome acts as a template of the RdRp, which is supposed to initiate the replication cycle at the 3′ end. Most studies that were aimed at understanding how the translation and replication cycles of positive-strand RNA viruses are coordinated were performed with the Picornaviridae members poliovirus and encephalomyocarditis virus (EMCV). As with HCV and pestiviruses, translation of the poliovirus and EMCV genomes is mediated by IRES elements (Racaniello, 2001). Studies with poliovirus led to the important finding that translation and replication of the viral RNA are mutually exclusive events. That is, during the mRNA phase, translation prevents the initiation of replication. At a yet undefined stage, translation initiation is blocked causing the release of ribosomes. Formation or activation of the initial replication complex then switches the genome to the replication mode promoting the synthesis of negative-strand RNA (Charini *et al*, 1994; Novak and Kirkegaard, 1994; Barton *et al.*, 1999; Barton *et al.*, 2001). According to a feed-back regulation model, which was originally proposed for the phage Qβ (Kolakowsky and Weissmann, 1971) and later applied to poliovirus, the transition from translation to RNA replication may be explained by the accumulation of viral proteins, which inhibit protein synthesis by binding to viral RNA motifs and/or cellular factors that are essential for translation. Along this line, the poliovirus $3CD^{pro}$ protein was indicated to associate with the so-called 5′-cloverleaf motif at the immediate 5′-terminus of the genome and to repress translation (Garmanik and Andino, 1998).

Such as the BVDV hairpin Ia motif, the poliovirus 5′-cloverleaf was shown to be essentially involved in both, negative and positive-strand RNA synthesis (Barton *et al.*, 2001; Herold and Andino, 2001) and to contribute to the efficiency of the poliovirus IRES (Simoes and Sarnow, 1991). Besides interacting with the 5′-cloverleaf, $3CD^{pro}$ was shown to associate also with the poly(A)$^{+}$ binding protein (pAb1p) that, in turn, binds to the A-rich domain near the 3′-terminus of the poliovirus genome. This protein complex, which also includes the cellular poly(rC) binding protein, was proposed to facilitate and to stabilize a circular conformation of the poliovirus viral genome, which was suggested to be important for the feed back inhibition of translation in favor of the initiation of RNA replication (Garmanik and Andino 1998; Herold and Andino, 2001; Walter *et al.*, 2002).

A speculative model on the initial steps of the BVDV replication pathway

As discussed in the earlier sections of this review, with all members of the Flaviviridae family, translation, processing of the polyprotein and replication were found to be closely linked processes that demand a sophisticated regulation. Considering the studies with picornaviruses and the fact that with all Flaviviridae members, cyclization of the viral genome was indicated to be an important prerequisite for replication, we postulate the following order of events at the early stage of the viral replication cycle. For this purpose, we discuss the pestivirus BVDV as a model, because BVDV was intensively studied with respect to the function of individual RNA elements as well as concerning the role of the NF/NFAR host proteins during the regulation of translation and RNA replication (see The role of RNA elements in the replication of pestiviruses and HCV and The role of NF/NFAR proteins in the replication of BVDV and HCV).

Thus, for the period of the mRNA phase of the BVDV genome, signals in the 5′ NTR (BVDV hairpin Ia) and 3′ NTR (BVDV 3′V) were indicated to act synergistically to facilitate IRES-mediated translation initiation. Along this line, the 3′ NTR contacts the 5′ end of the viral genome via RNA–RNA interactions that are supported by the NFAR proteins (Figure 4.4). The resulting circular conformation of the RNA may favor translation reinitiation of ribosomes after a completed round of polyprotein synthesis. Concomitant with the maturation of the viral proteins, the replication complex emerges at the 3′ C portion of the 3′ NTR while a major task of 3′ V involves blockage of the translation machinery at the ORF stop. As a conceivable scenario for the switch from translation to replication, the assembling replication complex may, at a certain stage, change the conformation of 3′ V in way that this affects also the 5′ NTR resulting in translation inhibition. Members of the NFAR proteins and/or de novo synthesized viral proteins such as NS5A may trigger this event (see The role of the pestiviral and HCV NS5A protein). Clearance of ribosomes from the viral RNA then would allow the initiation of replication, involving again signals at the 5′ and 3′ termini of the RNA such as 5′ NTR hairpin Ia and the 3′ NTR SLI (see The role of RNA elements in the replication of pestiviruses and HCV, and Figure 4.4).

Conclusion

Genetic studies of replication competent genomic and subgenomic RNA molecules of different Flaviviridae members revealed significant information on the identity and function of replicating RNA elements in the viral NTRs and in the viral ORF. Moreover, these studies defined the minimal number of viral protein coding units, the expression of which is necessary and sufficient for catalyzing the intracellular replication pathway of the viral RNA. However, for several viral proteins and virus-encoded enzymatic activities little is yet known on their tasks during the viral replication process, and, considering the strict *cis*-activity and tight membrane association of most viral proteins, the establishment of *in vitro* systems that would allow, for example, specific depletion and reconstitution studies, remains a challenge. In contrast, a rapidly emerging field concerns the characterization of host factors involved in viral replication, which is driven by the increasing number of transgenic or knockout cell lines and animals and the availability of gene array, RNAi knock down and proteome-technologies. In fact, the identification and characterization of cellular components of the viral RC is expected to accelerate unraveling the functional composition of the Flaviviridae replication complexes as well as to elucidate the functions of viral components.

References

Ackermann, M. and R. Padmanabhan. (2001). De novo synthesis of RNA by the dengue virus RNA-dependent RNA polymerase exhibits its temperature dependence at the initiation but not elongation phase. J. Biol. Chem. 276: 39926–37.

Agapov, E.V., C.L. Murray, I. Frolov, L. Qu, T.M. Myers, and C.M. Rice. (2004). Uncleaved NS2–3 is required for production of infectious bovine viral diarrhea virus. J. Virol. 78: 2414–25.

Agol, V. I., A. V. Paul, and E. Wimmer, E. (1999). Paradoxes of the replication of picornaviral genomes. Virus Res. 62: 129–47.

Ali, N. and A. Siddiqui. (1995). Interaction of polypyrimidine tract-binding protein with the 5′ noncoding region of the hepatitis C virus RNA genome and its functional requirement in internal initiation of translation. J. Virol. 69: 6367–75.

Ali, N. and A. Siddiqui. (1997). The La antigen binds 5′ noncoding region of the hepatitis C virus RNA in the context of the initiator AUG codon and stimulates internal ribosome entry site-mediated translation. Proc. Natl. Acad. Sci. USA *94*: 2249–54.

Alvarez, D.E., M.F. Lodeiro, S.J. Luduena, L.I. Pietrasanta, and A.V. Gamarnik. (2005). Long-range RNA–RNA interactions circularize the dengue virus genome. J. Virol. *79*: 6631–43.

Andino, R., G.E. Rieckhof, and D. Baltimore. (1990). A functional ribonucleoprotein complex forms around the 5′ end of poliovirus RNA. Cell *63*: 369–80.

Ansari, I.H., L.M. Chen, D. Liang, L.H. Gil, W. Zhong, and R.O. Donis. (2004). Involvement of a bovine viral diarrhea virus NS5B locus in virion assembly. J. Virol. *78*: 9612–23.

Anwar, A., N. Ali, R. Tanveer, and A. Siddiqui. (2000). Demonstration of functional requirement of polypyrimidine tract-binding protein by SELEX RNA during hepatitis C virus internal ribosome entry site-mediated translation initiation. J. Biol. Chem. *275*: 34231–5.

Appel, N., U. Herian, and R. Bartenschlager. (2005). Efficient rescue of hepatitis C virus RNA replication by trans-complementation with non-structural protein 5A. J. Virol. *79*: 896–909.

Banerjee, R. and A. Dasgupta. (2001). Specific interaction of hepatitis C virus protease/helicase NS3 with the 3′-terminal sequences of viral positive- and negative-strand RNA. J. Virol. *75*: 1708–21.

Bartholomeusz, A. and P. Thompson. (1999). Flaviviridae polymerase and RNA replication. J. Viral Hepatol. *6*: 261–70.

Barton, D.J., B.J. Morasco, and J.B. Flanegan. (1999). Translating ribosomes inhibit poliovirus negative-strand RNA synthesis. J. Virol. *73*: 10104–12.

Barton, D.J., B.J. O'Donnell, and J.B. Flanegan. (2001). 5′ cloverleaf in poliovirus RNA is a cis-acting replication element required for negative-strand synthesis. EMBO J. *20*: 1439–48.

Becher, P., M. Orlich, and H.J. Thiel. (1998). Complete genomic sequence of border disease virus, a pestivirus from sheep. J. Virol. *72*: 5165–73.

Becher, P., M. Orlich, and H.J. Thiel. (2000). Mutations in the 5′ nontranslated region of bovine viral diarrhea virus result in altered growth characteristics. J. Virol. *74*: 7884–94.

Beck, A.R., I.J. Miller, P. Anderson, and M. Streuli. (1998). RNA-binding protein TIAR is essential for primordial germ cell development. Proc. Natl. Acad. Sci. USA *95*: 2331–6.

Behrens, S.E., C.W. Grassmann, H.J. Thiel, G. Meyers, and N. Tautz. (1998). Characterization of an autonomous sub-genomic pestivirus RNA replicon. J. Virol. *72*: 2364–72.

Blackwell, J. L., and M.A. Brinton. (1995). BHK cell proteins that bind to the 3′ stem–loop structure of the West Nile virus genome RNA. J. Virol. *69*: 5650–8.

Blackwell, J.L. and M.A. Brinton. (1997). Translation elongation factor-1 alpha interacts with the 3′ stem–loop region of West Nile virus genomic RNA. J. Virol. *71*: 6433–44.

Blight, K. J., and C.M. Rice. (1997). Secondary structure determination of the conserved 98-base sequence at the 3′ terminus of hepatitis C virus genome RNA. J. Virol. *71*: 7345–52.

Blight, K.J., A.A. Kolykhalov, and C.M. Rice. (2000). Efficient initiation of HCV RNA replication in cell culture. Science *290*: 1972–4.

Blumenthal, T., and G.G. Carmichael. (1979). RNA replication: function and structure of Qbeta-replicase. Annu. Rev. Biochem. *48*: 525–548.

Brinton, M.A. (2002). The molecular biology of West Nile Virus: a new invader of the western hemisphere. Annu. Rev. Microbiol. *56*: 371–402.

Brinton, M.A. and J.H. Dispoto. (1988). Sequence and secondary structure analysis of the 5′-terminal region of flavivirus genome RNA. Virology *162*: 290–9.

Brinton, M.A., A.V. Fernandez, and J.H. Dispoto. (1986). The 3′-nucleotides of flavivirus genomic RNA form a conserved secondary structure. Virology *153*: 113–21.

Cahour, A., A. Pletnev, M. Vazielle-Falcoz, L. Rosen, and C.J. Lai. (1995). Growth-restricted dengue virus mutants containing deletions in the 5′ noncoding region of the RNA genome. Virology *207*: 68–76.

Charini, W.A., S. Todd, G.A. Gutman, and B.L. Semler. (1994). Transduction of a human RNA sequence by poliovirus. J. Virol. *68*: 6547–52.

Chen, C.J., M.D. Kuo, L.J. Chien, S.L. Hsu, Y.M. Wang, and J.H. Lin. (1997). RNA-protein interactions: involvement of NS3, NS5, and 3′ noncoding regions of Japanese encephalitis virus genomic RNA. J. Virol. *71*: 3466–73.

Cheney, I.W., S. Naim, V.C. Lai, S. Dempsey, D. Bellows, M.P. Walker, J.H. Shim, N. Horscroft, Z. Hong, and W. Zhong. (2002). Mutations in NS5B polymerase of hepatitis C virus: impacts on *in vitro* enzymatic activity and viral RNA replication in the sub-genomic replicon cell culture. Virology *297*: 298–306.

Chon, S.K., D.R. Perez, and R.O. Donis. (1998). Genetic analysis of the internal ribosome entry segment of bovine viral diarrhea virus. Virology *251*: 370–82.

Chung, R.T. and L.M. Kaplan. (1999). Heterogeneous nuclear ribonucleoprotein I (hnRNP-I/PTB) selectively binds the conserved 3′-terminus of hepatitis C viral RNA. Biochem. Biophys. Res. Commun. *254*: 351–62.

Collier, A.J., J. Gallego, R. Klinck, P.T. Cole, S.J. Harris, G.P. Harrison, F. Aboul-Ela, G. Varani, and S. Walker. (2002). A conserved RNA structure within the HCV IRES eIF3-binding site. Nat. Struct. Biol. *9*: 375–80.

Corver, J., E. Lenches, K. Smith, R.A. Robison, T. Sando, E.G. Strauss, and J.H. Strauss. (2003). Fine mapping of a cis-acting sequence element in yellow fever virus RNA that is required for RNA replication and cyclization. J. Virol. *77*: 2265–70.

Cui, T., R.J. Sugrue, Q. Xu, A.K. Lee, Y.C. Chan, and J. Fu. (1998). Recombinant dengue virus type 1 NS3 protein exhibits specific viral RNA binding and NTPase activity regulated by the NS5 protein. Virology *246*: 409–17.

De Nova-Ocampo, M., N. Villegas-Sepulveda, and R.M. del Angel. (2002). Translation elongation factor-1alpha, La, and PTB interact with the 3′ untranslated region of dengue 4 virus RNA. Virology *295*: 337–47.

Del Gatto-Konczak, F., C.F. Bourgeois, C. Le Guiner, L. Kister, M.C. Gesnel, J. Stevenin, and R. Breathnach. (2000). The RNA-binding protein TIA-1 is a novel mammalian splicing regulator acting through intron sequences adjacent to a 5′ splice site. Mol. Cell. Biol. *20*: 6287–99.

Deng, R. and K.V. Brock. (1993). 5′ and 3′ untranslated regions of pestivirus genome: primary and secondary structure analyses. Nucleic Acids Res. *21*: 1949–57.

Domitrovich, A.M., K.W. Diebel, N. Ali, S. Sarker, and A. Siddiqui. (2005). Role of La autoantigen and polypyrimidine tract-binding protein in HCV replication. Virology *335*: 72–86.

Egger, D., B. Wolk, R. Gosert, L. Bianchi, H.E. Blum, D. Moradpour, and K. Bienz. (2002). Expression of hepatitis C virus proteins induces distinct membrane alterations including a candidate viral replication complex. J. Virol. *76*: 5974–84.

Elazar, M., P. Liu, C.M. Rice, and J.S. Glenn. (2004). An N-terminal amphipathic helix in hepatitis C virus (HCV) NS4B mediates membrane association, correct localization of replication complex proteins, and HCV RNA replication. J. Virol. *78*: 11393–400.

Elghonemy, S., W.G. Davis, and M.A. Brinton. (2005). The majority of the nucleotides in the top loop of the genomic 3′-terminal stem–loop structure are cis-acting in a West Nile virus infectious clone. Virology *331*: 238–46.

El-Hage, N. and G. Luo. (2003). Replication of hepatitis C virus RNA occurs in a membrane-bound replication complex containing nonstructural viral proteins and RNA. J. Gen. Virol. *84*: 2761–9.

Evans, M.J., C.M. Rice, and S.P. Goff. (2004). Genetic interactions between hepatitis C virus replicons. J. Virol. *78*: 12085–9.

Friebe, P., V. Lohmann, N. Krieger, and R. Bartenschlager. (2001). Sequences in the 5′ nontranslated region of hepatitis C virus required for RNA replication. J. Virol. *75*: 12047–57.

Friebe, P. and R. Bartenschlager. (2002). Genetic analysis of sequences in the 3′ nontranslated region of hepatitis C virus that are important for RNA replication. J. Virol. *76*: 5326–38.

Friebe, P., J. Boudet, J.P. Simorre, and R. Bartenschlager. (2005). Kissing-loop interaction in the 3′ end of the hepatitis C virus genome essential for RNA replication. J. Virol. *79*: 380–92.

Frolov, I., M.S. McBride, and C.M. Rice. (1998). cis-acting RNA elements required for replication of bovine viral diarrhea virus-hepatitis C virus 5′ nontranslated region chimeras. RNA *4*: 1418–35.

Fukushi, S., K. Katayama, C. Kurihara, N. Ishiyama, F.B. Hoshino, T. Ando, and A. Oya. (1994). Complete 5′ noncoding region is necessary for the efficient internal initiation of hepatitis C virus RNA. Biochem. Biophys. Res. Commun. *199*: 425–32.

Gamarnik, A.V. and R. Andino. (1998). Switch from translation to RNA replication in a positive-stranded RNA virus. Genes Dev. *12*: 2293–304.

Gao, L., H. Aizaki, J.W. He, and M.M. Lai. (2004). Interactions between viral non-structural proteins and host protein hVAP-33 mediate the formation of hepatitis C virus RNA replication complex on lipid raft. J. Virol. *78*: 3480–8.

Gerber, K., E. Wimmer, and A.V. Paul. (2001). Biochemical and genetic studies of the initiation of human rhinovirus 2 RNA replication: identification of a cis-replicating element in the coding sequence of 2A(pro). J. Virol. *75*: 10979–90.

Ghetti, A., S. Pinol-Roma, W.M. Michael, C. Morandi, and G. Dreyfuss. (1992). hnRNP I, the polypyrimidine tract-binding protein: distinct nuclear localization and association with hnRNAs. Nucleic Acids Res. *20*: 3671–8.

Gil, A., P.A. Sharp, S.F. Jamison, and M.A. Garcia-Blanco. (1991). Characterization of cDNAs encoding the polypyrimidine tract-binding protein. Genes Dev. 5: 1224–36.

Goodfellow, I., Y. Chaudhry, A. Richardson, J. Meredith, J.W. Almond, W. Barclay, and D.J. Evans. (2000). Identification of a cis-acting replication element within the poliovirus coding region. J. Virol. *74*: 4590–600.

Gosert, R., K.H. Chang, R. Rijnbrand, M. Yi, D.V. Sangar, and S.M. Lemon. (2000). Transient expression of cellular polypyrimidine-tract binding protein stimulates cap-independent translation directed by both picornaviral and flaviviral internal ribosome entry sites in vivo. Mol. Cell. Biol. *20*: 1583–95.

Gosert, R., Egger, D., Lohmann, V., Bartenschlager, R., Blum, H. E., Bienz, K., and Moradpour, D. (2003). Identification of the hepatitis C virus RNA replication complex in Huh-7 cells harboring subgenomic replicons. J. Virol. *77*: 5487–92.

Grange, T., M. Bouloy, and M. Girard. (1985). Stable secondary structures at the 3′-end of the genome of yellow fever virus (17 D vaccine strain). FEBS Lett. *188*: 159–63.

Grassmann, C.W., O. Isken, and S.E. Behrens. (1999). Assignment of the multifunctional NS3 protein of bovine viral diarrhea virus during RNA replication: an *in vivo* and *in vitro* study. J. Virol. *73*: 9196–205.

Grassmann, C.W., O. Isken, N. Tautz, and S.E. Behrens. (2001). Genetic analysis of the pestivirus non-structural coding region: defects in the NS5A unit can be complemented in trans. J. Virol. *75*: 7791–802.

Grassmann, C.W., H. Yu, O. Isken, and S.E. Behrens. (2005). Hepatitis C virus and the related bovine viral diarrhea virus considerably differ in the functional organization of the 5′ non-translated region: implications for the viral life cycle. Virology *333*: 349–66.

Grosset, C., C.Y. Chen, N. Xu, N. Sonenberg, H. Jacquemin-Sablon, and A.B. Shyu. (2000). A mechanism for translationally coupled mRNA turnover: interaction between the poly(A) tail and a c-fos RNA coding determinant via a protein complex. Cell *103*: 29–40.

Gueydan, C., L. Droogmans, P. Chalon, G. Huez, D. Caput, and V. Kruys. (1999). Identification of TIAR as a protein binding to the translational regulatory AU-rich element of tumor necrosis factor alpha mRNA. J. Biol. Chem. *274*: 2322–6.

Guo, J.T., V.V. Bichko, and C. Seeger. (2001). Effect of alpha interferon on the hepatitis C virus replicon. J. Virol. *75*: 8516–23.

Hahm, B., Y.K. Kim, J.H. Kim, T.Y. Kim, and S.K. Jang. (1998). Heterogeneous nuclear ribonucleoprotein L interacts with the 3′ border of the internal ribosomal entry site of hepatitis C virus. J. Virol. *72*: 8782–8.

Hahn, C.S., Y.S. Hahn, C.M. Rice, E. Lee, L. Dalgarno, E.G. Strauss, and J.H. Strauss. (1987). Conserved elements in the 3′ untranslated region of flavivirus RNAs and potential cyclization sequences. J. Mol. Biol. *198*: 33–41.

Hahn, Y.S. (2003). Subversion of immune responses by hepatitis C virus: immunomodulatory strategies beyond evasion? Curr. Opin. Immunol. *15*: 443–9.

Harris, K. S., Xiang, W., Alexander, L., Lane, W. S., Paul, A. V., and Wimmer, E. (1994). Interaction of poliovirus polypeptide 3CDpro with the 5′ and 3′ termini of the poliovirus genome. Identification of viral and cellular cofactors needed for efficient binding. J. Biol. Chem. *269*: 27004–14.

Harris, C.E., R.A. Boden, and C.R. Astell. (1999). A novel heterogeneous nuclear ribonucleoprotein-like protein interacts with NS1 of the minute virus of mice. J. Virol. *73*: 72–80.

Hellen, C.U. and P. Sarnow. (2001). Internal ribosome entry sites in eukaryotic mRNA molecules. Genes Dev. *15*: 1593–612.

Herold, J. and R. Andino. (2001). Poliovirus RNA replication requires genome circularization through a protein–protein bridge. Mol. Cell. *7*: 581–91.

Honda, M., L.H. Ping, R.C. Rijnbrand, E. Amphlett, B. Clarke, D. Rowlands, and S.M. Lemon. (1996). Structural requirements for initiation of translation by internal ribosome entry within genome-length hepatitis C virus RNA. Virology *222*: 31–42.

Honda, M., M.R. Beard, L.H. Ping, and S.M. Lemon. (1999). A phylogenetically conserved stem–loop structure at the 5′ border of the internal ribosome entry site of hepatitis C virus is required for cap-independent viral translation. J. Virol. *73*: 1165–74.

Hresko, R.C. and M. Mueckler. (2000). A novel 68-kDa adipocyte protein phosphorylated on tyrosine in response to insulin and osmotic shock. J. Biol. Chem. *275*: 18114–20.

Intine, R.V., A.L. Sakulich, S.B. Koduru, Y. Huang, E. Pierstorff, J.L. Goodier, L. Phan, and R.J. Maraia. (2000). Control of transfer

RNA maturation by phosphorylation of the human La antigen on serine 366. Mol. Cell. *6*: 339–48.

Isken, O., C.W. Grassmann, R.T. Sarisky, M. Kann, S. Zhang, F. Grosse, P.N. Kao, and S.E. Behrens. (2003). Members of the NF90/NFAR protein group are involved in the life cycle of a positive-strand RNA virus. EMBO J. 22: 5655–65.

Isken, O., C.W. Grassmann, H. Yu, and S.E. Behrens. (2004). Complex signals in the genomic 3′ nontranslated region of bovine viral diarrhea virus coordinate translation and replication of the viral RNA. RNA *10*: 1637–52.

Ito, T. and M.M. Lai. (1997). Determination of the secondary structure of and cellular protein binding to the 3′-untranslated region of the hepatitis C virus RNA genome. J. Virol. *71*: 8698–706.

Ito, T. and M.M. Lai. (1999). An internal polypyrimidine-tract-binding protein-binding site in the hepatitis C virus RNA attenuates translation, which is relieved by the 3′-untranslated sequence. Virology *254*: 288–96.

Ito, T., S.M. Tahara, and M.M. Lai. (1998). The 3′-untranslated region of hepatitis C virus RNA enhances translation from an internal ribosomal entry site. J. Virol. 72: 8789–96.

Izumi, R.E., S. Das, B. Barat, S. Raychaudhuri, and A. Dasgupta. (2004). A peptide from autoantigen La blocks poliovirus and hepatitis C virus cap-independent translation and reveals a single tyrosine critical for La RNA binding and translation stimulation. J. Virol. *78*: 3763–76.

Jones, C.T., C.G. Patkar, and R.J. Kuhn. (2005). Construction and applications of yellow fever virus replicons. Virology *331*: 247–59.

Kaariainen, L. and T. Ahola. (2002). Functions of alphavirus nonstructural proteins in RNA replication. Prog. Nucleic Acid Res. Mol. Biol. *71*: 187–222.

Kamoshita, N., K. Tsukiyama-Kohara, M. Kohara, and A. Nomoto. (1997). Genetic analysis of internal ribosomal entry site on hepatitis C virus RNA: implication for involvement of the highly ordered structure and cell type-specific transacting factors. Virology *233*: 9–18.

Kapoor, M., L. Zhang, P.M. Mohan, and R. Padmanabhan. 1995a. Synthesis and characterization of an infectious dengue virus type-2 RNA genome (New Guinea C strain). Gene *162*: 175–80.

Kapoor, M., L. Zhang, M. Ramachandra, J. Kusukawa, K.E. Ebner, and R. Padmanabhan. 1995b. Association between NS3 and NS5 proteins of dengue virus type 2 in the putative RNA replicase is linked to differential phosphorylation of NS5. J. Biol. Chem. *270*: 19100–6.

Kedersha, N.L., M. Gupta, W. Li, I. Miller, and P. Anderson. (1999). RNA-binding proteins TIA-1 and TIAR link the phosphorylation of eIF-2 alpha to the assembly of mammalian stress granules. J. Cell Biol. *147*: 1431–42.

Khromykh, A.A., M.T. Kenney, and E.G. Westaway. (1998). trans-Complementation of flavivirus RNA polymerase gene NS5 by using Kunjin virus replicon-expressing BHK cells. J. Virol. *72*: 7270–9.

Khromykh, A.A., H. Meka, K.J. Guyatt, and E.G. Westaway. (2001). Essential role of cyclization sequences in flavivirus RNA replication. J. Virol. 75: 6719–28.

Khromykh, A.A., P.L. Sedlak, and E.G. Westaway. (1999). trans-Complementation analysis of the flavivirus Kunjin ns5 gene reveals an essential role for translation of its N-terminal half in RNA replication. J. Virol. *73*: 9247–55.

Khromykh, A.A., P.L. Sedlak, and E.G. Westaway. (2000). cis- and trans-acting elements in flavivirus RNA replication. J. Virol. *74*: 3253–63.

Khromykh, A.A. and E.G. Westaway. (1997). Sub-genomic replicons of the flavivirus Kunjin: construction and applications. J. Virol. *71*: 1497–505.

Kieft, J.S., K. Zhou, R. Jubin, M.G. Murray, J.Y. Lau, and J.A. Doudna. (1999). The hepatitis C virus internal ribosome entry site adopts an ion-dependent tertiary fold. J. Mol. Biol. *292*: 513–29.

Kieft, J.S., K. Zhou, A. Grech, R. Jubin, and J.A. Doudna. (2002). Crystal structure of an RNA tertiary domain essential to HCV IRES-mediated translation initiation. Nat. Struct. Biol. 9: 370–4.

Kim, J.H., K.Y. Paek, S.H. Ha, S. Cho, K. Choi, C.S. Kim, S.H. Ryu, and S.K. Jang. (2004). A cellular RNA-binding protein enhances internal ribosomal entry site-dependent translation through an interaction downstream of the hepatitis C virus polyprotein initiation codon. Mol. Cell. Biol. *24*: 7878–90.

Kim, Y.K., C.S. Kim, S.H. Lee, and S.K. Jang. (2002). Domains I and II in the 5′ nontranslated region of the HCV genome are required for RNA replication. Biochem. Biophys. Res. Commun. *290*: 105–12.

Kim, Y.K., S.H. Lee, C.S. Kim, S.K. Seol, and S.K. Jang. (2003). Long-range RNA–RNA interaction between the 5′ nontranslated region and the core-coding sequences of hepatitis C virus modulates the IRES-dependent translation. RNA *9*: 599–606.

Koch, J.O. and R. Bartenschlager. (1999). Modulation of hepatitis C virus NS5A hyperphosphorylation by non-structural proteins NS3, NS4A, and NS4B. J. Virol. *73*: 7138–46.

Kolakofsky, D. and C. Weissmann. (1971). Possible mechanism for transition of viral RNA from polysome to replication complex. Nat. New Biol. *231*: 42–6.

Kolupaeva, V.G., T.V. Pestova, and C.U. Hellen. (2000a). An enzymatic footprinting analysis of the interaction of 40S ribosomal subunits with the internal ribosomal entry site of hepatitis C virus. J. Virol. *74*: 6242–50.

Kolupaeva, V.G., T.V. Pestova, and C.U. Hellen. (2000b). Ribosomal binding to the internal ribosomal entry site of classical swine fever virus. RNA 6: 1791–807.

Kolykhalov, A.A., S.M. Feinstone, and C.M. Rice. (1996). Identification of a highly conserved sequence element at the 3′-terminus of hepatitis C virus genome RNA. J. Virol. *70*: 3363–71.

Kolykhalov, A.A., K. Mihalik, S.M. Feinstone, and C.M. Rice. (2000). Hepatitis C virus-encoded enzymatic activities and conserved RNA elements in the 3′ nontranslated region are essential for virus replication *in vivo*. J. Virol. *74*: 2046–51.

Kong, L.K. and P. Sarnow. (2002). Cytoplasmic expression of mRNAs containing the internal ribosome entry site and 3′ noncoding region of hepatitis C virus: effects of the 3′ leader on mRNA translation and mRNA stability. J. Virol. *76*: 12457–62.

Kummerer, B.M. and C.M. Rice. (2002). Mutations in the yellow fever virus non-structural protein NS2A selectively block production of infectious particles. J. Virol. *76*: 4773–84.

Lai, M.M. (1998). Cellular factors in the transcription and replication of viral RNA genomes: a parallel to DNA-dependent RNA transcription. Virology *244*: 1–12.

Le Guiner, C., F. Lejeune, D. Galiana, L. Kister, R. Breathnach, J. Stevenin, and F. Del Gatto-Konczak. (2001). TIA-1 and TIAR activate splicing of alternative exons with weak 5′ splice sites followed by a U-rich stretch on their own pre-mRNAs. J. Biol. Chem. *276*: 40638–46.

Le, S.Y., A. Siddiqui, and J.V. Maizel, Jr. (1996). A common structural core in the internal ribosome entry sites of picornavirus, hepatitis C virus, and pestivirus. Virus Genes *12*: 135–47.

Lee, H., H. Shin, E. Wimmer, and A.V. Paul. (2004). cis-acting RNA signals in the NS5B C-terminal coding sequence of the hepatitis C virus genome. J. Virol. *78*: 10865–77.

Lemm, J.A., T. Rumenapf, E.G. Strauss, J.H. Strauss, and C.M. Rice. (1994). Polypeptide requirements for assembly of functional Sindbis virus replication complexes: a model for the temporal regulation of minus- and plus-strand RNA synthesis. EMBO J. *13*: 2925–34.

Li, W., Y. Li, N. Kedersha, P. Anderson, M. Emara, K.M. Swiderek, G.T. Moreno, and M.A. Brinton. (2002). Cell proteins TIA-1 and TIAR interact with the 3′ stem–loop of the West Nile virus complementary minus-strand RNA and facilitate virus replication. J. Virol. *76*: 11989–2000.

Liao, H.J., R. Kobayashi, and M.B. Mathews. (1998). Activities of adenovirus virus-associated RNAs: purification and characterization of RNA binding proteins. Proc. Natl. Acad. Sci. USA *95*: 8514–9.

Lin, C., J.W. Wu, K. Hsiao, and M.S. Su. (1997). The hepatitis C virus NS4A protein: interactions with the NS4B and NS5A proteins. J. Virol. *71*: 6465–71.

Lindenbach, B.D. and C.M. Rice. (1997). trans-Complementation of yellow fever virus NS1 reveals a role in early RNA replication. J. Virol. *71*: 9608–17.

Lindenbach, B.D. and C.M. Rice. (1999). Genetic interaction of flavivirus non-structural proteins NS1 and NS4A as a determinant of replicase function. J. Virol. *73*: 4611–21.

Lindenbach, B.D. and Rice, C.M. (2001). Flaviviridae: the viruses and their replication. In: Virology, 4th ed., edited by B.N. Fields. Lippincott-Raven, Philadelphia, New York, pp. 991–1042.

Lindenbach BD, and C.M. Rice. (2003). Molecular biology of flaviviruses. Adv. Virus. Res. *59*: 23–61.

Liu, W.J., H.B. Chen, and A.A. Khromykh. (2003). Molecular and functional analyses of Kunjin virus infectious cDNA clones demonstrate the essential roles for NS2A in virus assembly and for a nonconservative residue in NS3 in RNA replication. J. Virol. *77*: 7804–13.

Lobert, P.E., N. Escriou, J. Ruelle, and T. Michiels. (1999). A coding RNA sequence acts as a replication signal in cardioviruses. Proc. Natl. Acad. Sci. USA *96*: 11560–5.

Lohmann, V., F. Korner, A. Dobierzewska, and R. Bartenschlager. (2001). Mutations in hepatitis C virus RNAs conferring cell culture adaptation. J. Virol. *75*: 1437–49.

Lohmann, V., F. Korner, J. Koch, U. Herian, L. Theilmann, and R. Bartenschlager. (1999). Replication of sub-genomic hepatitis C virus

RNAs in a hepatoma cell line. Science *285*: 110–3.

Lu, H.H. and E. Wimmer. (1996). Poliovirus chimeras replicating under the translational control of genetic elements of hepatitis C virus reveal unusual properties of the internal ribosomal entry site of hepatitis C virus. Proc. Natl. Acad. Sci. USA 93: 1412–7.

Lukavsky, P.J., G.A. Otto, A.M. Lancaster, P. Sarnow, and J.D. Puglisi. (2000). Structures of two RNA domains essential for hepatitis C virus internal ribosome entry site function. Nat. Struct. Biol. 7: 1105–10.

Lukavsky, P.J., I. Kim, G.A. Otto, and J.D. Puglisi. (2003). Structure of HCV IRES domain II determined by NMR. Nat. Struct. Biol. *10*: 1033–8.

Lundin, M., M. Monne, A. Widell, G. Von Heijne, and M.A. Persson. (2003). Topology of the membrane-associated hepatitis C virus protein NS4B. J. Virol. 77: 5428–38.

Luo, G., S. Xin, and Z. Cai. (2003). Role of the 5′-proximal stem–loop structure of the 5′ untranslated region in replication and translation of hepatitis C virus RNA. J. Virol. *77*: 3312–8.

Lytle, J.R., L. Wu, and H.D. Robertson. (2001). The ribosome binding site of hepatitis C virus mRNA. J. Virol. *75*: 7629–36.

Macdonald, A. and M. Harris. (2004). Hepatitis C virus NS5A: tales of a promiscuous protein. J. Gen. Virol. *85*: 2485–502.

Mandl, C.W., F.X. Heinz, E. Stockl, and C. Kunz. (1989). Genome sequence of tick-borne encephalitis virus (Western subtype) and comparative analysis of non-structural proteins with other flaviviruses. Virology *173*: 291–301.

Mandl, C.W., C. Kunz, and F.X. Heinz. (1991). Presence of poly(A) in a flavivirus: significant differences between the 3′ noncoding regions of the genomic RNAs of tick-borne encephalitis virus strains. J. Virol. *65*: 4070–7.

Mandl, C.W., H. Holzmann, C. Kunz, and F.X. Heinz. (1993). Complete genomic sequence of Powassan virus: evaluation of genetic elements in tick-borne versus mosquito-borne flaviviruses. Virology *194*: 173–84.

Mandl, C.W., H. Holzmann, T. Meixner, S. Rauscher, P.F. Stadler, S.L. Allison, and F.X. Heinz. (1998). Spontaneous and engineered deletions in the 3′ noncoding region of tick-borne encephalitis virus: construction of highly attenuated mutants of a flavivirus. J. Virol. 72: 2132–40.

Maraia, R.J. (2001). La protein and the trafficking of nascent RNA polymerase iii transcripts. J. Cell Biol. *153*: F13–8.

Markoff, L. (2003). 5′- and 3′-noncoding regions in flavivirus RNA. Adv. Virus. Res. *59*: 177–228.

McKnight, K.L. and S.M. Lemon. (1998). The rhinovirus type 14 genome contains an internally located RNA structure that is required for viral replication. RNA *4*: 1569–84.

Meyers, G., N. Tautz, P. Becher, H.J. Thiel, and B.M. Kummerer. 1996a. Recovery of cytopathogenic and noncytopathogenic bovine viral diarrhea viruses from cDNA constructs. J. Virol. *70*: 8606–13.

Meyers, G., H.J. Thiel, and T. Rumenapf. 1996b. Classical swine fever virus: recovery of infectious viruses from cDNA constructs and generation of recombinant cytopathogenic defective interfering particles. J. Virol. *70*: 1588–95.

Michel, Y.M., A.M. Borman, S. Paulous, and K.M. Kean. (2001). Eukaryotic initiation factor 4G-poly(A) binding protein interaction is required for poly(A) tail-mediated stimulation of picornavirus internal ribosome entry segment-driven translation but not for X-mediated stimulation of hepatitis C virus translation. Mol. Cell. Biol. *21*: 4097–109.

Mizutani, A., M. Fukuda, K. Ibata, Y. Shiraishi, and K. Mikoshiba. (2000). SYNCRIP, a cytoplasmic counterpart of heterogeneous nuclear ribonucleoprotein R, interacts with ubiquitous synaptotagmin isoforms. J. Biol. Chem. *275*: 9823–31.

Moormann, R.J., H.G. van Gennip, G.K. Miedema, M.M. Hulst, and P.A. van Rijn. (1996). Infectious RNA transcribed from an engineered full-length cDNA template of the genome of a pestivirus. J. Virol. *70*: 763–70.

Moradpour, D., V. Brass, E. Bieck, P. Friebe, R. Gosert, H.E. Blum, R. Bartenschlager, F. Penin, and V. Lohmann. (2004). Membrane association of the RNA-dependent RNA polymerase is essential for hepatitis C virus RNA replication. J. Virol. 78: 13278–84.

Morasco, B.J., N. Sharma, J. Parilla, and J.B. Flanegan. (2003). Poliovirus cre(2C)-dependent synthesis of VPgpUpU is required for positive- but not negative-strand RNA synthesis. J. Virol. *77*: 5136–44.

Moser, C., P. Stettler, J.D. Tratschin, and M.A. Hofmann. (1999). Cytopathogenic and noncytopathogenic RNA replicons of classical swine fever virus. J. Virol. *73*: 7787–94.

Mourelatos, Z., L. Abel, J. Yong, N. Kataoka, and G. Dreyfuss. (2001). SMN interacts with a

novel family of hnRNP and spliceosomal proteins. EMBO J. *20*: 5443–52.
Muylaert, I.R., R. Galler, and C.M. Rice. (1997). Genetic analysis of the yellow fever virus NS1 protein: identification of a temperature-sensitive mutation which blocks RNA accumulation. J. Virol. *71*: 291–8.
Myers, T.M., V.G. Kolupaeva, E. Mendez, S.G. Baginski, I. Frolov, C.U. Hellen, and C.M. Rice. (2001). Efficient translation initiation is required for replication of bovine viral diarrhea virus sub-genomic replicons. J. Virol. *75*: 4226–38.
Neddermann, P., A. Clementi, and R. De Francesco. (1999). Hyperphosphorylation of the hepatitis C virus NS5A protein requires an active NS3 protease, NS4A, NS4B, and NS5A encoded on the same polyprotein. J. Virol. *73*: 9984–91.
Neddermann, P., M. Quintavalle, C. Di Pietro, A. Clementi, M. Cerretani, S. Altamura, L. Bartholomew, and R. De Francesco. (2004). Reduction of hepatitis C virus NS5A hyperphosphorylation by selective inhibition of cellular kinases activates viral RNA replication in cell culture. J. Virol. *78*: 13306–14.
Novak, J.E. and K. Kirkegaard. (1994). Coupling between genome translation and replication in an RNA virus. Genes Dev. *8*: 1726–37.
Penin, F., V. Brass, N. Appel, S. Ramboarina, R. Montserret, D. Ficheux, H.E. Blum, R. Bartenschlager, and D. Moradpour. (2004). Structure and function of the membrane anchor domain of hepatitis C virus non-structural protein 5A. J. Biol. Chem. *279*: 40835–43.
Pestova, T. V., Shatsky, I. N., Fletc.her, S. P., Jackson, R. J., and C.U.Hellen. (1998). A prokaryotic-like mode of cytoplasmic eukaryotic ribosome binding to the initiation codon during internal translation initiation of hepatitis C and classical swine fever virus RNAs. Genes Dev. *12*: 67–83.
Pestova, T.V. and C.U. Hellen. (1999). Internal initiation of translation of bovine viral diarrhea virus RNA. Virology *258*: 249–56.
Piccininni, S., A. Varaklioti, M. Nardelli, B. Dave, K.D. Raney, and J.E. McCarthy. (2002). Modulation of the hepatitis C virus RNA-dependent RNA polymerase activity by the non-structural (NS) 3 helicase and the NS4B membrane protein. J. Biol. Chem. *277*: 45670–9.
Piecyk, M., S. Wax, A.R. Beck, N. Kedersha, M. Gupta, B. Maritim, S. Chen, C. Gueydan, V. Kruys, M. Streuli, and P. Anderson. (2000). TIA-1 is a translational silencer that selectively regulates the expression of TNF-alpha. EMBO J. *19*: 4154–63.
Poole, T.L., C. Wang, R.A. Popp, L.N. Potgieter, A. Siddiqui, and M.S. Collett. (1995). Pestivirus translation initiation occurs by internal ribosome entry. Virology *206*: 750–4.
Proutski, V., M.W. Gaunt, E.A. Gould, and E.C. Holmes. 1997a. Secondary structure of the 3′-untranslated region of yellow fever virus: implications for virulence, attenuation and vaccine development. J. Gen. Virol. *78 (Pt 7)*: 1543–9.
Proutski, V., E.A. Gould, and E.C. Holmes. 1997b. Secondary structure of the 3′ untranslated region of flaviviruses: similarities and differences. Nucleic Acids Res. *25*: 1194–202.
Pugachev, K.V., F. Guirakhoo, S.W. Ocran, F. Mitchell, M. Parsons, C. Penal, S. Girakhoo, S.O. Pougatcheva, J. Arroyo, D.W. Trent, and T.P. Monath. (2004). High fidelity of yellow fever virus RNA polymerase. J. Virol. *78*: 1032–8.
Racaniello, V.R. (2001). Picornaviridae: the viruses and their replication. In: Virology, 4th ed., edited by B.N. Fields. Lippincott-Raven, Philadelphia, New York, pp. 685–722.
Rauscher, S., C. Flamm, C.W. Mandl, F.X. Heinz, and P.F. Stadler. (1997). Secondary structure of the 3′-noncoding region of flavivirus genomes: comparative analysis of base pairing probabilities. RNA 3: 779–91.
Reed, K.E., A.E. Gorbalenya, and C.M. Rice. (1998). The NS5A/NS5 proteins of viruses from three genera of the family flaviviridae are phosphorylated by associated serine/threonine kinases. J. Virol. *72*: 6199–206.
Reusken, C.B., T.J. Dalebout, P. Eerligh, P.J. Bredenbeek, and W.J. Spaan. (2003). Analysis of hepatitis C virus/classical swine fever virus chimeric 5′ NTRs: sequences within the hepatitis C virus IRES are required for viral RNA replication. J. Gen. Virol. *84*: 1761–9.
Rice, C.M., E.M. Lenches, S.R. Eddy, S.J. Shin, R.L. Sheets, and J.H. Strauss. (1985). Nucleotide sequence of yellow fever virus: implications for flavivirus gene expression and evolution. Science *229*: 726–33.
Rijnbrand, R., P. Bredenbeek, T. van der Straaten, L. Whetter, G. Inchauspe, S. Lemon, and W. Spaan. (1995). Almost the entire 5′ non-translated region of hepatitis C virus is required for cap-independent translation. FEBS Lett. *365*: 115–9.
Rijnbrand, R., T. van der Straaten, P.A. van Rijn, W.J. Spaan, and P.J. Bredenbeek. (1997). Internal entry of ribosomes is directed by the 5′ noncoding region of classical swine fever

virus and is dependent on the presence of an RNA pseudoknot upstream of the initiation codon. J. Virol. *71*: 451–7.

Rijnbrand, R.C. and S.M. Lemon. (2000). Internal ribosome entry site-mediated translation in hepatitis C virus replication. Curr. Top. Microbiol. Immunol. *242*: 85–116.

Rijnbrand, R., P.J. Bredenbeek, P.C. Haasnoot, J.S. Kieft, W.J. Spaan, and S.M. Lemon. (2001). The influence of downstream protein-coding sequence on internal ribosome entry on hepatitis C virus and other flavivirus RNAs. RNA 7: 585–97.

Ruggli, N., Tratschin, J.D., Mittelholzer, C., and Hofmann, M.A. (1996). Nucleotide sequence of classical swine fever virus strain Alfort/187 and transcription of infectious RNA from stably cloned full-length cDNA. J. Virol. *70*: 3478–87.

Salonen, A., T. Ahola, and L. Kaariainen. (2005). Viral RNA replication in association with cellular membranes. Curr. Top. Microbiol. Immunol. *285*: 139–73.

Saunders, L.R. and G.N. Barber. (2003). The dsRNA binding protein family: critical roles, diverse cellular functions. FASEB J. *17*: 961–83.

Shi, P.Y., W. Li, and M.A. Brinton. (1996). Cell proteins bind specifically to West Nile virus minus-strand 3′ stem–loop RNA. J. Virol. *70*: 6278–87.

Shimakami, T., M. Hijikata, H. Luo, Y.Y. Ma, S. Kaneko, K. Shimotohno, and S. Murakami. (2004). Effect of interaction between hepatitis C virus NS5A and NS5B on hepatitis C virus RNA replication with the hepatitis C virus replicon. J. Virol. *78*: 2738–48.

Shirota, Y., H. Luo, W. Qin, S. Kaneko, T. Yamashita, K. Kobayashi, and S. Murakami. (2002). Hepatitis C virus (HCV) NS5A binds RNA-dependent RNA polymerase (RdRP) NS5B and modulates RNA-dependent RNA polymerase activity. J. Biol. Chem. *277*: 11149–55.

Shurtleff, A.C., D.W. Beasley, J.J. Chen, H. Ni, M.T. Suderman, H. Wang, R. Xu, E. Wang, S.C. Weaver, D.M. Watts, K.L. Russell, and A.D. Barrett. (2001). Genetic variation in the 3′ non-coding region of dengue viruses. Virology *281*: 75–87.

Simoes, E.A. and P. Sarnow. (1991). An RNA hairpin at the extreme 5′ end of the poliovirus RNA genome modulates viral translation in human cells. J. Virol. *65*: 913–21.

Sizova, D.V., V.G. Kolupaeva, T.V. Pestova, I.N. Shatsky, and C.U. Hellen. (1998). Specific interaction of eukaryotic translation initiation factor 3 with the 5′ nontranslated regions of hepatitis C virus and classical swine fever virus RNAs. J. Virol. *72*: 4775–82.

Spahn, C.M., J.S. Kieft, R.A. Grassucci, P.A. Penczek, K. Zhou, J.A. Doudna, and J. Frank. (2001). Hepatitis C virus IRES RNA-induced changes in the conformation of the 40s ribosomal subunit. Science *291*: 1959–62.

Spangberg, K., L. Goobar-Larsson, M. Wahren-Herlenius, and S. Schwartz. (1999). The La protein from human liver cells interacts specifically with the U-rich region in the hepatitis C virus 3′ untranslated region. J. Hum. Virol. *2*: 296–307.

Spangberg, K., L. Wiklund, and S. Schwartz. (2001). Binding of the La autoantigen to the hepatitis C virus 3′ untranslated region protects the RNA from rapid degradation *in vitro*. J. Gen. Virol. *82*: 113–20.

Sumiyoshi, H., C. Mori, I. Fuke, K. Morita, S. Kuhara, J. Kondou, Y. Kikuchi, H. Nagamatu, and A. Igarashi. (1987). Complete nucleotide sequence of the Japanese encephalitis virus genome RNA. Virology *161*: 497–510.

Svitkin, Y.V., A. Pause, M. Lopez-Lastra, S. Perreault, and N. Sonenberg. (2005). Complete translation of the hepatitis C virus genome *in vitro*: membranes play a critical role in the maturation of all virus proteins except for NS3. J. Virol. *79*: 6868–81.

Takegami, T., M. Washizu, and K. Yasui. (1986). Nucleotide sequence at the 3′ end of Japanese encephalitis virus genomic RNA. Virology *152*: 483–6.

Tanaka, T., N. Kato, M.J. Cho, and K. Shimotohno. (1995). A novel sequence found at the 3′-terminus of hepatitis C virus genome. Biochem. Biophys. Res. Commun. *215*: 744–9.

Tardif, K.D., K. Mori, and A. Siddiqui. (2002). Hepatitis C virus sub-genomic replicons induce endoplasmic reticulum stress activating an intracellular signaling pathway. J. Virol. *76*: 7453–9.

Tautz, N., T. Harada, A. Kaiser, G. Rinck, S. Behrens, and H.J. Thiel. (1999). Establishment and characterization of cytopathogenic and noncytopathogenic pestivirus replicons. J. Virol. *73*: 9422–32.

Tellinghuisen TL, Rice CM. (2003). Interaction between hepatitis C virus proteins and host cell factors. Curr. Opin. Microbiol. 5: 419–27.

Thurner, C., C. Witwer, I.L. Hofacker, and P.F. Stadler. (2004). Conserved RNA secondary structures in Flaviviridae genomes. J. Gen. Virol. 85: 1113–24.

Tian, Q., J. Taupin, S. Elledge, M. Robertson, and P. Anderson. (1995). Fas-activated serine/threonine kinase (FAST) phosphorylates TIA-1 during Fas-mediated apoptosis. J. Exp. Med. *182*: 865–74.

Tischendorf, J.J., C. Beger, M. Korf, M.P. Manns, and M. Kruger. (2004). Polypyrimidine tract-binding protein (PTB) inhibits Hepatitis C virus internal ribosome entry site (HCV IRES)-mediated translation, but does not affect HCV replication. Arch. Virol. *149*: 1955–70.

Tsuchihara, K., T. Tanaka, M. Hijikata, S. Kuge, H. Toyoda, A. Nomoto, N. Yamamoto, and K. Shimotohno. (1997). Specific interaction of polypyrimidine tract-binding protein with the extreme 3′-terminal structure of the hepatitis C virus genome, the 3′X. J. Virol. *71*: 6720–6.

Tsukiyama-Kohara, K., N. Iizuka, M. Kohara, and A. Nomoto. (1992). Internal ribosome entry site within hepatitis C virus RNA. J. Virol. *66*: 1476–83.

Tu, H., L. Gao, S.T. Shi, D.R. Taylor, T. Yang, A.K. Mircheff, Y. Wen, A.E. Gorbalenya, S.B. Hwang, and M.M. Lai. (1999). Hepatitis C virus RNA polymerase and NS5A complex with a SNARE-like protein. Virology *263*: 30–41.

Tuplin, A., J. Wood, D.J. Evans, A.H. Patel, and P. Simmonds. (2002). Thermodynamic and phylogenetic prediction of RNA secondary structures in the coding region of hepatitis C virus. RNA *8*: 824–41.

Vassilev, V.B., M.S. Collett, and R.O. Donis. (1997). Authentic and chimeric full-length genomic cDNA clones of bovine viral diarrhea virus that yield infectious transcripts. J. Virol. *71*: 471–8.

Wallner, G., C.W. Mandl, C. Kunz, and F.X. Heinz. (1995). The flavivirus 3′-noncoding region: extensive size heterogeneity independent of evolutionary relationships among strains of tick-borne encephalitis virus. Virology *213*: 169–78.

Walter, B.L., T.B. Parsley, E. Ehrenfeld, and B.L. Semler. (2002). Distinct poly(rC) binding protein KH domain determinants for poliovirus translation initiation and viral RNA replication. J. Virol. *76*: 12008–22.

Wang, C., M. Gale, Jr., B.C. Keller, H. Huang, M.S. Brown, J.L. Goldstein, and J. Ye. (2005). Identification of FBL2 as a geranylgeranylated cellular protein required for hepatitis C virus RNA replication. Mol. Cell. *18*: 425–34.

Wang, C., S.Y. Le, N. Ali, and A. Siddiqui. (1995). An RNA pseudoknot is an essential structural element of the internal ribosome entry site located within the hepatitis C virus 5′ noncoding region. RNA *1*: 526–37.

Wengler, G. and E. Castle. (1986). Analysis of structural properties which possibly are characteristic for the 3′-terminal sequence of the genome RNA of flaviviruses. J. Gen. Virol. *67 (Pt 6)*: 1183–8.

Westaway, E.G. J.M. Mackenzie, M.T. Kenney, M.K. Jones, and A.A. Khromykh. (1997). Ultrastructure of Kunjin virus-infected cells: colocalization of NS1 and NS3 with double-stranded RNA, and of NS2B with NS3, in virus-induced membrane structures. J. Virol. *71*: 6650–61.

Westaway, E.G. J.M. Mackenzie, and A.A. Khromykh. (2003). Kunjin RNA replication and applications of Kunjin replicons. Adv. Virus. Res. *59*: 99–140.

Wood, J., R.M. Frederickson, S. Fields, and A.H. Patel. (2001). Hepatitis C virus 3′X region interacts with human ribosomal proteins. J. Virol. *75*: 1348–58.

Xiao, M., J. Gao, W. Wang, Y. Wang, J. Chen, and B. Li. (2004). Specific interaction between the classical swine fever virus NS5B protein and the viral genome. Eur. J. Biochem. *271*: 3888–96.

Yamshchikov, V.F., G. Wengler, A.A. Perelygin, M.A. Brinton, and R.W. Compans. (2001). An infectious clone of the West Nile flavivirus. Virology *281*: 294–304.

Yanagi, M., M. St Claire, S.U. Emerson, R.H. Purcell, and J. Bukh. (1999). In vivo analysis of the 3′ untranslated region of the hepatitis C virus after *in vitro* mutagenesis of an infectious cDNA clone. Proc. Natl. Acad. Sci. USA *96*: 2291–5.

Ye, J., Wang, C., Sumpter, R., Jr., Brown, M. S., Goldstein, J. L., and M. Gale, Jr (2003). Disruption of hepatitis C virus RNA replication through inhibition of host protein geranylgeranylation. Proc. Natl. Acad. Sci. USA *100*: 15865–70.

Yi, M. and S.M. Lemon. (2003a). 3′ nontranslated RNA signals required for replication of hepatitis C virus RNA. J. Virol. *77*: 3557–68.

Yi, M. and S.M. Lemon. (2003b). Structure-function analysis of the 3′ stem–loop of hepatitis C virus genomic RNA and its role in viral RNA replication. RNA *9*: 331–45.

Yin, J., A.V. Paul, E. Wimmer, and E. Rieder. (2003). Functional dissection of a poliovirus cis-acting replication element [PV-cre(2C)]: analysis of single- and dual-cre viral genomes and proteins that bind specifically to PV-cre RNA. J. Virol. *77*: 5152–66.

You, S., B. Falgout, L. Markoff, and R. Padmanabhan. (2001). In vitro RNA synthesis from exogenous dengue viral RNA templates requires long range interactions between 5′- and 3′-terminal regions that influence RNA structure. J. Biol. Chem. *276*: 15581–91.

You, S., D.D. Stump, A.D. Branch, and C.M. Rice. (2004). A cis-acting replication element in the sequence encoding the NS5B RNA-dependent RNA polymerase is required for hepatitis C virus RNA replication. J. Virol. *78*: 1352–66.

Yu, H., C.W. Grassmann, and S.E. Behrens. (1999). Sequence and structural elements at the 3′-terminus of bovine viral diarrhea virus genomic RNA: functional role during RNA replication. J. Virol. *73*: 3638–48.

Yu, H., O. Isken, C.W. Grassmann, and S.E. Behrens. (2000). A stem–loop motif formed by the immediate 5′-terminus of the bovine viral diarrhea virus genome modulates translation as well as replication of the viral RNA. J. Virol. *74*: 5825–35.

Yu, L. and L. Markoff. (2005). The topology of bulges in the long stem of the flavivirus 3′ stem–loop is a major determinant of RNA replication competence. J. Virol. *79*: 2309–24.

Zeenko, V. V., Ryabova, L. A., Spirin, A. S., Rothnie, H. M., Hess, D., Browning, K. S., and Hohn, T. (2002). Eukaryotic elongation factor 1A interacts with the upstream pseudoknot domain in the 3′ untranslated region of tobacco mosaic virus RNA. J. Virol. *76*: 5678–91.

Zeng, L., B. Falgout, and L. Markoff. (1998). Identification of specific nucleotide sequences within the conserved 3′- SL in the dengue type 2 virus genome required for replication. J. Virol. *72*: 7510–22.

Zhang, J., O. Yamada, T. Sakamoto, H. Yoshida, T. Iwai, Y. Matsushita, H. Shimamura, H. Araki, and K. Shimotohno. (2004). Down-regulation of viral replication by adenoviral-mediated expression of siRNA against cellular cofactors for hepatitis C virus. Virology *320*: 135–43.

Zhao, B., E. Mackow, A. Buckler-White, L. Markoff, R.M. Chanock, C.J. Lai, and Y. Makino. (1986). Cloning full-length dengue type 4 viral DNA sequences: analysis of genes coding for structural proteins. Virology *155*: 77–88.

A New Approach to Dengue Epidemiology: Sequential Infection and ADE Hypothesis — the Story of a Dogma

5

Xavier Deparis

Abstract

Since the 20th century, dengue fever has become one of the leading causes of morbidity and mortality in tropical areas. Dengue virus is a mosquito-borne virus transmitted by *Aedes* mosquitoes. Four distinct serological types of this virus exist. According to the World Health Organization, a person infected by one of the four dengue viruses goes through a spectrum of illness ranging from classical dengue fever, a self-limiting illness characterized by high temperature, headache, myalgia, arthralgia to a more severe form, dengue hemorrhagic fever. For over 30 years, the pathogenesis of dengue was attributed to the presence of enhancing antibodies that are acquired during a primary infection and lead to an increase of infected cells, thereby an increase in viremia, during secondary infections. This hypothesis is termed "antibody-dependent enhancement" (ADE). Although extensively studied, the role of ADE in the pathogenesis of dengue remains unverified. A lot of proofs exist demonstrating that we need to transcend the ADE theory in order to progress in dengue fever knowledge. A lot of facts demonstrate that dengue is logically close to yellow fever, and it is time to carry out a global study in several countries to assess the validity of the ADE theory.

Introduction

During the past two decades, dengue fever became one of the leading causes of morbidity and mortality in tropical and subtropical areas throughout the world (Monath, 1988). The dengue virus, a mosquito-borne member of the family of Flaviviridae, circulates as four distinct serological types DEN-1, -2, -3 and -4.

According to the World Health Organization (WHO), anyone infected by dengue virus goes through different signs of illness ranging from classical dengue fever (DF), a self-limiting illness characterized by high temperature, headache, myalgia, arthralgia, and abdominal pain to a more severe form, dengue hemorrhagic fever (DHF). A clinical definition of DHF (WHO, 1997) requires the simultaneous presence of high temperature, hemorrhagic manifestations, thrombocytopenia and hemoconcentration. The principal pathophysiologic sign of DHF is plasma leakage due to an increase in vascular permeability. The WHO classification further subdivides DHF into four grades of severity, grades III and IV corresponding to dengue shock syndrome (DSS) that is characterized by circulatory failure and can become life-threatening because of profound hypovolemic shock.

For 30 years, the pathogenesis of DHF has been attributed to the presence of enhancing antibodies that are acquired during a primary infection and lead to an increase of infected cells, thereby an increase in viremia, during secondary infections (Halstead, 1982; Morens, 1994). This hypothesis is called "antibody-dependent enhancement" (ADE). Although extensively studied, the role of ADE in the pathogenesis of DHF remains unverified, and there is increasing evidence that other mechanisms are involved (Morens, 1994; Holmes, 1998). The difference of opinions has very important practical implications because of its relevance for the use of dengue vaccines: the major consequence of ADE hypothesis is the necessity to use a tetravalent vaccine explaining the fact that there is still no vaccine available (Holmes, 1998).

During the last 30 years, ADE was accepted as the major theory explaining dengue pathogenesis. Recently published data in evident contradiction with ADE theory demonstrated that it is time to take up the debate about dengue pathogenesis.

This work is based on a review of dengue articles which were published from the beginning of the 1920s up to our days. Our objective is to recapitulate most of the epidemiological data available in order to reconsider the validity of ADE hypothesis. This work begins with a historical review of dengue epidemics in the world, leading to the history of the elaboration of the ADE hypothesis. It is then followed by questions concerning the validity of the ADE hypothesis.

Historical review of dengue epidemics

Dengue and other arboviruses with similar ecology have had widespread distribution in the tropics for over 250 years (Morens *et al.*, 1986; Monath, 1988). Suspected dengue epidemics occurred in Africa, in Europe, in America, and in Asia before the 18th century (Levaditi *et al.*, 1938).

During the 18th century, the first colonists in the French West Indies died from yellow fever. After and before these yellow fever epidemics arose, the successive epidemics of the "fiebre amarilla frustra" occurred. The clinical description of the "fiebre amarilla frustra" corresponded to dengue fever (Frederiksen, 1955). The first epidemics considered clinically due to dengue occurred in Philadelphia in 1780 (Morens *et al.*, 1986), in Cairo (Egypt) and in India in 1779/80. In 1787 and 1788, two other epidemics occurred in Spain (Levaditi *et al.*, 1938; Gubler, 1997). The origin of the word dengue as the name of an illness is currently attributed to the Swahili phrase "ka dinga pepo," thought to have crossed over from Africa to the Caribbean in 1827. In Cuba, this phrase was popularly identified with the Spanish word dengue and then after, this word was used in order to distinguish the acute and benign febrile illnesses between yellow fever (Rigau-Perez, 1998a).

During the 19th century, dengue reached Asia, the Caribbean, India, South America, and Africa (Levaditi *et al.*, 1938; Joyeux, 1944; Morens, 1994). Dengue epidemics occurred in the Caribbean islands in 1824, 1826/28, 1850/54 and 1880 (Griffiths *et al.*, 1968; Spence *et al.*, 1969; Ventura and Hewitt, 1970). Between 1828 and 1850, dengue and yellow fever epidemics did not cease alternating in the main harbors located in the south of the United States of America, close to the Caribbean. In Goré island (Senegal), several dengue epidemics arose, after or before yellow fever epidemics, like in 1878 (Frederiksen, 1955).

In Asia, dengue was described for the first time in Malaysia in 1902 (Lee, 1994). At the beginning of the 20th century, the epidemiological observations reported that dengue was more frequent among Europeans than among the native inhabitants in the Asian region (Levaditi *et al.*, 1938; Joyeux, 1944; Halstead, 1974). Indeed, in the autochthonous population, dengue infection was very often asymptomatic among Philippine children, and the herd immunity level was high among adults (Halstead *et al.*, 1969a). Conversely, for example in the Manila area of Luzon in the 1920s and 1930s, it was well known that mostly all adult visitors from other countries caught classical dengue fever within a year or less, and if they remained longer they frequently had a second similar attack, suggesting that at least two immunologic types of virus were already present (Halstead *et al.*, 1969a; Hammon, 1973). When they arose, the attack rates of the dengue epidemics were very high, explaining the numerous reported cases and the relatively short duration of the epidemic (Levaditi *et al.*, 1938; Joyeux, 1944; Kuno, 1995). At that time, it was interesting to report that the autochthonous inhabitants in the mountains located in the North of Luzon were not immunized against dengue, certainly because the vector *Aedes* could not survive or could not be infected by the dengue virus because of the altitude (Hammon, 1973; Halstead, 1974).

Epidemics with fatal cases occurred in Queensland (Australia) in 1897 (Morens, 1994; Hober *et al.*, 1995) in Singapore in 1901 (George, 1987), in Cochin (China) in 1895/96 and in Indochina in 1907 (Morens, 1994), in the Caribbean from 1915 to 1918 (Rymzo *et al.*, 1976), in the South of the United States in Louisiana (30 000 dengue cases) and in Texas (600 000 cases), in Brazil in 1922 (Moreau *et al.*, 1973; Dietz *et al.*, 1990; Hober *et al.*, 1995; Reiter, 1996) and in Athens (Greece) in 1927/28 (Levaditi *et al.*, 1938; Morens, 1994; Hober *et al.*, 1995; Rosen, 1996).

The Greek epidemic which is known as the most severe, was the first one to be documented and retrospectively confirmed by laboratory diagnostic (Halstead and Papaevangelou, 1980b; Rosen, 1986a). The epidemic began at the end of autumn 1927, and about 20 000 cases occurred. In July 1928, at the revival of the epidemic, about 650 000 (Halstead and Papaevangelou, 1980b) to 1 000 000 (Rosen, 1986a) cases were reported; 1061 fatal cases were documented (Halstead and Papaevangelou, 1980b), which resulted in a case fatality rate equal to 0.08% (Joyeux, 1944).

After the beginning of the 20th century, activities related to World War II resulted in expanded geographic distribution of the dengue disease and in an increase of the distribution of *Aedes aegypti* in several regions of the world. The first dengue viruses, serotypes 1 and 2, were isolated by Sabin and Schlessinger in 1944 from soldiers who were infected in Calcutta (India), New Guinea, and Hawaii (Monath, 1988).

In the Pacific region, a large dengue epidemic probably due to serotype 1 occurred in 1930, particularly in the Fiji and Samoa islands (Gubler *et al.*, 1978; Kuno, 1995). Nosocomial transmission of dengue was reported during this outbreak (Kuno, 1995). A major regional pandemic of dengue-1 arose in most islands in 1942–45 (Morens *et al.*, 1987; Gubler, 1997). In Tahiti, the outbreak was explosive and serotype 1 was retrospectively identified as the cause of this epidemic. It was proven that dengue-2 had certainly circulated before 1944 in French Polynesia (Rosen *et al.*, 1954; Rosen, 1958a; Laigret *et al.*, 1967; Moreau *et al.*, 1973; Gubler *et al.*, 1978).

During the war, about 90 000 dengue cases were reported among the American soldiers including 4 fatal cases (Morens *et al.*, 1987). A dengue epidemic was also reported in Taiwan in 1942/43 (Kuo *et al.*, 1992). In Japan, a large outbreak occurred between 1942 and 1945. People stored water because they were afraid of the American air raids. Thousands of cases were reported. The only dengue vector at this time in Japan was *Aedes albopictus* (Hammon, 1973; Kuno, 1995).

The following epidemic occurred in Tahiti in 1964/65 due to a viral strain of serotype 3 coming from Asia where it caused a lot of hemorrhagic dengue cases. Fortunately, the epidemic was benign in the Pacific region (Laigret *et al.*, 1967; Gubler *et al.*, 1978). In 1971, a dengue-2 outbreak began in Tahiti and Fiji Islands and then reached the Pacific region. Severe dengue cases which were more frequently reported among adults were seen for the first time in the region (Reeves, 1972; Moreau *et al.*, 1973; Saugrain *et al.*, 1973; Gubler *et al.*, 1978).

Later, the dengue-1 epidemic in 1975/76 and the dengue-4 epidemic in 1979 occurred in French Polynesia without severe cases (Carme *et al.*, 1975; Parc *et al.*, 1981; Laudon *et al.*, 1992). In December 1988, a large dengue-1 outbreak followed by a dengue-3 epidemic in 1989 arose in Tahiti (Chungue *et al.*, 1992a; Mallet *et al.*, 1993). The dengue-3 epidemic was very severe with over 500 hospitalized cases and 13 deaths. Dengue hemorrhagic cases were observed among primary and secondary dengue infection (Chungue *et al.*, 1992b; Glaziou *et al.*, 1992; Laudon *et al.*, 1992). In 1996/97, a dengue-2 epidemic of lower severity in Tahiti occurred and then spread over the Pacific region (Chungue *et al.*, 1998; Deparis *et al.*, 1998a; Deparis *et al.*, 1998b). A dengue-1 outbreak arose in 2001 during which several deaths were reported in French Polynesia. In Australia, the province of North Queensland reported dengue outbreak in 1981/83, 1992/93 (dengue-2) and 1996/97 (Moloney *et al.*, 1998).

In Africa, a sylvatic cycle of dengue was discovered first in Senegal and then in the Côte d'Ivoire (Rodhain, 1991; Kuno, 1995; Dietz *et al.*, 1996; Rosen, 1996; Pinheiro *et al.*, 1998; Diallo *et al.*, 2003). In 1926/27, a large dengue epidemic has been reported in South Africa (Moreau *et al.*, 1973; Diallo *et al.*, 2003). An outbreak occurred in Burkina Faso in 1982 certainly due to a dengue virus coming from the Seychelles archipelago where a large epidemic arose before (Diallo *et al.*, 2003). The circulation of the four dengue serotypes was documented in Africa (Gubler *et al.*, 1986). Dengue seems more present in East Africa, in Somalia, in Sudan and in Djibouti where a dengue-2 epidemic was reported in 1991/92 (Gubler *et al.*, 1986; Jupp and Kemp, 1993; Rodier *et al.*, 1996).

In the Caribbean Islands and in America, dengue epidemics occurred in 1938/39 and 1945, the latter being the last epidemic reported in the United States. In Panama an outbreak of 1941/42 was due to a dengue-2 virus, but dengue-3 virus circulated in the region until 1953 (Rosen, 1974a). Successive epidemics occurred in the Caribbean region, in 1953 due to dengue-2, 1963/64 due to dengue-3, and 1969 due to a dengue-2 strain which circulated up until 1976 (Griffiths *et al.*, 1968; Spence *et al.*, 1969; Ventura and Hewitt, 1970; Rymzo *et al.*, 1976; Morens *et al.*, 1986). In 1977, an epidemic was reported in Porto Rico due to the dengue-3 viral strain already responsible of the 1963 outbreak in the Caribbean (Kuno, 1995). In the beginning of the 1970s, dengue epidemics due to dengue-1 and dengue-4 occurred in Brazil. It was likely that dengue became endemic in the Caribbean region because the seroprevalence level of the exposed population was often higher than 50% in several islands (Griffiths *et al.*, 1968; Spence *et al.*, 1969; Ventura and Hewitt, 1970; Rosen, 1974a; Rymzo *et al.*, 1976). In 1977, the arrival of the dengue-1 serotype caused large epidemic, as the arrival of the serotype 4 in 1981/82 (Henchal *et al.*, 1986; Morens *et al.*, 1986).

Whereas dengue cases had always been benign, the dengue-2 epidemic in Cuba in 1981 was particularly severe, with over 10 000 hospitalized cases and 150 deaths. After 1981, severe dengue cases were sporadically reported in El Salvador, Mexico, Nicaragua and Porto Rico between 1982 and 1988. Then, epidemics with severe cases arose in 1985 in Nicaragua (Harris *et al.*, 2000), in 1989/90 in Venezuela (Focks *et al.*, 1995), in Brazil in the state of Rio de Janeiro (Dietz *et al.*, 1990; Pinheiro and Chuit, 1998) and finally in 1991/92 in French Guyana (Reynes *et al.*, 1994; Pinheiro and Chuit, 1998). In 1994/95, dengue-3 was reintroduced to Nicaragua and Panama and then to Colombia and Mexico, followed by the occurrence of epidemics with severe dengue cases. Since then, severe dengue cases are regularly reported in Central and South America where the four dengue serotypes circulate (Reynes *et al.*, 1993; Reynes *et al.*, 1994; Division of disease prevention and control—PAHO, 1996; Pinheiro and Chuit, 1998).

In Asia, a sylvatic cycle of dengue was suspected since 1931 by Simmons in the Philippines, and this hypothesis was confirmed by Rudnick in the jungle of Malaysia in the 1950s and 1960s (Rosen *et al.*, 1954; Frederiksen, 1955; George, 1987; Rodhain, 1991; Kuno, 1995; Fauran, 1996; Rico-Hesse, 1997a; Pinheiro and Chuit, 1998; Wang *et al.*, 2000). Recently, several studies confirmed the sylvatic cycle of dengue in Indonesia and Sri Lanka (De Silva *et al.*, 1999; Wolfe *et al.*, 2001). In the years after World War II, epidemics of hemorrhagic fever had been documented among children all over southeast Asia. The first epidemic of dengue with hemorrhagic cases occurring in Asia was described in 1953 and 1956 in Manila, Philippines. Thus the hypothesis of a new disease (George, 1987; Lum *et al.*, 1993) was put forward until the discovery of the dengue-3 and dengue-4 serotypes (Hammon, 1973; Rosen and Gubler, 1974b; Lum *et al.*, 1993). Another dengue epidemic with hemorrhagic cases occurred in 1958 in Bangkok, Thailand (Hammon, 1973; Gubler, 1997). Subsequent to 1958, dengue hemorrhagic cases began to be observed in epidemics in several other urban areas in Southeast Asia including Singapore, Saigon, Penang, Kuala Lumpur and other smaller cities in Thailand, Malaya and Vietnam, apparently from extension through major ports and routes of travel affecting indigenous people (Hammon, 1973). Then in Calcutta, adults with primary or secondary dengue infections were reported to be heavily affected and for the first time after World War II Caucasians were involved (Johnson *et al.*, 1967; Hammon, 1973). In the 1960s, a study estimated the annual incidence rate among the American or European people living in Cambodia at 3% (Hahn and Chastel, 1970). Hammon highlighted that the host characteristics were rather similar in all places except in Bangkok where the modal age was lowest (about 3 years) compared to Singapore and Manila. In the middle of the 1960s, dengue infection became an important health problem in Southeast

Asia, but not in Africa and America. The disease was more often, but not exclusively, reported among children.

The beginning of the sequential infection theory

In 1964, during the WHO conference in Bangkok, the definitions of the dengue cases have been standardized in dengue fever (DF) cases and dengue hemorrhagic fever (DHF) cases. Among the DHF cases, two categories were distinguished, the DHF cases without shock and the DHF with shock, whereas numerous clinicians did not find that the shock was specific of dengue infection because it was often described in several other arboviruses infections (Bre, 1980; Rosen, 1986b; Martet *et al.*, 1990).

In 1965, scientists and physicians tried to explain the specific epidemiological setting of dengue in Thailand (Halstead and Yamarat, 1965b; Morens, 1994): (1) dengue hemorrhagic fever (DHF) seemed to be only a children's disease; (2) The average age of DHF was between 3 and 4 years.

Even if some DHF cases occurred during primary infections, the form of the age-specific hemorrhagic fever attack rate curves for Bangkok could only be explained if a second dengue infection caused hemorrhagic fever (Halstead and Yamarat, 1965b; Hammon, 1973).

Consequently, the secondary infection was interpreted as a risk factor of the occurrence of severe dengue cases. This epidemiological explanation is called the sequential infection theory. This last point was then supported by epidemiological observations made in Thailand in 1967 (Johnson *et al.*, 1967), which concluded that DHF is more frequent in secondary than in primary dengue infections. Most of the severe dengue cases occurred when dengue-2 was responsible of the secondary dengue infection, with 3 months to 5 years interval between primary and secondary infections (Halstead, 1970). The implication of an immunological mechanism was then suggested (Morens, 1994), and it seemed possible that severe dengue was a function of the sequence of dengue infection, in particular when the sequence of infection ended with serotype 2.

But, at the same time, this mechanism could not be applied in Cambodia, in the Caribbean or in South India where no severe dengue cases were reported despite the co-circulation of several dengue serotypes (Johnson *et al.*, 1967). Furthermore, this hypothesis could not explain the occurrence of fatal dengue cases among primary dengue infections as during the Ubon epidemic investigated in Thailand in 1964 (Halstead and Yamarat, 1965b; Hammon, 1973).

During the 1970s, a lot of publications explained the role of the sequential infection scheme supporting an immunological mechanism in order to explain the occurrence of the severe dengue cases. Between 1970 and 1977, *in vivo* experiments performed in monkeys (Marchette *et al.*, 1973) as well as a profusion of *in vitro* experiments were made and published which increased the credibility of the ADE hypothesis (Morens, 1994). Although extensively studied, the role of ADE in the DHF pathogenesis remained unverified *in vivo*. In some countries like in India (Broor *et al.*, 1997), in Africa (Digoutte *et al.*, 1980; Deubel and Drouet, 1997), or in Brazil (Nogueira *et al.*, 1995), the co-circulation of several dengue serotypes including dengue-2 were reported without DHF cases. Some physicians thought that ADE could be an Asian phenomenon which is not confirmed in other countries.

But in 1981 the Cuban dengue-2 epidemic appeared as an illustration of the sequential infection scheme and of the ADE mechanism. Four years after a first dengue-1 epidemic during which no DHF cases were reported, this dengue-2 epidemic broke out with over 10 000 hospitalized dengue cases, including 158 deaths (Bravo *et al.*, 1987). After the Cuban epidemic, in the 1980s, ADE was considered as the only explanation for the occurrence of severe cases. For certain scientists, DHF cases could occur only among children with secondary infection (Halstead, 1980a). In Southeast Asia, prospective studies showed that DHF arose in about 2% to 4% of the secondary infections (Halstead, 1982; Burke *et al.*, 1988; Thein *et al.*, 1997), whereas the DHF cases among the primary dengue infections were rare or nonexistent.

In the 1980s, many reports were published concluding that DHF occurred in secondary dengue cases after D1/D2, D3/D2 and D4/D2 sequences (Halstead, 1982; Sangkawibha *et al.*, 1984). In the regions in which the absence of DHF cases during secondary epidemics was reported, it was explained that epidemics occurred over 5 years after the first epidemic, and that the sequence of infection did not end by serotype 2 (Halstead, 1980a).

In the 1980s and 1990s, the Relative Risk (RR) of secondary versus primary severe dengue infection following the D1/D2, D3/D2 and D4/D2 sequences of dengue infection has been assessed respectively at 163, 157 and 57.5 (Sangkawibha *et al.*, 1984; Monath, 1994). However, many of the criteria that were used were not supported or were in conflict with others data which could question on their validity (Monath, 1988; Rosen, 1997; Deparis *et al.*, 1998b; Dechant *et al.*, 1999). Finally, in view of the number of published studies during the 1990s, it was accepted that all DSS cases were secondary cases with some exceptions (Halstead, 1993), and that DHF occurred notwithstanding the serotype responsible of the secondary infection (Rothman *et al.*, 1997; Rigau-Perez *et al.*, 1998b).

Last pieces of the puzzle

Why did DHF cases occur among children of less than 1 year and mainly during primary infection?

In 1988, the role of maternal antibodies was studied by *in vitro* experiments in order to explain the occurrence of DHF in babies (Kliks *et al.*, 1988), supporting the ADE hypothesis. We will show that *in vivo*, the ADE theory cannot explain the mechanism of the occurrence of severe dengue cases among newborn children.

Why do DHF cases not occur systematically during all sequential dengue epidemics?

As stated before, no DHF cases were reported during several dengue epidemics in endemic regions (Johnson *et al.*, 1967; Halstead, 1974; Anderson *et al.*, 1976). In America, before the 1981 dengue-2 epidemic in Cuba, dengue-2 serotype circulated but was not responsible for severe cases. During the 1995 dengue-2 epidemic in Peru, a prospective study conducted on all dengue cases in IQUITOS demonstrated the absence of severe dengue cases although more than half of the dengue cases were secondary (Watts *et al.*, 1999). This epidemic was due to dengue-2 American genotype. The severe Cuban epidemic was due to the introduction of a Southeast Asian genotype. The difference between Cuba and Peru was explained by the fact that the American genotype 2 was not virulent enough to induce severe disease even during secondary infections. By contrast, the Southeast Asian genotype in Cuba was more virulent and it was suggested that American dengue-2 genotype strains did not bear the necessary properties to cause severe disease (Leitmeyer *et al.*, 1999).

In Indonesia, only 11 out of 536 dengue cases were hospitalized among 1837 children despite the circulation of Asian strains of different serotypes during the epidemic of 1999 (Graham *et al.*, 1999). The scientists who agreed with the ADE hypothesis explained that the occurrence of DHF cases is related not only with the presence of enhancing antibodies acquired during a primary infection, but also with a virulent strain during the secondary infection. In other words, some strains were able to induce the ADE phenomenon and others were not.

Why did DHF occur among adults during certain epidemics?

Contrary to the situation in Thailand, reports of DHF among adults are frequent, especially in nonendemic regions (Bravo *et al.*, 1987; Kouri *et al.*, 1987; Diaz *et al.*, 1988; Huerre *et al.*, 1995; Deparis *et al.*, 1998a; Pinheiro and Chuit, 1998; Cunha *et al.*, 1999; Rigau-Perez *et al.*, 2001; Garcia-Rivera and Rigau-Perez, 2003). Generally, when the whole population is susceptible for dengue virus infection, several authors observed that dengue severity increased with the age of the patients and that children under 5 years often had dengue infections of lower severity (Kuberski *et al.*, 1977; Reed *et al.*, 1977; Gubler *et al.*, 1978; Fagbami *et al.*, 1988). Thus, during the great DHF/DSS epidemic in Venezuela in 1990, a higher lethality was observed among adults (Kuno, 1995; Pinheiro and Chuit, 1998).

Moreover, during the Cuban DEN-2 epidemic in 1981, all but three of the DHF/DSS cases occurred among children over 5 years (Morens *et al.*, 1986), and a study reported that 65% of the hospitalized DHF/DSS cases were over 5 years old (Kuno, 1995; Pinheiro and Chuit, 1998). In 1997, 16 years later and after the disappearance of the dengue-2 circulation (The first international conference on dengue, 2000), a second DEN-2 epidemic occurred, also caused by the DEN-2 virus identified in 1981. The lethality rate was 40 times higher than in 1981 and the fatal dengue cases occurred only among adults (Kouri *et al.*, 1998; Guzman *et al.*, 2000b; Valdes *et al.*, 2002).

How to explain the frequent primary DHF infections?

Rosen, who agreed in the beginning with the very attractive ADE hypothesis, carried out investigations in 1972 on the dengue-2 epidemic in Niue Island where half of the DHF cases were primary infections, including 12 deaths (Barnes and Rosen, 1974; Rosen, 1977; Rosen, 1986b). Since then, the data of severe primary cases have been reported but have rarely been published.

Scientists who agreed with the ADE hypothesis had never given an explanation on this fact, and the studies reporting severe primary dengue cases had been rarely referred to and often been forgotten. Considering these facts, it seems justified to ask if the link between ADE hypothesis and the severity of dengue is really so obvious.

To confirm ADE theory, there are three conditions:

1. The first condition is that ADE must be demonstrated *in vivo* in order to

prove that this phenomenon may play an important role in the occurrence of DHF cases.

2. The second condition is that DHF should only exceptionally occur in the course of a primary infection.
3. The third condition is that, as a result of the second condition, a constant proportion of secondary dengue cases be classified as severe dengue cases, whatever the epidemic or the country.

Analysis of the three conditions of validity of the ADE theory

First condition: has ADE ever been demonstrated *in vivo*?

It is obvious that the *in vitro* evidence for ADE could only be clinically relevant if it can be shown that immune enhancement of dengue virus replication also occurs in the intact host (Rosen, 1989a; Morens, 1994). But since many *in vitro* experiments often cited in favor of the ADE hypothesis were carried out with strains of dengue virus that had never been responsible for severe dengue cases in human beings, and also because there is a large number of inter-individual variations within the viral strains and the host (Morens *et al.*, 1991), the *in vivo* relevance of most *in vitro* experiments has to be critically discussed.

ADE and dengue virus in vivo

Two *in vivo* experiments have been performed on monkeys in the same laboratory in the 1970s (Marchette *et al.*, 1973; Halstead, 1979; Halstead, 1982; Rosen, 1989a). In the first study, monkeys with a secondary dengue infection (actively or passively acquired antibody to a heterologous dengue virus) had higher levels of viremia than if they had a primary infection. Higher viremia was only observed when monkeys had a secondary DEN-2 infection, not with the other serotypes (Halstead, 1982). The second experiment compared the level of viremia in two groups of five monkeys, according to the fact that they had received human sera with or without enhanced antibodies. Besides several methodological problems as discussed by Rosen (Rosen, 1977; Rosen, 1986b; Rosen, 1989a) and Monath (Monath, 1988), and the fact that these observations have never been duplicated (Monath, 1988), none of these monkeys developed a severe disease (Rosen, 1977; Halstead, 1982; Rosen, 1986b). Furthermore, the statistical analysis performed in these studies compared means by statistical sign tests, i.e. Chi-squared test (Halstead, 1979): "viremia titers in paired experiments were analyzed by the statistical sign test. Daily mean values of viremia in experimental and control animals were compared by the statistical sign test (Chi-squared test = 10.2)." We do not know exactly how the authors matched paired monkeys, thus it seems difficult to conclude that means of viremia were different. It would be necessary to use a more appropriate statistical analysis before concluding to such a difference, and if such a difference exists, we have to ask if a daily mean of viremia is really appropriate to describe and compare the evolution of the level of viremia over time and between the monkeys, because inter individual variations of viremia are very high among the monkeys, ranging from 1 to 1000 (Rosen *et al.*, 1985; Rosen, 1986b; Rosen, 1989a).

Several other studies were carried out concluding the absence of ADE mechanism. In Bangkok, a study of sequential dengue infections with the four dengue serotypes has been performed in 39 white-handed gibbons (Whitehead *et al.*, 1970). The authors found that viremia in secondary infection was lower than in primary and that a primary infection induced cross protective reaction. These results have been confirmed in vaccine studies performed in Rhesus monkeys and showing an evident cross protection after dengue sequential infection, particularly the cross protection acquired against a dengue-3 virus after a primary dengue-1 infection (Halstead *et al.*, 1973a; Halstead et Palumbo, 1973b). More interestingly in the 1930s, human plasma of dengue cases taken on the 3rd or 4th day of the disease, was injected to human volunteers including former dengue cases (Levaditi *et al.*, 1938; Joyeux, 1944; Hammon, 1973). The authors reported that the severity of dengue was lower among the population of the former dengue cases.

ADE and other flaviviruses

ADE has been observed *in vitro* during sequential flavivirus infections (Kuno *et al.*, 1993; Rosen *et al.*, 1989b; Halstead, 1982; Gould and Buckley, 1989; Gollins and Porterfield, 1984; Philpotts *et al.*, 1985; Fagbami *et al.*, 1988).

Cross-infection enhancement was proven *in vitro* with seven flaviviruses present in Africa: DEN-2, yellow fever, West Nile Virus, Wesselbron, Zika, Uganda S and Dakar bat viruses (Fagbami and Halstead, 1986; Fagbami *et al.*, 1987a; Fagbami *et al.*, 1988)

In particular, if ADE had an *in vivo* impact, these observations should imply that people in Africa or in South America, vaccinated or exposed to yellow fever would be exposed to high risk of DHF/DSS. Fortunately, it has not been the case. During vaccine studies, it has been concluded that there is no proof of the existence of the ADE phenomenon after the infection by a dengue-1 strain or a dengue-2 strain in persons who were previously immunized against yellow fever or tick-borne encephalitis viruses (Bancroft *et al.*, 1984; McKee *et al.*, 1987). In 1984, experimental vaccinations against dengue-2 virus in military persons previously vaccinated against yellow fever and in un-vaccinated military persons did not show any significant difference of the viremia levels in the two groups (Bancroft *et al.*, 1984).

Neutralizing antibodies against dengue-2 virus developed in 90% of the recipients with antibodies to yellow fever virus and 61% of those without (Bancroft *et al.*, 1984; Eckels *et al.*, 1985). There was no statistically significant difference in the frequency of complaints between dengue virus seroconverting volunteers with and without immunity of yellow fever and the authors concluded: "none of the mechanisms of ADE has been shown to apply to the PR-159/S-1 strain of dengue-2." This result has been recently confirmed *in vivo* using a hamster model (Xiao *et al.*, 2003). These data suggest that immunological cross-protection could provide a barrier for the spread of yellow fever and dengue (Halstead, 1969d; Monath, 1997; Wolfe *et al.*, 2001).

In 1973, an attenuated dengue-2 strain gave an efficient seroprotective level among vaccinated monkeys. This vaccine study concluded to the usefulness of a sequential scheme of vaccinations, a combination of yellow fever and JE vaccine, attenuated strains of Langat and dengue-2 viruses, because this scheme protected the vaccinated monkeys against 28 group B arboviruses (Price *et al.*, 1973).

In Asia, there was no report of the ADE phenomenon in sequential JEV and dengue virus infections. In a study, human anti-Japanese encephalitis sera failed to enhance dengue-2 infections *in vitro*. This result is corroborating the clinical findings of Putavana and co-workers who observed no case of dengue shock syndrome in humans sequentially infected with JEV and dengue virus (Putvana *et al.*, 1984).

ADE has been demonstrated *in vitro* for many other viruses of different families such as postikum virus (Fagbami *et al.*, 1987b), Ross River virus (Linn *et al.*, 1996), human respiratory syncitial virus (Gimenez *et al.*, 1996), equine infectious anemia virus (Raabe *et al.*, 1999), Ebola virus (Takeda *et al.*, 2001), coxsackie virus B4 (Hober *et al.*, 2001), TBE (Kreil and Eibl, 1997) and HIV (Takeda *et al.*, 1988; Takeda and Innis, 1990).

Concerning the potential risk of antibody-dependent enhancement in human HIV vaccine trials (Mascola *et al.*, 1993; Morens, 1994), a study reported that *in vitro* data are insufficient to determine the development of current HIV-1 vaccine candidates as there are several notable diseases for which *in vitro* ADE does not correlate with ADE *in vivo* (Morens, 1994). *In vivo* correlates of protection/enhancement are necessary to measure exactly the risk for ADE (Mascola *et al.*, 1993).

Finally, it is difficult to relate *in vitro* with *in vivo* experiments. It seems clear that the *in vivo* counterpart of ADE mechanism has never been demonstrated. The *in vivo* experiments rather show the absence of an unspecified role of the ADE in dengue pathogenesis.

ADE among infants of less than 1 year

The *in vitro* implication of maternal dengue antibodies on the occurrence of DHF/DSS in children less than 1 year old was demonstrated in 1988 (Kliks *et al.*, 1988). But once again these experiments confirmed the *in vitro* ADE phenomenon by using subneutralizing concentrations of dengue antibodies. Moreover, it seems that statistical analysis was open to criticism. Making a Fisher test, authors used a one tailed statistical comparison, although their study did not seem to be suitable for this approach. They would have used a two tailed comparison which, in this case, led to a nonsignificant difference. The second point is that the authors made 24 correlation tests in order to compare three groups of paired results. Without taking into account that they did not choose the best statistical tests, they found 4 significant results among the 24 correlation tests, but using a level of significance equal to 0.05. As these 24 correlation tests were not independent because they were both related to the same global comparison, it is necessary to use the Bonferroni correction: the significant level should have been equal to 0.0021 in order to keep a global alpha level equal to 0.05 (Armitage and Berry, 1994). Applying this statistical correction, there was no significant correlation.

In 1988, other experiments were carried out in Nigeria and Hawaii employing the same methodology. ADE was observed *in vitro* between dengue viruses, but also between dengue virus and West Nile virus (Fagbami *et al.*, 1988). Taking their results into account, the authors concluded that the risk of severe dengue epidemics was very high in Nigeria because several dengue virus serotypes circulated there. However, no severe dengue cases were reported since then.

In South America, in Africa, and in the Pacific region where the four dengue serotypes circulate for a long time, dengue infections were not particularly severe among children of less than 1 year (Fagbami *et al.*, 1988), as also demonstrated in the Philippines islands (Songco *et al.*, 1987).

Finally, how to explain the severity of dengue in newborn children?

Vertical transmission of dengue virus has already been reported (Poli *et al.*, 1991; Thaithmyanon *et al.*, 1994; Carles *et al.*, 2000; Rigau-Perez *et al.*, 2001). In French Polynesia, five cases of dengue virus transmission from mother to child at the end of pregnancy occurred during the 1989 dengue-3 outbreak. All of these children were classified as DHF, and among the five mothers, two had had a primary dengue infection (Poli *et al.*, 1991). In Thailand, a study reports vertical transmission from a mother who experienced severe DSS to her child who experienced a brief and uncomplicated dengue fever (Thaithmyanon *et al.*, 1994). During the dengue-1 epidemic in Puerto Rico in 1977, one case of fetal death was reported after dengue virus transmission from mother to child during the pregnancy (Morens *et al.*, 1986). In French Guiana, 38 cases of mothers infected by dengue fever throughout the three trimesters of pregnancy have been reported (Carles *et al.*, 2000). The consequences on the fetus were premature birth in eight cases, five *in utero* fetal deaths, four cases of acute fetal distress during labor and two cases of mother-to-child transmission. As in similar surveys set up in Cuba (Tudela Coloma *et al.*, 1980; Fernandez *et al.*, 1994), authors concluded that dengue fever during pregnancy is a serious risk of premature birth and fetal death and in case of infection close to term, there is a risk of hemorrhage for both the mother and the newborn.

The ADE theory cannot explain these clinical observations. A simple explanation of the peak of DHF in Thailand among children ranging between 6 months and 1 year old may be that in dengue, as in other infectious diseases such as measles, a peak of incidence of mild or severe cases occurs after the end of the protective effect of the maternal dengue antibodies.

ADE and higher viremia levels

Following the ADE hypothesis, viremia levels should be higher during secondary dengue infections. Studies that reported the magnitude of dengue viremia indicated variability in the amount of circulating dengue virus, according to the serotype and even to the different strains of the same serotype (Gubler *et al.*, 1979b; Gubler *et al.*, 1981a). Above all, none have given evidence of higher viremia in secondary infections compared to primary infections (Kuberski *et al.*, 1977; Gubler *et al.*, 1981b). On the contrary, several studies in Asia or in the Pacific region showed that viremia level was higher among the primary dengue infection cases than among the secondary dengue infection cases (Kuberski *et al.*, 1977; Eram *et al.*, 1979; Gubler *et al.*, 1979b; Gubler *et al.*, 1981a). These results paralleled the observations among monkeys (Whitehead *et al.*, 1970; Halstead *et al.*, 1973a). More recently, it has been shown that higher viremia levels were observed during DHF cases compared to DF cases but independently of the immune status of the dengue infection (Murgue *et al.*, 2000).

The escape mutants hypothesis

In 1977, a large dengue-1 epidemic occurred in Cuba, followed by a severe dengue-2 epidemic in 1981. Thirty years after a possible dengue-1 primary infection, a new severe dengue-2 epidemic occurred in Santiago de Cuba in 1997 due to the dengue-2 genotype III strain called Jamaica strain that was already responsible for the epidemic in 1981 (Kouri *et al.*, 1998; Guzman *et al.*, 2000b). In contrast to this observation, ADE hypothesis suggested that the immunologic enhancement mechanism was possible only in the 5 years following a primary dengue infection (Halstead, 1980a).

During the 1981 epidemic there would have been a high number of young infants with DEN-1 antibodies whereas in 1997 only individuals over 20 would have these antibodies. Later, the scientists who agreed with the ADE hypothesis suggested that during the 1977 DEN-1 epidemic, antibodies with neutralizing heterotypic activities were produced. These antibodies later interacted with dengue-2 viruses acquired in 1981, as well as but to a lesser extent because of the maturation affinity of dengue-1 antibodies over time, with dengue-2 viruses acquired in 1997. In these latest cases, the disease was mild or asymptomatic. Moreover, some dengue-2 viruses must have escaped the heterotypic antibodies, and this allowed their progeny to interact with enhancing antibodies (Guzman *et al.*, 2000a). Finally these heterotypic antibodies which do not prevent dengue infection serve to down-regulate the disease to mild illness or asymptomatic infection. A population of dengue-2 viruses that replicates in dengue-1 immune hosts escapes heterotypic neutralization. When inoculated into a new dengue-1 immune host, these viruses are free to interact with the more abundant infection-enhancing antibodies to produce severe diseases.

Yet, several authors highlighted that dengue-2 virus did not circulate between 1982 and 1996 in Cuba (Pinheiro and Chuit, 1998; The First International Conference on Dengue and Dengue Hemorrhagic Fever, 2000), preventing at the same time the escape mutant viruses to appear. Aside this first objection and following the hypothesis of escape mutants, the severity of the Jamaica strain should increase regularly over the time from 1981 to 1999, in the different countries of the Caribbean islands and South America where several dengue serotypes already circulated (Van der Stuyft *et al.*, 1998). Fortunately, this has not been the case, as we will show afterwards. The natural evolution of the Jamaica strain which circulated from Cuba in 1981 to South America demonstrated that the hypothesis of the escape mutants does not fit in the reality of dengue epidemiology (Anonymous, 1994; Reynes *et al.*, 1994; Da Cunha *et al.*, 1995; Nogueira *et al.*, 1995; Van der Stuyft *et al.*, 1998).

Thus, in answer to the first condition we may conclude that ADE has still not been demonstrated *in vivo*.

Second condition: are primary DHF cases absent or rare?

A lot of studies have reported fatal primary dengue cases in Asia (Halstead, 1965a; Halstead and Yamarat, 1965b; Halstead *et al.*, 1969c; Nimmannitya *et al.*, 1969; Hammon, 1973; Khai Ming *et al.*, 1974; Eram *et al.*, 1979; Li *et al.*, 1986; George, 1987; Kuo *et al.*, 1992; Lum *et al.*, 1996; Pancharoen and Thisyakorn, 2001), in Pacific region (Barnes and Rosen, 1974; Reed *et al.*, 1977; Gubler *et al.*, 1978), in the Caribbean islands and in South America (Pinheiro and Chuit, 1998) or in Greece (Rosen, 1986a). Numerous surveys reported severe primary dengue cases in Asia (Halstead, 1965a; Halstead and Yamarat, 1965b; Winter *et al.*, 1968; Halstead *et al.*, 1969a; Halstead *et al.*, 1969c; Nimmannitya *et al.*, 1969; Hammon, 1973; Eram *et al.*, 1979; Gubler *et al.*, 1979b; Nimmannitya, 1987a; Songco *et al.*, 1987; Hayes *et al.*, 1988; Hayes *et al.*, 1989; Qiu *et al.*, 1991; Rodhain, 1996a; Lum *et al.*, 1996; Capeding *et al.*, 1997; Rathavuth *et al.*, 1997; Richards *et al.*, 1997; Graham *et al.*, 1999), in the Pacific region (Barnes and Rosen, 1974; Gubler *et al.*, 1978; Rosen, 1986b; Poli *et al.*, 1991; Glaziou *et al.*, 1992), in south America (Reynes, 1996; Pinheiro and Chuit, 1998; Da Cunha *et al.*, 1999), in Greece in 1927/28 (Levaditi *et al.*, 1938; Rosen, 1986a) or among travelers coming back from the tropics (Morens *et al.*, 1987). It is likely that the number of studies reporting fatal or severe primary dengue cases is underestimated due mostly to the difficulty of publishing these data. We are going to illustrate our opinion with five examples.

First example: the Greek epidemic in 1927/28

Concerning the Greek epidemic, a study tried to demonstrate that the 1928 epidemic was due to a co-circulation of dengue-1 and dengue-2 strains, whereas it was difficult to identify the cause of the epidemic in 1927 (Halstead and Papaevangelou, 1980b). This hypothesis allowed a pathogenic role of sequential infection in the outbreak. But, following the observations of the epidemic in the 1930s (Levaditi *et al.*, 1938; Joyeux, 1944), it was reported that not any dengue case among the 20 000 cases reported in 1927 contracted the disease in 1928, and the authors concluded that they were protected. Rosen demonstrated finally that the circulation of a dengue-2 serotype occurred in Greece, but after 1928 (Rosen, 1986a). Thus, the 1927/28 dengue epidemic in Greece was only due to the serotype 1 and, consequently, the 1061 fatal dengue cases were all primary dengue cases. These data are in contradiction with the ADE theory.

Second example: the Ubon epidemic in Thailand

In 1964, an epidemic broke out in Ubon (Thailand); 94 children were hospitalized and eight deaths were reported, all due to primary dengue infection. A study conducted among 34 sera showed that 26 were primary dengue infections and dengue-1 virus was isolated in 12 sera (Halstead and Yamarat, 1965b; Hammon, 1973). Why should the age distribution of cases in Ubon be similar to that in Bangkok when the antibody responses of Ubon patients suggest the absence of previous dengue virus exposure?

Third example: the 1972 epidemic in Niue and the 1989 epidemic in Tahiti

Since 1972, an epidemic with primary fatal dengue cases was described in Niue by Rosen (Barnes and Rosen, 1974). In the South Pacific, dengue viruses from Asia were also responsible of severe epidemics. In Tahiti, during the severe 1989 dengue-3 epidemic, related to the introduction of a virus of Asian origin, 11% of the DHF cases were primary infections older than 1 year, including two fatal cases (Murgue *et al.*, 1999).

Fourth example: the 1994 epidemic in Saudi Arabia

In Saudi Arabia, dengue was for the first time confirmed during the 1994 epidemic and this was followed by a continual transmission of dengue virus. Between 1994 and 1999, dengue infection was confirmed by virus isolation in 207 out of 985 suspected cases, 90% occurred in 1994. Three dengue serotypes circulated: dengue-3 (13 isolates) in 1996 and 1997, dengue-2 (138 isolates) and dengue-1 (56 isolates) during the study period. All confirmed dengue patients were self limiting DF, except two fatal DHF cases among adults in 1994, both due to dengue-2 virus (Fakeeh and Zaki, 2001). These two cases were obviously primary infections and were due to a dengue-2 strain of African origin.

Fifth example: the 1998 Nicaraguan epidemic

In 1998 an epidemic involving both DEN-3 and DEN-2 viruses of Asian origin, occurred in Nicaragua. However, among the DHF cases, nine were primary infections representing 15% among the total number of primary dengue cases. Secondary infection was not significantly correlated with DHF/DSS (Harris *et al.*, 2000). Authors concluded that the epidemiology of dengue can differ according to region and viral serotype.

In answer to the second condition, it is obvious that the primary DHF cases are not rare, but rather underestimated and underreported.

Third condition: is a constant proportion of secondary dengue cases classified as severe dengue cases?

Seroepidemiological studies conducted in a global population and not only in hospitalized cases are able to demonstrate in Asia (Johnson *et al.*, 1967; Halstead *et al.*, 1969a; Hammon, 1973; Halstead, 1974; Anderson *et al.*, 1976; Hayes *et al.*, 1988; Chen *et al.*, 1996; Pande and Kabra, 1996; Broor *et al.*, 1997; Graham *et al.*, 1999), in South America, in the Caribbean islands (Griffiths *et al.*, 1968; Ventura and Hewitt, 1970; Rymzo *et al.*, 1976; Waterman *et al.*, 1985; Morens *et al.*, 1986; Herrera-Basto *et al.*, 1992; Halstead, 1993; Da Cunha *et al.*, 1995; Nogueira *et al.*, 1995; Dietz *et al.*, 1996; Division of disease prevention and control—PAHO, 1996; Hayes *et al.*, 1996; Ricco-Hesse *et al.*, 1997b; Van Der Stuyft *et al.*, 1998; Vasconcelos *et al.*, 1998; Da Cunha *et al.*, 1999), in Africa (Digoutte *et al.*, 1980; Gubler *et al.*, 1986; Hyams *et al.*, 1986; Boisier *et al.*, 1994; Rico-Hesse, 1997a), and in the Pacific region (Kuberski *et al.*, 1977; Reed *et al.*, 1977; Deparis *et al.*, 1998a; Deparis *et al.*, 1998b) that the secondary dengue cases may be as benign as the primary dengue cases.

It is interesting to point out the epidemiological setting in South America: Prior to 1981, dengue outbreaks were not accompanied by DHF in this region, despite of simultaneous or sequential transmission of more than one serotype (Anonymous, 1994; Armstrong and Ricco-Hesse, 2001). Dengue-2 virus from Southeast Asia was introduced to America, first to Cuba in 1981. Since then, this strain was responsible for DHF epidemics, contrary to the American dengue-2 genotype, as the dengue-2 variant of Puerto Rico present in the Caribbean since 1969 (Monath, 1994; Wang *et al.*, 2000; Armstrong and Ricco-Hesse, 2001; Rigau-Perez *et al.*, 2001).

Dengue-3 virus of Asian origin was then introduced in America in the 1990s and was responsible of a severe epidemic in Nicaragua and in Panama in 1994 (Division of disease prevention and control—PAHO, 1996; Fauran, 1996; Pinheiro and Chuit, 1998). This genotype spread further to Central and South America.

For scientists supporting the ADE theory, these observations led to the conclusions that DHF is always associated with secondary infections, but requires an infection with dengue viruses of Asian origin. This hypothesis might explain the absence of DHF cases among the numerous secondary infections in Peru in 1999, due to DEN-2 American genotype (Kochel *et al.*, 2002). Yet, it cannot explain that the same dengue-2 strain of Asian origin that was found in Cuba did not induce any DHF epidemic among the secondary dengue infections when it circulated in Bolivia, French Guyana, and in Brazil in the 1980s and 1990s (Anonymous, 1994; Reynes *et al.*, 1994; Da Cunha *et al.*, 1995; Nogueira *et al.*, 1995; Van Der Stuyft *et al.*, 1998).

In Brazil, a dengue-2 epidemic occurred in 1990/91 after a benign dengue-1 epidemic in 1986/87. The epidemiological setting paralleled the Cuban situation in 1981 (Pinheiro and Chuit, 1998). Contrary to Cuba, the morbidity and the lethality remained low, and in a megalopolis like Rio de Janeiro, only 462 DHF/DSS cases including eight deaths were reported. In 1990/91, a seroepidemiological survey was carried out in schoolchildren in Rio de Janeiro, during the epidemic (Da Cunha *et al.*, 1995). Among 450 blood samples, dengue virus infection was detected in 66% (297/450) and among them, secondary dengue infection occurred in 61% (181/297). Not any DHF/DSS case was reported and only asymptomatic or oligosymptomatic infections were detected in 56% of the studied population. But the circulating strain was the dengue-2 Jamaica strain coming from Cuba (Nogueira *et al.*, 1995).

The epidemic spread to other Brazilian states, and in 1994, in Fortaleza, State of Ceara, only 24 cases of DHF/DSS occurred including 11 fatal cases (Pinheiro and Chuit, 1998; Pontes *et al.*, 2000). During the dengue-2 epidemic in Belem in 1996/97, among 17 440 serum samples positive to dengue of whom primary and secondary serologic responses were found, but no hemorrhagic cases or deaths were reported in spite of the fact that since October 1997 dengue-2 had been isolated from patients with previous dengue-1 infection (Travassos Da Rosa *et al.*, 2000). In 2001/02, a large dengue-3 epidemic occurred in Rio de Janeiro due to a viral strain responsible in Asia for severe dengue cases. But beside the sequence of infection dengue-2 (present since 1990) and then dengue-3 in 2001, the DHF/DSS cases are rare.

The Jamaica strain also circulated in French Guyana during the 1986 dengue epidemic. At this time, no case of DHF occurred despite the co-circulation of other dengue serotypes (Reynes *et al.*, 1994; Fouque *et al.*, 1995; Pinheiro and Chuit, 1998). In 1992, the first epidemic of DHF cases occurred in French Guyana. Yet, this epidemic was due to the Jamaica strain highlighting that the presence of dengue virus strains of Asian origin may be a necessity but not a sufficient condition for a dengue outbreak with hemorrhagic cases (Reynes *et al.*, 1994; Pinheiro and Chuit, 1998).

The same phenomenon was reported in the Pacific region where the same strain induced severe dengue cases in Tahiti and in Niue and no severe cases in Tonga and Samoa archipelagoes (Rosen, 1986b).

Referring to the third condition, it is obvious that secondary or multiple dengue infections are not systematically fatal or severe.

The why and the how of a dogma

Even if the majority of articles published in the main medical journals since the 1970s support the ADE theory, the facts do not. The sequential infection scheme may be a dramatic epidemiological misleading which was maintained involuntarily by the scientific press (Seglen, 1997). It is notable that in all dengue reviews published recently, the data documenting DHF cases in primary infections or the absence of DHF in secondary epidemics as well as data demonstrating the absence of any correlation between factors related to severity and secondary infections are seldom referred.

The bias of the sequential infection and ADE theory: the influence of the age of the exposed population

For the past 30 years, the pathogenesis of dengue has been the subject of scientific arguments with important consequences. Still, the ADE hypothesis remains the only official reference for the World Health Organization, considering the necessity of a tetravalent dengue vaccine because of ADE.

However, despite the numerous publications devoted to this subject, the proof of the *in vivo* relevance of ADE in dengue virus infection remains to be produced (Bielefeldt-Ohmann, 1997). As discussed previously, ADE theory might explain the fact that severe cases in Asia are mostly secondary. In addition, the severe dengue-2 epidemic in Cuba in 1981convinced lots of scientists of the validity of the ADE theory.

However, the Cuban medical literature points out some disturbing observations (Sanchez Velga *et al.*, 1979; Estrada Gonzalez *et al.*, 1981; Estrada Gonzalez, 1983; Senado Dumoy *et al.*, 1984; Martinez *et al.*, 1993; Palacios Serrano *et al.*, 2001): Cuban physicians reported 54 cases of DHF and 63 cases of DSS among 250 dengue cases hospitalized during the dengue-2 epidemic in 1981. Among these cases, 112 were secondary and 138 were primary dengue infections. They concluded that in their clinical experience, secondary dengue infection has never been a risk factor of DHF/DSS (Senado Dumoy *et al.*, 1984). During the supposed benign epidemic of primary dengue-1 infections in 1977 in Cuba, several physicians noticed an epidemic of polyneuritis and polyradiculoneuritis (Sanchez Velga *et al.*, 1979; Estrada Gonzalez *et al.*, 1981; Estrada Gonzalez, 1983). These reports are for us sufficient to maintain the doubt about the validity of the ADE theory even during the 1977 and the 1981 dengue epidemics in Cuba.

Concerning the studies conducted in Asia and supporting the sequential infection scheme, as Monath (Monath, 1988) and Rosen (Rosen, 1977; Rosen, 1989a) wrote, these studies suffered from several epidemiological problems and have to take into account the age of the exposed population.

For example, in order to highlight the difficulty to conduct a reliable and representative epidemiological study, during the dengue epidemic in Puerto Rico in 1994–1995, 24 700 suspected dengue cases were reported. Among them, only 11 363 (46%) were tested. Among those tested, 5564 (49%) were positive, 1928 (17%) were negative and above all 3871 (34%) were indeterminate. Among the 2004 hospitalized suspected dengue cases, 139 were classified as DHF among them 45 positive, 8 negative and especially 52 indeterminate and 34 without diagnostic sample (Rigau-Perez *et al.*, 2001). Among the 936 dengue virus positive patients with detectable IgG, 173 (18.5%) were primary infections and 763 (81.5%) were secondary infections. The analysis of the results is not simple and we have to come to some hypothesis. If we supposed, according to the ADE theory, that there was not any primary DHF case, the maximum of secondary DHF cases is equal to 45. In total, there were about 9819 (= [24 700*0.49*0.815]−45) secondary DF cases. Thus, the maximum of the proportion of DHF/DSS cases among the secondary dengue cases was equal to 0.46% (45/9 819). We can conclude that the level of severity was low among the secondary dengue infection despite the high circulation of the dengue-2 strain Jamaica.

Studying the influence of the age structure of the susceptible population in Thailand, it was obvious that the transmission of dengue should be age dependent (Pongsumpun and Tang, 1987). Thus, before we conclude that severe dengue cases are more frequent in secondary or multiple dengue infections, it is necessary to adjust the incidence and prevalence of dengue infections to the age of the susceptible population (Halstead *et al.*, 1969b; George, 1987; Monath, 1988; Hayes *et al.*, 1996; Deparis *et al.*, 1998b). Age-stratified antibody prevalence studies show that most children in the urban centers of Southeast Asia have had at least one dengue infection by the age of 6 years (Russel *et al.*, 1968; Burke *et al.*, 1988). The problem of determining the importance of secondary infection as a risk factor for DHF/DSS is that there are no reliable data available in most endemic areas, as we showed with the Porto Rican study, and laboratory techniques required to determine whether a patient is experiencing a secondary or tertiary infection are not generally available in most laboratories (Kuno *et al.*, 1993; De Souza *et al.*, 2004).

To confuse the issue even more, cases of primary DSS have been documented in older children and adults in most DHF endemic areas (Kuno *et al.*, 1993). Moreover, because of the lack of a second blood sample, certain primary dengue infections are not diagnosed leading to underestimated dengue primary incidence rates (Rosen, 1986b; Rosen, 1989a), or can be classified as secondary infections if the sole serum sample has a very high level of IgM antibodies (Rosen, 1977; Rosen, 1986b; Rosen, 1989a; Barnes and Rosen, 1974; Souza *et al.*, 2004).

Some other difficulties can occur. In Thailand, studies conducted among hospitalized cases suffered from bias. Indeed, DF in hospitalized infants (0–2 years old) who experience more often primary dengue than older children, was more common because older children who have less severe dengue were less commonly admitted (Pancharoen and Thisyakorn, 2001). Another problem is due to the fact that dengue spread is characterized by the occurrence and the succession of several focuses limited in time and in place. The level of severity and the clinical aspects of the dengue cases vary according to the particular setting of each dengue focus (Halstead, 1965a; Halstead *et al.*, 1969b; Winter *et al.*, 1969; Kuno, 1995). But the calculations of relative risk made to prove the reality of the sequential infection scheme and the ADE theory did not take into account this phenomenon.

The assessment of the severity of the reported dengue cases

The real severity of a given risk factor of a given disease is usually assessed by its induced case fatality rates (CFR) and morbidity. Then, one of the major questions concerning the importance of the sequential infection scheme and the ADE theory should be: "Are the CFR and the level of morbidity among the secondary dengue infections higher than among the primary dengue infections?" The level of morbidity could be assessed, for example, through the proportion of dengue cases hospitalized in intensive care units.

Yet it is obvious that in epidemiological studies, it is rarely possible to determine the CFR or the level of morbidity of primary and secondary dengue cases. Taking into account the lethality and morbidity level of 408 hospitalized dengue cases in Tahiti from 1989 to 1996, we showed that the immunological dengue status, and thus, the ADE theory, was not related to dengue severity (Murgue *et al.*, 1999). We highlighted that some clinical dengue manifestations could not be reported as severe dengue cases using the actual WHO dengue classification because it is based only on the presence of hemorrhagic and shock manifestations. Indeed, in our study, among the dengue cases classified as DF, hepatitis, encephalitis, and opportunistic infections constituted as frequent and severe manifestations leading sometimes but not rarely to death or complications

(Murgue *et al.*, 1999). Yet, these manifestations could not be reported by the clinicians to the WHO because of the present dengue classification.

Under these conditions, it is difficult to evaluate the real gravity of dengue in the field, and even more difficult to assess the validity of the ADE hypothesis (Murgue *et al.*, 1999; Solomon *et al.*, 2000).

Propositions

First proposition: to reconsider the importance of viral virulence, viral load, the role of the host and the vector

Some former and recent results (Rosen and Gubler, 1974b; Gubler *et al.*, 1978; Deparis *et al.*, 1998b; Watts *et al.*, 1999; Murgue *et al.*, 2000; Corwin *et al.*, 2001; Murgue *et al.*, 2001) emphasized the important role of the virulence of the dengue strains, the viral load and the host factor and its interactions in the dengue severity. It seems necessary to pay more attention to the virus itself, to the host–virus interaction and to vector-related factors in a specific environment which have not been extensively considered (Rodhain, 1991; Sehgal, 1997; Holmes, 1998). Some examples may illustrate the importance of the vector–host–virus interaction.

In 1952, Sabin showed that *Aedes* previously infected by dengue viruses cannot be infected by yellow fever virus (Frederiksen, 1955), highlighting the role of the possible interactions between the vector and the arboviruses circulating in a specific place.

Several studies have shown that dengue transmission takes place from person to person starting in houses where infected mosquitoes are found. Distribution of severe cases of dengue appears to follow the same transmission mode (Halstead, 1965a; Halstead, 1969b; Turell *et al.*, 1987; Kuno, 1995). In a given house, one mosquito carrying a virulent strain of the virus can cause several severe cases of dengue among the occupants of the house or among their neighbors, if the mosquito moves about.

Former studies explained that *Ae. aegypti*, a relatively poorly efficient vector, selected the viral dengue strains which produce a higher level of viremia in infected persons. These dengue strains are likely to be the most virulent: for a given dengue strain, the higher the level of viremia, i.e. the level of virulence, the higher the probability to enter a specific epidemiological setting (Gubler and Rosen, 1976; Gubler *et al.*, 1978; Gubler *et al.*, 1979a; Rosen *et al.*, 1985; Freier and Rosen, 1987; Rosen, 1987; Kuno, 1995). Vaccine studies confirmed this result showing that monkeys developed a poor level of viremia when infected by an attenuated dengue-2 strain and noting that no *Aedes* became infected when they sting monkeys, whereas a large proportion of mosquitoes became infected when they bit the monkeys who were infected by a nonattenuated dengue-2 strain (Price *et al.*, 1973). In central Java, a second dengue-3 epidemic occurred without severe cases after a first dengue-3 epidemic with severe dengue cases. Authors reported that the second epidemic was characterized by a low level of viremia among the patients (Gubler *et al.*, 1981a; Gubler, 1989). In Jamaica, during the dengue-3 epidemic in 1963 and 1968/69, the cases were all asymptomatic or benign. The level of viremia of the patients was so low that viral cultures were very difficult to be carried out, whereas a new dengue-2 strain introduced in 1968/69 was easy to isolate (Griffiths *et al.*, 1968; Spence *et al.*, 1969; Ventura and Hewitt, 1970; Kuno, 1995). Thereby, viral strains able to induce high level of viremia could be characterized by a high power of spread and virulence (Gubler and Rosen, 1976; Gubler *et al.*, 1978; Focks *et al.*, 1995; Kuno, 1995).

A pathogenesis scheme explaining dengue severity as the result of the interactions between the virulence of the viral strain, the viral load, the characteristics of the host, and the role of the vector has a great advantage, considering the sequential infection scheme and the ADE theory: it is more classical and more simple, and it may explain the different aspects of dengue epidemiology whatever the regions where dengue virus circulates. This new scheme releases directly a new strategy against dengue fever: the vaccination with four monovalent vaccines.

Second proposition: to reconsider the WHO dengue classification

The WHO classification considers that the shock syndrome is specific of the severe form of dengue disease and that the other severe manifestations of dengue (hepatitis, encephalitis, etc.) are rare or unusual (WHO, 1997).

But other arboviruses, including yellow fever viruses, are able to induce a shock syndrome (Bre, 1980; Rosen, 1986b), and we showed previously in this chapter that dengue viruses, as other arboviruses, are able to induce severe and fatal hepatitis or meningoencephalitis (Murgue *et al.*, 1999).

Thus, we want to underline the necessity to improve the criteria of dengue severity in the WHO classification in order to improve the adequacy of the clinical classification to the field reality.

For example, several authors reported a large epidemic of polyneuritis and polyradiculoneuritis, including Guillain-Barré syndrome, during the epidemic in 1977 (Sanchez Velga *et al.*, 1979; Estrada Gonzalez *et al.*, 1981; Estrada Gonzalez, 1983). The same report has been made during the dengue-2 epidemic in 1981. But, as dengue neurologic syndromes cannot be reported as severe cases, the 1977 dengue-1 epidemic in Cuba is supposed to be completely benign. The same way, in Asia, the dengue encephalitis cases are likely underestimated because they are often reported as JEV infection (Solomon *et al.*, 2000).

Third proposition: to consider a new "old dengue pathogenesis scheme"

As Hammon said in view of the Ubon epidemic (Halstead and Yamarat, 1965b, Hammon, 1973), other causal factors than immunologic sensitization should be sought to explain the occurrence of DHF cases in Southeast Asia in the 1950s. Dr. Rudnick had given an interesting explanation.

At first, he showed the evidence of a dengue sylvatic cycle. Simmons had already suspected a jungle dengue cycle in 1931 in the Philippines islands (Rosen *et al.*, 1954; Frederiksen, 1955). Rudnick isolated dengue virus from a sentinel monkey in a jungle area of Malaya and demonstrated a seroconversion for dengue among several similar animals suggesting the existence of a sylvatic dengue cycle (Rosen *et al.*, 1954; George, 1987; Rodhain, 1991; Kuno, 1995; Rico-Hesse, 1997a; Pinheiro and Chuit, 1998; Wang *et al.*, 2000). Another jungle cycle involving monkeys was later discovered in Africa (Rodhain, 1991; Kuno, 1995; Dietz *et al.*, 1996; Rosen, 1996; Diallo *et al.*, 2003). Recent publications confirm the presence of a sylvatic dengue cycle in Indonesia (Wolfe *et al.*, 2001) and in Sri Lanka (De Silva *et al.*, 1999). In South America, it has been demonstrated that several monkey species might be infected and had viremia without clinical signs (Rosen, 1958b, de Thoisy *et al.*, 2004).

Secondly, Dr. Rudnick suggested that through repeated passages between monkey and *Aedes* in the jungle, in particular *Ae. albopictus*, dengue virus variants have been selected which differ significantly in their pathogenicity for humans, with unusual hemorrhagic potential (George, 1987; Rodhain, 1991; Kuno, 1995; Wang *et al.*, 2000).

Thirdly, urbanization and the concomitant introduction of *Aedes aegypti*, whose populations, sometimes of considerable size, were inclined from the beginning of the 20th century to replace those of *Aedes albopictus* in the urban areas (Rodhain, 1991; Rosen, 1996; Rodhain, 1996b), resulted in the appearance of severe, and sometimes hemorrhagic, dengue in endemo-epidemic form in the cities whereas at this time dengue remained endemic in the rural areas (Kuno, 1995).

Finally, following World War II, the trade of cynomolgus monkeys could be the key factors of the brutal change of the epidemiologic situation of dengue in southeast Asia (Hammon, 1973; Rosen *et al.*, 1985). At this time, dengue was endemic in the rural areas where low virulent dengue strains circulated for a long time with *Ae. albopictus*, and a non-negligible proportion of the human population living in the cities were not immune against dengue viruses. At the beginning of the 1950s, a tremendous export trade for cynomolgus monkeys began to develop from the jungle of the Philippine Islands, Thailand, Malaysia, and other areas, and tens of thousands of these monkeys were soon funneled through the major ports of Manila, Bangkok, Singapore, and other cities for shipment to the United States and elsewhere for poliovirus vaccine production. While the cages were held on the wharf in the port of exportation, these monkeys with jungle-acquired dengue viremia were fed with urban *Aedes aegypti*, and the hemorrhagic epidemics resulted (Rosen *et al.*, 1985).

Since the jungle strains were present simultaneously with the larger number of classical strains, a mixed urban epidemic situation occurred. Epidemics of dengue hemorrhagic fever in the 1950s in southeast Asia started each time in large, metropolitan port cities, then spread along travel routes, but have been limited so far to previously recognized endemic dengue areas of southeast Asia because people are immunized against dengue, especially in the rural areas (Hammon, 1973; Kuno, 1995). It was like the spread of yellow fever epidemic in Africa (George, 1987). Gradually, the urbanization and the rural migration increased making dengue hemorrhagic fever endemic everywhere in Southeast Asia nowadays (Kuno, 1995).

This hypothesis could explain the early limited local distribution in the exact areas where epidemics were first discovered (Hammon, 1973). This scenario did not need the sequential scheme or the ADE hypothesis but it uses several notions that we want to justify. It is suggested by Rudnick that the passage of dengue viruses through *Ae. aegypti*, a new vector arrived in Southeast Asia, resulted in the modification of the virulence of the viruses. Is the vector able to select dengue viruses?

As we stated above, it is well known that *Ae. aegypti* is a relatively poor vector compared to *albopictus*, *pseudoscutellaris* and *polynesiensis*, and that it selected the viral dengue strains which produce the higher level of viremia in the infected persons (Gubler and Rosen, 1976; Gubler *et al.*, 1978; Gubler *et al.*, 1979a; Rodhain and Hannoun, 1979; Kuno, 1985; Rosen *et al.*, 1985; Freier and Rosen, 1987; Rosen,

1987; Rodhain, 1991; Monath, 1994; Rodhain, 1996b; Rosen, 1996; Wang *et al.*, 2000). This phenomenon of selection is also reported with the monkeys in the jungle cycle of yellow fever (Rodhain and Hannoun, 1979; Rodhain, 1991; Galat and Galat-Luong, 1997).

Recently, in South America, a study demonstrated that viral dengue-2 strains of the southeast Asian genotype tend to infect and disseminate more efficiently in field-derived mosquitoes than did variants of the American dengue-2 genotype. This could explain the large spread of dengue-2 severe epidemics in South America because Asian dengue strains are responsible of a large proportion of the reported dengue hemorrhagic fever cases in this continent (Armstrong and Rico-Hesse, 2001). The phenomenon of selection of the circulated viral strains by a vector has been also reported in the Pacific region. In Niue Island, where the dengue vector is *Aedes cooki*, the virulence of the dengue strain has been very high, leading to fatal cases whereas the same dengue strain had circulated in the same time with its usual vector *Aedes aegypti* in the other islands without severe dengue cases (Rosen *et al.*, 1985; Rosen, 1986b). In the same way, during a large epidemic in the region, dengue strains of reduced virulence had circulated in South Pacific Islands of Tonga resulting in less severe disease than in the other archipelagoes. The only difference was that the dengue vector in Tonga islands was *Ae. polynesiensis*, a relatively high susceptible vector whereas a less susceptible one, *Ae. aegypti*, was the main vector elsewhere (Kuberski *et al.*, 1977; Wang *et al.*, 2000).

Could the epidemics in Southeast Asia in the 1950s result from increased virulence of jungle dengue strains entering the urban areas?

A first experiment demonstrated that changes in virulence according to different vectors and/or host-passage histories are possible for the Crimean Congo hemorrhagic fever virus and others, and it is probable that vectors can induce phenotypic changes that modulate viral virulence without major genotypic changes (Gonzalez *et al.*, 1995). During another experiment, dengue viruses were successively passaged through mice in a brief period of time, making them more virulent and cause paralysis and death among most of the animals inoculated with late passages. This change in virulence is then a permanent characteristic of the viruses (Halstead, 1965a). During several epidemics, in Athens in 1927/28, Niue in 1972, Cuba in 1981, and Tahiti in 1996/97, the virulence of the circulating viral strains has been modified over time. This could explain the higher severity of the dengue cases occurring at the end of the epidemics (Levaditi *et al.*, 1938; Joyeux, 1944; Barnes and Rosen, 1974; Rosen, 1977; Rosen, 1986a; Rosen, 1986b; Bravo *et al.*, 1987; Kouri *et al.*, 1987; Deparis *et al.*, 1998a). The acceleration of the passages, vector to host and from host to vector during the epidemic, could explain the selection and then the modification of the virulence of the dengue strains (Rosen, 1977; Rosen, 1986b).

Finally, taking into account this new "old dengue pathogenesis scheme," could it explain the absence of yellow fever in Asia?

It was known that Asian *Aedes species* are vectors for yellow fever viruses because they can infect monkeys (Frederiksen, 1955). Sabin demonstrated that Rhesus monkeys were protected against a fatal yellow fever infection after having been previously infected by dengue virus

(Frederiksen, 1955). Then, the presence of antibodies to dengue virus among wild animals may have led to cross-reactive immunity preventing yellow fever from establishing a stable sylvatic cycle in Asia (Wolfe *et al.*, 2001). During the 19th century, it was noted that indentured laborers from India and the British troops who had served in India were less susceptible to contracting YF during the YF outbreaks (Monath, 1997). Until the end of the 19th century, yellow fever epidemics and dengue epidemics were never superposed on each other in time and in space (Frederiksen, 1955). Moreover, Sabin showed in 1952 that after a previous dengue infection, *Aedes* can not be infected by the yellow fever virus which weakens the probability of the circulation of the yellow fever virus in the Asian continent (Frederiksen, 1955). The high endemicity of dengue in Asia and a cross-reactive immunity between yellow fever and dengue combined with the low affinity of several African *Aedes* to dengue virus (Gubler *et al.*, 1979a) may explain the absence of yellow fever in Asia and the low level of endemicity of dengue in Africa (Monath, 1997; Wolfe *et al.*, 2001).

Conclusion

For 30 years the progress in dengue knowledge is prevented by the sequential infection scheme and the ADE hypothesis. The actual WHO classification of dengue severity is a perfect illustration and reflection of the lack of clinical and prognostic usefulness of the ADE theory.

In this chapter, we tried to show that a lot of evidence exists demonstrating that it is necessary to modify the ADE theory in order to progress in dengue fever knowledge. A lot of facts demonstrate that the epidemiology and pathogenicity of dengue virus is in many points comparable to yellow fever virus (Frederiksen, 1955; Rosen *et al.*, 1989b; Rodhain, 1991). It is time to carry out a global study in several countries in order to assess the validity of the sequential infection scheme and thus, the ADE theory. It is the role of international organizations like the WHO to order and manage this kind of epidemiological study.

We thank Bernadette Murgue and Léon Rosen for their help.

References

Acevedo, J., Casanova, M.F., Antonini, A.C., and Morales, H. (1982). Acute polyneuritis associated with dengue. Lancet. *828*, 1357.

Adebajo, A.O. (1996). Dengue arthritis. Br. J. Rheumatol. 35, 909–910.

Anderson, K.E., Joseph, S.W., Nasution, R., Sunoto, Butler, T., Van Peenen, P.F.D., Irving, G.S., Saroso, J.S., and Watten, R.H. (1976). Febrile illnesses resulting in hospital admission: a bacteriological and serological study in Jakarta, Indonesia. Am. J. Trop. Med. Hyg. 25, 116–121.

Anonymous. (1994). Dengue in the Americas. Weekly Epidemiological Report. 69, 177–184.

Armitage, P. and Berry, G. (1994). Statistical Methods in Medical Research. (Oxford: Blackwell science Ltd).

Armstrong, P.M., and Rico-Hesse, R. (2001). Differential susceptibility of *Aedes aegypti* to infection by the American and Southeast Asian Genotypes of dengue type 2 virus. Vector Borne Zoonotic Dis. *1*, 159–168.

Bancroft, W.H., Mc N Scott, R., Eckels, K.H., Hoke, C.H., Simms, T.E., Jesrani, K.D.T., Summers, P.L., Dubois, D.R., Tsoulos, D., and Russell, P.K. (1984). Dengue virus type 2 vaccine; reactogenicity and immunogenicity in soldiers. J. Infect. Dis. *149*, 1005–1010.

Barnes, W.J.S., and Rosen, L. (1974). Fatal hemorrhagic disease and shock associated with primary dengue infection on a Pacific Island. Am. J. Trop. Med. Hyg. *24*, 495–506

Baruah, H.C., Mohapatra, P.K., Kire, M., Pegu, D.K., and Mahanta, J. (1996). Hemorrhagic manifestations associated with dengue virus infection in Nagaland. J. Commun. Dis. *28*, 301–303

Bielefeldt-Ohmann, H. (1997). Pathogenesis of dengue virus diseases: missing pieces in the jigsaw. Trends Microbiol. 5, 409–412

Boisier, P., Morvan, J., Laventure, S., Charrier, N., Martin, E., Ouledi, A., and Roux, J. (1994).

Epidémie de dengue 1 sur l'île de la grande Comore (république fédérale islamique des Comores). Mars–mai (1993). Ann. Soc. Belge Méd. Trop. *74*, 217–229.

Bravo, J.R., Guzman, M.G., and Kouri, G.P. (1987). Why dengue hemorrhagic fever in Cuba? 1. Individual risk factors for dengue hemorrhagic fever/dengue shock syndrome. Trans. R. Soc. Trop. Med. Hyg. 81, 816–820

Bre, S P. (1980). Les arboviroses: un domaine où la santé publique et la recherche appellent une coopération internationale. Med. Trop. *40*, 485–491.

Broor, S., Dar, L., Sengupta, S., Chakraborty, M., Wali, J.P., Biswas, A., Kabra, S.K., Jain, Y., and Seth, P. (1997). Recent dengue epidemic in Delhi, India. In Factors in the emergence of arbovirus diseases, J.F. Saluzzo and B. Dodet, ed. (Paris, France: Elsevier), pp 123–126.

Burke, D.S., Nisalak, A., Johnson, D.E., and Mc N Scott, R. (1988). A prospective study of dengue infections in Bangkok. Am. J. Trop. Med. Hyg. 38, 172–180.

Capeding, M.R.R., Paladin, F.J.E., Miranda, E.G. and Navarro, X.R. (1997). Dengue surveillance in Metro Manila. Southeast Asian J. Trop. Med. Public Health. *28*, 530–534.

Carles, G., Talarmin, A., Peneau, C., and Bertsch, M. (2000). Dengue et grossesse. Etude de 38 cas en Guyane française. J. Gynecol. Obstet. Biol. Reprod. 29, 758–762.

Carme, B., Chamouard, J.F., and Delebecque, P. (1977). Lutte contre la dengue à Tahiti en 1975 (2ème partie). Med. Océanienne. *7*, 1–7.

Chen, W.J., Chen, S.L., Chien, L.J., Chen, C.C., King, C.C., Harn M.R., Hwang K.P., and Fang, J.H. (1996). Silent transmission of the dengue virus in southern Taiwan. Am. J. Trop. Med. Hyg. 55, 12–16.

Chungue, E., Marche, G., Plichart, R., Boutin, J.P., and Roux, J.F. (1989). Comparison of immunoglobulin G enzyme-linked immunosorbent assay (IgG-ELISA) and haemagglutination inhibition (HI) test for the detection of dengue antibodies. Prevalence of dengue IgG-ELISA antibodies in Tahiti. Trans. R. Soc. Trop. Med. Hyg. *83*, 708–711.

Chungue, E., Burucoa, C., Boutin, J.P., Phillippon, G., Laudon, F., Plichart, R., Barbazan, P., Cardines, R., and Roux J.F. (1992a). Dengue-1 epidemic in French Polynesia, 1988–1989: surveillance and clinical, epidemiological, virological and serological findings in 1752 documented clinical cases. Trans. R. Soc. Trop. Med. Hyg. 86, 193–197.

Chungue, E., Glaziou, P., Spiegel, A., Martin, P.M.V., and Roux, J.F. (1992b). Estimation of dengue infection attack rate in a cohort of children during a dengue 3 outbreak in Tahiti (1989–1990). Southeast Asian J. Trop. Med. Public Health. *23*, 157–158.

Chungue, E., Deparis, X., and Murgue, B. (1998). Dengue in French Polynesia: major features, surveillance, molecular epidemiology and current situation. Public Health Dialog. *5*, 154–162.

Corwin, A.L., Larasati, R.P., Bangs, M.J., Wuryadi, S., Arjoso, S., Sukri, N., Listyaningsih, E., Hartati, S., Namursa, R., Anwar, Z., Chandra, S., Loho, B., Ahmad, H., Campbell, R., and Porter, K.R. (2001). Epidemic dengue transmission in southern Sumatra, Indonesia. Trans. R. Soc. Trop. Med. Hyg. *95*, 257–265.

Da Cunha, R.V., Dias, M., Nogueira, R.M.R., Chagas, N., Miagostovich, M.P., and Schatzmayr, H.G. (1995). Secondary dengue infection in schoolchildren in a dengue endemic area in the state of Rio de Janeiro, Brazil. Rev. Inst. Med. Trop. 37, 517–521.

Da Cunha, R.V., Schatzmayr, H.G., Miagostovich, M.P., Barbosa, A.M.A., Paiva, F.G., Miranda, R.M.O., Ramos, C.C.F., Coelho, J.C.O., and Dos Santos, F.B. (1999). Dengue epidemic in the State of Rio de Janeiro Grande do Norte, Brazil, in (1997). Trans. R. Soc. Trop. Med. Hyg. *93*, 247–249

De Silva, A.M., Dittus, W.P.J., Amerasinghe P.H., and Amerasinghe F.P. (1999). Serologic evidence for an epizootic dengue virus infecting toque macaques (*Macaca sinica*) at Polonnaruwa, Sri Lanka. Am. J. Trop. Med. Hyg. *60*, 300–306.

De Souza, V.A., Fernandes, S., Araujo, E.S., Tateno, A.F., Oliveira, O.M.N.P.F., Oliveira, R., and Pannuti, C.S. (2004). Use of an immunoglobulin G avidity test to discriminate between primary and secondary dengue virus infections. J. Clin. Microbiol. 42, 1782–1784.

Dechant, E.J., Rigau-Perez, J.G., and the Puerto Rico association of epidemiologists. (1999). Hospitalizations for suspected dengue in Puerto Rico, 1991–1995: estimation by capture-recapture methods. Am. J. Trop. Med. Hyg. *61*, 574–578.

Del Valle Diaz, S., Pinera Martinez, M., and Guash Saent-Felix, F. (2001). Hepatitis reactiva por virus del dengue hemorragico. Rev. Cubana Med. Trop. 53, 28–31.

Deparis, X., Murgue, B., Roche, C., Cassar, O., and Chungue, E. (1998a). Changing clinical and biological manifestations of dengue during the dengue-2 epidemic in French Polynesia in 1996/97. Description and analysis in a prospective study. Trop. Med. Int. Health. *3*, 859–865.

Deparis, X., Roche, C., Murgue, B., and Chungue, E. (1998b). Possible dengue sequential infection: spread of the infection in a neighborhood during the 1996/97 epidemic in French Polynesia. Trop. Med. Int. Health. *3*, 866–871.

Deubel, V., and Drouet, M.T. (1997). Biological and molecular variations of yellow fever virus strains. In Factors in the emergence of arbovirus diseases, J.F. Saluzzo and B. Dodet, ed. (Paris, France: Elsevier), pp. 157–165.

Diallo, M., Ba, Y., Sall, A., Diop, O., Ndione, J., Mondo, M., Girault, L., and Mathiot, C. (2003). Amplification of the sylvatic cycle of dengue virus type 2, Senegal, 1999–2000: entomologic findings and epidemiologic considerations. Emerg. Infect. Dis. *9*, 362–367.

Diaz, A., Kouri, G., Guzman, M.G., Lobaina, L., Bravo, J., Ruiz, A., Ramos, A., and Martinez, R. (1988). Description of the clinical picture of dengue hemorrhagic fever/dengue shock syndrome in adults. Bull. Pan Am. Health Organ. *22*, 133–143.

Dietz, V., Gubler, D.J., Rigau-Perez, J.G., Pinheiro, F., Schatzmayr, H.G., Bailey, R., and Gunn, R.A. (1990). Epidemic dengue-1 in Brazil, 1986: evaluation of a clinically based dengue surveillance system. Am. J. Epidemiol. *131*, 693–701.

Dietz, V., Gubler D.J., Ortiz, S., Kuno, G., Casta-Velez, A., Sathier, G.E., Gomez, I., and Vergne, E. (1996). The 1986 dengue and dengue hemorrhagic fever epidemic in Puerto Rico: epidemiologic and clinical observations. Puerto Rico Health Sci. J. *15*, 201–210.

Digoutte, J.P., Salaun, J.J., Robin, Y., Bres, P., and Cagnard, V.J.M. (1980). Les arboviroses mineures en Afrique centrale et occidentale. Med. Trop. *40*, 523–533.

Division of disease prevention and control, communicable diseases program, HCP/HCT, PAHO. (1996). Dengue and dengue hemorrhagic fever (1996). Epidemiological bulletin/ PAHO. *17*, 12–4.

Eckels, K.H., Kliks, S.C., Dubois, D.R., Wahl, L.M., and Bancroft W.H. (1985). The association of enhancing antibodies with seroconversion in humans receiving a dengue-2 live-virus vaccine. J. Immunol. *135*, 4201–4203.

Eram, S., Setyabudi, Y., Sadono, I., Sutrisno, S., Gubler, D.J., and Sulianti Saroso, J. (1979). Epidemic dengue hemorrhagic fever in rural Indonesia. II. Clinical studies. Am. J. Trop. Med. Hyg. *28*, 711–716.

Estrada Gonzalez, R., Goyenechea, A, and Herrera, C. (1981). Brote de polirradiculoneurtis aguda tipo Landry-Guillain-Barre-Strohl durante una epidemia de dengue. Rev. Cub. Hig. Epid. *19*, 252–265.

Estrada Gonzalez, R. (1983). Sobre los syndromes neurologicos que han ocurrido durante nuestra dos recientes epidemias por virus dengue y sus posibles interrelaciones. Rev. Cub. Hig. Epid. *21*, 105–113.

Fagbami, A.H., and Halstead, S.B. (1986). Antibody-mediated enhancement of Wesselbron virus in P388D1 cells. Afr. J. Med. Med. Sci.*15*, 103–107.

Fagbami, A.H., Halstead, S.B., Marchette N.J., and Larsen K. (1987a). Cross-infection enhancement among African flaviviruses by immune ascitic fluid. Cytobios. *49*, 49–55.

Fagbami, A.H., Halstead, S.B., Marchette, N., and Larsen, K. (1987b). Potiskum virus: enhancement of replication in a macrophage-like cell line. Acta Virol. *31*, 463–467.

Fagbami, A.H., Halstead, S.B., Marchette N.J., and Larsen K. (1988). Heterologous flavivirus infection-enhancing antibodies in sera of Nigerians. Am. J. Trop. Med. Hyg. *38*, 205–207.

Fakeeh, M., and Zaki, A.M. (2001). Virologic and serologic surveillance for dengue fever in Jeddah, Saudi Arabia, 1994–1999. Am. J. Trop. Med. Hyg. *65*, 764–7.

Familusi, J.B., Moore, D.L., Fomufod, A.K., and Causey O.R. (1972). Virus isolates from children with febrile convulsions in Nigeria. Clin. Ped. *11*, 272–276.

Fauran, P. (1996). Prévision et prévention des épidémies de dengue. Bull. Soc. Path. Ex. *89*, 123–127

Fernandez, R., Rodriguez, T., Borbonet, F., Vazquez, S., Guzman M.G., and Kouri, G. (1994). Estudio de la relacion dengue-embarazo en un grupo de madres cubanas. Rev. Cubana Med. Trop. *46*, 76–78.

Focks, D.A., Daniels, E., Haile, D.G., and Keesling, J.E. (1995). A simulation model of the epidemiology of urban dengue fever: literature analysis, model development, preliminary validation, and samples of simulations results. Am. J. Trop. Med. Hyg. *53*, 489–506.

Fouque, F., Reynes, J.M., and Moreau, J.P. (1995). Dengue in French Guyana, 1965–1993. Bull. Pan Am. Health Organ. *29*, 147–155.

Frederiksen, H. (1955). Historical evidence for interference between dengue and yellow fever. Am. J. Trop. Med. Hyg. *4*, 483–491.

Freier, J.E., and Rosen, L. (1987). Vertical transmission of dengue viruses by mosquitoes of the *Aedes scutellaris* group. Am. J. Trop. Med. Hyg. 37, 640–647.

Galat, G., and Galat-Luong, A. (1997). Circulation des virus en milieu tropical, socio-écologie des primates et équilibre des écosystèmes. Cahiers Santé. 7, 81–87.

Garcia Arotzarene, J., Corteguera, M.E., Gonzalez Valdes, J., and Riveron Corteguera, R. (1988). Manifestaciones pleuropulmonares en el curso de la fiebre hemorragica del dengue: estudio de 212 pacientes. Rev. Cubana Pediatr. *60*, 375–396.

Garcia-Rivera E.J., and Rigau-Perez JG. (2003). Dengue severity in the elderly in Puerto Rico. Pan. Am. J. Public Health. *13*, 362–368.

George, R. (1987). Dengue hemorrhagic fever in Malaysia: a review. Southeast Asian J. Trop. Med. Public Health. 18, 278–283.

George, R., Liam, C.K., Chua, C.T., Lam, S.K., Pang, T, Geethan, R., and Foo L.S. (1988). Unusual clinical manifestations of dengue virus infection. Southeast Asian J. Trop. Med. Public Health. *19*, 585–590.

Gimenez, H.B., Chisholm, S., Dornan, J., and Cash P. (1996). Neutralizing and enhancing activities of human respiratory syncytial virus-specific antibodies. Clin. Diagn. Lab. Immunol. 3, 280–286.

Glaziou, P., Chungue, E., Gestas, P., Soulignac, O., Couter, J.P., Plichart, R., Roux, J.F., and Poli, L. (1992). Dengue fever and dengue shock syndrome in French Polynesia. Southeast Asian J. Trop. Med. Public Health. 23, 531–532.

Gollins, S.W., and Porterfield, J.S. (1984). Flavivirus infection enhancement in Macrophages: radioactive and biological studies on the effect of antibody on viral fate. J. Gen. Virol. 65, 1261–1272.

Gonzalez, J.P., Wilson, M.L., Cornet, J.P., and Camicas, J.L. (1995). Host-passage-induced phenotypic changes in Crimean-Congo hemorrhagic fever virus. Res. Virol. *146*, 131–140.

Gould, E.A., and Buckley, A. (1989). Antibody-dependent enhancement of yellow fever and Japanese encephalitis virus neurovirulence. J. Gen. Virol. *70*, 1605–1608.

Graham, R.R., Juffrie, M., Tan, R., Hayes, C.G., Laksono, I., Ma'Roef, C., Sutaryo, E., Porter, K.R., and Halstead, S.B. (1999). A prospective seroepidemiologic study on dengue in children four to nine years of age in Yogyakarta, Indonesia I. Studies in 1995–1996. Am. J. Trop. Med. Hyg. *61*, 412–419

Griffiths, B.B., Grant, L.S., Minott, O.D., and Belle, E.A. (1968). An epidemic of dengue-like illness in Jamaica-1963. Am. J. Trop. Med. Hyg. *17*, 584–589.

Gubler, D.J., and Rosen, L. (1976). Variation among geographic strains of *Aedes albopictus* in susceptibility to infection with dengue viruses. Am. J. Trop. Med. Hyg. *25*, 318–325.

Gubler, D.J., Reed, D., Rosen, L., and Hitchcock, J.C. Jr. (1978). Epidemiologic, clinical and virologic observations on dengue in the kingdom of Tonga. Am Trop. Med. Hyg. *27*, 581–9.

Gubler, D.J., Nalim, S., Tan, R., Saipan, H., and Sulianti Saroso, J. (1979a). Variation in susceptibility to oral infection with dengue viruses among geographic strains of *Aedes aegypti*. Am. J. Trop. Med. Hyg. *28*, 1045–1052.

Gubler, D.J., Suharyono, W., Lubis, I., Eram, S., and Sulianti Saroso, J. (1979b). Epidemic dengue hemorrhagic fever in rural Indonesia. I. Virological and epidemiological studies. Am. J. Trop. Med. Hyg. *28*, 701–710.

Gubler, D.J., Suharyono, W., Lubis, I., Eram, S., and Gunarso S. (1981a). Epidemic dengue-3 in central Java associated with low viremia in man. Am. J. Trop. Med. Hyg. *30*, 1094–1099.

Gubler, D.J., Suharyono, W., Tan, R., Abidin, M., and Sie, A. (1981b). Viremia in patients with naturally acquired dengue infection. Bull. World Health Organ. *59*, 623–630

Gubler, D.J., Sather, G.E., Kuno, G., and Cabral J.R.(1986). Dengue 3 virus transmission in Africa. Am. J. Trop. Med. Hyg. *35*, 1280–1284.

Gubler, D.J. (1989). Surveillance for dengue and dengue hemorrhagic fever. Bull. Pan Am. Health Organ. 23, 397–404.

Gubler, D.J. (1997). The emergence of dengue/dengue hemorrhagic fever as a global public health problem. In Factors in the emergence of arbovirus diseases, J.F. Saluzzo and B. Dodet, ed. (Paris, France: Elsevier), pp. 83–90

Guzman, M.G., Kouri, G., Morier, L., Soler, M., and Fernandez A. (1984) A study of fatal hemorrhagic dengue cases in CUBA (1981). Bull. Pan Am. Health Organ. *18*, 213–220.

Guzman, M.G., Kouri, G., and Halstead, S.B. (2000a). Do escape mutants explain rapid increases in dengue case-fatality rates within epidemics? Lancet. 355, 1902–1903

Guzman, M.G., Kouri, G., Valdes, L., Bravo J., Alvarez, M., Vazques, S., Delgado, I., and Halstead, S.B. (2000b). Epidemiologic studies on dengue in Santiago de Cuba (1997). Am. J. Epidemiol. *152*, 793–799.

Hahn, H., and Chastel, C. (1970). Dengue in CAMBODIA in (1963). Nineteen nine laboratory-proved cases. Am. J. Trop. Med. Hyg. *19*, 106–9.

Halstead, S.B. (1965a). Dengue and hemorrhagic fevers of Southeast Asia. Yale J. Biol. Med. *37*, 434–454.

Halstead, S.B., and Yamarat, C. (1965b). Recent epidemics of hemorrhagic fever in Thailand. A.J.P.H. *55*, 1386–1395.

Halstead, S.B., Nimmannitya, S., and Margiotta, M.R. (1969a). Dengue and Chikungunya virus infection in man in Thailand, 1962–1964. Part II: observations on disease in outpatients. Am. J. Trop. Med. Hyg. *18*, 972–983.

Halstead, S.B., Scanlon J.E., Umpaivit, P., and Udomsakdi, S. (1969b). Dengue and Chikungunya virus infection in man in Thailand, 1962–1964. Part IV: epidemiologic studies in the Bangkok metropolitan area. Am. J. Trop. Med. Hyg. *18*, 997–1021.

Halstead, S.B., Udomsakdi, S., Scanlon J.E., and Rohitayodhin, S. (1969c). Dengue and Chikungunya virus infection in man in Thailand, 1962–1964. Part V: epidemiologic observations outside Bangkok. Am. J. Trop. Med. Hyg. *18*, 1022–1033.

Halstead, S.B., Udomsakdi, S., Singharaj, P., and Nisalak, A. (1969d). Dengue and Chikungunya virus infection in man in Thailand, 1962–1964. Part III: clinical, epidemiologic, and virologic observations on disease in non-indigenous white persons. Am. J. Trop. Med. Hyg. *18*, 984–996.

Halstead, S.B. (1970). Observations related to pathogenesis of dengue hemorrhagic fever. Yale J. Biol. Med. *42*, 350–362.

Halstead, S.B., Casals, J., Shotwell, H., and Palumbo, N. (1973a). Studies on the immunization of monkeys against dengue. I. Protection derived from single and sequential virus infection. Am. J. Trop. Med. Hyg. *22*, 365–374.

Halstead, S.B., and Palumbo, N. (1973b). Studies on the immunization of monkeys against dengue. II. Protection following inoculation of combinations of viruses. Am. J. Trop. Med. Hyg. *22*, 375–81.

Halstead, S.B. (1974). Etiologies of the experimental dengue of Siler and Simmons. Am. J. Trop. Med. Hyg. *23*, 974–982.

Halstead, S.B. (1979). *In vivo* enhancement of dengue virus infection in rhesus monkeys by passively transferred antibody. J Infect. Dis. *140*, 527–533.

Halstead, S.B. (1980a). Dengue hémorragique—problème de santé publique et domaine de recherche. Bull OMS. *58*, 375–397.

Halstead, S.B., and Papaevangelou, G. (1980b). Transmission of dengue 1 and 2 viruses in Greece in (1928). Am. J. Trop. Med. Hyg. *29*, 635–637.

Halstead, S.B., Porterfield, J.S., and O'Rourke, E.J. (1980c). Enhancement of dengue virus infection in monocytes by flavivirus antisera. Am. J. Trop. Med. Hyg. *29*, 638–642.

Halstead, S.B. (1982). Immune Enhancement of viral Infection. Prog. Allergy. *31*, 301–364.

Halstead, S.B. (1993). Global epidemiology of dengue: health systems in disarray. Trop. Med. *35*, 137–146.

Halstead, S.B., Streit, T.G., Lafontant, J.G., Putvatana, R., Russell, K, Sun W., Kanesa-Thasan, N., Hayes C.G., and Watts, D.M. (2001). Haiti: absence of dengue hemorrhagic fever despite hyperendemic dengue virus transmission. Am. J. Trop. Med. Hyg. *65*, 180–183.

Hammon, W.M.D. (1973). Dengue hemorrhagic fever do we know its cause? Am. J. Trop. Med. Hyg. *22*, 82–91.

Harris, E., Videa, E., Perez, L., Sandoval, E., Tellez, Y., De Los Angeles Perez, M., Cuadra, R., Rocha, J., Idiaquez, W., Alonso, R., Delgado, M., Campo, L., Acevedo, F., Gonzalez, A., Aamador, J., and Balmaseda, A. (2000). Clinical, epidemiologic, and virologic features of dengue in the 1998 epidemic in Nicaragua. Am. J. Trop. Med. Hyg. *63*, 5–11.

Hayes, C.G., Manaloto, C.R., Gonzales, A., and Ranoa, C.P. (1988). Dengue infections in the Philippines: Clinical and virological findings in 517 hospitalized patients. Am. J. Trop. Med. Hyg. *39*, 110–115.

Hayes, C.G., O'Rourke, T.F., Fogelman, V., Leavengood, D.D., Crow, G., and Albersmeyer, M.M. (1989). Dengue fever in American military personnel in the Philippines: clinical observations on hospitalized patients during a 1984 epidemic. Southeast Asian J. Trop. Med. Public Health. *20*, 1–8.

Hayes, C.G., Philips, I.A., Callahan, J.D., Griebenow, W.F., Hyams, C., Wu, S.J., and Watts, D.M. (1996). The epidemiology of dengue virus infection among urban, jungle, and rural populations in the Amazon region of Peru. Am. J. Trop. Med. Hyg. *55*, 459–463.

Henchal, E.A., Repik, P.M., McCown, J.M., and Brandt, W.E. (1986). Identification of an antigenic and genetic variant of dengue-4 virus from the Caribbean. Am. J. Trop. Med. Hyg. *35*, 393–400.

Herrera-Basto, E., Prevots, D.R., Zarate, M.L., Silva, J.L., and Sepulveda-Amor, J. (1992). First reported outbreak of classical dengue fever at 1700 meters above sea level in Guerrero state, Mexico, June (1988). Am. J. Trop. Med. Hyg. *46*, 649–53.

Hober, D., Roulin, G., Deubel, V., and Wattre, P. (1995). La dengue : une maladie virale en pleine expansion. Méd. Mal. Infect. *25*, 888–895.

Hober, D., Chehadeh, W., Bouzidi, A., and Wattre, P. (2001). Antibody-dependent enhancement of coxsackievirus B4 infectivity of human peripheral blood mononuclear cells results in increased interferon-alpha synthesis. J. Infect. Dis. *184*, 1098–1108.

Holmes, E.C. (1998). Molecular epidemiology of dengue virus—the time for big science. Trop. Med. Int. Health. *3*, 855–856.

Hommel, D., Talarmin, A., Deubel, V., Reynes, J.M., Drouet, M.T., Sarthou, J.L., and Hulin, A. (1998). Dengue encephalitis in French Guiana. Res. Virol. *149*, 235–238.

Huerre, M., Camprasse, M.A., and Laille, M. (1995). Problèmes posés par les épidémies de dengue. L'exemple des foyers d'Océanie. Situation mondiale actuelle et dans les DOM-TOM. Méd. Mal. Infect. *25*, 688–695.

Hyams, K.C., Oldfield, E.C., McNair, S.R., Bourgeois, A.L., Gardiner, H., Pazzaglia, G., Moussa, M., Saleh, A.S., Dawi, O.E., and Daniell, F.D. (1986). Evaluation of febrile patients in Port Sudan, SUDAN: isolation of dengue virus. Am. J. Trop. Med. Hyg. *35*, 860–865.

Innis, B.L., Myint, K.S.A., Nisalak, A., Ishak, K.G., Nimmannitya, S., Laohapand, T., Tanprasertsuk, S., Pongritsakda, V., and Thisyakorn, U. (1990). Acute liver failure is one important cause of fatal dengue infection. Southeast Asian J Trop. Med. Public Health. *21*, 695–696.

Isarangkura, P.B., Pongpanich, B., Pintadit, P., Phanichyakarn P., and Valyasevi, A. (1987). Hemostatic derangement in dengue hemorrhagic fever. Southeast Asian J. Trop. Med. Public Health. *18*, 331–39.

Johnson, K.M., Halstead, S.B., and Cohen, S.N. (1967). Hemorrhagic fevers of Southeast Asian and South America: a comparative appraisal. Progr. Med. Virol. *9*, 106–127.

Joyeux, C. (1944). Précis de Médecine coloniale (Paris: Editions Masson et Compagnie).

Jupp, P.G., and Kemp, A. (1993). The potential for dengue in South Africa: vector competence tests with dengue 1 and 2 viruses and 6 mosquito species. Trans. R. Soc. Trop. Med. Hyg. *87*, 639–643

Kabra, S.K., Jain, Y., Pandrey, R.M., Madhulika Singhal, T., Tripathi, P., Broor, S., Seth, P., and Seth, V. (1999). Dengue hemorrhagic fever in children in the 1996 Delhi epidemic. Trans. R. Soc. Trop. Med. Hyg. *93*, 294–298.

Khai Ming, C., Thein, S., Thaung, U., Tin, U., Myint, K.S., Swe, T., Halstead, S.B., and Diwan, A.R. (1974). Clinical and laboratory studies on hemorrhagic fever in Burma, 1970–72. Bull. World Health Organ. *51*, 227–235.

Kliks, S.C., Nimmannitya, S., Nisalak, A., and Burke D.S. (1988). Evidence that maternal dengue antibodies are important in the development of dengue hemorrhagic fever in infants. Am. J. Trop. Med. Hyg. *38*, 411–419.

Kochel, T.J., Watts, D.M., Halstead, S.B., Hayes, C.G., Espinoza, A., Felices, V., Caceda, R., Bautista, C.T., Montoya, Y., Douglas, S., and Russell K.L. (2002). Effect of dengue-1 antibodies on American dengue-2 viral infection and dengue hemorrhagic fever. Lancet. *360*, 310–312.

Kouri, G.P., Guzman, M.G., and Bravo, J.R. (1987). Why dengue hemorrhagic fever in Cuba? 2. An integral analysis. Trans. R. Soc. Trop. Med. Hyg. *81*, 821–823.

Kouri, G., Guzman, M.G., Valdes, L., Carbonel, I., Del Rosario, D., Vazquez, S., Laferte, J., Delgado, J., and Cabrara, M.C. (1998). Reemergence of dengue in Cuba: a 1997 epidemic in Santiago de Cuba. Emerg. Infect. Dis. *4*, 89–92.

Kreil, T.R., and Eibl, M.M. (1997). Pre- and Post exposure protection by passive immunoglobulin but no enhancement on infection with a flavivirus in a mouse model. J. Virol. *71*, 2921–2927.

Kuberski, T., Rosen L., Reed, D., and Mataika, J. (1977). Clinical and laboratory observations on patients with primary and secondary dengue type 1 infection with hemorrhagic manifestations in Fiji. Am. J. Trop. Med. Hyg. *26*, 775–83.

Kuno, G., Gubler, D.J., and Oliver, A. (1993). Use of 'original antigenic sin' theory to determine serotypes of previous dengue infections. Trans. R. Soc. Trop. Med. Hyg. *87*, 103–105.

Kuno, G. (1995). Review of the factors modulating dengue transmission. Epidemiol. Rev. *17*, 321–35.

Kuo, C.H., Tai, D.I., Chang-Chien, C.S., Lan, C.K., Chiou, S.S., Liaw, Y.F. (1992). Liver biochemical tests and dengue fever. Am. J. Trop. Med. Hyg. *47*, 265–270.

Laigret, J., Rosen, L., Scholammer, G. (1967). Sur une épidémie de dengue survenue à Tahiti en (1964). Relations avec les "fièvres hémorragiques" du sud-est asiatique. Bull. Soc. Pathol. Exot. 60, 339–352.

Laudon, F., Cardines, R., Svarc, M. (1992). Situation de la dengue en Polynésie Française. Bulletin d'Informations Sanitaires et Epidémiologiques de la Direction de la Santé publique de Polynésie Française. 2, 1–3.

Lee, H.L. (1994). Research on dengue vectors: an overview. In First International Congress of the Parasitology and Tropical Medicine, pp. 48–55.

Leitmeyer, K.C., Vaughn, D.W., Watts, D.M., Salas, R., Villalobos De Chacon, I., Ramos, C., and Rico-Hesse, R. (1999). Dengue virus structural differences that correlate with pathogenesis. J. Virol. *73*, 4738–4747.

Levaditi, C., Lepine, P., Blanc, G., Bijl, J.P., Galloway, I.A., Gastinel, P., Gratia, A., Haber, P., Hornus, G., Kreis, B., Lecomte du Nouy, P., Levaditi, J., Mathis, C., Mesrobeanu, J., Nicolau, S., Plotz, H., Vieuchange, J., and Wohlwill F. (1938). Les ultra virus des maladies humaines. (Paris: Librairie Maloine).

Li, F.S., Yang, F.R., Song, J.C., Gao, H., Tang, J.Q., Zou, C.H., Hu, B.N., Wen, S.R., and Qiu, F.X. (1986). Etiologic and serologic investigations of the 1980 epidemic of dengue fever on Hainan island, China. Am. J. Trop. Med. Hyg. *35*, 1051–1054.

Linn, M.L., Aaskov, J.G., Suhrbier, A. (1996). Antibody-dependent enhancement and persistence in macrophages of an arbovirus associated with arthritis. J. Gen. Virol. *77*, 407–11.

Lopez-Velez, R., Perez-Casas, C., Vorndam, A.V., and Rigau, J. (1996). Dengue in Spanish travelers returning from the tropics. Eur. J. Clin. Microbiol. Infect. Dis. *15*, 823–825.

Lum, L.C.S., Lam, S.K., George, R., and Devi, S. (1993). Fulminant hepatitis in dengue infection. Southeast Asian J Trop. Med. Public Health. *24*, 467–471.

Lum, L.C.S., Lam, S.K., Choy, Y.S., George, R., and Harun, F. (1996). Dengue encephalitis: a true entity? Am. J. Trop. Med. Hyg. *54*, 256–259.

Mallet, E.C., Gestas, P., Pilorget, H., and Bataille, H. (1993). La Dengue hémorragique avec choc chez l'enfant en Polynésie Française. Bull. Soc. Path. Ex. 86, 450–454.

Marchette, N.J., Halstead, S.B., Falkler, W.A., Jr., Stenhouse, A., and Nash, D. (1973). Studies on the pathogenesis of dengue infection in monkeys. J. Infect. Dis. *128*, 23–30.

Martet, G., Coue, J.C., and Lecamus, J.L. (1990). Epidémiologie et prophylaxie des fièvres hémorragiques virales. Med. Trop. *50*, 331–337.

Martinez, E., Guzman, M.G., Valdes, M., Soler, M., and Kouri, G. (1993). Fiebre del dengue y dengue hemorragico en infantes con infeccion primaria. Rev. Cubana Med. Trop. *45*, 97–101.

Mascola, J.R., Mathieson, B.J., Zack, P.M., Walker, M.C., Halstead S.B., and Burke D.S. (1993). Summary report: workshop on the potential risks of antibody-dependent enhancement in human HIV vaccine trials. AIDS Res. Hum. Retroviruses. *9*, 1175–1184.

Mazaud, R., Salaun J.J., Montabone, H., Goube, P., and Bazillo, R. (1971). Troubles neurologiques et sensoriels aigus dans la dengue et la fièvre à Chinkungunya. Bull. Soc. Path. Ex. *64*, 22–30.

McKee, K.T. Jr., Bancroft, W.H., Eckels, K.H., Redfield, R.R., Summers, P.L., and Russell, P.K. (1987). Lack of attenuation of a candidate dengue 1 vaccine (45AZ5) in human volunteers. Am. J. Trop. Med. Hyg. *36*, 435–442.

Moloney, J.M., Skelly, C., Weinstein, P., Maguire, M., and Ritchie, S. (1998). Domestic *Aedes aegypti* breeding site surveillance limitations of remote sensing as a predictive surveillance tool. Am. J. Trop. Med. Hyg. *59*, 261–264.

Monath, T.P. (1988). Dengue. In The Arboviruses: ecology and epidemiology. Vol. II, T.P. Monath, ed. (Boca Raton, Florida USA: CRC Press), pp. 224–260.

Monath, T.P. (1994). Yellow fever and dengue—the interactions of virus, vector and host in the re-emergence of epidemic disease. Seminars in Virology. *5*, 133–145.

Monath, T.P. (1997). Epidemiology of yellow fever: current status and speculations on future trends. In Factors in the emergence of arbovirus diseases, J.F. Saluzzo and B. Dodet, ed. (Paris, France: Elsevier), pp. 143–156.

Moreau, J.P., Rosen, L., Saugrain, J., and Lagraulet, J. (1973). An epidemic of dengue on Tahiti associated with hemorrhagic manifestations. Am. J. Trop. Med. Hyg. *22*, 237–241.

Morens, D., Rigau-Perez, J.G., Lopez-Correa, R.H., Moore, C.G., Ruiz-Tiben, E.E., Sather, G.E., Chiriboga, J., Eliason, D.A., Casta-Velez, A., Woodall, J.P., and the dengue outbreak investigation group. (1986). Dengue in Puerto Rico, 1977: public health response to characterize and control an epidemic of multiple serotypes. Am. J. Trop. Med. Hyg. 35, 197–211.

Morens, D.M., Sather, G.E., Gubler, D.J., Rammohan, M., and Woodall, J.P. (1987). Dengue shock syndrome in an American traveler with primary dengue 3 infection. Am. J. Trop. Med. Hyg. 36, 424–426.

Morens, D.M., Marchette, N.J., Chu M.C., and Halstead, S.B. (1991). Growth of dengue type 2 virus isolates in human peripheral blood leukocytes correlates with severe and mild dengue disease. Am. Trop. Med. Hyg. *45*, 644–651.

Morens, D.M. (1994). Antibody-dependent enhancement of infection and the pathogenesis of viral disease. CID. *19*, 500–512.

Murgue, B., Deparis, X., Chungue, E., Cassar, O., and Roche C. (1999). Dengue: a re-evaluation of disease severity based on an analysis of 403 laboratory-confirmed cases. Trop. Med. Int. Health. *4*, 765–773.

Murgue, B., Roche, C., Deparis, X., and Chungue, E. (2000). Viremia in children hospitalized during the 1996/1997 dengue 2 outbreak in French Polynesia. J. Med. Virol. *60*, 455–462.

Nimmannitya, S., Halstead, S.B., Cohen, S.N., and Margiotta, M.R. (1969). Dengue and Chikungunya virus infection in man in Thailand, 1962–1964. Part I: observations on hospitalized patients with hemorrhagic fever. Am. J. Trop. Med. Hyg. *18*, 954–971.

Nimmannitya, S. (1987a). Dengue hemorrhagic fever in Thailand. Southeast Asian J. Trop. Med. Public Health. *18*, 291–294.

Nimmannitya, S., Thisyakron, U., and Hemrichart, V. (1987b). Dengue hemorrhagic fever with unusual manifestations. Southeast Asian J. Trop. Med. Public Health. *18*, 398–406.

Nogueira, R.M.R., Miagostovich, M.P., Schatzmayr, H.G., Moraes, G.C., Cardoso, F.M.A., Ferreira, J., Cerqueira, V., and Pereira M. (1995). Dengue type 2 outbreak in the south of the state of BAHIA, Brazil: laboratorial and epidemiological studies. Rev. Inst. Med. Trop. *37*, 507–510.

Palacios Serrano, H., Vargas Caballero, M.E., and Aguirre Portuondo, T.M. (2001). Dengue hemorragico en dengue primario. Rev. Cubana Med. Trop. *53*, 59–62.

Pancharoen, C., and Thisyakorn, U. (2001). Dengue virus infection during infancy. Trans. R. Soc. Trop. Med. Hyg. *95*, 307–308.

Pande, J.N., and Kabra, S.K. (1996). Dengue hemorrhagic fever and shock syndrome. Natl. Med. J. India. *9*, 256–258.

Parc, F., Pichon, G., Tetaria, C., Louis, F., and Laigret, J. (1981). La Dengue due au virus de type 4 en Polynésie Française. I. Epidémiologie générale. Aspects cliniques particuliers. Med. Trop. *41*, 93–96.

Patey, O., Ollivaud, L., Breuil, J., and Lafaix, C. (1993). Unusual neurologic manifestations occurring during dengue fever infection. Am. J. Trop. Med. Hyg. *48*, 793–802.

Paul, C., Dupont, B., and Pialoux, G. (1990). Polyradiculonévrite aiguë secondaire à une dengue. Presse Med. *19*, 1503.

Phillpotts, R.J., Stephenson J.R., and Porterfield J.S. (1985). Antibody-dependent enhancement of tick-borne encephalitis virus infectivity. J. Gen. Virol. *66*, 1831–1837.

Pinheiro, F., and Chuit, R. (1998). Emergence of dengue hemorrhagic fever in the Americas. Infect. Med. *15*, 244–251.

Poli, L., Chungue, E., Soulignac, O., Gestas, P., Kuo, P., and Papouin-Rauzy, M. (1991). Dengue materno-fœtale. A propos de 5 cas observés pendant l'épidémie de Tahiti (1989). Bull. Soc. Path. Ex. *84*, 513–521.

Pongsumpun, P., and Tang, I.M. (1987). A realistic age structured transmission model for dengue hemorrhagic fever in Thailand. Southeast Asian J. Trop. Med. Public Health. *32*, 336–340.

Pontes, R.J.S., Freeman, J., Oliveira-Lima J.W., Hodgson, C., and Spielman A. (2000). Vector densities that potentiate dengue outbreaks in a Brazilian city. Am. J. Trop. Med. Hyg. *62*, 378–383.

Price, W.H., Thind, I.S., and O'Leary, W. (1973). The attenuation of the 26th mouse brain passage of New Guinea C strain of dengue-2 virus for use in the sequential immunization procedure against group B arboviruses. Am. J. Trop. Med. Hyg. *22*, 92–99.

Putvana, R., Yoksan, S., Chayayodhin, T, Bamarapravati, N., and Halstead S.B. (1984). Absence of dengue-2 infection enhancement in human sera containing Japanese encephalitis antibodies. Am. J. Trop. Med. Hyg. *33*, 288–294.

Qiu, F.X., Chen, Q.Q., Ho, Q.Y., Chen, W.Z., Zhao, Z.G., and Zhao, B.W. (1991). The first epidemic of dengue hemorrhagic fever in the peoples republic of China. Am. J. Trop. Med. Hyg. *44*, 364–370.

Raabe, M.L., Issel, C.J., and Montelaro, R.C. (1999). In vitro antibody-dependent enhancement assays are insensitive indicators of *in vivo* vaccine enhancement of equine infectious anemia virus. Virology. *259*, 416–427.

Rathavuth, H., Vaughn, D.W., Minn, K., Nimmannitya, S., Nisalak, A., Raengsakulrach, B., Rorabaugh, M.L., Yuvatha, K., and Sophal, O. (1997). Hemorrhagic fever in Cambodia in caused by dengue viruses: Evidence for transmission of all four serotypes. Southeast Asian J. Trop. Med. Public Health. *28*, 120–125.

Reed, D., Maguire, T., and Mataika, J. (1977). Type 1 dengue with hemorrhagic disease in FIJI: epidemiologic findings. Am. J. Trop. Med. Hyg. 26, 784–791.

Reeves, W.C. (1972). Can the war to contain infectious diseases be lost? Am. J. Trop. Med. Hyg. *21*, 251–259.

Reiter, P. (1996). La dengue dans les Amériques. Bull. Soc. Path. Ex. *89*, 95–97.

Reynes, J.M., Laurent, A., Deubel, V., Telliam, E., and Moreau, J.P. (1993). Etude rétrospective, clinique et virologique des cas de dengue hémorragique de l'épidémie de dengue 1991–1992 en Guyane Française. Bull. Soc. Path. Ex. *86*, 467–469.

Reynes, J.M., Laurent, A., Deubel, V., Telliam, E., and Moreau, J.P. (1994). The first epidemic of DHF in French Guyana. Am. J. Trop. Med. Hyg. *51*, 545–553.

Reynes, J.M. (1996). La dengue en Guyane Française. Historique et actualités. Bull. Soc. Path. Ex. *89*, 98–101.

Richards, A.L., Bagus, R., Baso, S.M., Follows, G.A., Tan, R., Graham, R.R., Sandjaja, B., Corwin, A.L., and Punjabi, N. (1997). The first reported outbreak of dengue hemorrhagic fever in IRIAN JAYA, Indonesia. Am. J. Trop. Med. Hyg. *57*, 49–55.

Rico-Hesse, R. (1997a). Molecular evolution and pathogenesis of dengue type 2 viruses. In Factors in the emergence of arbovirus diseases, J.F. Saluzzo and B. Dodet, ed. (Paris, France: Elsevier), pp. 103–108.

Rico-Hesse, R., Harrison, L.M., Salas, R.A., Tovar, D., Nisalak, A., Ramos, C., Boshell, J., De Mesa, M.T.D., Nogueira, R.M.R., and Travassos Da Rosa, A. (1997b). Origins of dengue type 2 viruses associated with increased pathogenicity in the Americas. Virology. *230*, 244–251.

Rigau-Perez, J.G. (1998a). The early use of breakbone fever (Quebranta Huesos, 1771) and dengue (1801) in Spanish. Am. J. Trop. Med. Hyg. 59, 272–274.

Rigau-Perez, J.G., Clark, G.G., Gubler, D.J., Reiter, P., Sanders, E.J., and Vorndam, A.V. (1998b). Dengue and dengue hemorrhagic fever. Lancet. 352, 971–977.

Rigau-Perez, J.G., Vance Vorndam, A., and Clark, G.G. (2001). The dengue and dengue hemorrhagic fever epidemic in Puerto Rico, 1994–1995. Am. J. Trop. Med. Hyg. *64*, 67–74.

Rodhain, F., and Hannoun, C. (1979). Ecologie dynamique des systèmes virus-vecteur. Rev. Epidém. Santé Publ. *27*, 399–408.

Rodhain, F. (1991). The role of the monkeys in the biology of dengue and yellow fever. Comp. Immun. Microbiol. Infect. Dis. *14*, 9–19.

Rodhain, F. (1996a). L'écologie d'*Aedes aegypti* en Afrique et en Asie. Bull. Soc. Path. Ex. *89*, 103–106.

Rodhain, F. (1996b). Problèmes posés par l'expansion d'*Aedes albopictus*. Bull. Soc. Path. Ex., *89*, 137–141.

Rodier G.R., Gubler, D.J., Cope, S.E., Cropp, C.B., Soliman, A.K., Polycarpe, D., Abdourhaman, M.A., Parra, J.P., Maslin, J., and Arthur, R.R. (1996). Epidemic of dengue-2 in the city of Djibouti 1991–1992. Am. J. Trop. Med. Hyg. *90*, 237–240.

Rosen, L., Roseboom, L.E., Sweet, B.J., Sabin, A.B. (1954). The transmission of dengue by *Aedes polynesiensis* Marks. Am. J. Trop. Med. Hyg. *3*, 878–882.

Rosen, L. (1958a). Dengue antibodies in residents of the Society Islands, French Oceania. Am. J. Trop. Med. Hyg. *7*, 403–405.

Rosen, L. (1958b). Experimental infection of new world monkeys with dengue and yellow fever viruses. Am. J. Trop. Med. Hyg. *7*, 406–410.

Rosen, L. (1974a). Dengue type 3 infection in Panama. Am. Trop. Med. Hyg. *23*, 1205–1206.

Rosen, L., and Gubler, D.J. (1974b). The use of mosquitoes to detect and propagate dengue viruses. Am. Trop. Med. Hyg. *23*, 1153–1160.

Rosen, L. (1977). The emperor's new clothes revisited, or reflections on the pathogenesis of dengue hemorrhagic fever. Am. J. Trop. Med. Hyg. *26*, 337–342.

Rosen, L., Roseboom, L.E., Gubler, D.J., Lien, J.C., and Chaniotis, B.N. (1985). Comparative susceptibility of mosquito species and strains to oral and parenteral infection with dengue and Japanese encephalitis viruses. Am. Trop. Med. Hyg. *28*, 603–615.

Rosen, L. (1986a). Dengue in Greece in 1927 and 1928 and the pathogenesis of dengue hemorrhagic fever: new data and a different conclusion. Am. J. Trop. Med. Hyg. *35*, 642–653.

Rosen, L. (1986b). La pathogenèse de la dengue hémorragique : discussion critique des hypothèses actuelles. Bull. Soc. Path. Ex. *79*, 342–349.

Rosen, L. (1987). Sexual transmission of dengue viruses by *Aedes albopictus*. Am. J. Trop. Med. Hyg. 37, 398–402.

Rosen, L. (1989a). Disease exacerbation caused by sequential dengue infections: myth or reality? Rev. Infect. Dis. *11*, 840–842.

Rosen, L., Khin, M.M., and U, T. (1989b). Recovery of virus from the liver of children with fatal dengue: reflections on the pathogenesis of the disease and its possible analogy with that of yellow fever. Res. Virol. 140, 351–360.

Rosen, L. (1996). Dengue hemorrhagic fever. Bull. Soc. Path. Ex. *89*, 91–94.

Rothman, A.L., Green, S., Vaughn, D.W., Kalayanarooj, S., Nimmannitya, S., Innis, B.L., Stephens, H.A.F., Rico-Hesse, R., Suntayakorn, S., Nisalak, A., Sudiro, T.M., Lew, R., and Ennis, F.A. (1997). Dengue hemorrhagic fever. In Factors in the emergence of arbovirus diseases, J.F. Saluzzo and B. Dodet, ed. (Paris, France: Elsevier), pp. 109–116.

Row, D., Weinstein, P., and Murray-Smith, S. (1996). Dengue fever with encephalopathy in Australia. Am. J. Trop. Med. Hyg. *54*, 253–255.

Russel, P.K., Yuill, T.M., Nisalak, A., Udomsakdi, S., Gould, D.J., Winter, P.E., Nantapanich, S. (1968). An insular outbreak of dengue hemorrhagic fever. II virologic and serologic studies. Am. J. Trop. Med. Hyg. *17*, 600–608.

Rymzo, W.T., Cline, B.L., Kemp, G.E., Sather, G.E., Craven, P.C. (1976). Dengue outbreaks in Guanica–Ensenada and Villalba, Puerto Rico, 1972–1973. Am. J. Trop. Med. Hyg. *25*, 136–145.

Sainte-Foie, S., Niel, L., Moreau, J.P., Ast-Ravallec, and Chippaux, A. (1993). Un cas de polyradiculonévrite associé à une dengue chez une patiente originaire de la Guyane. Bull. Soc. Path. Ex. *86*, 117–119.

Sanchez Velga, F., Diaz Rey, J.M., Perez Vizcaino, D., and Goyenechea Hernandez, A. (1979). El dengue: estudio de sus posibles complicaciones. Rev. Cub. Ped. *51*, 249–53.

Sangkawibha, N., Rojanasuphot, S., Ahandrik, S., Viriyapongse, S., Jatanasen, S., Salitul, V., Phanthumachinda, A B., and Halstead, S.B. (1984). Risk factors in dengue shock syndrome: a prospective epidemiologic study in Rayong, Thailand. Am. J. Epidemiol. *120*, 653–669.

Saugrain, J., Moreau, J.P., and Rosen, L. (1973). L'épidémie de dengue de Tahiti en 1971: évolution de la tendance hémorragique et comparaisons avec les épidémies précédentes. Bull. Soc. Pathol. Exot. *66*, 381–385.

Schwartz, E., Mendelson, E., Sidi, Y., and Hashomer, T. (1996). Dengue fever among travelers. Am. J. Med. *101*, 516–520.

Seglen, P.O. (1997). Why the impact factor of journals should not be used for evaluating research. Br. Med. J. *314*, 498–502.

Sehgal, R. (1997). Dengue fever and El Nino. Lancet. *349*, 729–730.

Senado Dumoy, J., Capote Perez, C.A., and Sbarch Sharrager, N.N. (1984). Fiebre hemorragica dengue: patogenia y fisiopatologia. Cub. Med. *23*, 374–380.

Solomon, T., Nguyen Minh Dung, Vaughn D.W., Kneen, R., Le Thi Thu Thao, Raengsakulrach, B., Ha Thi Loan, Day N.P.J., Farrar J., Myint Ksa, Warrell M.J., James W.S., Nisalak A., and White N.J. (2000). Neurological manifestations of dengue infection. Lancet. *355*, 1053–1059.

Songco, R.S., Hayes, C.G., Leus C.D., and Manaloto, C.O. (1987). Dengue fever/dengue hemorrhagic fever in Filipino children: clinical experience during the 1983/1984 epidemic. Southeast Asian J. Trop. Med. Public Health. *18*, 284–290.

Spence, L., Jonkers, A.H., and Casals, J. (1969). Dengue type 3 virus isolated from an Antiguan patient during the 1963/64 Caribbean epidemic. Am. J. Trop. Med. Hyg. *18*, 584–587.

Takeda, A., Tuazon, C.U., and Ennis, F.A. (1988). Antibody-enhanced infection by HIV-1 via Fc receptor-mediated entry. Science. *242*, 580–583.

Takeda, A., and Ennis, F. (1990). FcR-mediated enhancement of HIV-1 infection by antibody. AIDS Res. Hum. Retroviruses. *6*, 999–1004.

Takeda, A., Watanabe, S., Okazaki, K., Kida, H., and Kawaoka, Y. (2001). Infectivity-enhancing antibodies to Ebola virus glycoprotein. J. Virol. *75*, 2324–2330.

Thaithmyanon, P., Thisyakorn, U., Deerojnawong, J., and Innis B.L. (1994). Dengue infection complicated by severe hemorrhage and vertical transmission in a parturient woman. CID. *18*, 248–9.

The first international conference on dengue and dengue hemorrhagic fever. Proceedings. (2000). (Chiang Mai, Thailand: Saengtharatip, S. Press).

Thein, S., Aung, M.M., Shwe, T.U., Aye, M., Zaw, A., Aye, K., Aye, K.M., and Aaskov, J. (1997). Risk factors in dengue shock syndrome. Am. J. Trop. Med. Hyg. *56*, 566–572.

Thisyakorn, U., Thisyakorn, C., Limpitikul, W., and Nisalak A. (1999). Dengue infection with central nervous system manifestations. Southeast Asian J. Trop. Med. Public Health. *30*, 504–506.

de Thoisy, B., Dussart, P., Kazanji, M. (2004). Wild terrestrial rainforest mammals as potential reservoirs for flaviviruses (yellow fever, dengue 2 and St Louis encephalitis viruses) in French Guiana. Trans. R. Soc. Trop. Med. Hyg. *98*, 409–12.

Travassos Da Rosa, A., Vasconcelos, P., Travassos Da Rosa, E., Rodrigues, S., Mondet, B., Cruz, A., Sousa, M., and Travassos Da Rosa, J. (2000). Dengue epidemic in Belem, Para, Brazil, 1996–97. Emerg. Infect. Dis. *6*, 298–301.

Tudela Coloma, J.M., Portuondo Ferrer, M., Joa Mesa, T., Bello Machado, P., Ayra Perez, C., and Toirac Lamarque, A. (1980). Dengue y embarazo. Consecuencia sobre el producto. Rev. Cub. Obstet. Ginec. *6*, 347–53.

Turell, M.J., Mather, T.N., Spielman, A., and Bailey, C. (1987). Increased dissemination of dengue-2 virus in *Aedes aegypti* associated with concurrent ingestion of microfilariae *Brugia malayi*. Am. J. Trop. Med. Hyg. *37*, 197–201.

Valdes, L.G., Mizhrahi, J.V., and Guzman, MG. (2002). Impactico economico de la epidemia de dengue 2 in Santiago de Cuba (1997). Rev. Cubana Med. Trop.. *54*, 220–227.

Van Der Stuyft, P., Gianella, A., Pirard, M., Holzman, A., Peredo, C., Boelaert, M., and Vorndam, V. (1998). Short communication: Dengue serotype 2 subtype III (Jamaica) epidemic in Santa Cruz, Bolivia. Trop. Med. Int. Health. *3*, 857–858.

Vasconcelos, P.F.C., Wellington O.L., J., Travassos Da Rosa, A.P.A., Timbo, M.J., Travassos Da Rosa, E.S., Lima, H.R., and Travassos Da Rosa, R.J.F.S. (1998). Epidemia de dengue em fortaleza, Ceara inquérito soro-epidemiologico aleatorio. Rev. Saude Publica. *32*, 447–454.

Venot, C., Texereau, M., Le Guenno, B., Bourgoin, A., Beby, A., Breux, J.P., and Agius, G. (1997). Un cas de dengue d'importation à forme hémorragique en France métropolitaine. Méd. Mal. Infect. *27*, 1037–1039.

Ventura, A.K., and Hewitt, C.M. (1970). Recovery of dengue-2 and dengue-3 viruses from man in JAMAICA. Am. J. Trop. Med. Hyg.*19*, 712–715.

Vidal Lopez, B., Martinez Torres, E., and Peramo Gomez, S.T. (1985). Alteraciones endocardicas en el dengue hemorragico. Rev. Cub. Ped. *57*, 146–51.

Wang, E., Ni, H., Xu, R, Barrett, A.D.T., Watowich, S.J., Gubler, D.J., and Weaver, S.C. (2000). Evolutionary relationship of endemic/epidemic and sylvatic dengue viruses. J. Virol. *74*, 3227–3234.

Waterman, S.H., Novak, R.J., Sather, G.E., Bailey, R.E., Rios, I., and Gubler, D.J. (1985). Dengue transmission in two Puerto Rican communities in (1982). Am. J. Trop. Med. Hyg. *34*, 625–632.

Watts, D.M., Porter, K.R., Putvatana, P., Vasquez, B., Calampa, C., Hayes, C.G., and Halstead, S.B. (1999). Failure of secondary infection with American genotype dengue 2 to cause dengue hemorrhagic fever. Lancet. *354*, 1431–1434.

Whitehead, R.H., Chaicumpa, V., Olson, L.C., and Russell, P.K. (1970). Sequential dengue virus infections in the white-handed gibbon (*Hylobates Lar*). Am. J. Trop. Med. Hyg. 19, 94–102.

Winter, P.E., Yuill, T.M., Udomsakdi, S., Gould, D., Nantapanich, S., and Russell, P.K. (1968). An insular outbreak of dengue hemorrhagic fever. I Epidemiologic observations. Am. J. Trop. Med. Hyg. *17*, 590–599.

Winter, P.E., Nantapanich, S., Nisalak, A., Udomsakdi, S., and Dewey, R.W. (1969). Recurrence of epidemic dengue hemorrhagic fever in an insular setting. Am. J. Trop. Med. Hyg. *18*, 573–579.

Wolfe, N.D., Kilbourn, A.M., Karesh, W.B., Rahman, H.A., Bosi, E.J., Cropp, B.C., Andau, M., Spielman, A., and Gubler, D.J. (2001). Sylvatic transmission of arboviruses among Bornean orangutans. Am. J. Trop. Med. Hyg. *64*, 310–16.

World Health Organization. Technical Advisory Committee on Dengue Hemorrhagic Fever for the Southeast Asian and Western Pacific Regions. (1997). Guide for Diagnosis, Treatment and Control of Dengue Hemorrhagic Fever. (Geneva: World Health Organization).

Xiao, S., Guzman, H., Da Rosa, A.P., Zhu, H., and Tesh, RB. (2003). Alteration of clinical outcome and histopathology of yellow fever virus infection in a hamster model by previous infection with heterologous flaviviruses. Am. J. Trop. Med. Hyg. *68*, 695–703.

Functional Role of prM Glycoprotein in Dengue Virus Replication

6

Nopporn Sittisombut, Poonsook Keelapang, and Prida Malasit

Abstract

Proteolytic cleavages of the prM protein during the replication of flaviviruses are intimately involved with the production and maturation of virion. Cleavage at the amino-terminus of prM by signal peptidase is coordinately regulated with the viral protease cleavage of anchored C protein to optimize the production of virions with nucleocapsid. Internal cleavage of prM on immature virion by furin allows rearrangement of the receptor-binding and viral fusion protein, E, enabling E to mediate envelope-membrane fusion upon exposure to low pH environment. Despite its importance in the generation of infectious virion, cleavage of prM is variable among different flaviviruses, partly due to the influence of charged residues within and proximal to the furin consensus sequence at the pr-M cleavage junction. Generation and characterization of pr-M junction chimeric and mutant viruses begin to unravel the structural basis for partial prM cleavage consistently observed in dengue viruses as well as the influence of cleavage on virus export. The remaining prM on extracellular dengue virion potentially serves as target for anti-prM antibodies, many of which exhibit infection-enhancing activity on dengue infection of Fc receptor bearing cells *in vitro*, but not strongly neutralizing effect. Several previous studies indicating the host-protective effect of anti-prM antibodies may reflect either their cross-reactivity with the E protein, or the contribution of remaining prM on post-binding step of the multiplication cycle.

Introduction

During the replication of flaviviruses, translation of a single-stranded viral RNA genome generates a polyprotein, which is cleaved by cellular and viral-encoded proteases into the three structural proteins (capsid [C], premembrane [prM], and envelope [E]) required for the production of new viral particles as well as seven non-structural proteins (Lindenbach and Rice, 2001). Binding of the C protein with genomic RNA generates the nucleocapsid that is enclosed in the envelope of viral particles. Of the two envelope proteins, prM is the class I membrane protein of about 19–23 kDa that serves as a chaperone for the receptor-binding, class II viral fusion protein, E, and, together with E, functions in the formation of viral particles (Lindenbach and Rice, 2001). The prM protein is cleaved from the polyprotein at the C-prM and prM-E junctions by signal peptidase, associates noncovalently with E, and, through the lateral association of the prM-E heterodimers, participates in viral assembly and budding into the lumen of endoplasmic reticulum. Intracellular virions remain noninfectious until shortly before release and are then converted to an infectious form by the internal cleavage of prM at the pr-M junction. Mechanistically, cleavage of prM dissociates the prM-E heterodimer and allows E to undergo structural arrangements, resulting in an acquisition of the virion's abilities to induce cell fusion and efficiently infect susceptible cells (Stadler *et al.*, 1997). Despite the importance of prM cleavage to the infectivity of virions, it is intriguing that this cleavage is not complete in dengue and a few other flaviviruses. In this review, we examine the structural basis for partial prM cleavage and the influence of the cleavages at the pr-M and other junctions on virus production, transport and other biological properties.

Recent studies revealed the influence of the proteolytic cleavages at the amino-terminus of prM (the C-prM junction) on the production of infectious flaviviral particles (Lobigs and Lee, 2004) as well as at the pr-M junction on the proportion of noninfectious, recombinant particles of different sizes (Allison *et al.*, 2003). Considerable sequence variations of the prM protein among flaviviruses (Chang, 1997) and within dengue antigenic complex (Diaz *et al.*, 2002) are observed which may differentially affect the functioning of prM and other viral properties. A notable example is an incomplete internal prM cleavage consistently detected in dengue and a few other flaviviruses. Characterization of the pr-M junction chimeric and mutant dengue viruses begins to unravel how sequence variation affects the pr-M cleavage and virus export (Keelapang *et al.*, 2004). The basis for understanding the dynamic interaction between prM and E prior to and after this critical cleavage at the pr-M junction is provided by structural analyses of the immature and mature particles of dengue virus (Kuhn *et al.*, 2002; Zhang *et al.*, 2003), West Nile virus (Mukhopadhyay *et al.*, 2003), as well as the subviral particles of tick-borne encephalitis virus (Ferlenghi *et al.*, 2001). These and other studies on the mechanism of envelope-membrane fusion mediated by the E protein upon exposure to low pH (Bressanelli *et al.*, 2004; Modis *et al*, 2004) together with further studies on the functional consequences of prM retention on virus replication will lead to a better understanding on why dengue and other flaviviruses evolve to retain some prM proteins on the extracellular virions.

Structural aspects of prM in virion assembly and maturation

Assembly of flavivirus particles in the endoplasmic reticulum requires co-expression of the envelope proteins prM and E (Konishi and Mason, 1993). Following cleavages of the polyprotein by signal peptidase, prM and E rapidly associate in the endoplasmic reticulum. The prM-E heterodimers are subsequently incorporated into immature virions by lateral interaction, forming "lattice-like" multimeric structures of increasing size (Wang *et al.*, 1999). Heterodimeric interaction between prM and E is important for proper folding of E (Konishi and Mason, 1993) whereas folding of prM does not require the presence of E (Lorenz *et al.*, 2002). Sites located within the predicted α-helical regions of the stem and the membrane-spanning region of E are required for stabilization of prM-E heterodimer (Allison *et al.*, 1999).

In virus-infected cells, two types of noninfectious particles are initially formed: the large, rapidly sedimenting, nucleocapsid-containing viral particles (immature virions), and the smaller, slowly sedimenting subviral particles without the C protein and viral RNA (Russell *et al.*, 1980; Westaway, 1980). During their transport along the secretory system, the envelope proteins on the surface of these particles are N-glycosylated and, just prior to release, the prM protein is cleaved internally into the soluble pr peptide and the particle-associated M (approx. 8–8.5 kDa) protein by a *trans*-Golgi-resident proprotein convertase furin (Stadler *et al.*, 1997). Depending on the extent of pr-M cleavage, the extracellular particles in different flaviviruses may contain varying proportions of prM and M. Cleavage of prM results in an acquisition of infectivity for the large, extracellular particles (mature virions) (Elshuber *et al.*, 2003) simultaneous with the rearrangement of envelope proteins from the forms of prM-E heterodimer to E homodimers (Wengler and Wengler, 1989).

Structural analyses of the cell-associated, immature virions of dengue virus and yellow fever virus reveal the rough surface formed by the organization of three prM-E heterodimers into each of the 60 icosahedrally arranged spikes that project a portion of prM out of the envelope (Zhang *et al.*, 2003). The protruding portion of prM strategically covers the fusion loop at the tip of domain II of the E protein, which also tilts outward (about 25°) relative to the lipid bilayer (Zhang *et al.*, 2003). In marked contrast, the extracellular virions of dengue virus and West Nile virus display a smooth surface under which a head-to-tail dimer of E lies parallel to the lipid bilayer (Kuhn *et al.*, 2002; Mukhopadhyay *et al.*, 2003). The overall surface is formed by 30 sets of three parallel E dimers arranged in a herringbone pattern over a layer of nonexposed M protein (Kuhn *et al.*, 2002; Mukhopadhyay *et al.*, 2003). A different arrangement (T=1) of head-to-tail E dimers is found on the surface of recombinant, extracellular subviral particles of tick-borne encephalitis virus (Ferlenghi *et al.*, 2001). The most distinctive feature from the comparison of the two types of dengue virions is the positioning of prM on top of the fusion loop in the immature virions. This configuration fits well with the previous findings that prM prevents E from undergoing premature conformational changes and oligomeric rearrangement induced by acidic pH of the *trans*-Golgi network (Allison *et al.*, 1995; Guirakhoo *et al.*, 1992; Heinz *et al.*, 1994; Stiasny *et al.*, 1996). Drastic changes of the surface texture from the rough type into the smooth type reflect a highly efficient prM cleavage of the virus preparations employed in these studies, and also the extent of which vertical and lateral rearrangements of the E protein can take place in the mature virions after prM cleavage.

Proteolytic cleavages of prM

The two types of proteolytic processing involving prM are associated with virion assembly and subsequent maturation event. The prM molecule is liberated from the polyprotein by signal peptidase cleavage at both the C-prM junction and prM-E junction. Extensive studies of the C-prM cleavage in Murray Valley encephalitis virus and other viruses reveal the complexity of this cleavage and the influence of cleavage efficiency on the production of infectious virus particles (Lobigs, 1993; Yamshchikov and Compans, 1993; Stocks and Lobigs, 1995; Stocks and Lobigs, 1998; Amberg and Rice, 1999; Lobigs and Lee, 2004; Lee *et al.*, 2000). During the transport of virions through the secretory pathway, an internal cleavage of prM is mediated by the cellular enzyme furin, a member of the proprotein convertase family (Stadler *et al.*, 1997). Cleavage at the pr-M junction allows the rearrangement of E, resulting in an acquisition of the capacity to mediate envelope-membrane fusion upon exposure to acidic environment of the endosome of newly infected cells. An internal cleavage of prM is critical for the generation of infectious virions.

Signal peptidase cleavage at C-prM junction and prM-E junction

Prior to cleavage from the polyprotein, prM is connected on the amino-terminal side to the C protein by a short hydrophobic sequence that serves as the signal sequence for the translocation of prM into the lumen of endoplasmic reticulum and also as the membrane anchor for the cytoplasmically located C protein. Cleavage of prM from the signal sequence occurs on the luminal side by signal peptidase whereas cleavage of C from the cytoplasmic end of this sequence requires the viral protease NS2B-3. Earlier studies establish that, in contrast to the co-translational cleavage of other signal sequences, cleavage at the C-prM junction is inefficient in the absence of viral protease cleavage of the C protein (Lobigs, 1993; Yamshchikov and Compans, 1993; Stocks and Lobigs, 1995; Amberg *et al.*, 1994). While the hydrophobic signal sequences in different flaviviruses are rather heterogeneous with regards to length and amino acid sequence, they all appear to lack polar residues in a short segment just proximal to the C-prM junction (Stocks and Lobigs, 1998). Introduction of polar side chains into the C-prM cleavage proximal segment of yellow fever virus and Murray Valley encephalitis virus by *in vitro* mutagenesis enhances the efficiency of C-prM cleavage, but reciprocally reduces the production of infectious virus particles (Lee *et al.*, 2000; Lobigs and Lee, 2004). The incorporation of nucleocapsid into the viral particles is markedly decreased in the C-prM junction mutants; the majority of secreted viral particles contains only prM and E glycoproteins and is smaller in size than the infectious particles (Lobigs and Lee, 2004). Conversely, enhanced cleavage at the amino-terminus of prM by signal peptidase results in an inefficient cleavage of the C protein from the hydrophobic anchor (Lobigs and Lee, 2004). The importance of the cleavage-proximal segment of the signal sequence on the efficiency of C-prM junction is further revealed by an analysis of spontaneous revertant viruses with increased replicative ability and infectious particle production. Many of the revertants contain additional single point mutations at the C-prM junction close to the intended mutations; these revertants display both reduced C-prM cleavage and efficient cleavage of C protein (Lobigs and Lee, 2004). These results indicate that, with the wild type signal sequence, the C-prM cleavage remains inefficient as long as the C protein is connected to the signal peptide, and the production of noninfectious subviral particle composed of only prM and E is minimized. Cleavage of the C protein from the signal peptide by the viral protease then releases the C-prM junction from the cryptic configuration for efficient cleavage by signal peptidase, allowing assembly and budding of the viral particles to proceed in the vicinity of functional C protein. Cleavages at the lumenal and cytoplasmic ends of the hydrophobic signal sequence for prM by signal peptidase and viral protease, respectively, are therefore coordinately regulated to ensure the optimal production of infectious viral particles (Lobigs and Lee, 2004). An additional protein required for the cleavage of anchored C protein from the cytoplasmic end of the signal sequence by the viral protease NS2B-3 is the nonstructural protein

NS2A (Kummerer and Rice, 2002), but the mechanism remains unknown.

Cleavage of the carboxy-terminus of prM, the prM-E junction, is also mediated by signal peptidase. A rather long hydrophobic stretch at the carboxy-terminus of prM is divided by the conserved arginine (or lysine) residue into two domains: the transmembrane segment for anchoring of prM to the membrane, and the signal sequence for translocation of E protein (Chang, 1997; Markoff *et al.*, 1994). The simultaneous presence of the two domains is required for membrane insertion of prM as well as an efficient cleavage at the prM-E junction by signal peptidase (Markoff *et al.*, 1994).

Furin cleavage at pr-M junction

Internal cleavage of flaviviral prM protein by furin was first demonstrated in tick-borne encephalitis virus (Stadler *et al.*, 1997). Cleavage by furin and other proprotein convertases within the mammalian subtilisin and Kex2p-like serine endoprotease superfamily generally requires a cluster of basic amino acids with the consensus sequence, Arg-Xaa-(Lys/Arg)-Arg (where Xaa is any amino acid), just proximal to the substrate cleavage site (reviewed in Molloy *et al.*, 1999; Nakayama, 1997; Steiner, 1998; Thomas, 2002; Zhou *et al.*, 1999). This requirement reflects the presence of separate binding pockets that accommodate each of the positively charged side chains corresponding to the P1, P2 and P4 cleavage positions of the substrate through charge-charge interaction (Henrich *et al.*, 2003; Holyoak *et al.*, 2003). A furin consensus sequence is present in the prM protein of tick-borne encephalitis virus and other members of the genus *Flavivirus* (Table 6.1). Cell fusing agent virus represents an exceptional case as two furin consensus sequences are located just five residues apart (Table 6.1). In addition to these indispensable residues of the furin consensus sequence, from 1–4 basic residues can be found at other positions (P3, P5 and other more proximal locations) of the pr-M cleavage junction that appear to be quite conserved, especially among phylogenetically related viruses within the dengue, tick-borne encephalitis and Japanese encephalitis antigenic complexes. Analysis of the crystal structure of the catalytic domain of furin complexed with peptide inhibitor and modeling indicate that the positively charged side chains at the substrate cleavage positions P3, P5 and P6 can potentially interact with acidic residues surrounding the substrate binding pockets on the surface of enzyme molecule and these nonconsensus residues may contribute to the enzyme-substrate interaction (Henrich *et al.*, 2003). In several members of the tick-borne encephalitis antigenic complex, all members of the JE antigenic complex with known prM sequence, and an insect flavivirus (Kamiti river virus), some of the nonconsensus basic residues form the minimal furin recognition motif, Arg-Xaa-Xaa-Arg, that is known to be cleaved by furin in a few cellular precursor proteins (Brennan and Nakayama, 1994; Hatsuzawa *et al.*, 1992; Klimpel *et al.*, 1992; Molloy *et al.*, 1992; Morsy *et al.*, 1994). Among flaviviruses, the four serotypes of dengue virus are unique with the presence of two highly conserved acidic residues at the cleavage positions P3 and P7. Similarly, the cell fusing agent virus contains an acidic residue at the position P3 within the second furin consensus sequence (Table 6.1). The occurrence of P3 acidic residue is rare among cellular and viral furin substrates (Molloy *et al.*, 1999; Rockwell *et al.*, 2002); examples include HIV-1 gp160 and the amino-terminal set of the tandem consensus sequence within the hemagglutinin precursor of Hong Kong H5N1 influenza A viruses (Hallenberger *et al.*, 1992; Subbarao *et al.*, 1998; Hatta *et al.*, 2001). The presence of additional charged residues surrounding the furin consensus sequence is suggestive of the plasticity in which different flaviviruses could modify the pr-M junction for interacting with host enzymes in their own niche.

Up to seven members of the proprotein convertase family have been identified in human and other vertebrates (Rockwell *et al.*, 2002; Thomas, 2002). These homologs share the general organization of protein domains, but they differ in size, fine substrate specificities, the number and location of potential glycosylation sites, the presence of carboxy-terminal membrane spanning and cytoplasmic domains, the requirement for a specific chaperone 7B2, and their associations with the constitutive or regulated secretory systems. The heterogeneity of the proprotein convertase family appears to be less extensive in insects as only three furin-related genes were detected in the genome of drosophila and anopheles mosquito (Riehle *et al.*, 2002). In human, Furin, PC7, PC5/6 and PACE4 have a broad tissue distribution. PC1/3 and PC2 are restricted to the neuroendocrine system whereas PC4 expression is limited to testicular and ovarian germ cells (Thomas, 2002). Furin is synthesized as inactive precursor and two successive auto-proteolytic cleavages of the self-inhibitory propeptide are needed for functional activation. The final cleavage of the propeptide is promoted by the mildly acidic environment of the trans-Golgi/endosome compartment (Thomas, 2002). Furin localizes to the trans-Golgi network, but can be sorted to endosome and the cell surface, and then recycled (Molloy *et al.*, 1999; Thomas, 2002). Cleavage of precursor protein by furin occurs at a pH range of 5–8 with a strict requirement for calcium (Thomas, 2002). While the involvement of ubiquitous furin in cleavage of prM has been established (Stadler *et al.*, 1997), the role of other members of the proprotein convertase family in different hosts and their relevant tissues remains to be investigated.

Cleavage of prM is required for the infectivity of flaviviruses. Failure to cleave prM in a TBEV mutant with an engineered loss of P2 arginine was shown to result in the release of totally noninfectious virions that could be revived only temporarily by trypsin digestion (Elshuber *et al.*, 2003). While this study clearly establishes the importance of prM cleavage in the generation of infectious virus and of the P2 basic residue in determining cleavage efficiency, it leaves open the functional role of other charged residues at nonconsensus positions in the process of prM cleavage and other steps of the virus replication cycle. In several earlier studies, blocking of prM cleavage by treating flavivirus-infected cells with acidotropic amines or bafilomycin, which raises the pH of the endosomal compartment, generates virions with higher proportion of prM that are less infectious and do not induce cell fusion at acidic pH (Guirakhoo *et al.*, 1992; Heinz *et al.*, 1994; Randolph *et al.*, 1990). The reduction of specific infectivity of the immature virions released from amine-treated cells, and also of the intracellular virions, as compared with the mature forms is not uniform, but varies form 6–8 folds for dengue virus (Randolph *et al.*, 1990), 10 folds for Murray Valley encephalitis virus (Guirakhoo *et al.*, 1992), 50 folds for tick-borne encephalitis virus (Heinz *et al.*, 1994), to 62 folds for West Nile virus (Wengler and Wengler, 1989). Such variable effects may be due to the differences in doses and schedules of inhibitors, the diversity of cells employed, the variation of intracellular level and functional characteristics of proprotein convertases that are involved, and the inconsistent sensitivity of the methods used for comparing viral infectivity. However, these findings also raise a possibility that structural differences inherent to flaviviruses, particularly at the pr-M junction, affect prM cleavability and thereby contribute to the variable responses to the inhibitory effect of acidotropic amines/bafilomycin.

Variations of pr-M junction cleavage in flaviviruses

For many flaviviruses, such as tick-borne encephalitis virus and Japanese encephalitis virus, cleavage of prM is mostly complete (for examples see Guirakhoo *et al.*, 1991; Schalich *et al.*, 1996; and Konishi *et al.*, 2001). Previously documented partial prM cleavage in certain preparations of tick-borne encephalitis virus (Allison *et al.*, 2003; Gehrke *et al.*, 2003), Langat virus (Guirakhoo *et al.*, 1991; Iacono-Connors *et al.*, 1996), Japanese encephalitis virus (Kimura-Kuroda and Yasui, 1983), and Kunjin virus (Khromykh *et al.*, 1998) appears to be exceptional cases. In contrast, cleavage of dengue pr-M junction was consistently found to be inefficient in mosquito cells (Falconar, 1999; Henchal *et al.*, 1985; Randolph *et al.*, 1990; Murray *et al.*, 1993; He *et al.*, 1995; Anderson *et al.*, 1997; Roehrig *et al.*, 1998; Sriburi *et al.*, 2001; Chang *et al.*, 2003; Keelapang *et al.*, 2004), Vero cells (Randolph *et al.*, 1990; Murray *et al.*, 1993; He *et al.*, 1995; Wang *et al.*, 1999) and LLC-MK2 cells (Crabtree *et al.*, 2003) as crude and gradient-purified preparations of infectious virions liberated from these cells always contain some prM molecules. Up to 40–60 prM molecules/virion were detected in a dengue serotype 2 strain purified from ^{35}S-methionine- and ^{35}S-cysteine-labeled, infected C6/36 culture fluid by co-immunoprecipitation with anti-E monoclonal antibody (Keelapang *et al.*, 2004). Incomplete prM cleavage in dengue is not limited to infectious virus preparations; recombinant subviral particles secreted from mammalian cells, including COS-1 cells (Chang *et al.*, 2003), following transfection with dengue prM/E-expressing plasmids also contain intact prM molecules (Chang *et al.*, 2003). However, there appears to be a tendency for subviral particles of other flaviviruses to contain uncleaved prM as well (Gehrke *et al.*, 2003; Konishi *et al.*, 2001). Interestingly, a very low level of prM cleavage is observed in cell fusing agent virus, which shares the P3 acidic residue within the furin consensus sequence of prM with dengue virus, and this P3 acidic residue has been thought to adversely affect prM cleavage efficiency (Crabtree *et al.*, 2003). Nevertheless, other flaviviruses that do not contain P3 acidic residue, such as Murray Valley encephalitis virus (Guirakhoo *et al.*, 1992; Lobigs and Lee, 2004), or even contain P3 basic residue, such as Kamiti River virus (Crabtree *et al.*, 2003), can also manifest partial prM cleavage. As all of these flaviviruses contain at least a set of furin consensus sequence, it is clear that the presence of a furin consensus sequence does not ensure complete prM cleavage. Variation of charged residues surrounding the furin consensus sequence may, on the other hand, influence prM cleavage efficiency among members of the genus *Flavivirus*.

In an attempt to determine whether sequence variations of the nonconsensus positions of the pr-M junction among flaviviruses affect cleavage efficiency, the 13-amino-acid region proximal to the pr-M junction of dengue virus was substituted with those of tick-borne encephalitis virus, yellow fever virus, and Japanese encephalitis virus to generate the TBEV-, YFV- and JEV-pr-M junction chimeric viruses with dengue background (Keelapang *et al.*, 2004). Comparison of the levels of prM and M in these chimeras with the parent dengue type 2 virus revealed that prM cleavage is enhanced to near completion in the JEV chimera. Cleavage is modestly increased in the YFV chimera, but slightly reduced in the TBEV chimera (Keelapang *et al.*, 2004). The differences in prM cleavage observed in these pr-M junction chimeras attest to the influence of the cleavage-proximal, nonconsensus positions (and their variations) on prM cleavage efficiency. An extensive prM cleavage in the JEV chimera also implies that the cleavage-modifying effect of structural components other than the 13-amino-acid, cleavage-proximal segment of prM would be minimal for dengue virus.

The greatest level of prM cleavage enhancement detected in the JEV chimera correlates with the highest positive charge content, the presence of an additional furin minimal motif, and the lack of negative charges within the pr-M junction proximal segment derived from Japanese encephalitis virus (Table 6.1). In a subsequent study, dengue mutants were generated which received only the basic residues at the positions P8, P10 and P13 (in effect also harboring the furin minimal motif), or lacked the two acidic residues at the positions P3 and P7 (that are highly conserved in dengue) (Songjaeng, 2004). Enhancement of prM cleavage to the same level as the JEV chimera occurs only in the mutant lacking the acidic residues (N.S. and P.K., unpublished data). The presence of multiple basic residues at the positions P8, P10 and P13 (and an additional furin minimal motif) minimally influences cleavage of the pr-M junction. The effect of individual charged residue within the cleavage positions P7-P1′ of dengue pr-M junction was then determined by generating a set of alanine-scanning mutants using C6/36 cell line. The P1 and P2 Ala mutants were not viable whereas the low-titer P4 Ala mutant acquired compensatory changes that resulted in the loss of P7 negative charge or the acquisition of P3 positive charge, confirming the importance of the P1 and P2 basic residues for virus viability, and revealing the requirement of P4 basic residue for efficient replication. A comparison among viable mutants revealed extensive prM cleavage in the P3 Glu→Ala mutant, but not the P7 Glu→Ala mutant (N.S. and P.K., unpublished data), strongly implicating the cleavage inhibitory effect of the negative charge at the P3 position of dengue pr-M junction. This result agrees well with the previously proposed role of the P3 negative charge (Crabtree *et al.*, 2003). On the contrary, cleavage-facilitating influence of the P6 His and P5 Arg residues is revealed by reduced prM cleavage in the P6 His→Ala and P5 Arg→Ala mutants. Charged residues at these nonconsensus positions of dengue pr-M junction may, therefore, exert contrasting influences on prM cleavage and the observed partial cleavage likely reflects the net result of such interaction. Structurally, the opposing effects of the P3, P5 and P6 charged residues on dengue pr-M cleavage could be explained by the extended interaction between these acidic and basic residues with the conserved acidic residues surrounding the substrate binding sites on the surface of furin molecule (Henrich *et al.*, 2003; Holyoak *et al.*, 2003). Depending on the charge of each particular residue, such interaction may affect cleavage efficiency by either weakening or strengthening the interaction between the P1, P2 and P4 basic residues of the substrate and the corresponding substrate-binding pockets of the enzyme.

It is intriguing that cleavage of prM in the YFV and TBEV pr-M junction chimeras is not enhanced to the same extent as in the JEV chimera despite an absence of the inhibitory P3 acidic residue in both viruses. This may reflect different ways in which residues at other nonconsensus positions of the pr-M junction could affect cleavage. Both chimeras contain an acidic residue (P10 Asp and P8 Glu, respectively) more upstream of the cleavage junction that, in the absence of strong inhibitory effect of the P3 negative charge, may also reduce cleavage efficiency. In other viral glycoproteins, such as the HA_0 precursor of some influenza A viruses and possibly HIV-1 gp160, furin consensus sequence is located in an exposed loop that is accessible to the enzyme (Steinhauer, 1999; Chen *et al.*, 1998; Moulard *et al.*, 1998; Oliva *et al.*, 2002). For HA_0 precursor, cleavage is reduced by the presence of nearby N-glycan group, but the polybasic amino acid insertion or the substitution of uncharged residue with basic residue in the loop reverses such inhibitory effect, possibly by improving the accessibility of N-glycan-blocked target sequence (Kawaoka and Webster, 1989; Steinhauer, 1999). If dengue pr-M junction resides in a similar loop as previously proposed (Keelapang *et al.*, 2004), the deletion of an amino acid in the TBEV chimera, which corresponds to the P5 position in other flaviviruses, may alter the length and/or configuration of the loop and adversely affect access or positioning of the target sequence in the loop. Shortening of the loop may also bring the P8 Glu residue in closer proximity to the consensus positions allowing this residue to affect cleavage. The prM cleavage defect in a TBEV mutant with an engineered loss of P2 arginine (19) coincides with further shortening of the cleavage proximal sequence.

An additional feature of the pr-M junction distinguishes the TBEV chimera from the JEV counterpart. The TBEV chimera lacks a minimal furin motif, Arg-Xaa-Xaa-Arg, which is present at positions P10 and 13 of the pr-M junction of Japanese encephalitis virus and two related viruses, and at more cleavage proximal positions in two other members of the Japanese encephalitis antigenic complex (Table 6.1). The presence of minimal furin motif adjacent to a consensus sequence is reminiscent of the duplicated furin consensus sequence, which has been observed in the HA_0 precursor of some avian influenza A strains (as tandem repeat) and in HIV-1 gp160 (with short intervening sequence) (Chakrabarti *et al*, 1987; Horimoto and Kawaoka, 2001; Rohm *et al.*, 1995; Subbarao *et al.*, 1998). For influenza HA_0 precursor and a recombinant HIV-1 gp140, which represents gp160 engineered to contain a tandem repeat of furin consensus sequence as well as a deletion of the transmembrane and intracytoplasmic domains, duplicated furin motif results in significant cleavage enhancement at the physiologic, cleavage-proximal site in both cases (Basak *et al.*, 2001; Binley *et al.*, 2002). Although the minimal furin motif from the pr-M junction of Japanese encephalitis virus did not affect cleavage efficiency when tested in dengue prM background, its influence has to be further determined in the native environment where the P3 and P7 acidic residues are absent. While the furin minimal motif is present in at least four members of the tick-borne encephalitis antigenic complex, it is replaced by the sequence, Arg-Cys-Gly-Lys, in tick-borne encephalitis virus and six others (Table 6.1). Such altered minimal motif is unlikely to be recognizable by human furin because of furin's strict structural requirement for arginine at the P1 position (Henrich *et al.*, 2003). Even the

Table 6.1 Comparison of pr-M junction sequence of flaviviruses.

Virus	Amino acid sequence [a]		↓			P1–P13 charge [b]				Accession number
						+ ve	– ve	Net	Set [c]	
DENV type 1	YGTCS	QTG**E**H	**R**R**D**KR	SVALA	PHVGL					P33478
DENV type 3	YGTCN	QAG**E**H	**R**R**D**KR	SVALA	PHVGM					NC 001475
DENV type 2	YGTCT	TMG**E**H	**R**R**E**KR	SVALV	PHVGM	4	2	+2	1	U87411
DENV type 4	YGTCT	QSG**ER**	**R**R**E**KR	SVALT	PHSGM					M14931
TBEV-N	YG**R**CG	**K**Q**E**GS	-RTRR	SVLIP	SHAQG	5	1	+4	1	U27495
LIV	YG**R**CG	**K**Q**E**GS	-RTRR	SVLIP	THAQG					M59376
TBEV-S	YG**R**CG	**K**Q**E**GS	-RTRR	SVLIP	SHAQG					X03870
TBEV-V	YG**R**CG	**K**Q**E**GS	-RTRR	SVLIP	SHAQG					L40361
OHFV	YG**R**CG	**K**Q**E**GT	-RSRR	SVLIP	SHAQK					AY193805
LGTV	YG**R**CG	**R**R**E**GS	-RSRR	SVLIP	SHAQR					P29837
ALKV	YG**R**CG	**K**PVGG	-RSRR	SVSIP	VHAHS					AF331718
KFDV	YG**R**CG	**K**PAGG	-RNRR	SVSIP	VHAHS					X74111
POWV	YG**R**CG	**R**QAGS	-RGKR	SVVIP	THAQK					L06436
DTV	YG**R**CG	**R**QAGF	-RGRR	SVVIP	THAQK					NC003218
YFV	YG**K**C**D**	SAG**R**S	**R**RSRR	AIDLP	THENH	6	1	+5	1	AF094612
JEV	YG**R**CT	**R**T**R**HS	**K**RSRR	SVSVQ	THGES	7	0	+7	2	M55506
MVEV	YG**R**CT	**R**A**R**HS	**K**RSRR	SITVQ	THGES					AF161266
KUNV	YG**R**CT	**K**T**R**HS	**R**RSRR	SLTVQ	THGES					D00246

WNV	YG**R**CT	**K**T**R**HS	**R**RSRR	SLTVQ	THGES	NC 001563
SLEV	YG**R**CT	**R**MGHS	**R**RSRR	SISVQ	HHGDS	M16614
RBV	V**K**YPL	C**K**PGG	HRLKR	SLSIT	EHPSS	AF144692
MMLV	V**E**YST	CNPTV	**E**RAKR	SLVIQ	DHPHS	NC004119
MODV	V**E**YPL	C**KK**GS	NRVRR	AVNIA	SHPEI	AJ242984
APOIV	MVTYP	TC**KR**N	TRTRR	DVTIQ	DHPPS	AF160193
CFAV	TL**R**Y**R**	**R**CVV**K**	**K**R**E**KR	SREPP	KADLL	M91671
KRV	HL**R**YH	**R**CVPQ	VR**R**RR	APQPQ	VSNQV	AY149904
TABV	QWG**KR**	**E**T**E**HM	HRTRR	SVTET	THESS	AF285080
YKSV	ITTNL	LTGLM	**K**R**K**RR	SSVSC	ELLLT	NC005039
Cleavage position	15 11	6	1	1′ 6′	10′	

a, Amino acids in bold represent additional charged residues between the cleavage positions P1 and P13. b, Charge characteristics of the four viruses employed in the construction of pr-M chimeric viruses. c, Set of nonoverlapping Arg-Xaa-(Lys/Arg)-Arg and Arg-Xaa-Xaa-Arg sequences. Arrow indicates pr-M cleavage site. Underline indicates the 13-amino-acid segment employed in the construction of pr-M junction chimeric viruses. Box indicates the 8-amino-acid segment employed in the generation of alanine-scanning mutants. DV, dengue virus; TBEV-N, tick-borne encephalitis virus Neudoerfl strain; LIV, louping ill virus; TBEV-S, tick-borne encephalitis virus Sofjin strain; TBEV-V, tick-borne encephalitis virus Vasilchenko strain; OHFV, Omsk hemorrhagic fever virus; LGTV, Langat virus; ALKV, Alkhurma virus; KFDV, Kyasanur Forest disease virus; POWV, Powassan virus; DTV, Deer tick virus; YFV, yellow fever virus; JEV, Japanese encephalitis virus; MVEV, Murray Valley encephalitis virus; KUNV, Kunjin virus; WNV, West Nile virus; SLEV, St. Louis encephalitis virus; RBV, Rio Bravo virus; MMLV, Montana myotis leukoencephalitis virus; MODV, Modoc virus; APOIV, Apoi virus; CFAV, mosquito cell fusing agent virus; KRV, Kamiti River virus; TABV, Tamana bat virus; YKSV, Yokose virus. Listing of selected members of the genus *Flavivirus* follows the phylogenetic relationship of the whole genome based on Charlier *et al*. (2002), de Lamballerie *et al*. (2002), Crabtree *et al*. (2003), and Lin *et al*. (2003).

minimal furin motif is a relatively poor target for furin than the consensus sequence (Hatsuzawa *et al.*, 1992; Klimpel *et al.*, 1992; Molloy *et al.*, 1992). Whether the altered furin minimal motif of the pr-M junction of tick-borne encephalitis virus exerts any effect on prM cleavage in other insect cells remains to be elucidated.

Comparison of the nucleotide sequence of the pr-M junction indicates that incomplete prM cleavage is a highly conserved characteristic among members of dengue antigenic complex as the P3 Glu (or P3 Asp) residue is almost invariant (Keelapang *et al.*, 2004). Among the rare dengue strains in which variations of pr-M junction are detected, only two strains substitute the P3 acidic residue with uncharged ones (Table 6.2). In two other strains, the P6 His residue is replaced with arginine, and in one other the P5 Arg is substituted with glycine. Because of the strong inhibitory effect of the P3 acidic residue, prM cleavage is unlikely to be affected in the last three strains.

Effects of alteration of pr-M cleavage

Internal cleavage of viral glycoprotein precursor by proprotein convertases during their transport through the secretory pathway is an obligatory step in the multiplication of many enveloped RNA viruses (Flint *et al.*, 2000). Unlike other viruses that bud from the cell membrane, flaviviruses assemble in the rough endoplasmic reticulum and cleavage of prM takes place subsequently on immature virion rather than on individual precursor molecule. Cleavage serves to terminate the chaperone function of prM for E, allowing E to dimerize and capable of further rearrangement required for mediating envelope-host membrane fusion, rather than activates the potential functional capability of prM. Cleavage of prM is not a prerequisite for virus export as complete failure of prM cleavage, either by engineering the pr-M junction (Elshuber *et al.*, 2003) or treatment with acidotropic amines/bafilomycin (Guirakhoo *et al.*, 1992; Heinz *et al.*, 1994; Randolph *et al.*, 1990), still allows transport of noninfectious virus particles to the extracellular compartment. However, alteration of prM structure and cleavage may affect virion formation, stability and transport in other ways.

Particles secreted from flavivirus-infected cells are not homogeneous; large, rapidly sedimenting particles with high infectivity are present along with smaller, slowly sedimenting, noninfectious subviral particles that are devoid of genomic RNA (Russell *et al.*, 1980; Westaway, 1980). Formation of both types of particles has been thought to reflect lateral interaction of the prM-E heterodimer on the intracellular membrane (Allison *et al.*, 2003) as subviral particles can be generated and secreted from cells expressing only prM and E proteins (Konishi and Fujii, 2002; Kroeger and McMinn, 2002; Mason *et al.*, 1991; Konishi *et al.*, 1992; Pincus *et al.*, 1992; Fonseca *et al.*, 1994; Allison *et al.*, 1995b; Schalich *et al.*, 1996; Hunt *et al.*, 2001; Konishi *et al.*, 2001; Chang *et al.*, 2003). The factor(s) governing the pathways leading to the formation of large or small particles in flavivirus-infected cells is not known. A detailed study on the heterogeneity of viral particles secreted from a mammalian cell line expressing prM and E proteins of tick-borne encephalitis virus revealed two distinct size classes of the recombinant prM-E particles with diameters of about 35 and 55 nm, which correspond to the subviral particles and the virion-size particles, respectively (Allison *et al.*, 2003). The proportion of these two classes of secreted, recombinant particles is affected by

Table 6.2 Variation of the pr-M junction of dengue viruses (Modified from Keelapang *et al.*, 2004).

		Amino acid sequence [a]					P1–P13 charge				
Type	Strain			↓			+ve	−ve	Net	Set [b]	Accession number
1	Mochizuki	YGTCS	QTGEH	RR**G**KR	SVALA	PHVGL	4	1	+3	1	BAB72261
2	16681	YGTCT	TMGEH	RR**Q**KR	SVALV	PHVGM	4	1	+3	1	AAA73185
2	D2–04	YGT**R**T	TTGEH	**G**REKR	SVALV	PHVGM	4	2	+2	1	P30026
3	68784	YGTCN	QAGE**R**	RRDKR	SVALT	PHSGM	5	2	+3	1	AAK74146
3	BR74886/02	YGTCN	QAGE**R**	RRDKR	SVALA	PHVGM	5	2	+3	1	AY679147
Cleavage position		15 11	6	1	1′ 6′	10′					

a, Amino acids in bold represent variation of charged residues between position P1 and P13 of the pr-M junction. b, Set of nonoverlapping Arg-Xaa-(Lys/Arg)-Arg and Arg-Xaa-Xaa-Arg sequences. Arrow indicates pr-M cleavage site.

prM cleavability: the virion-size particle is markedly enriched when prM cleavage is abolished by an engineered deletion of the P2 arginine residue (Allison *et al.*, 2003). Although the proportion of the intracellular particles has not been determined, the finding is suggestive of the stabilizing effect of uncleaved prM when present in the virion-size, recombinant particles. Whether this effect of prM represents its native function operating during the assembly, transport and maturation of infectious viral particles in infected cells (in the presence of C, viral RNA, and viral protease NS2B-3), and is capable of influencing the proportion of infectious particles and subviral particles are still unknown.

While cleavage of prM on immature flavivirus virion is not required for transport through the secretory pathway (Elshuber *et al.*, 2003; Guirakhoo *et al.*, 1992; Heinz *et al.*, 1994; Randolph *et al.*, 1990), alteration of furin target sequence in the pr-M junction can affect virus export. In the study of pr-M junction chimeric viruses, comparison of the changes in intracellular virus and the extracellular virus levels using the one-step kinetic model revealed delayed accumulation of infectious particles of the JEV chimera, but not others, in the extracellular compartment (Keelapang *et al.*, 2004). Based on previous studies of HIV gp140, which displayed similar export defect as a result of the tandem repeat of furin minimal motif and a consensus motif (Binley *et al.*, 2002; Staropoli *et al.*, 2000), the delay has been thought to reflect complexing of the minimal motif sequence at the pr-M junction with furin and the retention of resultant virion-enzyme complex in the *trans*-Golgi network (Keelapang *et al.*, 2004). Because the delay of virus export is concurrent with greatly enhanced cleavage of prM in this chimera, it is also possible that enhanced cleavage of prM causes premature expression of fusion-competent virions, which, in the presence of sufficiently low pH in the Golgi apparatus and secretory vesicles, results in fusion of virus envelope to intracellular membrane and the apparent export retardation. Such fusion is possible as fusion of flavivirus (and also alphavirus) envelope to membrane can be mediated by the E glycoprotein upon exposure of the mature virion to low pH even in the absence of E-receptor interaction (Jardetzky and Lamb, 2004; Stiasny *et al.*, 2002; Modis *et al.*, 2004; Bressanelli *et al.*, 2004; Gibbons *et al.*, 2004). The latter possibility appears unlikely as a dengue pr-M junction mutant with the P3 Glu→Ala mutation exhibits as extensive prM cleavage as the JEV pr-M junction chimera but without the reduction of virus titer and replication kinetics (N.S. and P.K., unpublished data). Thus, the physical contact between the proprotein convertase enzyme(s) with the target sequence(s) at the pr-M junction and the result of enzyme-substrate interaction may affect on flavivirus export. During the evolution of dengue virus, the export delay may present a strong selective pressure since no dengue virus has been found that contains the furin minimal motif in the cleavage–proximal region as commonly observed in the Japanese encephalitis antigenic complex and in some other flaviviruses.

Comparison of the specific infectivity of the JEV chimera with the parent strain did not reveal significant increase of the infectivity of the chimeric virus over the parent despite an enhancement of prM cleavage (Keelapang *et al.*, 2004). This indicates that optimal level of dengue virus infectivity has been achieved without complete prM cleavage, and that other virus structural component(s) is(are) able to compensate for any possible infectivity-reducing consequences of incomplete prM cleavage.

Possible roles of prM in antibody-mediated infection enhancement and fusion

Conservation of the cleavage inhibitory P3 acidic residue at the pr-M junction of dengue viruses strongly suggests that it is advantageous for this group of virus to retain some prM on the extracellular virions. The advantage must be sufficiently substantial so as to withstand the selective pressure imposed by specific antibody response during their passage in human. Host protective effect of anti-prM antibodies has been documented for dengue virus (Bray and Lai, 1991; Falconar, 1999; Kaufman *et al.*, 1989; Vazquez *et al.*, 2002), but how these antibodies exert their effect is not yet clear. PrM is not known to involve in receptor binding and monoclonal antibodies specific for prM do not, or only weakly, exhibit neutralizing activity (Henchal *et al.*, 1985; Roehrig *et al.*, 1998; Falconar, 1999; Men *et al.*, 2004), even by the one displaying broad cross-protective effect in passive immunization study (Kaufman *et al.*, 1989). Weak neutralization of dengue infectivity by some anti-prM monoclonal antibodies (Falconar, 1999) and anti-prM-peptide sera (Vazquez *et al.*, 2002) could be due to their cross-reactivity with E protein.

In contrast to displaying only weak and inconsistent neutralizing activity, an anti-prM monoclonal antibody (2H2) enhances dengue virus infection of Fc receptor-bearing U937 and P388D1 cell lines to the similar extent as some strongly enhancing anti-E monoclones (Henchal *et al.*, 1985; Randolph *et al.*, 1990). A comparison of ten murine monoclonal anti-dengue prM antibodies with anti-dengue E antibodies indicates that infection-enhancement and lack of potent neutralization are common properties of anti-prM antibodies (P.K. and N.S., unpublished data). These results not only suggest that prM constitutes an alternative target for infection-enhancing antibodies but also imply that prM-containing, extracellular dengue virions are infectious. Anti-prM antibody is regularly generated during dengue infection in human (Se-Thoe *et al.*, 1999), enhancement of dengue infection of cells expressing Fc receptors mediated by anti-prM antibodies could contribute to efficient virus replication *in vivo*. Modeling indicates that infection enhancement is an important factor in maintaining the four serotypes of dengue virus in nature (Burke and Monath, 2001; Ferguson *et al.*, 1999). By serving as a target for infection-enhancement, but not neutralization, prM may play an important role in dengue virus evolution, especially the conservation of the four serotypes in nature.

Structural analysis of immature dengue virions indicates that a portion of dengue prM projects out of the surface of immature virion and covers the fusion peptide at the distal end of the E glycoprotein (Zhang *et al.*, 2003). This critical positioning of prM precludes further rearrangement of E that is required for fusion, and provides a basis for the lack of fusion activity of the intracellular virions and the prM-containing, extracellular virions (Guirakhoo *et al.*, 1991). The presence of some prM molecules in infectious dengue preparations poses a structural problem as the E homodimers and the pr-M heterodimers assume quite different configuration on dengue virions (Kuhn *et al*, 2002; Zhang *et al.*, 2003). It is not yet clear whether all dengue virions retain some prM molecules, or there exist two subpopulations of intermingling M-only virions and prM-only virions. The prM-only virions in the second model would, however, be fusion-incompetent and, thus, noninfectious. Retention of some prM in the first model would fit better with the implication that the prM-containing virions are infectious. Until the structure of extracellular dengue virions is fully resolved, it would be important to realize that these static structural models may be too simplistic. As furin can be sorted to the cell surface, shed, or recycled to endosome (Molloy *et al.*, 1999; Thomas, 2002), dengue virions may encounter furin and, thereupon, be further modified, while they are either free in the extracellular compartment, bound to the cell surface, or endocytosed.

The presence of prM on extracellular virions may possibly affect fusion in a different way. Upon encountering acidic environment, the E dimers rearrange into trimer (Allison *et al.*, 1995a), which then insert the hydrophobic fusion loop into host cell membrane (Bressanelli *et al.*, 2004; Modis *et al*, 2004). It has been thought that rearrangement of E homodimers in the acidic environment requires an outward expansion of the envelope to adequately accommodate molecular movements (Kuhn *et al.*, 2002). Acid-induced movement of envelope glycoproteins and expansion of virions are observed in a related alphavirus (Haag *et al.*, 2002; Hammar *et al.*, 2003). By virtue of its carboxy-terminal transmembrane anchor and the amino-terminal projecting portion, prM may, in addition to directly preventing E of the same prM-E heterodimer to engage in fusion, collectively hinder or retard the general outward expansion of the virion envelope (Kuhn *et al.*, 2002). If this is the case, the anti-prM antibodies could possibly interfere with dengue virus replication by reinforcing this obstructive effect of prM on fusion and cause further retardation or even failure of the fusion event. The retardation of virus replication by non- or weakly neutralizing anti-prM antibodies *in vivo* may result in inefficient virus replication beneficial to the host.

Another aspect of prM that may contribute to the multiplication of dengue virus is the potential of prM to induce apoptosis. A nine-amino-acid segment at the carboxy-terminal portion of the ectodomain of the M protein from all four dengue serotypes, as well as Japanese encephalitis virus, West Nile virus and yellow fever virus, has been shown to induce apoptosis in human and murine cell lines when expressed as fusion proteins with green fluorescent protein (Catteau *et al.*, 2003b). Apoptosis induction requires the export of the M ectodomain from the Golgi apparatus to plasma membrane and is associated with the disruption of mitochondrial transmembrane potential and the activation of caspase 3-like activity (Catteau *et al.*, 2003a). It is not yet clear whether apoptosis can be induced in infected cells by virion-bound prM, and whether prM cleavage alters the apoptosis induction potential of the intracellular virions.

Conclusion and future directions

Cleavage of prM represents a critical step in the replication of flaviviruses; several of these viruses evolve to minimize the failure to cleave prM internally by maintaining an optimal number of the basic residues at the pr-M junction. Partial cleavage of prM in dengue virus and a few others has been conserved during evolution, possibly by also altering the structure and function of other viral components to compensate for the possible reduction of infectivity. It will be important to define such compensatory modifications in structural and function terms as well as the distribution of prM at both the single virion level and the virus population level. Insights form the study of the variations of prM level and their effect on the envelope-membrane fusion and infectivity will be important to our better understanding of the fusion event and its control. It will also be interesting to define the possible *in vivo* effects of maintaining prM on the virion surface against the continuous onslaught of various components of the immune system. The knowledge gained from these studies will undoubtedly benefit the design of next generations of vaccine candidates and the search for drugs against dengue hemorrhagic fever.

References

Allison, S. L., J. Schalich, K. Stiasny, C.W. Mandl, C. Kunz, and F.X. Heinz. (1995a). Oligomeric rearrangement of tick-borne encephalitis virus envelope proteins induced by an acidic pH. J. Virol. *69*: 695–700.

Allison, S. L., K. Stadler, C. W. Mandl, C. Kunz, and F. X. Heinz. (1995b). Synthesis and secretion of recombinant tick-borne encephalitis virus protein E in soluble and particulate form. J. Virol. *69*:5816–5820.

Allison, S. L., K. Stiasny, K. Stadler, C.W. Mandl, and F.X. Heinz. (1999). Mapping of functional elements in the stem-anchor region of tick-borne encephalitis virus envelope protein E. J. Virol. *73*: 5605–5612.

Allison, S.L.,Y.J. Tao, G. O'Riordain, C.W. Mandl, S.C. Harrison, and F.X. Heinz. (2003). Two distinct size classes of immature and mature subviral particles from tick-borne encephalitis virus. J. Virol. *77*: 11357–11366.

Amberg, S. M., A. Nestorowicz, D. W. McCourt, and C. M. Rice. (1994). NS2B-3 proteinase-mediated processing in the yellow fever virus structural region: *in vitro* and *in vivo* studies. J. Virol. *68*:3794–3802.

Amberg, S. M., and C. M. Rice. (1999). Mutagenesis of the NS2B-NS3-mediated cleavage site in the flavivirus capsid protein demonstrates a requirement for coordinated processing. J. Virol. *73*:8083–8094.

Anderson, R., S. Wang, C. Osiowy, and A.C. Issekutz. (1997). Activation of endothelial cells via antibody-enhanced dengue virus infection of peripheral blood monocytes. J. Virol. *71*: 4226–4232.

Basak, A., M. Zhong, J. S. Munzer, M. Chretien, and N.G. Seidah. (2001). Implication of the proprotein convertases furin, PC5 and PC7 in the cleavage of surface glycoproteins of Hong Kong, Ebola and respiratory syncytial viruses: a comparative analysis with fluorogenic peptides. Biochem. J. 353: 537–545.

Binley, J.M., R.W. Sanders, A. Master, C.S. Cayanan, C.L. Wiley, L. Schiffner, B. Travis, S. Kuhmann, D.R. Burton, S.-L. Hu, W.C. Olson, and J.P. Moore. (2002). Enhancing the proteolytic maturation of human immunodeficiency virus type 1 envelope glycoproteins. J. Virol. *76*: 2606–2616.

Bray, M., and C.J. Lai. (1991). Dengue virus premembrane and membrane proteins elicit a protective immune response. Virology *185*: 505–508.

Brennan, S.O., and K. Nakayama. (1994). Cleavage of proalbumin peptides by furin reveals unexpected restrictions at the P2 and P'1 sites. FEBS Lett. *347*: 80–84.

Bressanelli, S., K. Stiasny, S.L Allison, E.A Stura, S. Duquerroy, J. Lescar, F.X Heinz, and F. A Rey. (2004). Structure of a flavivirus envelope glycoprotein in its low-pH-induced membrane fusion conformation. EMBO J. *23*: 728–738.

Burke, D.S., and T.P. Monath. (2001). Flaviviruses, p. 1043–1125. *In* D.M. Knipe, P.M. Howley, D.E. Griffin, R.A. Lamb, M.A. Martin, B. Roizman, and S.E. Strauss (ed.), Fields virology, 4th ed. Lippincott Williams & Wilkins, Philadelphia, Pa.

Catteau, A., G. Roue, V.J. Yuste, S.A. Susin, and P. Despres. (2003a). Expression of dengue ApoptoM sequence results in disruption of mitochondrial potential and caspase activation. Biochimie 85: 789–793.

Catteau, A., O. Kalinina, M.-C. Wagner, V. Deubel, M.-P. Courageot, and P. Despres. (2003b). Dengue M protein contains a proapoptotic sequence referred to as ApoptoM. J. Gen. Virol. 84: 2781–2793.

Chang, G.-J (1997). Molecular biology of dengue viruses, p 175–198. In D.J. Gubler and G. Kuno (ed.), Dengue and Dengue Hemorrhagic Fever. CAB International, New York, N.Y.

Chang, G.-J.J., A.R. Hunt, D.A. Holmes, T. Springfield, T.-S. Chiueh, J.T. Roehrig, and D. J. Gubler. (2003). Enhancing biosynthesis and secretion of premembrane and envelope proteins by the chimeric plasmid of dengue virus type 2 and Japanese encephalitis virus. Virology *306*: 170–180.

Chakrabarti, L., M. Guyader, M. Alizon, M.D. Daniel, R.C. Desrosiers, P. Tiollais, and P. Sonigo. (1987). Sequence of simian immunodeficiency virus from macaque and its relationship to other human and simian retroviruses. Nature *328*: 543–547.

Charlier, N., P. Leyssen, C.W.A. Pleij, P. Lemey, F. Billoir, K. Van Laethem, A.-M. Vandamme, E. De Clercq, X. de Lamballerie, and J. Neyts. (2002). Complete genome sequence of

Montana Myotis leukoencephalitis virus, phylogenetic analysis and comparative study of the 3′ untranslated region of flaviviruses with no known vector. J. Gen. Virol. *83*: 1875–1885.

Chen, J., K.H. Lee, D.A. Steinhauer, S.J. Stevens, J.J. Skehel, and D.C. Wiley. (1998). Structure of the hemagglutinin precursor cleavage site, a determinant of influenza pathogenicity and the origin of the labile conformation. Cell *95*: 409–417.

Crabtree, M.B., R. C. Sang, V. Stollar, L. M. Dunster, and B. R. Miller. (2003). Genetic and phenotypic characterization of the newly described insect flavivirus, Kamiti River virus. Arch. Virol. *148*: 1095–1118.

Diaz, F.J., J. A. Farfan-ale, K.E. Olson, M.A. Lorono-pino, D.J. Gubler, C.D. Blair, W.C. Black IV, and B.J. Beaty. (2002). Genetic variation within the premembrane coding region of dengue viruses from the Yacatan peninsula of Mexico. Am. J. Trop. Med. Hyg. *67*: 93–101.

Elshuber, S., S.L. Allison, F.X. Heinz, and C. W. Mandl. (2003). Cleavage of protein prM is necessary for infection of BHK-21 cells by tick-borne encephalitis virus. J. Gen. Virol. *84*:183–191.

Falconar, A.K.I. (1999). Identification of an epitope on the dengue virus membrane (M) protein defined by cross-protective monoclonal antibodies: design of an improved epitope sequence based on common determinants present in both envelope (E and M) proteins. Arch. Virol. *144*: 2313–2330.

Ferguson, N., R. Anderson, and S. Gupta. (1999). The effect of antibody-dependent enhancement on the transmission dynamics and persistence of multiple-strain pathogens. Proc. Natl. Acad. Sci. USA 96: *790–794.*

Ferlenghi, I., M. Clarke, T. Ruttan, S. L. Allison, J. Schalich, F. X. Heinz, S. C. Harrison, F. A. Rey, and S. D. Fuller. (2001). Molecular organization of a recombinant subviral particle from tick-borne encephalitis virus. Mol. Cell *7*: 593–602.

Flint, S.J., L.W. Enquist, R.M. Krug, V.R. Racaniello, and A.M. Skalka. (2000). Intracellular transport of viral components: prelude to assembly, p. 418–421. *In* S.J. Flint, L.W. Enquist, R.M. Krug, V.R. Racaniello, and A.M. Skalka (ed.), Principles of virology. American Society for Microbiology, Washington, D.C.

Fonseca, B. A., S. Pincus, R. E. Shope, E. Paoletti, and P. W. Mason. (1994). Recombinant vaccinia viruses co-expressing dengue-1 glycoproteins prM and E induce neutralizing antibodies in mice. Vaccine *12*:279–285.

Gehrke, R., M. Ecker, S.W. Aberle, S.L. Allison, F.X. Heinz, and C.W. Mandl. (2003). Incorporation of tick-borne encephalitis virus replicons into virus-like particles by a packaging cell line. J. Virol. *77*: 8924–8933.

Gibbons, D.L., M.-C. Vaney, A. Roussel, A. Vigouroux, B. Reilly, J. Lepault, M. Kielian, and F.A. Rey. (2004). Conformational change and protein–protein interactions of the fusion protein of Semliki Forest virus. Nature 427: *320–325.*

Guirakhoo, F., F.X. Heinz, C.W. Mandl, H. Holzmann, and C. Kunz. (1991). Fusion activity of flaviviruses: comparison of mature and immature (prM-containing) tick-borne encephalitis virions. J. Gen. Virol. *72*: 1323–1329.

Guirakhoo, F., R.A. Bolin, and J.T. Roehrig. (1992). The Murray Valley encephalitis virus prM protein confers acid resistance to virus particles and alters the expression of epitopes within the R2 domain of E glycoprotein. Virology *191*: 921–931.

Haag, L., H. Garoff, L. Xing, L. Hammar, S-T. Kan, and R. Holland Cheng. (2002). Acid-induced movements in the glycoprotein shell of an alphavirus turn the spikes into membrane fusion mode. EMBO J. *21*: 4402–4410.

Hallenberger, S., V. Bosch, H. Angliker, E. Shaw, H.-D. Klenk, and W. Garten. (1992). Inhibition of furin-mediated cleavage activation of HIV-1 glycoprotein gp160. Nature *360*: 358–361.

Hammar, L., S. Markarian, L. Haag, H. Lankinen, A. Salmi, and R. Holland Cheng. (2003). Prefusion rearrangements resulting in fusion peptide exposure in Semliki Forest Virus. J. Biol. Chem. *278*: 7189–7198.

Hatsuzawa, K., M. Nagahama, S. Takahashi, K. Takada, K. Murakami, and K. Nakayama. (1992). Purification and characterization of furin, a Kex2-like processing endoprotease, produced in Chinese Hamster Ovary cells. J. Biol. Chem. *267*: 16094–16099.

Hatta, M., Gao, P., Halfmann, P. and Kawaoka, Y. (2001). Molecular basis for high virulence of Hong Kong H5N1 influenza A viruses. Science *293*: 1840–1842.

He, R.-T., B.L. Innis, A. Nisalak, W. Usawattanakul, S. Wang, S. Kalayanarooj, and R. Anderson. (1995). Antibodies that block virus attachment to Vero cells are a major component of the human neutralizing antibody response against dengue virus type 2. J. Med. Virol. *45*: 451–461.

Heinz, F. X., K. Stiasny, G. Puschner-Auer, H. Holzmann, S.L. Allison, C.W. Mandl, and C.

Kunz. (1994). Structural changes and functional control of the tick-borne encephalitis virus glycoprotein E by the heterodimeric association with protein prM. Virology *198*: 109–117.

Henchal, E.A., J.M. McCown, D.S. Burke, M.C. Seguin, and W.E. Brandt. (1985). Epitope analysis of antigenic determinants on the surface of dengue-2 virions using monoclonal antibodies. Am. J. Trop. Med. Hyg. *34*: 162–169.

Henrich, S., A. Cameron, G.P. Bourenkov, R. Kiefersauer, R. Huber, I. Lindberg, W. Bode, and M.E. Than. (2003). The crystal structure of the proprotein processing proteinase furin explains its stringent specificity. Nature Struct. Biol. *10*: 520–526.

Horimoto, T., and Y. Kawaoka. (2001). Pandemic threat posed by avian influenza A viruses. Clin. Microbiol. Rev. *14*: 129–149.

Holyoak, T., M.A. Wilson, T.D. Fenn, C.A. Kettner, G.A. Petsko, R.S. Fuller, and D. Ringe. (2003). 2.4 Å resolution crystal structure of the prototypical hormone-processing protease Kex2 in complex with an Ala-Lys-Arg boronic acid inhibitor. Biochemistry *42*: 6709–6718.

Hunt, A. R., C. B. Cropp, and G. J. J. Chang. (2001). A recombinant particulate antigen of Japanese encephalitis virus produced in stably-transformed cells is an effective noninfectious antigen and subunit immunogen. J. Virol. Meth. *97*:133–149.

Iacono-Connors, L.C., J.F. Smith, T.G. Ksiazek, C.L. Kelley, and C.S. Schmaljohn. (1996). Characterization of Langat virus antigenic determinants defined by monoclonal antibodies to E, NS1 and preM and identification of a protective, non-neutralizing preM-specific monoclonal antibody. Virus Res. *43*: 125–136.

Jardetzky, T.S., and R.A. Lamb. (2004). A class act. Nature *427*: 307–308.

Kaufman, B.M., P.L. Summers, D.R. Dubois, W. Houston Cohen, M.K. Gentry, R.L. Timchak, D.S. Burke, and K.H. Eckels. (1989). Monoclonal antibodies for dengue virus prM glycoprotein protect mice against lethal dengue infection. Am. J. Trop. Med. Hyg. *41*: 576–580.

Kawaoka, Y., and R. G. Webster. (1989). Interplay between carbohydrate in the stalk and the length of the connecting peptide determines the cleavability of influenza virus hemagglutinin. J. Virol. *63*: 3296–3300.

Keelapang, P., R. Sriburi, S. Supasa, N. Punyadee, A. Songjaeng, A., Jairungsri, C. Puttikunt, W. Kasinrerk, P. Malasit, and N. Sittisombut. (2004). Alterations of pr-M cleavage and virus export in pr-M junction chimeric dengue viruses. J. Virol. *78*: 2367–2381.

Khromykh, A.A., A.N. Varnavski, and E.G. Westaway. (1998). Encapsidation of the flavivirus Kunjin replicon RNA by using a complementation system providing Kunjin virus structural proteins in *trans*. J. Virol. *72*: 5967–5977.

Kimura-Kuroda, J., and K. Yasui. (1983). Topographical analysis of antigenic determinants on envelope glycoprotein V3 (E) of Japanese encephalitis virus, using monoclonal antibodies. J. Virol. *45*: 124–132.

Klimpel, K.R., S.S. Molloy, G. Thomas, and S.H. Leppla. (1992). Anthrax toxin protective antigen is activated by a cell surface protease with the sequence specificity and catalytic properties of furin. Proc. Natl. Acad. Sci. USA *89*: 10277–10281.

Konishi, E., and A. Fujii. (2002). Dengue type 2 virus subviral extracellular particles produced by a stably transfected mammalian cell line and their evaluation for a subunit vaccine. Vaccine *20*:1058–1067.

Konishi, E., A. Fujii, and P.W. Mason. (2001). Generation and characterization of a mammalian cell line continuously expressing Japanese encephalitis virus subviral particles. J. Virol. *75*: 2204–2212.

Konishi, E., and P.W. Mason. (1993). Proper maturation of the Japanese encephalitis virus envelope glycoprotein requires cosynthesis with the premembrane protein. J. Virol. *67*:1672–1675.

Konishi, E., S. Pincus, E. Paoletti, R. E. Shope, T. Burrage, and P. W. Mason. (1992). Mice immunized with a subviral particle containing the Japanese encephalitis virus prM/M and E proteins are protected from lethal JEV infection. Virology *188*:714–720.

Kroeger, M. A., and P. C. McMinn. (2002). Murray Valley encephalitis virus recombinant subviral particles protect mice from lethal challenge with virulent wild-type virus. Arch. Virol. *147*:1155–1172.

Kuhn, R.J., W. Zhang, M. G. Rossmann, S. V. Pletnev, J. Corver, E. Lenches, C. T. Jones, S. Mukhopadhyay, P. R. Chipman, E. G. Strauss, T. S. Baker, and J. H. Strauss. (2002). Structure of dengue virus: Implications for flavivirus organization, maturation, and fusion. Cell *108*: 717–725.

Kummerer, B. M., and C. M. Rice. (2002). Mutations in the yellow fever virus nonstructural protein NS2A selectively block production of infectious particles. J. Virol. *76*:4773–4784.

de Lamballerie, X., S. Crochu, F. Billoir, J. Neyts, P. de Micco, E. C. Holmes, and E. A. Gould. (2002). Genome sequence analysis of Tamana bat virus and its relationship with the genus Flavivirus. J. Gen. Virol. *83*: 2443–2454.

Lee, E., C. E. Stocks, S.M. Amberg, C.M. Rice, and M. Lobigs. (2000). Mutagenesis of the signal sequence of yellow fever virus prM protein: enhancement of signalase cleavage *in vitro* is lethal for virus production. J. Virol. *74*: 24–32.

Lin, D., L. Li, D. Dick, R.E. Shope, H. Feldmann, A.D.T. Barrett, and M.R. Holbrook. (2003). Analysis of the complete genome of the tick-borne flavivirus Omsk hemorrhagic fever virus. Virology *313*: 81–90.

Lindenbach, B.D., and C.M. Rice. (2001). *Flaviviridae*: the viruses and their replication, p. 991–1041. *In* D.M. Knipe, P.M. Howley, D.E. Griffin, R.A. Lamb, M.A. Martin, B. Roizman, and S.E. Strauss (ed.), Fields virology, 4th ed. Lippincott Williams & Wilkins, Philadelphia, Pa.

Lobigs, M. (1993). Flavivirus premembrane protein cleavage and spike heterodimer secretion require the function of the viral proteinase NS3. Proc. Natl. Acad. Sci. USA *90*:6218–6222.

Lobigs, M., and E. Lee. (2004). Inefficient signalase cleavage promotes efficient nucleocapsid incorporation into budding flavivirus membranes. J. Virol. *78*: 178–186.

Lorenz, I.C., S.L. Allison, F.X. Heinz, and A. Helenius. (2002). Folding and dimerization of tick-borne encephalitis virus envelope proteins prM and E in the endoplasmic reticulum. J. Virol. *76*: 5480–5491.

Markoff, L., A. Chang, and B. Falgout. (1994). Processing of flavivirus structural glycoproteins: stable membrane insertion of premembrane requires the envelope signal peptide. Virology *204*: 526–540.

Mason, P.W., S. Pincus, M.J. Fournier, T.L. Mason, R.E. Shope, and E. Paoletti. (1991). Japanese encephalitis virus-vaccinia recombinants produce particulate forms of the structural membrane proteins and induce high levels of protection against lethal JEV infection. Virology *180*: 294–305.

Men, R., T. Yamashiro, A.P. Goncalvez, C. Wernly, D.J. Schofield, S.U. Emerson, R.H. Purcell, and C.-J. Lai. (2004). Identification of chimpanzee Fab fragments by repertoire cloning and production of a full-length humanized immunoglobulin G1 antibody that is highly efficient for neutralization of dengue type 4 virus. J. Virol. *78*: 4665–4674.

Modis, Y., S. Ogata, D. Clements, and S.C. Harrison. (2004). Structure of the dengue virus envelope protein after membrane fusion. Nature *427*: 313–319.

Molloy, S.S., E.D. Anderson, F. Jean, and G. Thomas. (1999). Bi-cycling the furin pathway: from TGN localization to pathogen activation and embryogenesis. Trends Cell. Biol. *9*: 28–35.

Molloy, S.S., P. A. Bresnahan, S. H. Leppla, K.R. Klimpel, and G. Thomas. (1992). Human furin is a calcium-dependent serine endoprotease that recognizes the sequence Arg-X-X-Arg and efficiently cleaves anthrax toxin protective antigen. J. Biol. Chem. *267*: 16396–16402.

Morsy, J., W. Garten, and R. Rott. (1994). Activation of an influenza virus A/Turkey/Oregon/71 HA insertion variant by the subtilisin-like endoprotease furin. Virology *202*: 988–991.

Moulard, M., L. Chaloin, S. Canarelli, K. Mabrouk, H. Darbon, and L. Challoin. (1998). Retroviral envelope glycoprotein processing: structural investigation of the cleavage site. Biochemistry *37*: 4510–4517.

Mukhopadhyay, S., B.-S. Kim, P.R. Chipman, M.G. Rossmann, and R.J. Kuhn. (2003). Structure of West Nile virus. Science *302*:248.

Murray, J.M., J.G. Aaskov, and P.J. Wright. (1993). Processing of the dengue virus type 2 proteins prM and C-prM. J. Gen. Virol. *74*: 175–182.

Nakayama, K. (1997). Furin: a mammalian subtilisin/Kex2p-like endoprotease involved in processing of a wide variety of precursor proteins. Biochem. J. *327*: 625–635.

Oliva, R., M. Leone, L. Falcigno, G. D'Auria, M. Dettin, C. Scarinci, C. Di Bello, and L. Paolillo. (2002). Structural investigation of the HIV-1 envelope glycoprotein gp160 cleavage site. Chemistry *8*: 1467–1473.

Pincus, S., P. W. Mason, E. Konishi, B. A. Fonseca, R. E. Shope, C. M. Rice, and E. Paoletti. (1992). Recombinant vaccinia virus producing the prM and E proteins of yellow fever virus protects mice from lethal yellow fever encephalitis. Virology *187*:290–297.

Randolph, V. B., G. Winkler, and V. Stollar. (1990). Acidotropic amines inhibit proteolytic processing of flavivirus prM protein. Virology *174*: 450–458.

Riehle, M.A., S.F. Garczynski, J.W. Crim, C.A. Hill, and M.R. Brown. (2002). Neuropeptides and peptide hormones in *Anopheles gambiae*. Science *298*: 172–175.

Rockwell, N.C., D.J. Krysan, T. Komiyama, and R.S. Fuller. (2002). Precursor processing by kex2/furin proteases. Chem. Rev. *102*: 4525–4548.

Roehrig, J.T., R.A. Bolin, and R.G. Kelly. (1998). Monoclonal antibody mapping of the envelope glycoprotein of the dengue 2 virus, Jamaica. Virology *246*: 317–328.

Rohm, C., T. Horimoto, Y. Kawaoka, J. Suss, and R.G. Webster. (1995). Do hemagglutinin genes of highly pathogenic avian influenza viruses constitute unique phylogenetic lineages? Virology *209*: 664–670.

Russell, P. K., W. E. Brandt, and J. Dalrymple. (1980). Chemical and antigenic structure of flaviviruses, p. 503–529. *In* R. W. Schlesinger (ed.), The togaviruses: biology, structure, replication. Academic Press, New York, N.Y.

Schalich, J., S. L. Allison, K. Stiasny, C. W. Mandl, C. Kunz, and F. X. Heinz. (1996). Recombinant subviral particles from tick-borne encephalitis virus are fusogenic and provide a model system for studying flavivirus envelope glycoprotein functions. J. Virol. *70*:4549–4557.

Se-Thoe, S.Y., M.M. Ng, and A.E. Ling. (1999). Retrospective study of Western blot profiles in immune sera of natural dengue virus infections. J. Med. Virol. *57*: 322–330.

Songjaeng, A. (2004). M.Sc. thesis. Chiang Mai University, Chiang Mai, Thailand.

Sriburi, R., P. Keelapang, T. Duangchinda, S. Pruksakorn, N. Maneekarn, P. Malasit, and N. Sittisombut. (2001). Construction of infectious dengue 2 cDNA clones using high copy number plasmid. J. Virol. Meth. *92*: 71–82.

Stadler, K., S.L. Allison, J. Schalich, and F.X. Heinz. (1997). Proteolytic activation of tick-borne encephalitis virus by furin. J. Virol. *71*: 8475–8481.

Staropoli, I., C. Chanel, M. Girard, and R. Altmeyer. (2000). Processing, stability, and receptor binding properties of oligomeric envelope glycoprotein from a primary HIV-1 isolate. J. Biol. Chem. *275*: 35137–35145.

Steiner, D.F. (1998). The proprotein convertases. Curr. Opin. Chem. Biol. *2*: 31–39.

Steinhauer, D.A. (1999). Role of hemagglutinin cleavage for the pathogenicity of influenza virus. Virology *258*: 1–20.

Stiasny, K., S.L. Allison, A. Marchler-Bauer, C. Kunz, and F.X. Heinz. (1996). Structural requirements for low-pH-induced rearrangements in the envelope glycoprotein of tick-borne encephalitis virus. J. Virol. *70*: 8142–8147.

Stiasny, K., S. L. Allison, J. Schalich, and F. X. Heinz. (2002). Membrane interactions of the tick-borne encephalitis virus fusion protein E at low pH. J. Virol. *76*:3784–3790.

Stocks, C. E., and M. Lobigs. (1995). Posttranslational signal peptidase cleavage of the flavivirus C-prM junction *in vitro*. J. Virol. *69*:8123–8126.

Stocks, C. E., and M. Lobigs. (1998). Signal peptidase cleavage at the flavivirus C-prM junction: dependence on the viral NS2B-3 protease for efficient processing requires determinants in C, the signal peptide, and prM. J. Virol. *72*: 2141–2149.

Subbarao, K., A. Klimov, J. Katz, H. Regnery, W. Lim, H. Hall, M. Perdue, D. Swayne, C. Bender, J. Huang, M. Hemphill, T. Rowe, M. Shaw, X. Xu, K. Fukuda, and N. Cox. (1998). Characterization of an avian influenza A (H5N1) virus isolated from a child with a fatal respiratory illness. Science *279*: 393–396.

Thomas, G. (2002). Furin at the cutting edge: from protein traffic to embryogenesis and disease. Nat. Rev. Mol. Cell. Biol. *3*: 753–766.

Vazquez, S., M.G. Guzman, G. Guillen, G. Chinea, A.B. Perez, M. Pupoa, R. Rodriguez, O. Reyes, H.E. Garay, I. Delgado, G. Garcia, and M. Alvarez. (2002). Immune response to synthetic peptides of dengue prM protein. Vaccine *20*: 1823–1830.

Wang, S., R. He, and R. Anderson. (1999). PrM- and cell-binding domains of the dengue virus E protein. J. Virol. *73*: 2547–2551.

Wengler, G., and G. Wengler. (1989). Cell-associated West Nile flavivirus is covered with E+pre-M protein heterodimers which are destroyed and reorganized by proteolytic cleavage during virus release. J. Virol. *63*: 2521–2526.

Westaway, E.G. (1980). Chemical and antigenic structure of flaviviruses, p. 531–581. *In* R. W. Schlesinger (ed.), The togaviruses: biology, structure, replication. Academic Press, New York, N.Y.

Yamshchikov, V. F., and R. W. Compans. (1993). Regulation of the late events in flavivirus protein processing and maturation. Virology *192*:38–51.

Zhang, Y., J. Corver, P.R. Chipman, W. Zhang, S.V. Pletnev, D. Sedlak, T.S. Baker, J. H. Strauss, R.J. Kuhn, and M.G. Rossmann. (2003). Structures of immature flavivirus particles. EMBO J. *22*: 2604–2613.

Zhou, A., G. Webb, X. Zhu, and D.F. Steiner. (1999). Proteolytic processing in the secretory pathway. J. Biol. Chem. *274*: 20745–20748.

Diagnosis of Dengue Virus Infection

7

María G. Guzmán, Delfina Rosario, and Gustavo Kourí

Abstract

The availability of a rapid, sensitive, specific and economical laboratory diagnostic test to confirm a dengue infection is of utmost importance in the management of dengue fever and dengue hemorrhagic fever and in the prevention and control of outbreaks. In addition, travelers returning from endemic countries also constitute a diagnostic challenge to physicians.

Virus detection throughout virus isolation, antigen and genome detection allows a confirmatory serotype diagnosis. Antidengue IgM detection or high antidengue IgG titer in a serum sample or seroconversion in a paired of sera are the most routinely applied serological methods for dengue diagnosis.

In spite of the advance of dengue diagnosis, some problems still warrant the timely development of new solutions. For instance, virus isolation is time consuming; PCR requires specific laboratory equipments and facilities as well as extensive evaluation of the different protocols under field conditions; IgM antibody detection requires proper timing and is confounded by false positive reactions and the long persistence of IgM antibodies, commercial kits still need to be critically evaluated, and the costs and availability of these kits and other reagents need to be addressed. In this chapter, an update of dengue diagnosis, its needs and challenges are discussed.

Introduction

Dengue fever (DF) occurs in epidemic form when the virus is being introduced into an environment in which it can establish itself due to the presence of *Aedes aegypti* and of a susceptible population. Explosive epidemics followed by endemic establishment of the virus are frequently observed. When a different dengue serotype is introduced into a region, dengue hemorrhagic fever/ dengue shock syndrome (DHF/DSS) epidemic can occur. Considering the duration of the disease (usually 5 to 7 days), the severity of the clinical picture observed in some individuals, and the epidemiological burden of this illness, clinical diagnosis and laboratory surveillance of dengue is of crucial importance (Guzman *et al.*, 2004).

Early differential diagnosis of dengue infection is desired, as it is the prerequisite for appropriate interventions. Also, it can prevent unnecessary treatments against other diseases with similar clinical presentation.

Thus, the availability of a rapid, sensitive, specific, and economical laboratory diagnostic test to confirm dengue infection is of greatest importance in the management of dengue fever, DHF/DSS, and in the prevention and control of outbreaks. Finally, due to the increased mobility of the society, many nonimmune travelers visit endemic areas and are exposed to virus infection. Travelers returning from endemic countries also constitute a diagnostic challenge for physicians (Schwartz *et al.*, 1996, Schwartz *et al.*, 2000).

Dengue virus

After the description of the yellow fever virus, dengue fever was the second human disease whose etiology was identified as a "filtrable virus" (Ashburn and Craig, 1907). In 1944, Kimura and Hotta were the first to isolate the virus by inoculating serum from acutely ill patients into suckling mice (Kimura and Hotta, 1944). In 1944 to 1945, Sabin and colleagues recovered seven strains from American soldiers stationed in Hawaii, New Guinea, and India (Sabin, 1952). Initially, the virus strains were recovered by intracutaneous inoculation in human volunteers. Later, some of the human isolates were adapted to mice. Hemagglutination inhibition assay (HI) allowed the classification of two antigenically different viruses, named dengue 1 and dengue 2. The Hawaiian dengue 1 and New Guinea dengue 2 strains were considered as prototypes. Finally, two more serotypes, named dengue 3, and dengue 4 were isolated from patients with hemorrhagic disease during the epidemic in Manila in 1956 (Hammon *et al.*, 1960, Hammon *et al.*, 1964). In 1954, Smithburn reported the existence of dengue complex-specific antigenic determinants using neutralization assays. Later studies proposed a single unique complex of dengue viruses (Calisher *et al.*, 1989, De Madrid and Porterfield, 1974). Finally, studies employing monoclonal antibodies confirmed that the dengue viruses form an antigenic complex (Henchal *et al.*, 1983).

Since dengue viruses are maintained in cycles involving hematophagous arthropod vectors and vertebrate hosts (Sabin, 1950, Schelesinger, 1977), they are considered as arthropod-borne viruses. Dengue viruses were first classified as members of group B of arboviruses and specifically as members of the Togaviridae family (Fenner *et al.*, 1974). In 1985, the genus *Flavivirus* was reclassified into the family of Flaviviridae (Westaway *et al.*, 1985).

As other members of the *Flavivirus* genus, dengue viruses are spherical, lipid enveloped viruses that contain a positive single strand RNA genome of approximately 11 kilobases in length. They are sensitive to ether, chloroform and other lipid solvents. They are labile and readily inactivated at temperatures above 30°C and in acid conditions and remain infectious for years at −70°C (Heinz *et al.*, 1990, Innis 1995, Monath *et al.*, 1996). The RNA of positive polarity encodes three structural proteins (capsid, C, membrane M and its precursor prM, and the envelope E) and seven nonstructural proteins (NS1-NS2a-NS2b-NS3-NS4a-NS4b-NS5). The genomic RNA has a type I cap at its 5′ end (m7GpppAmp) and lacks a 3′-terminal poly(A) tract (Chambers *et al.*, 1990, Mandl *et al.*, 1989).

The C protein (12–14 kd) forms the structural component of the viral nucleocapsid. Its protein homology is low; however, hydrophobic and hydrophilic amino acid regions are conserved (Mandl *et al.*, 1988). The cleavage of prM glycoprotein (18–19 kd) is presumably linked to virus maturation as prM and M proteins are found in intracellular and extracellular virions. The prM protein is part of immature virions and its proteolytic cleavage generates the mature virions. Antibodies to prM protein have been shown to protect mice against lethal dengue infection (Kaufman *et al.*, 1989). Also, synthetic peptides of M and prM proteins are able to induce neutralizing antibody and protection in immunized mice (Vazquez *et al.*, 2002).

The E protein (53–54 kd) is involved in virion assembly, receptor binding, and membrane fusion. It is the viral hemagglutinin and the primary target for neutralizing antibodies. Consequently, cross-neutralization data reflect differences in the antigenic structure of this protein. Cross-HI tests using polyclonal antibodies provide evidences for the presence of group, serocomplex, and type-specific determinants. Subcomplex, subtype, strain, and even substrain specificity have been found using monoclonal antibodies. Being the major component of the virus surface, E protein plays a dominant role in the induction of neutralizing antibodies and immune protective response but also in the immune-enhancement phenomenon. Active immunization of animals using native or recombinant proteins and passive protection experiments with E monoclonal antibodies support this (Brandriss *et al.*, 1986, Guzman *et al.*, 2003a, Kaufman *et al.*, 1987Mune *et al.*, 2003).

Three major antigenic epitope clusters (A, B, C) have been characterized on the E protein of flaviviruses. These antigenic domains correspond roughly to domains II (finger-like), III (immunoglobulin-like) and I (central). Although the three antigenic domains contain epitopes involved in neutralization and hemagglutination inhibition, only domain A contains flavivirus cross-reactive epitopes. This domain derives from two regions (residues 50–125 and 200–250) and contains variable regions as well as conserved sequences. Particularly residues 98–111 are possibly involved in acid-catalyzed fusion. Domain B comprises residues 300–395. Studies with Tick-borne encephalitis virus (TBE) suggest that this region is involved in tissue tropism and receptor binding. Fusion proteins of domain B induce neutralizing antibody development in immunized mice and protection after challenge (Hermida *et al.*, 2004, Rey *et al.*, 1995, Rice, 1996, Roehrig, 1997).

The NS1 glycoprotein (39–41 kDa) exists as cell-associated, cell surface and extracellular forms. Type-specific, complex-specific and group-reactive epitopes have been identified on NS1 protein. NS1-specific antibodies can bind to the surface of infected cells and mediate complement-dependent cell lysis. Immunization with native and recombinant NS1 protein protects mice from lethal dengue encephalitis (Mackenzie *et al.*, 1996, Schlesinger *et al.*, 1987, Srivastava *et al.*, 1995, Zhang *et al.*, 1988).

NS3 (68–70 kDa) has a protease activity and a nucleotide triphosphatase/helicase activity. It plays an important role in the viral RNA replication. Passive immunization of mice with specific anti-NS3 monoclonal antibodies was reported to prolong survival after intracerebral challenge with dengue virus. However, the primary immunological role of NS3 seems to be its function as target for cytotoxic T cells (Kurane *et al.*, 1999, Mathew *et al.*, 1998, Tan *et al.*, 1990).

NS5 is the largest (103–105 kDa) and most conserved of the flavivirus proteins. It is presumed to play a role as the viral RNA polymerase (Chambers, 1990, Devappa *et al.*, 2003). Diagnostic tools for dengue infection take in advantage the characteristics of the virus and the humoral immune response to the different dengue proteins.

Human infection

After an incubation period of 4 to 10 days, dengue infection can produce an asymptomatic infection, undifferentiated fever, classical DF or the severe form, DHF/DSS. Early studies demonstrated that human volunteers reinoculated with the same dengue serotype were immune to the homologous virus. Contrarily, results of inoculation with a heterologous virus were found to depend on the interval after first infection. Immunity was demonstrated only during the first two months after the second infection and cross protection was demonstrated between 2 and 9 months (Sabin, 1952). It is accepted that dengue antibodies probably persist for life. Nine of the Siler's and Simmon's volunteers had dengue antibodies 42 to 48 years after dengue virus inoculation (Halstead, 1974).

Two main forms of dengue infection have been characterized, primary and secondary infection. The first is induced by the bite of an infested mosquito in an individual without previous exposition to any flavivirus. Secondary infection mostly occurs in individuals previously infected by any of the four dengue serotypes, but also in individuals immune to other flaviviruses. Although tertiary and quaternary dengue infections are possible, not much data are available.

After a primary flavivirus infection, IgM antibodies are initially produced followed by IgG antibodies. A second infection with a heterologous serotype results mainly in the production of IgG antibodies that are highly cross reactive. In some of these cases, IgM response can be absent (Dittmar *et al.*, 1979; Edelman *et al.*, 1973). The kinetics of dengue IgM antibody has been studied in patients with primary and secondary dengue infection. In these patients, the disappearance of fever and viremia with the simultaneous appearance of IgM antibody is also observed.

Primary dengue infection is characterized by elevations of anti-dengue IgM, three to five days after the onset of fever. IgM antibody generally persists for 30 to 60 days. IgG levels also become elevated after 9–10 days and remain detectable for life (Gubler, 1996).

During secondary infection, IgM levels generally rise more slowly and reach lower levels than in primary infection, while IgG levels rise rapidly from one to two days after the onset of fever to higher levels than observed in primary or past dengue infection and remain at these levels for 30–40 days (Gubler 1996, Innis 1997). Although not well studied, it is known that specific IgA and IgE antibodies are developed both during primary and secondary infection. IgA is broadly cross-reactive to the four dengue virus serotypes and IgE has been related with disease severity (Balmaseda *et al.*, 2003, Groen *et al.*, 1999, Koraka *et al.*, 2003, Summers *et al.*, 1984, Vazquez *et al.*, 2004). More details are presented below.

In humans, viremia is generally detected in all cases at the same time as symptoms appear, being no longer detectable at the time of defervescence (Nisalak *et al.*, 1970, Rothman, 1997, Vaughn *et al.*, 2000).

Objectives of dengue diagnosis

Accurate and efficient diagnosis of dengue infection is important for clinical care, surveillance support, pathogenesis studies, and vaccine research. Diagnosis is also important for case confirmation (DF or DHF/DSS), to differentiate dengue from other diseases such as leptospirosis, rubella and other flavivirus infections and for the clinical management and evaluation of patients with severe disease (Guzman *et al.*, 1996; Vorndam *et al.*, 1997).

In conjunction with the clinical and epidemiological surveillance, the early detection of dengue circulation or an increase in dengue activity provides to health authorities useful information on time, location, virus serotype and disease severity (PAHO, 1994). The utility of good dengue diagnostic tools is critical for laboratory confirmation of DHF/DSS. This includes the registration of case fatalities, the determination of the strains involved, and the estimation of the total incidence following epidemics (Guzman *et al.*, 1990, Guzman *et al.*, 1995, PAHO, 1994).

Finally, dengue diagnosis is of major importance for research on host, virus and vector characteristics, for determining the epidemiological conditions influencing the pathogenesis of the disease and for vaccine evaluation (phase 1–3 studies) (Durbin *et al.*, 2001, Jirakanjanakit, *et al.*, 1999, Kanesa-thasan, *et al.*, 2001)

Clinical samples

Virus detection throughout virus isolation, antigen, and genome detection allows a confirmatory serotype diagnosis. Detection of anti-dengue IgM, high anti-dengue IgG titer or of seroconversion in a pair of sera are the most frequently applied serological methods for dengue diagnosis (Table 7.1).

Which clinical sample is suitable for dengue diagnosis and which methodology be employed, depends of the phase of the disease. In most clinical states, serum is the sample of choice. If collected during the first four to five days after the onset of fever, it is useful for virus detection. Acute and convalescent serum samples are useful for serological diagnosis. Sterile serum without anticoagulant is useful for dengue virus detection. Samples must be rapidly transported at 4°C to the laboratory and must be processed as soon as possible.

Tissue specimens from fatal cases are also useful for virus detection. The tissue samples should be frozen at –70°C immediately. Liver, spleen, lymph nodes, and other tissues are useful. Hemolysis of serum, repeated freezing, thawing, and bacterial contamination of clinical samples must be avoided. Finally, peripheral blood

Table 7.1 Main methods currently employed for dengue diagnosis.

	Clinical sample	Methods
Serology	IgM detection in monosera collected after five-six days of onset of fever	MAC/ELISA
	Seroconversion of antibody in paired sera collected during the acute and convalescent phase of illness Detection of high level of antibody in monosera collected in the first days after onset	ELISA, HI assay, PRNT
Viral isolation	Virus detected in serum, plasma, homogenized tissues (liver, spleen, lymph nodes, others)	Direct inoculation of mosquitoes, insect and mammalian cell cultures, intracranial inoculation of suckling mice
Antigen detection	Serum Fresh and paraffin embedded-tissues	NS1 detection by ELISA Immunohistochemical technique for antigen detection in tissues
Genome detection	Serum, plasma, fresh and paraffin embedded-tissues	RT/PCR, Realtime PCR

leukocytes have been recommended by some authors (McNair *et al.*, 1980).

Sera for serological studies can be stored at 4°C for short and at –20°C for longer periods. Blood collected on filter paper is also useful for serological diagnosis including IgM and IgG detection by ELISA, neutralizing and hemagglutination inhibition antibody detection. More recently, blood on filter paper has been also used for RNA detection (Guzman *et al.*, 1982, Prado *et al.*, 2005, Vazquez *et al.*, 1991). Clinical samples must be accompanied by the clinical and epidemiological patient and specimen information. Name, age, gender, address, day of onset of fever, travel history, presence of similar cases, day of sample collection, sample type, a summary of the clinical and epidemiological data of the patient in conjunction with the name and data of the physician and hospital must accompany the specimen.

Serological dengue diagnosis

Currently, serological studies are the most applied methods for dengue diagnosis. The employed techniques are less expensive, relatively easy to perform and to apply into routine work. Depending on the test and the studied sample, they allow to confirm a dengue infection or to suspect a recent infection. Different serological markers are being used, the anti-dengue IgM detection being the most frequently employed as a marker of recent infection. Yet, some studies suggest that anti-dengue IgA and IgE can also be used. The presence of anti-dengue IgG in a serum sample indicates a past infection. However, the presence of a high IgG titer in a serum sample, seroconversion or fourfold increase in a paired sera from a dengue suspected case, are criteria for a recent or confirmed dengue infection respectively.

IgM detection

Up to now, the detection of anti-dengue IgM is the best indicator of an active or recent infection. Contrary to IgG, IgM detection provides information from a single serum sample. Since IgM antibodies are produced first, their presence provides the earliest serological evidence of an ongoing infection.

The following topics deserve special interest: (1) time of serum sample collection, (2) false positive and false negative reactions, (3) Usefulness of IgM detection for serotyping, (4) Cross-reaction with other dengue serotypes and flaviviruses, (5) formats for dengue IgM detection.

The development of the IgM antibody capture ELISA (MAC-ELISA) provided a relatively simple and rapid test to those endemic countries where laboratory facilities are limited. It requires only a single serum dilution, no serum treatment and provides definite results from single serum samples. Considering these advantages, it has been applied for dengue diagnosis and serological surveillance since the 1980s. Most of the dengue IgM studies have been performed using this assay. MAC-ELISA uses a solid phase coated with anti-human μ chain immunoglobulins that captures human IgM, followed by a viral antigen and indicator antibodies (Burke 1983). Both dengue antigens prepared in mice and cell cultures have been employed although the former are the most widely applied (Cardosa *et al.*, 1992).

The majority of confirmed dengue cases (60–78%) have detectable IgM (Gubler, *et al.*, 1988, Innis *et al.*, 1989, Lam *et al.*, 1987). Some studies show that 98% of the patients developed IgM antibodies between six to ten days after the onset of infection and all sera from patients tested between 10 and 30 days have IgM (PAHO, 1994). There is a variation among the amount of produced IgM, in the rapidity with which IgM develops, and in the duration that detectable IgM antibodies persists after infection. Some patients show IgM at day two or three while others do have detectable IgM at seven or eight days after infection. Different studies have demonstrated that some patients have detectable IgM antibody by 60 to 90 days post infection (Gubler *et al.*, 1988, Nogueira *et al.*, 1992). Innis observed that anti-dengue IgM levels remain longer positive in patients with primary infections compared to patients with secondary infections (Innis *et al.*, 1989). Studies done using sera of dengue cases from Nicaragua, Panama and Costa Rica, demonstrated that 56% of the samples were positive at day one to five increasing to 95% at days six to ten after onset of fever (Guzman *et al.*, 1996, Guzman *et al.*, 2002, Guzman *et al.*, 2004). In general, the use of IgM detection assays for scientific purposes is recommended in serum samples collected after days five or six after the onset of fever.

Compared to HI assay, 1.7% of the studied samples show a false positive response and 10% a false negative response by MAC-ELISA. During a period of two years (1998–1999), 0.23% of false positive reactions in 14806 studied serum samples received through the Cuban dengue surveillance system were observed. No dengue circulation was reported in this period (Guzman *et al.*, 2002).

The specificity of MAC-ELISA is in general broadly reactive among dengue and other flaviviruses and it is not recommended as an identification method of the infecting dengue serotype. Different studies have suggested that IgM antibodies detected by ELISA show high or complete specificity in cases of yellow fever, Zika and Wesselsbron infection whereas extensive cross-reactions were noted in dengue infection (Saluzzo *et al.*, 1986). Other authors have proven that anti-dengue and anti-Japanese encephalitis IgM antibodies were clearly distinguished using acute or convalescent sera (Innis *et al.*, 1989). Although relatively specific in distinguishing dengue from other flavivirus infections, MAC-ELISA is not suitable for the discrimination of dengue serotypes. In a series of serum samples of dengue patients from Nicaragua, Panama and Costa Rica, a serotype-specific IgM response was observed in only 15% and 16% of DF and DHF cases respectively and in 17% and 14% of the primary and secondary cases (Guzman *et al.*, 2004). In a more recent study, Vazquez demonstrated that in primary DF, in secondary DF, and DHF cases, anti-dengue IgM is cross reactive against the four dengue serotypes (Vazquez *et al.*, 2004). Similar results have been reported previously (Delgado *et al.*, 2002).

In 84 confirmed cases of DF, 11% cross-reacted by MAC-ELISA using the yellow fever antigen. Contrarily, in 21 individuals vaccinated with the 17D yellow fever vaccine, all of them were negative for dengue IgM demonstrating a high specificity of IgM antibodies against yellow fever (Vazquez *et al.*, 2003a).

Immunofluorescence has been reported to detect specific IgM antibodies in serum (Vathanophas *et al.*, 1973). However, the fluorescence was often weak, and the assay only allows the study of a low number of samples.

Because MAC-ELISA is easy and relatively simple, it has been widely applied for dengue surveillance providing the most rapid method for the detection of dengue virus infections in endemic countries. In addition, IgM serosurveys in nonendemic areas allow a rapid screening for dengue virus circulation considering that those positive samples represent recent infections.

Although most laboratories use "in house" tests, commercial kits for IgM detection have appeared on the market in the last years. Different formats such as the MAC-ELISA, the immunochromatographic test, the dipstick dot blot enzyme-linked immunosorbent assay, the AuBioDOT IgM capture, and the dot blot enzyme-linked immunosorbent assay have been employed (Cardosa *et al.*, 1995, Cuzzubo *et al.*, 2001, Kittigul *et al.*, 1998, Laferte *et al.*, 1998, Lam *et al.*, 1996, Palmer *et al.*, 1999, Sang *et al.*, 1998, Vazquez *et al.*, 2003b, Wu *et al.*, 1997). Some of these formats provide a diagnosis in minutes being most suitable for individuals or a small number of clinical samples. Some others such as the IgM capture ultramicroELISA (UME/dengue) are widely employed in the serological surveillance of some countries in Latin America allowing a rapid screening for dengue IgM antibodies. A recent review showed the existence of more than 50 commercial kits for IgM and IgG detection mainly based on ELISA, dot blot and immunochromatographic techniques (WHO/TDR, 2004).

IgA detection

The presence of virus-specific IgA is also a reliable indicator for recent virus replication on mucosal surfaces for many virus infections. Recently, Talarmin detected anti-dengue IgM and IgA antibodies in 178 sera from patients with DF (Talarmin *et al.*, 1998). 100% sensitivity and specificity was obtained. IgA antibodies were detected at day six following the onset of fever until day 25. In average, IgM antibodies were detected on day 3.8 and IgA antibodies on day 4.6.

Using immunofluorescence, the kinetics of anti-dengue IgA was measured in serum samples from a patient with primary infection (Groen *et al.*, 1999). During the first two months after infection, specific IgA antibody titers largely paralleled those of the specific IgM antibody levels detected by MAC-ELISA. 113 days after infection, IgA antibodies were no longer detected contrary to IgM that still was present. The level of IgG antibodies remained the same during the three months of study. Although these authors also postulated the diagnostic value of IgA serum detection, a higher percentage of IgA detection was observed in acute serum samples from secondary cases (62%) when compared to primary cases (17%). Similar results were obtained by Balmaseda when the authors applied an ELISA for the detection of anti-dengue IgA in sera (Balmaseda *et al.*, 2003). In a recent study, Vazquez and co-workers, demonstrated significantly higher levels of anti-dengue IgA in acute sera of patients with a second infection independently of the severity of the disease (DF or DHF) (Vazquez *et al.*, 2004).

IgE detection

Koraka studied the kinetic of total and dengue-specific immunoglobulin IgE in serial samples from 168 patients (Koraka *et al.*, 2003). Higher levels of both total and specific IgE were noted in dengue cases compared to controls. No differences were observed in acute samples from patients with primary or secondary infection although dengue-specific IgE was higher in sera from DHF cases. In a similar approach, other authors demonstrated higher levels of dengue-specific IgE in secondary than in primary infection independently of the clinical picture (DF or DHF) (Vazquez *et al.*, 2004). The implications of these observations in terms of diagnosis, clinical prognosis, and pathogenetic role deserve careful study.

IgG detection

Serodiagnosis of most patients by the IgG response requires the use of paired acute/

convalescent sera to obtain confirmatory results. The diagnosis depends upon the demonstration of rising titers of dengue antibodies between acute and convalescent sera or the presence of high titer dengue antibodies in secondary dengue infection using HI assay or ELISA. Very often, the collection of paired sera is not possible or both serum samples are not well spaced to show a rise in titers of diagnostic value. Thus, in many cases nonconclusive results are obtained.

Historically, HI assay, as described by Clarke and Casals has been the election method for dengue diagnosis (Clarke and Casals, 1958). This assay, while sensitive and reproducible, requires treatment of sera with acetone or kaolin to remove inhibitors and with goose red cells to eliminate the nonspecific agglutinins. It also requires convalescent sera. HI assay is satisfactory for identifying flavivirus infections; however, it lacks the specificity to distinguish the type of flavivirus, particularly in patients who have been infected by several flaviviruses (Innis *et al.*, 1989). In general, a test serum must be tested with a battery of several different flavivirus antigens to avoid inadvertent selection of the incorrect antigens. Conversely, the presence of serum HI antibodies does not prove prior infection with the test antigen virus; they only constitute evidence for past infection with an unspecified flavivirus (Burke *et al.*, 1987).

Although HI assay has been used traditionally, ELISA gained acceptance as a faster and more convenient alternative (Gubler 1996). ELISA for dengue antibody detection is inexpensive and simple to perform, being a good screening test with broad cross-reactivity and high sensitivity. Several ELISA assays for arbovirus antibody detection have been described. In the double antibody sandwich ELISA, viral antigen is captured by affinity-purified antibodies adsorbed to the solid phase; human serum is later added and binding of antibodies is detected by an antihuman conjugate. Other methods include direct antibody assay using microplates coated with gradient-purified virus or virus infected cultured cells as antigens (Miagostovich *et al.*, 1999, Oseni *et al.*, 1983, Tadeu *et al.*, 1987).

Good correlation between HI and IgG-ELISA results have been demonstrated previously (Chungue *et al.*, 1989, Figueiredo el at., 1989), emphasizing the usefulness of this assay for dengue surveillance. An ELISA inhibition method (EIM) has been widely applied for both dengue diagnosis and seroprevalence studies showing an elevated sensitivity and specificity when compared to HI assay (Fernandez *et al.*, 1990, Vazquez *et al.*, 1989). A monoclonal antibody based capture ELISA for the detection of anti-arboviral IgG was developed by Johnson and co-workers (Johnson *et al.*, 2000). The system allowed testing IgG against some arboviruses from Alphavirus, *Flavivirus* and Bunyavirus genera, and IgG results correlated with those obtained by plaque reduction neutralization technique (PRNT). Previously, Burke reported the usefulness of an epitope-blocking immunoassay using monoclonal antibodies for identifying individuals immune to dengue and Japanese encephalitis (Burke *et al.*, 1987), and Summers studied the antibody response to dengue 2 vaccine using radioimmunoassay method (Summers *et al.*, 1984). This latter system employs purified antigen but was 10 to 500 times more sensitive than neutralization or HI assays.

As an alternative of IgG-ELISA, immunoperoxidase monolayer assay (IPMA) measures antigen-antibody interactions in virus infected cells that are cultured and sequentially reacts with a specific antibody and with horseradish peroxidase-labeled antiglobulin chromogenic substrate. IPMA has been employed by some authors with good results (Soliman *et al.*, 1997). In addition, a dot enzyme immunoassay for dengue IgG detection has been applied (Cardosa *et al.*, 1988a, Cardosa *et al.*, 1988b, Young, 1989).

HI assay and IgG ELISA are cross-reactive within the flavivirus group and may show higher antibody titers to homologous antigens only in carefully timed samples from primary infections (Gubler 1988). The complement fixation test is considered to be serotype-specific within the dengue group, but this test relies on short-lived antibodies and is rarely used (Vorndam *et al.*, 2002). Complement fixation (CF) antibodies generally appear later than HI antibodies, are more specific in primary infections, and usually persist for short periods (Herrera *et al.*, 1985, Vazquez *et al.*, 1986, Vorndam *et al.*, 1997, Vorndam *et al.*, 2002).

The virus neutralization, mainly through the PRNT is the election method for specific serotype determination (Russell *et al.*, 1967). In general, neutralizing antibodies rise at about the same time than HI and ELISA antibody titers but more quickly than CF antibody titers and persist for more than 48 years (Halstead *et al.*, 1974). Since the first descriptions of a dengue plaque assay (Georgiades *et al.*, 1965, Schulze *et al.*, 1963), several protocols using agar, agarose, carboxymethycellulose and different cell cultures such as African green monkey kidney cell line, hamster kidney cells (BHK21), rhesus monkey kidney cells ($LLCMK_2$), porcine kidney cells (PS) have been employed (De Madrid *et al.*, 1969, Hotta *et al.*, 1966, Morens *et al.*, 1985a, Morens *et al.*, 1985b, Stim *et al.*, 1970a, Stim *et al.*, 1970b, Yuill *et al.*, 1968). At present, Vero and BHK21 cell lines are mainly used (Rao, 1976, Morens *et al.*, 1985a). Specifically, a simplified plaque reduction neutralization assay using semi-micro method in BHK21 cell suspension was described by Morens (Morens *et al.*, 1985a.). This protocol employs a small volume of sera and general reagents, allowing the screening of neutralizing antibodies to the four serotypes. Results can be interpreted at five to nine days after inoculation. Finally, it allows testing a high number of samples in one experiment. This method with some modifications (Alvarez *et al.*, 2005) has been widely employed for dengue serosurvey allowing the clear distinction in homotypic and heterotypic dengue antibodies and the classification of the type of the infection (Guzman *et al.*, 1990, Guzman *et al.*, 1991, Guzman *et al.*, 2000). In the PRNT, serum dilutions are tested against a constant amount of each of the four dengue viruses calibrated to give a defined number of plaques. This procedure is laborious, time consuming, and requires a good standardization.

In an attempt to improve the neutralization assay, the immuno-peroxidase technique has been developed to visualize virus-infected cells in infectivity assays and neutralization tests with dengue viruses (Churdboonchart *et al.*, 1984, Jirakanjanakit *et al.*, 1997, Okuno *et al.*, 1979) and for detection of antibody enhancement activity in human monocytes (Kamasanttaya, *et al.*, 1987). A microfocus reduction neutralization test (based on peroxidase-antiperoxidase technique) for determining neutralizing antibody response in persons given a dengue vaccine has been successfully evaluated (Jirakanjanakit *et al.*, 1997). In addition, a microneutralization test based on an ELISA format has been reported by Vorndam *et al.*, 2000. The reduction in virus growth due to neutralization by antibody was measured optically. Good correlation with PRNT was obtained when sera from patients with a primary infection were tested, however in samples from patients with a secondary infection, no consistent results were obtained.

Other serological tests such as single radial Hemolysis and hemadsorption immunosorbent technique have been employed for dengue antibody detection (Chan *et al.*, 1985; Guzman *et al.*, 1985).

Classification in primary or secondary infection

The HI assay has been widely applied to classify dengue infections as primary or secondary. The WHO definition of secondary dengue virus infection depends upon the analysis of paired sera, whereas one of the samples with a titer of 1/2560 or higher is confirmatory of a second infection (PAHO, 1994).

It has been shown that simultaneous dengue IgM and IgG measurements by ELISA can be used to classify infections as primary or secondary. The distribution of the ratios of anti-dengue IgM and IgG demonstrate that primary and secondary cases defined by HI assay could be clearly discriminated (Innis *et al.*, 1989). Using a ratio of IgM/IgG of 1.78 or higher to define a primary case, 98% of primary infections and 96% of secondary infections were correctly classified. In acute sera from dengue cases, levels of anti-dengue IgG of 1/1280 or higher are criteria for a second dengue infection if detected by ELISA Inhibition Method (Vazquez *et al.*, 1997).

Usefulness of different clinical samples for serological diagnosis

Both IgM and IgG ELISA have shown a good correlation to HI assay for finger prick collection, even after long-term storage at 4°C (Garcia *et al.*, 1997, Vazquez *et al.*, 1991). Comparison of anti-dengue IgM and IgG results both in serum and in blood collected on filter paper have shown figures of 99% coincidence, 98.1% sensitivity, and 98.5% specificity (Vazquez *et al.*, 1991, Vazquez *et al.*, 1998). In addition, blood collected on filter paper has been widely applied in serosurvey studies using both HI and PRNT assays (Guzman *et al.*, 1982, Guzman *et al.*, 1990, Guzman *et al.*, 2000).

Most body fluids contain antibodies, although at much lower levels than in blood. However, because of the obvious advantage and convenience of noninvasive samples such as saliva, some studies have evaluated different tests to detect dengue IgM and IgG antibodies. Cuzzubo demonstrated that salivary IgG levels correlated well with serum HI antibody titers allowing the distinction between primary and secondary infection (Cuzzubo *et al.*, 1998). With a positive result defined as either salivary IgM or IgG level above the cutoff value, an overall sensitivity of 92% was obtained. In a similar approach, Balmaseda studied the presence of specific dengue IgM and IgA in sera and saliva in confirmed dengue cases (Balmaseda *et al.*, 2003). IgM was detected in saliva with a sensitivity of 90.3% and specificity of 92% demonstrating that salivary IgM is a useful serological marker for dengue infection. Although detection of IgA in saliva was not found to be a useful tool for dengue diagnosis, its high sensitivity in serum (94.4% of the IgM-positive cases) suggested that IgA might be another feasible alternative for dengue diagnosis.

Antibodies against dengue viral proteins

Most of the serological tests quantify dengue antibodies; however, none of them can demonstrate the antibody response to each of the viral proteins at the same time.

Previous studies using Western blotting demonstrated that primary dengue cases showed low titers of IgG antibodies to E, NS3 and NS5 proteins in sera collected during the convalescent phase. In secondary dengue cases, IgG to E protein in acute sera and high titers of IgG antibodies to E, NS1, NS3, NS5 and C proteins were observed during the convalescent phase. Results obtained in this study suggested that this test was not appropriate to detect IgM antibodies (Churdboonchart *et al.*, 1991). More recent investigations applying the same methodology suggested that patients with past dengue infection retain antibody towards the E protein but not all patients retain antibodies against other structural and nonstructural proteins (Se-Thoe *et al.*, 1999). On the other hand, Valdes observed that the antibody response in acute-phase samples from secondary cases was greater than that in primary cases, including the intensity of the reaction (Valdes *et al.*, 2000). Anti NS1, NS3 and NS5 were detected mainly in secondary cases.

NS1-specific IgM, IgA, and IgG antibodies from patients with DF and DHF were studied using an indirect ELISA (Shu *et al.*, 2000). A strong anti-NS1 antibody response in all convalescent sera was observed. Results showed that significant NS1-specific IgG was induced, whereas 75% and 60% of primary DF patients versus 40% and 90% of secondary DF patients produced IgM and IgA antibodies respectively. The serotype specificity of NS1-specific IgM, IgA, and IgG were found to be 80%, 67% and 75% for primary infections and 50%, 22% and 30% for secondary infections in positive samples of DF cases. A similar pattern was found in DHF patients. In addition, these authors obtained a good correlation between NS1 serotype-specific IgG ELISA and PRNT suggesting that the serotype-specific IgG ELISA could replace PRNT for seroepidemiologic studies to differentiate Japanese encephalitis and dengue virus infections and for dengue virus serotyping (Shu *et al.*, 2002). Finally, they proposed that this assay could be of value to differentiate primary and secondary dengue virus infections (Shu *et al.*, 2003b).

Virus isolation

In the past, virus isolation methods were slow, labor intensive and insensitive; however, in the recent years methods for virus isolation have improved with the introduction of mosquito cell cultures and mosquito inoculation. The use of mosquito larvae inoculation has provided a more sensitive and rapid method of virus isolation for dengue viruses (Lam *et al.*, 1986).

The first biological system successfully employed for dengue virus isolation was the Swiss albino mouse (Kimura *et al.*, 1944, Sabin *et al.*, 1945). Dengue strains adapted to mice causing flaccid paralysis after intracerebral injection. Suckling mice, intracerebrally inoculated develop encephalitis after 4 to 7 days depending on the dengue serotype and the adaptation of the virus. Some strains fail to adapt to mice. Dengue virus can be found in high titer in the brain of the inoculated mice. This system has been widely used in the past for dengue virus isolation but was replaced by other more rapid and sensitive methods (Guzman *et al.*, 1984). Currently, infant mice brain is used for dengue antigen preparation using the method developed by Clarke and Casals, 1958. In addition, mice are widely employed for vaccine testing and experimental research.

Although dengue viruses were successfully propagated in mammalian cell lines, the employment for virus isolation of cells derived from vertebrates was rapidly replaced by mosquito cell lines. In many isolates the early passages do not produce a cytopathic effect (CPE) or plaques in vertebrate cell cultures. Moreover, recovery and identification is time consuming and requires several passages to obtain adequate titers. BHK21, Vero and LLCMK2 cells have been employed for dengue virus isolation with higher sensitivity than suckling mice but lower sensitivity than mosquito cell cultures.

Early studies performed by Singh demonstrated that a cell line derived from the larvae of *Aedes albopticus* was more suitable for arbovirus multiplication than from *Aedes aegypti* (Singh *et al.*, 1968). Later studies demonstrated the higher sensitivity of *Aedes albopticus* (C6/36) and *Aedes pseudoscutelarris* (AP61) cell lines for dengue virus isolation (Hebert *et al.*, 1980, Race *et al.*, 1978, Race *et al.*, 1979, Tesh, 1979). Some dengue virus strains produce a marked syncytial cytopathic effect often visible four days after inoculation. Formation of syncytia eventually involves the whole cell sheet followed by retraction of syncytia to form dense clumps and leaving large holes in the cell monolayer (Varma *et al.*, 1974). A comparative study on the isolation of dengue viruses from human sera using the C6/36 clone of *Aedes albopticus*, AP61 and TRA-284 from *Toxorhynchites amboinensis* cell lines showed the highest isolation rates in the TRA-284 cells (Kuno *et al.*, 1985). Castillo and Morier obtained two clones from C636 and AP61 cell lines respectively (Castillo *et al.*, 1994 and Morier at al., 1995). Both cell clones showed good results in terms of sensitivity for dengue virus replication and virus isolation. In an attempt to increase the isolation frequency, Rodriguez applied the rapid centrifugation assay for dengue 2 virus isolation (Rodriguez *et al.*, 2000). These methods allowed isolation of virus from 16.6% more samples than the conventional method and it shortened the time for dengue virus detection. In addition, it allowed the isolation of dengue virus in 42.8% of tissue samples from fatal cases.

The direct dengue inoculation of live adult *Aedes albopticus* was first reported in 1974/76 (Gubler *et al.*, 1976, Kuberski *et al.*, 1976a, Kuberski *et al.*, 1976b, Rosen *et al.*, 1974). Later, a nonbiting mosquito, *Toxorhynchites amboinensis* was employed because of its safety and better survival (Rosen, 1981). Five species of colonized *Toxorhynchites* mosquitoes were compared for their susceptibility to parental infection with the four dengue serotypes being *Tx. amboinensis, Tx. brevipalpis, Tx. rutilis, Tx. splendens* equally susceptible to infection (Rosen *et al.*, 1985). The use of *Toxorynchites* mosquitoes or male *Aedes albopticus* is safe; the mosquitoes can be inoculated intrathoracically or intracerebrally with serum or tissue extract and dengue antigen can be detected by immunofluorescence, complement fixation test, antigen-capture ELISA and PCR (Kuberski *et al.*, 1977a, Kuberski *et al.*, 1977b, Lanciotti *et al.*, 1992, Schoepp *et al.*, 1984). A mosquito larvae inoculation has also been developed for dengue virus isolation (Pang *et al.*, 1983).

Direct mosquito inoculation is by far the most sensitive culture method, however it requires a large colony of mosquitoes, and it also demands a certain amount of dexterity to inoculate the insects. In general, for routine diagnosis, mosquito cell culture inoculation is preferred. Mosquito cell cultures are relatively easy to maintain, grow at room temperature and can be keep for at least 14 days without a change of medium. At present, the C6/36 cell line is probably the most employed system for routine dengue virus isolation.

The development of specific dengue monoclonal antibodies (Gentry *et al.*, 1982, Henchal *et al.*, 1982, Henchal *et al.*, 1983) represented an important advance in dengue diagnosis. They have been extensively used for identifying field isolates of dengue viruses both from mosquito cell cultures and from direct mosquito inoculation through indirect immunofluorescence (Gubler *et al.*, 1984, Soler *et al.*, 1985, Soler *et al.*, 1988). In general, most of the isolates are easily identified. However, in some occasions two or three passages in the mosquito cell line are required in order to increase the virus titer and facilitate virus identification.

Flow cytometry has been recently reported to be a useful method for dengue identification. In one study, flow cytometry allowed the identification of virus 10 hours earlier than the immunofluorescence assay when an anti-NS1 monoclonal antibody was used (Kao *et al.*, 2001). In another study, it was applied to antigen detection in infected human mononuclear leukocytes, C636 and Vero cell cultures (Sydow *et al.*, 2000). Other methods have been tested in an attempt to improve dengue virus isolation. A method involving antibody-dependent enhancement of infectivity in the mouse macrophage-like cell line P388D1 and the human promonocyte cell line (HL-CZ) have been used to isolate dengue viruses. Although good results were obtained, these methods have not been extensively applied (Cardosa, 1987, Liu *et al.*, 1991).

Antigen detection

Direct detection of dengue antigens in blood samples is extremely difficult. One reason for this difficulty is the presence of antibody in acute phase sera of persons with a second infection. Different dengue antigen detection methods have been evaluated. Early studies using counter-immunoelectrophoresis showed figures of 55% of antigen detection in acute-phase sera collected at days one to nine (Chunboorchart *et al.*, 1979, Chunboorchart *et al.*, 1974). Dengue antigen was detected in peripheral blood leukocytes of dengue confirmed cases using a direct fluorescent antibody; however, a low sensitivity was obtained (Waterman *et al.*, 1985). In a different approach, Monath evaluated a monoclonal radioimmunoassay (RIA) for the detection of dengue antigen in blood samples from patients (Monath *et al.*, 1986). In this study, dengue antigen was more frequently detected in cases of primary infection (54%) than in second infection (16%). More recently, Malergue reported the usefulness of an amplified fluorogenic ELISA for the detection and identification of dengue 3 virus (Malergue *et al.*, 1995). This assay uses biotinylated mouse IgG antibody directed against dengue antigens captured by an anti-dengue monoclonal antibody coated onto polystyrene microplate wells. This system showed a sensitivity and specificity of 90% and 98%, respectively, compared to virus isolation results. In another approach, dengue antigens were detected in acute sera and peripheral blood mononuclear cells (PBMC) from dengue infected patients using a biotin-streptavidin ELISA. The detection rate in PBMC was six times greater than in sera. Particularly, in severe clinical cases, the presence of dengue antigen in PBMC increased from 36% to 100% (Kittigul *et al.*, 1997). Finally, a commercial kit based on an antigen EIA system has been produced (Globio Blue and Globio Red kit). The sensitivity and specificity of the Blue kit (dengue antigen detection) were 84% and 89%, respectively, and of the Red kit (serotype identification) 91% and 93%, respectively (Globio Blue and Red Kit for antigen detection, Globio Corp., Beverly, MA, USA).

Young described the development of a capture ELISA for the detection of dengue NS1 protein (Young *et al.*, 2000). As much as 15 μg/ml levels of NS1 were found in acute-phase sera taken from patients secondarily, dengue 2 infected but was not detected in the convalescent sera. Contrarily, these authors could not demonstrate levels of NS1 protein in acute sera from serologically confirmed primary infection. More recently, Alcon reported the detection of NS1 protein from the first day after fever onset up to day nine (Alcon *et al.*, 2002). This protein could be detected even when viral RNA was negative or in the presence of IgM antibodies. Contrary to previous reports, they found similar NS1 concentrations in serum specimens obtained from patients experiencing primary or secondary dengue virus infections. Finally, Libraty reported that soluble NS1 in plasma correlate with viremia and those higher levels were observed in patients with DHF than in those with DF suggesting that high levels of NS1 could be employed as a prognosis marker for the severe disease (Libraty *et al.*, 2002a).

Very recently, an immunochip based on quartz crystal microbalance (QCM) coating, using two monoclonal antibodies that act specifically against the E and NS1 proteins, was developed for the detection of dengue antigens. The sensitivity of the immunochip was 100-fold higher than the conventional sandwich ELISA method and the time required for detection was shorter (1 hour) (Su *et al.*, 2003).

An antigen capture ELISA was developed for detecting dengue and Japanese encephalitis viruses in infected mosquitoes. The results indicated that this system could readily screen dengue antigen in individual and pooled mosquitoes and could be extended to dengue surveillance for screening of large number of potentially infected vectors (Sithiprasasna *et al.*, 1994).

Finally, immunohistochemical techniques have been applied to dengue antigen detection in tissues samples collected from fatal cases allowing a definitive diagnosis of dengue infection from autopsy material. Using a combination of polyclonal and monoclonal antibodies on paraffin-embedded tissues, a dengue diagnosis has been confirmed in human liver samples from fatal cases both in fresh and archive material (Hall *et al.*, 1991, Pelegrino *et al.*, 1997, Sarmiento *et al.*, 1995).

Genome detection

Over the past 10 years, a number of molecular amplification assays have been developed for the detection of flavivirus and particularly for dengue virus genome detection. Most of these assays utilize the reverse transcriptase- polymerase chain reaction (RT-PCR). RT-PCR for flavivirus has become an important tool for diagnosis, entomological surveillance, and for molecular epidemiological studies, being a very useful research tool in pathogenesis, antiviral drug, and vaccine studies (Rosario *et al.*, 2001). Modifications of the standard RT-PCR (TaqMan), using fluorescence-labeled oligonucleotide probes for detection have been developed recently. In addition, another amplification format, nucleic acid sequence based amplification (NASBA), has been developed and utilized for the detection of dengue virus and others flavivirus de medical importance (Lanciotti, 2003).

Dengue virus RNA has been detected by RT-PCR directly from clinical samples such as serum, plasma, and autopsy fresh and paraffin-embedded tissues. Supernatant of dengue virus-infected mosquito cell cultures and infected mosquito larvae have also been employed (De Paula 2002, Guzman *et al.*, 1999, Lucia *et al.*, 1994, Sariol *et al.*, 1999, Rosario *et al.*, 2001).

Several RT-PCR protocols for flavivirus and dengue RNA detection have been developed; many of them apply a combination of four serotype-specific oligonucleotide primer pairs in a single reaction tube or use a consensus flavivirus or dengue oligonucleotide primer pair followed by a second amplification with serotype-specific oligonucleotides. These procedures also vary in the genomic location of primers (C/prM, E, NS1, E/NS1, prM/E, NS5, NS5/3′), and in their specificity and sensitivity. Some of these protocols allow the detection of less than 50–100 dengue virus plaque forming units (Chang *et al.*, 1994, Chow *et al.*, 1997, Harris *et al.*, 1998, Lanciotti *et al.*, 1992, Meiyu *et al.*, 1997, Sudiro *et al.*, 1997). Table 7.2 summarizes some of the PCR protocols used in flavivirus and/or dengue diagnosis. A second round of amplification by PCR (nested PCR), hybridization with specific probes, and enzymatic restriction are some of the main technologies used for the specific identification with an increase in the sensitivity and the specificity of the assays. The combination of PCR and ELISA has allowed the quantification of the virus present in the original sample.

In the Americas, the protocol developed by Lanciotti *et al.*, 1992, has been widely applied. Modifications to this protocol as well as new protocols have also been used (Chang *et al.*, 1994, Figueiredo

Table 7.2 List of some protocols employed for dengue PCR diagnosis.

References	Genome	RT-PCR	Hybridization	nPCR	ELISA	RE
Deubel *et al.*, 1990	E	Den/C	Den/S			
Henchal *et al.*, 1991	NS1	Den/C				
Eldadah *et al.*, 1991	C/E	Den/S				
Lanciotti *et al.*, 1992	C-prM	Den/C		Den/S		
Trent and Chang, 1992	NS5	Fla/C		Fla/S		
Tanaka *et al.*, 1993	3′NCR	Fla/C				
Fulop *et al.*, 1993	NS5	Fla/C				
Morita *et al.*, 1994	E/NS3	Den/S				
Pierre *et al.*, 1994	NS5–3′	Fla/C	Fla/S			
Chang *et al.*, 1994	NS5	Fla/C			Den/S	
Yenchitsomanus *et al.*, 1996	E, NS1	Den/C		Den/S		
Sudiro *et al.*, 1997	3′NCR	Den/S				
Meiyu *et al.*, 1997	NS1	Fla/C		Den/S		
Chow *et al.*, 1997	NS3	Den/C			Den/S	
De Paula *et al.*, 2002	NS1	Den/C				Den/S

Genome, RNA genome region amplified. Den/C, Consensus primers of dengue viruses. Den/S, Serotype-specific primers of dengue virus. Fla/C, Consensus primers of flavivirus. Fla/S, specific primers of some flavivirus. Hybridization, Dot or Southern blot Hybridization. nPCR, nested PCR. ER, enzymatic restriction.

et al., 1997, Harris *et al.*, 1998, Khawsak *et al.*, 2003, Kuno *et al.*, 1998a, Rosario *et al.*, 1998).

Some of the commonly employed RNA extraction protocols are guanidinium thiocyanate-silica (Chomczynki and Sacchi, 1987), silica method (Sudiro *et al.*, 1997) and, more recently described, a simple protocol using Trizol Reagent (Gibco BRL). In addition, the commercial QIAamp viral RNA mini Kit (QIAGEN) represents a very simple, complete, and efficient RNA extraction method as described by Wang (Wang *et al.*, 2000).

Currently many commercial kits for RT-PCR containing one or two-step protocols are available. de Paula employed five commercial kits and demonstrated a higher sensitivity using one-step protocols compared to two-steps RT-PCR approaches (De Paula *et al.*, 2004). When correctly applied, PCR has remarkable advantages as a tool for dengue diagnosis. Its application allows the detection of dengue virus in stored samples even after long periods. Alvarez and later on Sariol demonstrated the presence of dengue 2 virus in serum and autopsy tissue samples from DHF/DSS cases that had been stored for more than 15 years (Alvarez *et al.*, 1996; Sariol *et al.*, 1999). In addition, they were able to classify the isolates by genomic sequencing.

Alternatively, different authors have applied PCR for entomological surveillance. Chow and Kow, identified dengue viruses in *Aedes aegypti* and *Aedes albopticus* mosquito pools during a 1-year surveillance period in Singapore, allowing the application of vector control measures (Chow *et al.*, 1998 and Kow *et al.* in 2001). Infected *Aedes aegypti* were detected as early as six weeks before the recognition of a dengue outbreak in 1995 and 1996. These authors recommend this method as an early warning monitoring system for dengue outbreaks.

PCR also allows the detection of concurrent infections by multiple serotypes both in serum samples and in isolates (Laille *et al.*, 1991, Loroño-Pino *et al.*, 1999, Maneekarn *et al.*, 1993). Laille and coworkers were able to detect dual viremia by dengue 1 and dengue 3 viruses in six DF patients during the 1989 New Caledonia epidemic. Similarly, Loroño-Pino demonstrated that 5.5% out of 292 samples showed evidence for concurrent dengue infection with two or more serotypes (Loroño-Pino *et al.*, 1999). An important advantage of RT-PCR in comparison with viral isolation is that it allows the confirmation of dengue virus infection both in secondary as well as in primary infection, especially when plasma or serum specimens are collected before the defervescence (Sangasang *et al.*, 2003).

Traditional detection of amplified DNA relies upon electrophoresis of the nucleic acid in the presence of ethidium bromide and visual or densitometric analysis of the resulting bands after irradiation by ultraviolet light. Alternatively, Southern blot detection of amplicons using hybridization with a labeled oligonucleotide probe is time consuming and requires multiple PCR product handling steps, further risking the spread of amplicons throughout the laboratory (Holland *et al.*, 1991). In general, many classical PCR protocols bear the risk of false positive results by contamination through handling, and of false negative results due to the presence of inhibitors of the enzymes or to the presence of RNA nucleases in the sample (Wilson *et al.*, 1997). Many safety measures are implemented into the protocols in order to minimize these risks, including the introduction of internal and external controls. Recently, an external control to routine diagnosis laboratories revealed figures of sensitivity ranging from 33 to 80% for PCR assay (Lemmer *et al.*, 2004).

Most diagnostic PCR assays reported to date have been used in a qualitative, or "yes/no" format. The first quantitative PCR for dengue virus evolved from the PCR-ELISA format to a quantitative-competitive PCR (qcPCR). PCR allows determining the amount of template in two ways: as a relative or as an absolute quantification. Relative quantification describes changes of the amount of a target sequence compared with its level in a related matrix. Absolute quantification represents the exact number of nucleic acid targets present in the sample in relation to a specific unit. PCR-ELISA works on the basis of biotin or digoxigenin-labeled primers or digoxigenin incorporated into the amplicon that can be detected using the standard ELISA format. Some of these protocols have shown excellent results in dengue diagnosis.

Lanciotti's group monitored viremia in dengue patients by combining RT-PCR with an ELISA DIG detection kit (Lanciotti *et al.*, 1992; Chow *et al.*, 1997; Murgue *et al.*, 2000). A very accurate approach to absolute quantification by PCR is the use of competitive co-amplification of an internal control nucleic acid of known concentration and a wild-type target nucleic acid of unknown concentration. Applying this format or modifying the original qualitative RT-PCR, two very interesting studies were published allowing the determination of viral load and its implication in the pathogenesis of DHF (Sudiro *et al.*, 2001; Wang *et al.*, 2003).

The implementation of nucleic acid technology into routine diagnosis has represented an important step forward in the introduction of real time technology, which makes the detection of amplified products relatively easy (Mackay *et al.*, 2002; Niesters, 2001). This technology is able to quantify target nucleic acid in a single sample over a larger dynamic range than most other quantitative technology. Since both amplification and detection can currently be performed automatically, the most labor intensive and critical step is still the efficient nucleic acid extraction from the clinical samples (Mackay *et al.*, 2002). Technological improvements, from automated sample isolation to real time amplification technology, have made possible the development and introduction of detection systems for most viruses of clinical interest, such as flaviviruses and specifically dengue viruses.

Several studies have been performed in order to determine risk factors for the development of DHF and to define possible prognostic markers. Libraty found a significant correlation of the severity of the disease with the duration and magnitude of viremia (Libraty *et al.*, 2002b), measured by a novel quantitative RT-PCR fluorogenic system developed by Houng (Houng *et al.*, 2001). Both polymerase chain reaction based on TaqMan principle and NASBA can currently be used together for real-time detection and generate results in a short turn-around time. In 1999, a group from the Bernhard-Nocht-Institute (Hamburg, Germany) applied a fully automated amplification protocol (based on TaqMan principle) to measure virus-specific DNA amplification (Laue *et al.*, 1999). This method has a high specificity and sensitivity; it eliminates the possibility of cross-contamination and allows the determination of viral load. Similar results were observed by later studies (Callahan *et al.*, 2001). These investigators applied a TaqMan RT/PCR assay to identify dengue viruses, reporting a high sensitivity and specificity when compared to viral isolation in C6/36 cells. Results were obtained in less than two hours. Wu applied a NASBA assay to amplify the four dengue serotypes (Wu *et al.*, 2001). These investigators employed a set of universal primers and serotype-specific capture probes for typing. Nucleic acid was amplified without thermo cycling and the product was detected by electrochemoluminescence using probe hybridization. The assay showed a sensitivity of 98.5% and a specificity of 100% when compared to viral isolation in C6/36 cell line. Shu reported the development of group- and serotype-specific one-step SYBR green I-based real-time RT-PCR (Shu *et al.*, 2003a). The system employs a set of primers against consensus nucleotide sequences in the core region being the range of the detection limit for each serotype from 10 to 4.1 pfu/mL.

Recently, a novel real-time system was developed as one-step, single tube, real-time accelerated reverse transcription loop-mediated isothermal amplification (RT-LAMP) assay for detecting the E gene of West Nile virus. Using RT-LAMP assay, DNA amplification is effectuated in less than one hour and figures of 10-fold higher sensitivity compared to RT-PCR, with a detection limit of 0.1 pfu (Parida *et al.*, 2004).

Real-time PCR does not require any post-amplificational manipulation of the amplicon and is therefore considered a "closed" or homogeneous system. It includes a reduced turnaround time, minimized risk of contamination, and the ability to closely scrutinize the assay's performance. In addition, it has been shown that real-time PCR is cost effective when implemented in high throughput laboratories. However, these new technologies have some disadvantages in comparison with conventional PCR. The most important being the inability to monitor the amplicon size without opening the system, the incompatibility of some platforms with some fluorogenic chemistries, and the relatively restricted multiplex capabilities of current applications. In addition, the start-up expense of real-time PCR may be prohibitive when use in low-throughput laboratories.

Other important applications of genome detection technology

One of the most important scientific applications of PCR is the study of genetic strain variability in order to identify the origin of epidemics and to reveal markers of virulence. In conjunction with nucleotide sequencing or restriction enzyme analysis, PCR has allowed the classification of dengue serotypes into genotypes (Chungue *et al.*, 1993, Chungue *et al.*, 1995, Deubel *et al.*, 1993, Lanciotti *et al.*, 1994, Lanciotti *et al.*, 1997, Rico-Hesse, 1990, Usuku *et al.*, 2001). Rico-Hesse and co-workers have developed a method to compare dengue 2 genomes directly from plasma patient. They found some amino acid and nucleotide changes at the E protein gene as well as within the untranslated region that may represent primary determinants of DHF (Leitmeyer *et al.*, 1999, Rico-Hesse *et al.*, 1997). In other studies, Kuno established the genetic relationship among viruses of the *Flavivirus* genus. They proposed that two branches of virus evolved from the putative ancestor: nonvector and vector-borne virus clusters. According to their theory, tick-borne and mosquito-borne viruses emerged from the latter cluster (Kuno *et al.*, 1998b). Recent reports suggest that intra-serotype recombination of dengue virus occurs (Holmes *et al.*, 2000,

Twiddy *et al.*, 2002, Twiddy *et al.*, 2003, Worobey *et al.*, 1999). The consequences of these findings remain to be defined.

Recently, a differential display RT-PCR assay provided insight into the complex responses of transcriptional machinery of permissive and apoptotic human endothelial-like cells in the pathogenesis of dengue and its complications induced by the virulent dengue virus type 2 (Liew *et al.*, 2004). Applying real-time RT-PCR, the replication kinetic of the ChimericVax DEN tetravalent vaccine viruses in *Aedes aegypti* was studied demonstrating a low transmission potential by mosquitoes of the ChimericVax-Den vaccine viruses (Johnson *et al.*, 2004).

Bottlenecks in molecular diagnosis

Some problems remain to be solved before molecular diagnostic can be widely implemented into routine diagnostic (Table 7.3). Recently, a biosensor has been added to the NASBA technique, allowing the rapid detection of dengue virus RNA in only 15 minutes (Baeumner *et al.*, 2002). In an alternative approach, a fluorogenic RT-PCR system was developed for quantification and identification of dengue viruses using conserved and serotype-specific 3′ noncoding sequences. This system is able to detect 20–50 pfu per ml serum and showed a sensitivity of 92.8% and specificity of 92.4% compared to virus isolation in cell cultures (Houng *et al.*, 2001).

Micro- and macro-array technology will play a chimeric role in the real-time PCR of future research, but for now, there is a significant potential for routine, research, and commercial interests to redesign existing systems for greater sophistication, flexibility and the ability to generate high quality quantitative data. The development of real-time PCR assays that can discriminate as many targets as conventional multiplex PCR assay, whilst producing quantitative data at a greatly increased speed, will consolidate these as a routine tool for the laboratory of the future (Mackay *et al.*, 2002).

Laboratory confirmation of a clinical dengue infection

A definite diagnosis of dengue infection depends on the isolation of the virus, detection of the RNA in serum or tissues, and of the detection of the viral antigens.

Since the presence of anti-dengue IgM antibodies in the peripheral blood may be due to an infection up to three months earlier, positive samples for IgM antibody alone are thus not confirmatory for a current infection and are reported only as probable dengue cases. For a serological

Table 7.3 Bottlenecks in dengue molecular diagnosis.

Problem identified	Possible solution
Quantitative technology	Real-time amplification technology
Internal and external control	Well defined universal internal controls and quality control programs
Sample preparation still matrix dependent	New extraction technology
Hand-on time turn-around time results	Automation
No commercial kit available	Well characterized home-brewed assay

confirmation of dengue, a four-fold increase in the specific antibody titer should be obtained (Gubler, 1998, Guzman *et al.*, 2004, Rigau-Perez *et al.*, 1998).

The use of these diagnostic tests for dengue allows the classification of suspected clinical cases into probable or confirmed cases. A positive IgM serology, or a reciprocal HI antibody $\geq$1280 or equivalent IgG titer by ELISA are criteria indicating probable dengue infection. Dengue virus isolation, dengue antigen demonstration, or positive PCR, and a fourfold or higher change in reciprocal IgG in paired sera, is used to confirm dengue infection. Both probable and confirmed cases are reportable to health authorities (PAHO, 1994).

Conclusions

In spite of the advances in dengue virus diagnosis, some problems still warrant the timely development of new solutions. For instance, virus isolation is time consuming; PCR requires specific laboratory equipment and facilities as well as extensive evaluation of the different protocols under field conditions; IgM antibody detection requires proper timing and is confounded by false positive reactions and the long persistence of IgM antibodies, commercial kits still need to be critically evaluated, and the costs and availability of these kits and other reagents need to be addressed.

Inexpensive, sensitive, and specific tests, easy to perform that allow the early diagnosis of dengue virus infection are still needed. Specifically, the following aspects require the greatest attention:

To develop: (1) tests for early clinical diagnosis; (2) serological tests able to differentiate dengue from other flavivirus infections and for dengue serotype identification; (3) easy and inexpensive protocols for genomic characterization and viral load; (4) modifications of existing protocols that simplify specimen handling and transportation; (5) recombinant antigens as tools for test evaluation and to produce these for serological tests; and (6) tools for prognosis marker detection.

In addition to these specific items, it is also necessary to implement mechanisms for greater reagent availability, for sharing standard reagents (antigens, monoclonal antibodies, cell cultures, positive and negative control sera), for the standardization of protocols in endemic regions, for improving the quality and quantity of proficiency test, and for enhanced exchange of information and experiences between endemic areas (Guzman *et al.*, 2003b).

References

Alcon, S., Talarmin, A., Debruyne, M., Falconar, A., Deubel., V., and Flemand, M. (2002). Enzyme-linked immunosorbent assay specific to dengue virus type 1 nonstructural protein NS1 reveals circulation of the antigen in the blood during the acute phase of disease in patients experiencing primary or secondary infections. J. Clin. Microbiol. *40*, 376–381

Alvarez, M., Guzman, M.G., Rosario, D., Vazquez, S., Pelegrino, J.L., Sariol, C., and Kouri, G. (1996). Secuenciacion directa a partir de un producto de PCR de una muestra de suero de la epidemia de FHD de (1981). Rev. Cubana. Med. Trop. *48*, 53–55

Alvarez, M., Rodriguez-Roche, R., Bernado, L., Morier, L., and Guzmán, M.G. (2005). Improved dengue virus plaque formation on BHK21 and LLCMK2 cells: evaluation of some factors. Dengue Bulletin *29*, 49–57

Ashburg, P.M., and Craig, C.F. (1907). Experimental investigations regarding the etiology of dengue fever. J. Infect. Dis. *4*, 440–475

Baeumner, A.J., Schlesinger, N.A., Slutzki, N.S., Romano, J., Lee, E.M., and Montagna, R.A. (2002). Biosensor for dengue virus detection, sensitive, rapid and serotype specific. Anal. Chem. *74*, 1442–1448

Balmaseda, A., Robleto, G., Flores, C., Videa, E., Tellez, Y., Saborio, S., Perez, L., Sandoval, E., Rodriguez, Y., Guzman, M.G., and Harris, E. (2003). Diagnosis of dengue virus infection by detection of specific IgM and IgA antibodies in serum and saliva. Clin. Diagn. Lab. Immunol. 10, 327–322

Brandriss, M.W., Schelesinger, J.J., Walsh, E.E., and Briselli, M. (1986). Lethal 17D Yellow fever encephalitis in mice. I Passive protection by monoclonal antibodies to the envelope proteins of 17D Yellow fever and Dengue 2 virus. J. Gen. Virol. 67, 229–234

Burke, D.S., Nisalak, A., and Gentry, M.K. (1987). Detection of flavivirus antibodies in human serum by epitope-blocking immunoassay. J. Med. Virol. 23, 165–173

Burke, D.S. (1983). Rapid methods in the laboratory diagnosis of dengue virus infections. Pang T, Pathmanathan R, eds. Proceedings of the International Conference on dengue/dengue hemorrhagic fever. Kuala Lumpur, Malaysia, University of Malaya, 72–84

Calisher, C.H., Karabatsos, N., Dalrymple, J.M., Shope, R.E., Porterfield, J.S., Westaway, E.G. and Brandt, W.F. (1989). Antigenic relationships between flaviviruses as determined by cross-neutralization tests with polyclonal antisera. J. Gen. Virol. 70, 37–43

Callahan, J.D., Wu, S.J., Dion-Schultz, A., Mangold, B.E., Peruski, L.F., Watts, D.M., Porter, K.R., Murphy, G.R., Suharyono, W., King, C.C., Hayes, C.G., and Temenak, J.J. (2001). Development and evaluation of serotype and group-specific fluorogenic reverse transcriptase PCR (Taqman) assays for dengue virus. J. Clin. Microbiol. 39, 4119–4124

Cardosa, M.J., Baharudin, F., Hamid, S., Hooi, T.P., and Nimmanitya, S. (1995). A nitrocellulose membrane based IgM capture enzyme immunoassay for etiological diagnosis of dengue virus infections. Clin. Diagn. Virol. 3, 343–350

Cardosa, J., Hooi, P., Nimmannitya, S., Nisalak A., and Innis B. (1992). IgM capture ELISA for detection of IgM antibodies to dengue virus, comparison of 2 formats using hemagglutinins and cell culture derived antigens. Southeast Asian J. Trop. Med. Public Health 23, 726–729

Cardosa, M.J., Hooi, T.P., and Shaari, N.S (1988a). Development of a dot enzyme immunoassay for dengue 3, a sensitive method for the detection of antidengue antibodies. J. Virol. Methods 22, 81–88

Cardosa, M.J., Sham, S.N., Tio, P.H., and Lim S.S. (1988b). A dot enzyme immunoassay for dengue 3 virus, comparison with the haemagglutination inhibition test. Southeast Asian J. Trop. Med. Public Health 19, 591–594

Cardosa, M.J. (1987). Dengue virus isolation by antibody-dependent enhancement of infectivity in macrophages. Lancet, *January 24*, 193–194

Castillo, A., and Morier, L. (1994). Obtencion de una sublinea de C6/36 adaptada a crecer en medio libre de suero y su utilizacion para estudios sobre virus del dengue. Rev. Cub. Med. Trop. 46, 139–143

Chambers, T.J., Hahn, C.S., Galler, R., and Rice, C.M. (1990). Flavivirus genome organization, expression and replication. Annu. Rev. Microbiol. 44, 649–688

Chan, Y.C., Tan, H.C., Tan, S.H., and Balachandran, K. (1985). The use of the single radial haemolysis technique in the serological diagnosis of dengue and Japanese encephalitis virus infections. Bull. WHO 63, 1043–1053

Chang, G.J., Trent, D.W., Vorndam, V., Vergne, E., Kinney, R.M., and Mitchell, C.J. (1994). An integrated target sequence and signal amplification assay, reverse transcriptase-PCR-enzyme-linked immunosorbent assay, to detect and characterize flaviviruses. J. Clin. Microbiol. 32, 477–483

Chomczynki, P., and Sacchi, N.(1987). Single-step method of RNA isolation by acid guanidinium thiocyanate-phenol-chloroform extraction. Anal. Biochem. 162, 156–159.

Chow, V.T.K., Yong, R.Y.Y., and Chan, Y.C. (1997). Automated type specific ELISA probe detection of amplified NS3 gene products of dengue viruses. J. Clin. Pathol. 50, 346–349

Chow, V.T.K., Chan, Y.C., Yong, R., Lee, K.M., Lim, L.K., Chung, Y.K., Lam-Phua, S.G., and Tann, B.T. (1998). Monitoring of dengue viruses in field-caught *Aedes aegypti* and *Aedes albopictus* mosquitoes by a type-specific polymerase chain reaction and cycle sequencing. Am. J. Trop. Med. Hyg. 58, 578–586

Chow, V.T.K., Yong, R.Y.Y., Ngoh, B.L., and Chan, Y.C. (1997). Automated type specific ELISA probe detection of amplified NS3 gene products of dengue viruses. J. Clin. Pathol. 50, 346–349

Chungue, E., Cassar, O., Drouet, M.T., Guzman, M.G., Laille, M., Rosen, L., and Deubel, V. (1995). Molecular epidemiology of dengue-1 and dengue-4 viruses. J. Gen. Virol. 76, 1877–1884

Chungue, E., Deubel, V., Cassar, O., Laille, M., and Martín, P.M.V. (1993). Molecular epidemiology of dengue 3 viruses and genetic relatedness among dengue 3 strains isolated from patients with mild or severe form of dengue fever in French Polynesia. J. Gen. Virol. 74, 2765–2770

Chungue, E., Marche, G., Plichart, R., Boutin, J.P., and Roux, J. (1989). Comparison of immunoglobulin G enzyme-linked immunosorbent assay (IgG-ELISA) and haemagglutination inhibition (HI) test for the detection of dengue antibodies. Prevalence of dengue IgG-ELISA antibodies in Tahiti. Trans. R. Soc. Trop. Med. Hyg. *83*, 708–711

Churdboonchart, V., Bhamarapravati, N., Peamprampercha, S., and Sirinavin, S. (1991). Antibodies against dengue viral proteins in primary and secondary dengue hemorrhagic fever. Am. J. Trop. Med. Hyg. *44*, 481–493

Churdboonchart, V., Kamsattaya, K., Yoksan, S., Sinarachatanant, P., and Bhamarapravati, N. (1984). Application of peroxidase-antiperoxidase (PAP) staining for detection and localization of dengue 2 antigen. I. In an endogenous peroxidase containing systems. Southeast Asian J. Trop. Med. Public Health *15*, 547–553

Churdboonchart, V., Bhamarapravati, N., Harisdangkul, V., McNair Scott, R., Futrakul, P., Chiengsong, R., and Nimmanitaya, S. (1979). Rapid detection of dengue viral antigens by counterimmunoelectrophoresis. Am. J. Trop. Med. Hyg. *71*, 102–108

Churdboonchart, V., Harisdangkul, V., and Bhamarapravati, N. (1974). Countercurrent immunoelectrophoresis for rapid diagnosis of dengue hemorrhagic fever. Lancet *October* 5, 841

Clarke, D.H., and Casals, J. (1958). Techniques for hemagglutination and hemagglutination inhibition with arthropod-borne viruses. Am. J. Trop. Med. Hyg. *7*, 561–573

Cuzzubo, A.J., Endy, T.P., Nisalak, A., Kalayanarooj, S., Vaughn, D.W., Ogata, S.A., Clements, D.E., and Devine, L.D. (2001). Use of recombinant envelope proteins for serological diagnosis of dengue virus infection in an immunochromatographic assay. CDLI 8, 1150–1155

Cuzzubo, A.J., Vaughn, D.W., Nisalak, A., Suntayakorn, S., Aaskov, J., and Devine, P.L. (1998). Detection of specific antibodies in saliva during dengue infection. J. Clin. Microbiol. *36*, 3737–3739

Delgado, I., Vazquez, S., Bravo, J.R., and Guzman, M.G. (2002). Prediccion del serotipo del virus dengue mediante la respuesta de anticuerpos IgM. Rev. Cubana Med. Trop. *54*, 113–117

De Madrid, A.T., and Porterfield, J.S. (1974). The flaviviruses (group B arboviruses), a cross-neutralization study. J. Gen. Virol. *23*, 91–96

De Madrid, A.T., and Porterfield, J.S. (1969). A simple micro-culture method for the study of group B arboviruses. Bull. WHO *40*, 113–121

De Paula, S.O., Melo Lima, C., Torres, M.P., Garbin Pereira, M.R., and Lopes da Fonseca, B.A. (2004). One-step RT-PCR protocols improve the rate of dengue diagnosis compared to two-step RT-PCR approaches. J. Clin. Virol. *30*, 297–301

De Paula, S.O., Malta Lima, D., and Lopes da Fonseca, B. (2002). Detection and identification of Dengue-1 virus in clinical simples by a nested-PCR followed by restriction enzyme digestion of amplicons. J. Med. Virol. *66*, 529–534.

Deubel, V., Nogueira, R.M., Drouet, M.T., Seller, H., Reynes, J.M., and Ha, D.Q. (1993). Direct sequencing of genomic cDNA fragments amplified by the polymerase chain reaction for molecular epidemiology of dengue-2 viruses. Arch. Virol. *129*, 197–210

Deubel, V., Laille, M., Hugnot, J.P., Chungue, E., Guesdon, J.L., and Drouet, M.T. (1990). Identification of dengue sequence by genomic amplification , rapid diagnosis of dengue virus serotype in peripheral blood. J. Virol. Methods 30 ,41–54.

Devappa, P., and Satchidanandam, V. (2003). Architecture of the flaviviral replication complex. J. Biol. Chem. *278*, 24386–24398

Dittmar, D., Cleary, T.J., and Castro, A. (1979). Immunoglobulin G and M-specific enzyme-linked immunosorbent assay for detection of dengue antibodies. J. Clin. Microbiol. *9*, 498–502

Durbin AP, Karron RA, Sun W, Vaughn DW, Reynols MJ, Perreault JR, Thumar B, Men R, Lai CJ, Elkins WR, Chanock RM, Murphy BR, and Whitehead SS. (2001). Attenuation and immunogenicity in humans of alive dengue virus type-4 vaccine candidate with a 30 nucleotide deletion in its 3′-untranslated region. Am. J. Trop. Med. Hyg. 65, 405–413

Edelman, R.A., Nisalak, A., Pariyanoda, S., Udomsaki, S., and Johnsen, D.O. (1973). Immunoglobulin response and viremia in dengue vaccinated gibbons repeatedly challenged with Japanese encephalitis virus. Am. J. Epidemiol. *97*, 208–218

Eldadah, Z.A., Asher, D.M., Godec, M.S., Pomeroy, K.L., Goldfarb, L.G., Feinstone, S.M., Levitan, H., Gibbs, C.J., and Gajdusek, D.C. (1991). Detection of flaviviruses by reverse-transcriptase polymerase chain reaction. J. Med. Virol. *33*, 260–267

Fenner, F., Pereira, H.G., Porterfield, J.S., Joklik, W.K., and Downie, A.W. (1974). Family and generic names for viruses approved by the International Committee on Taxonomy of Viruses. Intervirology 3, 193–198

Fernandez, R., and Vazquez, S. (1990). Serological diagnosis of dengue by ELISA Inhibition Method (EIM). Mem. Inst. *Oswaldo Cruz*, Rio de Janeiro *85*, 347–351

Figueiredo, L.T.M., Batista, W.C., and Igarashi, A. (1997). A simple reverse transcription-polymerase chain reaction for dengue type 2 virus identification. Mem. Inst. *Oswaldo Cruz*, Rio de Janeiro 92, 395–398

Figuereido, L.T.M., Simoes, M.C., and Cavalcante, S.M.B. (1989). Enzyme immunoassay for the detection of dengue IgG and IgM antibodies using infected mosquito cell as antigen. Trans. R. Soc. Trop. Med. Hyg. *83*, 702–707

Fulop, L., Barrett, A.D.T., Phillpotts, R., Martin, K., Leslie, D., and Titball, R.W. (1993). Rapid identification of flaviviruses based on conserved NS5 gene sequences. J. Virol. Methods *44*, 179–188

Garcia, M., Cabezas, C., Callahan, J., Benedicta, Y., Gutierrez, V., Ortiz, A., and Anaya, E. (1997). Determinacion de IgG y anticuerpos totales contrra el virus dengue en muestras obtenidas en papel de filtro. Rev. Med. Exp. INS *14*, 45–49

Georgiades, J., Stim, T.B., McCollum, R.W., and Henderson, J.R. (1965). Dengue virus plaque formation in Rhesus monkey kidney cultures. Proc. R. Soc. Exp. Biol. Med. *118*, 385–388

Gentry, M.K., Henchal., E.A., McCown, J.M., Brandt, W.E., and Dalrymple, J.M. (1982). Identification of distinct antigenic determinants on dengue 2 virus using monoclonal antibodies. Am. J. Trop. Med. Hyg. *31*, 548–555

Groen, J., Velzing, J., Copra C., Balenthien, E., Deubel, V., Vorndam, V., and Osterhaus, A.D.M.E. (1999). Diagnostic value of dengue virus-specific IgA and IgM serum antibody detection. Microbes Infect. *1*, 1085–1090

Gubler, D.J. (1998). Dengue and dengue hemorrhagic fever. Clin. Microbiol. Rev. *11*, 480–496

Gubler, D.J. (1996). Serological diagnosis of dengue/dengue hemorrhagic fever. Dengue Bull. *29*,20–23

Gubler, D.J., and Sather, G. (1988). Laboratory diagnosis of dengue and dengue hemorrhagic fever. Proceedings of the International Symposium of Yellow fever and dengue, Rio de Janeiro, Brazil, May 15–19, 1988.

Gubler, D.J., Kuno, G., Sather, G.E., Velez, M., and Oliver, A. (1984). Mosquito cell cultures and specific monoclonal antibodies in surveillance for dengue viruses. Am. J. Trop. Med. Hyg. 33, 158–165

Gubler, D.J., and Rosen, L. (1976). A simple technique for demonstrating transmission of dengue virus by mosquitoes without the use of vertebrate hosts. Am. J. Trop. Med. Hyg. *25*, 146–150

Guzman, M.G., and Kouri, G. (2004). Dengue diagnosis, advances and challenges. Int. J. Infect. Dis. *8*, 69–80

Guzman, M.G., Rodriguez, R., Rodriguez-Roche, R., Hermida, L., Alvarez, M, Lazo, L., Mune, M., Rosario, D., Valdes, K., Vazquez, S., Martinez, R., Serrano, T., Paez, J., Espinosa, R., Pumariega, T., and Guillen, G. (2003a). Induction of neutralizing antibodies and partial protection from viral challenge in *Macaca fascicularis* immunized with recombinant dengue 4 envelope glycoprotein expresses in *Pichia pastoris*. Am. J. Trop. Med. Hyg. 69, 123–128

Guzmán, M.G., Pelegrino, J.L., Pumariega, T., Vazquez, S., Kouri, G., and Arias, J. (2003b). Control externo de la calidad del diagnostico serologico del dengue. Rev Panamericana de la Salud *14*, 371–376.

Guzman, M.G., and Kouri, G. (2002). Dengue, an update. Lancet Infect. Dis. *2*, 33–42

Guzman, M.G., Kouri, G., Valdes, L., Bravo, J., Alvarez, M., Vazquez, S., Delgado, I., and Halstead, S.B. (2000). Epidemiological studies on dengue in Santiago de Cuba, 1997 Am. J. Epidemiol. 152; 793–799

Guzman, M.G., Alvarez, M., Rodriguez, R., Rosario, D., Vazquez, S., Valdes, L., Cabrera, M.V., and Kouri, G. (1999). Fatal dengue haemorrhagic fever in Cuba (1997). Int. J. Infect. Dis. 3 130–135.

Guzman, M.G., and Kourí, G. (1996). Advances in dengue diagnosis. Clin. Diagn. Immunol. *3*, 621–627

Guzman, M.G., Kouri, G., Bravo, J., Soler, M., and Martinez, E. (1991). Sequential infection as risk factor for dengue hemorrhagic fever/dengue shock syndrome (DHF/DSS) during the 1981 dengue hemorrhagic fever Cuban epidemic. Mem. Inst. Oswaldo Cruz *86*, 367

Guzman, M,.G., Kouri, G., Bravo, J., Soler, M., Vazquez, S., and Morier, L. (1990). Dengue hemorrhagic in Cuba, *1981*, a retrospective seroepidemiologic study. Am. J. Trop. Med. Hyg. *42* 179–184.

Guzman, M.G., Vazquez, S., Bravo, J.R., Monteagudo, R., and Kouri, G. (1985). Utilidad de la hemolisis radial para el diagnostico del dengue. Rev. Cubana Med. Trop. *37*, 238–245

Guzman, M.G., Kouri, G., Soler, M., Morier, L., and Vazquez, S. (1984). Aislamiento del virus dengue 2 en sueros de pacientes utilizando el ratón lactante y cultivo de células LLCMK2. Rev. Cubana Med. Trop. *36*, 4–10

Guzman, M.G., Kouri, G., and Bravo, J. (1982). Normalizacion de la toma de muestra de sangre en papel de filtro para la serologia del dengue. Rev. Cubana Med. Trop. *34*, 114–118

Hall, W.C., Crowell, T.P., Watts, D.M., Barros, V.L.R., Kruger H., Pinheiro, F., and Peters, C.J. (1991). Demonstration of yellow fever and dengue antigens in formalin-fixed paraffin-embedded human liver by immunohistochemical analysis. Am. J. Trop. Med. Hyg. *45*, 408–417

Halstead, S.B. (1974). Etiologies of the experimental dengues of Siler and Simmons. Am. J. Trop. Med. Hyg. *23*, 974–982

Hammon, W. McD, Rudnick, A., and Sather, G.E. (1960). Viruses associated with epidemic hemorrhagic fevers of the Philippines and Thailand. Science *131*, 1102–1103

Hammon, W. McD, and Sather, G.E. (1964). Virological findings in the 1960 hemorrhagic fever epidemic (dengue) in Thailand. Am. J. Trop. Med. Hyg. *13*, 629–641

Harris, E., Roberts, T.G., Smith, L., Selle, J., Kramer, L.D., Valle, S., Sandoval, E., and Balmaseda, A. (1998). Typing of dengue viruses in clinical specimens and mosquitoes by single-tube multiplex reverse transcriptase PCR. J. Clin. Microbiol. 36, 2634–2639

Hebert, S.J., Bowan, K.A., Rudnick, A., and Burton, J.J.S. (1980). A rapid method for the isolation and identification of dengue viruses employing a single system. Malaysian J. Pathol. 3, 67–68

Heinz, F.X., and Roehrig, J.T. (1990). Flaviviruses. Eds M.H.V. van Regemortel and A.R. Neurath. Immunochemistry of viruses. II The bases for serodiagnosis and vaccines. Elsevier Science Publishers (Biomedical division). Chapter 14, pp. 289–305

Henchal, E.A., Polo, S., Vordam, V., Yaemsiri, C., Innis, B.L., and Hoke, C.H. (1991). Sensitivity and specificity of a universal primer set for the rapid diagnosis of dengue virus infections by the polymerase chain reaction and nucleic acid hybridization. Am. J. Trop. Med. Hyg. *45*, 418–428.

Henchal, E.A., McCown, J.M., Seguin, M.C., Gentry, M.K., and Brandt, W.E. (1983). Rapid identification of dengue virus isolates by using monoclonal antibodies in an indirect immunofluorescence assay. Am. J. Trop. Med. Hyg. 32, 164–169

Henchal., E.A., Gentry, M.K., McCown, J.M., and Brandt, W.E. (1982). Dengue virus-specific and flavivirus group determinants identified by immunofluore4scence with monoclonal antibodies. Am. J. Trop. Med. Hyg. *31* 830–836

Hermida, L., Rodríguez, R., Lazo, L., Silva, R., Zulueta, A., Chinea, G., Lopez, C., Guzman, M.G., and Guillen, G. (2004). A dengue-2 envelope fragment inserted within the structure of P64k meningococcal protein carrier enables a functional immune response against the virus in mice. J. Virol. Methods *115*, 41–49

Herrera, M., Vazquez, S., and Fernandez, A. (1985). Determinacion de anticuerpos fijadores de complemento en pacientes con fiebre hemorragica del dengue. Rev. Cubana Med. Trop. *37*, 195–202

Holland, P.M., Abramson, R.D., Watson, R., and Gelfand, D.H. (1991). Detection of specific polymerase chain reaction product by utilized the 5′-3′exonuclease activity of Thermus aquaticus. Proc. Natl. Acad. Sci. USA *88*, 7276–80.

Holmes, E.C., and Burch, S.S. (2000). The causes and consequences of genetic variation in dengue virus. Trends Microbiol. *8*, 74–76

Hotta, S., Fujita, N., and Maruyama, T. (1966). Research on dengue in tissue culture. I Plaque formation in an established monkey kidney cell line culture. Kobe J. Med. Sci. *12*, 179–187

Houng, H.S.H., Chen, R.C.M., Vaughn, D.W., and Kanesathasan, N. (2001). Development of a fluorogenic RT-PCR system for quantitative identification of dengue virus serotypes 1, 2, 3 and 4, using conserved and serotype-specific 3′non-coding sequences. J. Virol. Methods 95,19–32

Innis, B. (1997). Antibody responses to dengue virus infection. In, Gubler DJ, Kuno, G., editors. Dengue and dengue haemorrhagic fever. New York, CAB International, pages 221–243

Innis, B. (1995). Dengue and Dengue Hemorrhagic Fever. Exotic Viral Infections. Edited by J.S. Porterfield. Published by Chapman and Hall, London. Chapter 5, 103–144

Innis, B.L., Nimmannityia, S., Kusalerdchariya, S., Chongswasdi V., Suntayakorn, S., Puttisri, P., and Hoke, C.H. (1989). An enzyme-linked immunosorbent assay to characterise dengue infections where dengue and Japanese encephalitis cocirculate. J. Trop. Med. Hyg. *40*, 418–427

Jirakanjanakit, N, Khin, M.M., Yoksan, S., Bhamarapravati, N. (1999). The use of Toxorhynchites splendens for identification and quantitation of serotypes contained in the tetravalent live attenuated dengue vaccine. Vaccine *17*, 597–601

Jirakanjanakit, N., Sanohsomneing, T., Yoksan, S., and Bhamarapravati, N. (1997). The micro-focus reduction neutralization test for determining dengue and Japanese encephalitis neutralizing antibodies in volunteers vaccinated against dengue. Trans. R. Soc. Trop. Med. Hyd. *91*, 614–617

Johnson, B.W., Chambers, T.V., Crabtree, M.B., Guirakhoo, F., Monath, T. and Miller, BR. (2004). Analysis of the replication kinetic of the ChimericVax-DEN 1, 2, 3, 4, tetravalent virus mixture in *Aedes aegypti* by real-time reverse transcription-polymerase chain reaction. Am. J. Trop. Med. Hyg. *70*,89–97

Johnson, A.J., Martin, D.A., Karabatsos, N., and Roehrig, J.T. (2000). Detection of anti-arboviral immunoglobulin G using a monoclonal antibody based capture enzyme-linked immunosorbent assay. J. Clin. Microbiol. *38*, 1827–1831

Kamasanttaya, K., Churdboonchart, V., Yoksan, S., and Bhamarapravati, N. (1987). Application of peroxidase-antiperoxidase (PAP) staining for detection and localization of dengue 2 viral antigen. II Observations for the antibody enhancement activity in human monocytes. Southeast Asian J. Trop. Med. Public Health *18*, 137–141

Kanesa-thasan N, Sun W, Kim-Ahn G, Van Albert S, Putnak JR, King A, Raengsakulsrach B, Christ-Schmidt H, Gilson K, Zahradnik JM, Vaughn DW, Innis BL, Saluzzo JF, and Hoke CH Jr. (2001). Safety and immunogenicity of attenuated dengue virus vaccines (Aventis Pasteur) in human volunteers. Vaccine *19*, 3179–3188

Kao, C.L., Wu, M.C., Chiu, Y.H., Lin, J.L., Yueh, Y.Y., Chen, L.K., and King, C.C. (2001). Flow cytometry compared to indirect immunofluorescence for rapid detection of dengue virus type 1 after amplification in tissue culture. J. Clin. Microbiol. *39*, 3672–3677

Kaufman, B.M., Summers, P.L., Dubois, D.R., Cohen, W.H., Gentry, M.K., Timchak, R.L., Burke, D.S., and Eckels, K.H. (1989). Monoclonal antibodies for dengue virus prM glycoprotein protect mice against lethal dengue infection. Am. J. Trop. Med. Hyg. *41*, 576–580

Kaufman BM, Summers PL, Debois Dr, Eckels KH (1987). Monoclonal antibodies against dengue 2 E glycoprotein protect mice against lethal dengue infection. Am. J. Trop. Med. Hyg. 36, 427–434

Khawsak, P., Phantana, S., and Chansiri, K. (2003). Determination of dengue virus serotypes in Thailand using PCR based method. Southeast Asian J. Trop. Med. Public Health *34*, 781–785.

Kimura, R., and Hotta, S. (1944). On the inoculation of dengue virus into mice. Nippon Igakku *3379*, 629–633

Kittigul, X, Suthachana S., Kittigul, C., and Pengruangrojanachai V (1998). Immunoglobulin M-capture biotin-streptavidin enzyme-linked immunosorbent assay for detection of antibodies to dengue viruses. Am. J. Trop. Med. Hyg. 59, 352–356

Kittigul, L., Meethien, N., Sujirarat, D., Kittigul, C., and Vasanavat, S. (1997). Comparison of dengue virus antigens in sera and peripheral blood mononuclear cells from dengue infected patients. Asian Pacific J. Allergy Immunol. *15*, 187–191

Koraka, P., Murgue, B., Setiati, T.E., Deparis, X., Suharti, C., van Gorp, E.C.M., Hack, C.E., Osterhaus, A.D.M.E. and Groen, J. (2003). Elevated levels of total and dengue virus specific immunoglobulin E in patients with varying disease severity. J. Med. Virol. *70*, 91–98

Kow, C.Y., Koon, L.L., and Yin, P.F. (2001). Detection of dengue viruses in field caught male *Aedes aegypti* and *Aedes albopictus* (Diptera, *Culicidae)* in Singapore by type-specific PCR. J. Med. Entomol. 38, 475–479

Kuberski, T.T., and Rosen, L. (1977a). Identification of dengue viruses using complement fixing antigen produced in mosquitoes. Am. J. Trop. Med. Hyg. *26*, 538–543

Kuberski, T.T., and Rosen, L. (1977b). A simple technique for the detection of dengue antigen in mosquitoes by immunofluorescence. Am. J. Trop. Med. Hyg. *26*, 533–537

Kuno G. (1998a). Universal diagnostic RT-PCR protocol for arboviruses. J. Virol. Methods *72*, 27–41

Kuno, G., Chang, G.J., Tsuchiya, K.R., Karabatsos, N., and Cropp, C.B. (1998b). Phylogeny of the genus *Flavivirus*. J. Virol. *72*, 73–83

Kuno, G., Gubler, D.J., Velez, M., and Oliver, A. (1985). Comparative sensitivity of three mosquito cell lines for isolation of dengue viruses. Bull. WHO 63, 279–286

Kurane, I., Zeng, L., Brinton, M.A., and Ennis, F.A. (1999). Definition of an epitope on NS3 recognized by human CD4+ cytotoxic T lymphocyte clones cross-reactive for dengue virus types 2, 3 and 4. Virology *240*, 169–174

Laferte, J., Pelegrino, J.L., Guzman, M.G., Gonzalez, G., Vazquez, S., and Hermida, C. (1992). Rapid diagnosis of dengue virus infection using a novel 10ul IgM antibody capture ultramicroELISA assay (MAC UMELISA Dengue). Adv. Modern Biotechnol. *1*, 19.4

Laille, M., Deubel, V., and Flye, F.(1991). Demonstration of concurrent dengue 1 and dengue 3 infection in six patients by the polymerase chain reaction. J. Med. Virol. *34*, 51–54

Lam, S.K., Fong, M.Y., Chungue, E., Dormisingham, S., Igarashi, M.A., Khin, Z.T., Kyaw, Z.T., Nisalak, A., Roche, C., Vaughn, D.W., and Vorndam, V. (1996). Multicentre evaluation of dengue IgM dot enzyme immunoassay. Clin. Diagn. Virol. 7, 93–98

Lam, SK., Devi, S., and Pang, T. (1987). Detection of specific IgM in dengue infection. Southeast Asian J. Trop. Med. Public Health *18*, 532–538

Lam, S.K., Chew, C.B., Poon, G.K., Ramalingam, S., Show, S.C., and Pang, T. (1986). Isolation of dengue viruses by intracerebral inoculation of mosquito larvae. J. Virol. Methods *14*, 133.

Lanciotti, R.S. (2003). Molecular amplification assays for detection of flavivirus. Adv. Virus Res. 61 ,67–99.

Lanciotti, R.S., Gubler, D.J., and Trent, D.W. (1997). Molecular evolution and phylogeny of dengue-4 viruses. J. Gen. Virol. *78*, 2279–2286

Lanciotti, R.S., Lewis, J.G., Gubler, D.J., and Trent, D.W. (1994). Molecular evolution and epidemiology of dengue-3 viruses. J. Gen. Virol. *75*, 65–75

Lanciotti, R.S., Calisher, C.H., Gubler, D.J., Chang, G.J., and Vorndam, V. (1992). Rapid detection and typing of dengue viruses from clinical samples by using Reverse Transcriptase-Polymerase Chain Reaction. *30*, 545–551

Laue, T., Emmerich, P., and Schmitz, H. (1999). Detection of dengue virus RNA in patients after primary or secondary dengue infection by using the TaqMan automated amplification system. J. Clin. Microbiol. *37*, 2543–2547

Leitmeyer, K., Vaughn, D., Watts, D.M., Salas, R., Villalobos de Chacon, I., Ramos, C., and Rico-Hesse, R. (1999). Dengue virus structural differences that correlate with pathogenesis. J. Virol. *73*, 4738–4747

Lemmer, K., Mantke, O., Bae, H.G., Groen, J., Drosten, C., and Niedrige. External quality control assessment in PCR diagnosis of dengue virus infections. (2004). J. Clin. Virol. 30 , 291–6.

Libraty, D.H., Young, P.R., Pickering, D., Endy, T.P., Kalayanarooj, S., Green, S., Vaughn, D.W., Nisalak, A., Ennis, F.A., and Rothman, A.L. (2002a). High circulating levels of the dengue virus nonstructural protein NS1 early in dengue illness correlate with the development of dengue hemorrhagic fever. J. Infect. Dis. *186*, 1165–1168.

Libraty, D.H., Endy, T.P., Houng, H.S.H., Green, S., Kalayanarooj, S., Suntayakorn, S., Chansiriwongs, W., Vaughn, D.W., Nisalak, A., Ennis, F.A., and Rothman, A.L. (2002b). Differing influences of virus burden and immune activation on disease severity in secondary dengue 3 virus infections. J. Infect. Dis. *185*, 1213–1221

Liew, K.J., and Chow, V.T. (2004). Differential display RT-PCR analysis of ECV304 endothelial-like cells infected with dengue virus type 2 reveals messenger RNA expression profiles of multiple human genes involved in know and novel roles. J. Med. Virol. *72(4)*, 597–609.

Liu, W.T., Chen, C.L., Lee, S.S.J., Chan, C.C., Lo, F.L.L. and Ko, Y.C. (1991). Isolation of dengue virus with a human promonocyte cell line. Am. J. Trop. Med. Hyg. *44*, 494–499

Loroño-Pino, M.A., Cropp, C.B., Farfan, J.A., Vorndam, A.V., Rodriguez-Angulo, E.M., Rosado Paredes, E.P., Flores, L.F., Beaty, B.J., and Gubler, G.J. (1999). Common occurrence of concurrent infections by multiple dengue virus serotypes. Am. J. Trop. Med. Hyg. *61*, 725–730

Lucia, H.L., and Kangwanpong, D. (1994). Identification of dengue virus-infected cells in paraffin-embedded tissue using in situ polymerase chain reaction and DNA hybridization. J. Virol. Methods *48*, 1–8

Mackay, I.M., Arden, K.E., and Nitche, A. (2002). Real-time PCR in virology. Nucleic Acids Res. 30 1292–1305

Mackenzie, J.M., Jones, M.K., and Young, P.R. (1996). Immunolocalization of the dengue virus nonstructural glycoprotein NS1 suggests a role in viral RNA replication. Virology *220*, 232–240

Malergue, F., and Chungue, E. (1995). Rapid and sensitive streptavidin-biotin amplified fluorogenic enzyme-linked immunosorbent assay for direct detection and identification of dengue viral antigens in serum. J. Med. Virol. *47*, 43–47

Mandl, C.W., Heinz, F.X., Stockl, E., and Kunz, C. (1989). Genome sequence of tick-borne encephalitis virus (Western subtype) and comparative analysis of nonstructural proteins with other flaviviruses. Virology *173*, 291–301

Mandl, C.W., Heinz, F.X., and Kunz, C. (1988). Sequence of the structural proteins of tick-borne encephalitis virus (Western subtype) and comparative analysis with other flaviviruses. Virology *166*, 197–205

Maneekarn, N., Morita, K., Tanaka, M., Igarashi, A., Usawattanakul, W., Sirisanthana, V.,

Innis, B.L., Sittisombut, N., Nisalak, A., and Nimmanitya, S. (1993). Applications of Polymerase Chain reaction for identification of dengue viruses isolated from patient sera. Microbiol. Immunol. *37*, 41–47

Mathew, A., Kurane, I., Green, S., Stephens, H.A., Vaughn, D. W., Kalayanarooj, S., Suntayakorn, S., Chandanayingyong, D., Ennis, F. A. and Rothman, A. L. Predominance of HLA-restricted cytotoxic T lymphocyte responses to serotype-cross-reactive epitopes on nonstructural proteins following natural secondary virus infection. J. Virol. *72*, 3999–4004

McNair Scott R., Nisalak, A., Cheamudon, U., Seridhoranakulo, S., and Nimmannitya, S. (1980). Isolation of dengue viruses from peripheral blood leukocytes of patients with hemorrhagic fever. J. Infect. Dis. *141*, 1–6

Meiyu. F., Huosheng. C., Cuihua. C., Xiaodong. T., Lianhua. J., Yifei. P., Weijun. C, and Huiyu. G. (1997). Detection of flaviviruses by reverse transcriptase-polymerase chain reaction with the universal primer set. Microbiol. Immunol. *41*, 209–213

Miagostovich, M.P., Nogueira, R.M.R., dos Santos, F.B., Schatzmayr, H.G., Araujo, E.S.M., and Vorndam, V. (1999). Evaluation of an IgG enzyme-linked immunosorbent assay for dengue diagnosis. J. Clin. Virol. *14*, 183–189

Monath, T.P. and Heinz, F.X. 1996). Flaviviruses. Fields Virology, Third edition. Edited by B.N. Fields, D.M. Knipe, P.M. Howley. Lippincott-Raven Publishers, Philadelphia. Chapter 31, pages 961–1032

Monath, T.P., Wands, J.R., Hill, L.J., Gentry, M.K. and Gubler, D.J. (1986). Multisite monoclonal immunoassay for dengue viruses, detection of viraemic human sera and interference by heterologous antibody. J. Gen. Virol. *67*, 639–650

Morens, D.M., Halstead, S.B., Repik, P.M., Putvatana, R., and Raybourne, N. (1985a). Simplified plaque reduction neutralization assay for dengue viruses by semimicro methods in BHK21 cells, comparison of the BHK suspension test with standard plaque reduction neutralization. J. Clin. Microbiol. *22*, 250–254

Morens, D.M., Halstead, S.B., and Larsen, L.K. (1985b). Comparison of dengue virus plaque reduction neutralization by macro and semimicro methods in LLCMK2 cells. Microbiol. Immunol. *29*, 1197–1205

Morier, L., Castillo, A., Rodriguez, R., and Guzman, M.G. (1995). Utilidad de la sublinea celular CLA-1 para el aislamiento del virus dengue. Rev. Cubana Med. Trop. *47*, 217–218

Morita, K., Maemoto, T., Honda, S., Onishi, K., Murata, M., Tanaka, M., and Igarashi, A. (1994). Rapid detection of virus genome from imported Dengue Fever and dengue Hemorrhagic fever patients by direct polymerase chain reaction. J. Med. Virol. *44*, 54–58

Mune, M., Rodriguez, R., Soto, Y., Rodriguez Roche, R., Marquez, G., Garcia, J., Guillen, G., and Guzman, M.G. (2003). Carboxy-terminally truncated dengue 4 virus envelope glycoprotein expressed in *Pichia pastoris* induce neutralizing antibodies and resistance to dengue 4 virus challenge in mice. Arch. Virol. *148*, 2267–2273

Murgue, B., Roche, C., Chungue, E., and Deparis, X. (2000). Prospective study of the duration and magnitude of viremia in children hospitalised during the 1996–1997 dengue-2 outbreak in French Polynesia. J. Med. Virol. *60*, 432–438.

Niesters, H.G.M. (2002). Clinical virology in real time. J. Clin. Virol. 25 ,S3–312.

Nisalak, A., Halstead, S.B., Singharaj, P., Udomsakdi, S., Nye, S.W., and Vinijchaikul, K. (1970). Observations related to pathogenesis of dengue hemorrhagic fever. III Virologic studies of fatal disease. Yale J. Biol. Med. *42*, 293–310

Nogueira RMR, Miagostovich MP, Cavalcanti SMB, Marzochi KBF and Schatzmayr HG. (1992). Levels of IgM antibodies against dengue virus in Rio de Janeiro, Brazil. Res. Virol. *143*, 423–427

Okuno, Y., Fukunaga, T., Srisupaluck, S., and Fukai, K. (1979). A modified PAP (peroxidase-antiperoxidase) staining technique using sera from patients with dengue hemorrhagic fever (DHF), 4 step staining technique. Biken J. *22*, 131–135

Oseni, R.A., Donaldson, M.D., Dalglish, D.A., and Aaskov, J.G. (1983). Detection by ELISA of IgM antibodies to Toss River virus in serum from patients with suspected epidemic polyarthritis. Bull WHO *61*, 703

PAHO. (1994). Dengue and Dengue Hemorrhagic Fever in the Americas, guidelines for prevention and control. Scientific Publication, No. 548

Palmer, C.J., King, S.D., Cuadrado, R.R., Perez, e., Baum, M., and Ager, A.L. (1999). Evaluation of the MRL diagnostics dengue fever virus IgM capture ELISA and the PanBio Rapid Immunochromatographic Test for diagnosis of dengue fever in Jamaica. J. Clin. Microbiol. *37*, 1600–1601

Pang T, Lam SK, Chew CB, Poon GK, Ramalingam S. (1983). Detection of dengue viruses by immunofluorescence following in-ovulation of mosquito larvae. Lancet *i (8336)*, 1271

Parida, M., Posadas, G., Inoue, S,., Hasebe, F., and Morita, K. (2004). Real-time reverse transcription loop-mediated isothermal amplification for rapid detection of West Nile virus. J. Clin. Microbiol. *42*, 257–63

Pelegrino, J.L., Arteaga, E., Rodríguez, A.J., Gonzalez, E., Frontela, M.C., and Guzmán, M.G. (1997). Normalización de técnicas inmunohistoquimicas para la detección de antígenos del virus dengue en tejidos embebidos en parafina. Rev. Cubana Med. Trop. *49*, 100–107

Pierre, V., Drouet, M.T., and Deubel, V. (1994). Identification of mosquito-borne flavivirus sequences using universal primers and reverse transcription/polymerase chain reaction. Res. Virol. *145*, 93–104

Prado I., Rosario, D., Bernardo, L., Alvarez, M., Rodriguez, R., Vazquez, S., and Guzman, M.G. PCR detection of Dengue virus using Dried Whole Blood spotted on Filter Paper. J. Virol. Methods *125*, 75–81

Race, M.W., Williams, M.C. and Agostini, C.F.M. (1979). Dengue in the Caribbean, virus isolation in a mosquito (*Aedes pseudoscutellaris*) cell line. Trans. R. Soc. Trop. Med. Hyg. *73*, 18–22

Race, M.W., Fortune, R.A.J., and Varma, M.G.R. (1978). Isolation of dengue viruses in mosquito cell cultures under field conditions. Lancet, 48–49

Rao., B.L. (1976). Plaque formation of dengue viruses in Vero cell culture under carboxymethylcellulose overlay. Indian J. Med. Res. *64*, 1709–1712

Rey, F.A., Heinz, F.X., Mandl, C., Kunz, C., and Harrison, S.C. (1995). The envelope glycoprotein from tick-borne encephalitis virus at 2A resolution. Nature *375*, 291–298

Rice, C.M. (1996). Flaviviridae, the viruses and their replication. Fields Virology, Third edition. Edited by B.N. Fields, D.M., Knipe, P.M. Howley. Lippincott, Raven Publishers, Philadelphia. Chapter 30, pages 931–959

Rico-Hesse R, Harrison LM, Salas RA, Tovar D, Nisalak A, Ramos C, Boshell J, Mesa MTR, Nogueira RMR, and Rosa ATD. (1997). Origins of dengue type 2 viruses associated with increased pathogenicity in the Americas. Virology *230*, 244–251

Rico-Hesse, R. (1990). Molecular evolution and distribution of dengue viruses type 1 and 2 in nature. Virology *174*, 479–493

Rigau, J.G., Clark, G.G., Gubler, D.J., Reiter, P., Sanders, E.J., and Vorndam, V. (1998). Dengue and dengue haemorrhagic fever. Lancet *352*, 971–977

Rodriguez Roche, R., Alvarez, M., Guzman, M.G., Morier, L., and Kouri, G. (2000). Comparison of rapid centrifugation assay with conventional tissue culture method for isolation of dengue 2 virus in C6/36-HT cells. J. Clin. Microbiol. *38*, 3508–3510

Roehrig, J.T. Immunochemistry of dengue viruses. (1997). In Gubler DJ, Kuno G, editors. Dengue and dengue hemorrhagic fever. Wallingford. CAB International 199–219,

Rosario, D., Alvarez, M., Vazquez, S., Amin, N., Rodríguez, R., Valdes, K., and Guzman, M.G. (2001). Application of Molecular Methods to the diagnosis and characterization of a dengue outbreak in Cuba. Rev. Biotecnología Aplicada *18*, 1–4

Rosario, D., Alvarez, M., Diaz, J., Contreras, R., Vazquez, S., Rodriguez, R., Guzmán, M.G. (1998). Rapid detection and typing of Dengue viruses from clinical samples using reverse transcriptase- polymerase chain reaction. Pan Am. J. Public Health *4*, 1–5.

Rosen, L, and Shroyer, D.A. (1985). Comparative susceptibility of five species of Toxorhynchites mosquitoes to parenteral infection with dengue and other flaviviruses. Am. J. Trop. Med. Hyg. *34*, 805–809

Rosen, L. (1981). The use of *Toxorhynchites* mosquitoes to detect and propagate dengue and other arboviruses. Am. J. Trop. Med. Hyg. *30*, 177–183

Rosen, L., and Gubler, D. (1974). The use of mosquitoes to detect and propagate dengue viruses. Am. J. Trop. Med. Hyg. *23*, 1153–1160

Rothman, A. (1997). Viral pathogenesis of dengue infections. In Gubler DJ, Kuno G, editors. Dengue and dengue hemorrhagic fever. Wallingford. CAB International 245–271

Russell, P.K., and Nisalak, A. (1967). Dengue virus identification by the plaque reduction neutralization test. J. Immunol. *99*, 291–296

Sabin, A.B. (1952). Research on dengue during World War II. Am. J. Trop. Med. Hyg. *1*, 30–50

Sabin, A.B. (1950). The dengue group and its family relationships. Bact. Tev. *14*, 225

Sabin, A.B., and Schlesinger, R.W. (1945). Production of immunity to dengue with virus modified by propagation in mice. Science *101*, 640

Saluzzo, J.F., Sarthou, J.L., Cornet, M., Digoutte, J.P. and Monath, T.P. (1986). Interet du titrage par ELISA des IgM specifiques pour le diagnostic et la surveillance de la circulation selvatique des flavivirus en Afrique. Ann. Inst. Pasteur/Virol. 137E, 155–170

Sang, C.T., Hoon, L.S., Cuzzubbo, A., and Devine, P. (1998). Clinical evaluation of a rapid immunochromatographic test for the diagnosis of dengue virus infection. CDLI 5, 407–409

Sangasang, A., Wibulwattanakij, S., Chanama, S., O-rapinpatipat, A., Anuegoonpipat, A., Anantapreecha, S., Sawanpanyalert, P., and Kurane, I. (2003). Evaluation of RT-PCR as a tool for diagnosis of secondary dengue virus infection. Jpn J. Infect. Dis. 56 205–209

Sariol, C., Pelegrino, J.L., Martinez, A., Arteaga, E., Kouri, G., and Guzman, M.G. (1999). Detection and genetic relationship in seventeen-year old paraffin embedded samples of Cuba. Am. J. Trop. Med. Hyg., *60*, 994–1000

Sarmiento, L., Rodriguez, G., and Boshell, J. (1995). Diagnostico immunohistoquimico del dengue en cortes de parafina. Biomedica *15*, 10–15

Schelesinger, J.J., Brandriss, M.W., and Walsh, E.E. (1987). Protection of mice against dengue 2 virus encephalitis by immunization with the dengue 2 virus non-structural glycoprotein NS1. J. Gen. Virol. *68*, 853–857

Schlesinger, R.W. (1977). Dengue viruses. In, Virology Monographs. Wien, Sprinter-Verlag, 109

Schwartz, E., Mileguir, F., G Rossman, Z., and Mendelson, E. (2000). Evaluation of ELISA-based serodiagnosis of dengue fever in travelers. J. Clin. Virol. *19*, 169–173

Schwartz, E., Mendelson, E. And Sidi, Y. (1996). Dengue fever among travellers. Am. J. Med. *101*, 516–520

Schoepp, R.J. and Beaty, B.J. (1984). Titration of dengue viruses by immunofluorescence in microtiter plates. J. Clin. Virol. *20*, 1017–1019

Se-Thoe, S.Y., Ng, M.M.L., and Ling, A.E. (1999). Retrospective study of Western Blot profiles in immune sera of natural dengue virus infections. J. Med. Virol. *57*, 322–330

Shu, P.Y., Chang, S.F., Kuo, Y.C., Yueh, Y.Y., Chien, L.J., Sue, C.L., Lin, T.H., and Huang, J.H. (2003a). Development of group-and serotype-specific one step SYBR green I-bases real-time reverse transcription-PCR assay for dengue virus. Clin. Microbiol. 41 2408–16.

Shu, P.Y., Chen, L.K., Chang, S.F., Yueh, Y.Y., Chow, L., Chien, L.J., Chin, C., Lin, T.H., and Huang, J.H. (2003b). Comparison of capture immunoglobulin M (IgM) and IgG Enzyme-Linked Immunosorbent Assay (ELISA) and nonstructural protein NS1 serotype-specific IgG ELISA for differentiation of primary and secondary dengue virus infections. CDLI *10*, 622–630

Shu, P.Y., Chen, L.K., Chang, S.F., Yueh, Y.Y., Chow, L., Chien, L.J., Chin, C., Yang, H.H., Lin, T.H., and Huang, J.H. (2002). Potential application of nonstructural protein NS1 serotype-specific immunglobulin G Enzyme-Linked Immunosorbent Assay in the sero-epiedmiologic study of dengue virus infection, correlation of results with those of the plaque reduction neutralization test. J. Clin. Microbiol. *40*, 1840–1844

Shu, P.Y., Chen, L.K., Chang, S.F., Yueh, Y.Y., Chow, L., Chien, L.J., Chin, C., Lin, T.H., and Huang, J.H. (2000). Dengue NS1-specific antibody responses, isotype distribution and serotyping in patients with dengue fever and dengue hemorrhagic fever. J. Med. Virol. *62*, 224–232

Shulze IT, Schlesinger RW. (1963). Plaque assay of dengue and other group B arthropod-borne viruses under methylcellulose overlay media. Virology *19*, 40–48

Singh, K.R.P., and Devi Paul, S. (1968). Multiplication of arboviruses in cell lines from *Aedes albopictus* and *Aedes aegypti*. Curr. Sci. *3*, 65–67

Sithiprasasna, R., Strickman, D., Innis, B.L., and Linthicum, K.J. (1994). ELISA for detecting dengue and Japanese encephalitis viral antigen in mosquitoes. Ann. Trop. Med. Parasitol. *88*, 397–404

Smithburn, K.C. (1954). Antigenic relationships among certain arthropod-borne viruses as revealed by neutralization tests. J. Immunol. *72*, 376–388

Soler, M., Guzman, M.G., Mune, M., and Kouri, G. (1988). Identificacion mediante la tecnica de inmunofluorescencia indirecta de varias cepas de dengue aisladas durante la epdiemia de Nicaragua en (1985). Rev. Cubana Med. Trop. *40*, 5–12

Soler, M., Guzman, M.G., Morier, L., and Kouri, G. (1985). Utilizacion de los anticuerpos monoclonales para la identificacion, mediante la tecnica de inmunofluorescencia indirecta de varias cepas d dengue aisladas durante la epidemia de fiebre hemorragica, Cuba (1981). Rev. Cubana Med. Trop. 37, 246–251

Soliman, A.K., Watts, D.M., Salib, A.W., Shehata, A.E.D., Arthur, R.R., and Botros, B.A.M. (1997). Application of an immunoperoxidase monolayer assay for the detection of arboviral antibodies J. Virol. Methods 65, 147–151

Srivastava, A.K., Putnak, J.R., Warren, R.L., and Hoke, C.H. (1995). Mice immunized with a dengue type 2 virus E and NS1 fusion protein made in *Escherichia coli* are protected against lethal dengue virus infection. Vaccine *13*, 1251–1258

Stim, T.B. (1970a). Dengue virus plaque development in Simian cell systems. II. Agar variables and effect of chemical additives. Appl. Microbiol. *19*, 757–762

Stim, T.B. (1970b). Dengue virus plaque development in Simian cell systems. I. Factors influencing virus adsorption and variables in the agar overlay medium. Appl. Microbiol. *19*, 751–756

Su, C.C., Wu, T.Z., Chen, L.K., Yang, H.H., and Tai, D.F. (2003). Development of immunochips for the detection of dengue viral antigens. Analyt. Chim. Acta *479*, 117–123

Sudiro, T.M., Zivny, J., Ishiko, H., Green, S., Vaughn, D.W., Kalayanarooj, S., Nisalak, A., Norman, J.E., Ennis, F.A., and Rothman, A.L. (2001). Analysis of plasma viral RNA levels during acute dengue virus infection using quantitative competitor reverse transcription-polymerase chain reaction. J. Med. Virol. 63 ,29–34.

Sudiro, T.M., Ishiko, H., Green, S., Vaughn, D.W., Nisalak, A., Kalayanarooj, A.L., Rothman, A.L., Raengsakulrach, B., Janus, J., Kurane, I., and Ennis, F.A. (1997). Rapid diagnosis of dengue viremia by reverse transcriptase-polymerase chain reaction using 3′-non-coding region universal primers. Am. J. Trop. Med. Hyg. 56, 424–429

Summers, P.L., Eckels, K.H., Dalrymple, J.M., Scott, R.M., and Boyd, V.A. (1984). Antibody response to dengue 2 vaccine measured by two different radioimmunoassay methods. J. Clin. Microbiol. *19*, 651–659

Sydow, F.F., Santiago, M.A., Neves-Souza, P.C., Cerqueira, D.I.S., Gouvea, A.S., Lavatori, M.F.H., Bertho, A.L., and Kubelka, C.F. (2000). Comparison of dengue infection in human mononuclear leukocytes with mosquito C6/36 and mammalian Vero cells using flow cytometry to detect virus antigen. Mem Inst *Oswaldo Cruz*, Rio de Janeiro 95, 483–489

Tadeu, L., Figueiredo, M., and Shope, R. (1987). An enzyme immunoassay for dengue antibody using infected cultured mosquito cells as antigen. J. Virol. Methods *17*, 191–198

Talarmin, T., Labeau, B., Lelarge, J., and Sarthou, J.L. (1998). Immunoglobulin capture enzyme-linked immunosorbent assay for the diagnosis of dengue fever. J. Clin. Microbiol. *36*, 1189–1192

Tan, C.H.C, Yap, E.H., Singh, M., Deubel, V., and Chan, Y.C. (1990). Passive protection studies in mice with monoclonal antibodies directed against the non-structural protein NS3 of dengue 1 virus. J. Gen. Virol. *71*, 745–749

Tanaka, M. (1993). Rapid identification of flavivirus using the polymerase chain reaction. J. Virol. Methods *41*, 311–322

Tesh, R.B. (1979). A method for the isolation and identification of dengue viruses using mosquito cell cultures. Am. J. Trop. Med. Hyg. *28*, 1053–1059

Trent, D.W., and Chang, G.J. (1992). Detection and identification of Flavivirus by reverse transcriptase-polymerase chain reaction. In, Becker Y, Darai G, editors. Frontiers in Virology, Diagnosis of human virus by PCR Technology. Springer-Verlag, Berlin, 355–371

Twiddy, S.S., Holmes, E.C. (2003). The extend of homologous recombination in members of the genus Flavivirus. J. Virol. *84*, 429–440

Twiddy, S.S., Farrar, J., Chau, N.V., Wills, B., Gould, E.A., Gritsun, T., Lloyd, G., and Holmes, E.C. (2002). Phylogenetic relationships and differential selection pressures among genotypes of dengue-2 virus. Virology 2002; *298*, 63–72

Usuku, S., Castillo, L., Sugimoto, C., Noguchi, Y., Yogo, Y., and Kobayashi, N. (2001). Phylogenetic analysis of dengue-3 viruses prevalent in Guatemala during 1996–1998. Arch. Virol. *146*, 1381–1390

Valdes, K., Alvarez, M., Pupo, M., Vazquez, S., Rodriguez, R., and Guzman, M.G. (2000). Human dengue antibodies against structural and nonstructural proteins. CDLI *7*, 856–857

Varma, M.G.R., Pudney, M. and Leake, C.J. (1974). Cell lines from larvae of *Aedes (Stegomyia) malayensis* Colless and *Aedes (S) pseudoscutellaris* (Theobald) and their infection with some arboviruses. Trans. R. Soc. Trop. Med. Hyg. 68; 374–382

Vathanophas, K., Hammon, W.M., Atchison, R.W. and Sather, G.F. (1973). Attentep type specific diagnosis of dengue virus infection by the indirect fluorescence antibody method directed at differentiating IgM and IgG responses. Proc. Soc. Exp. Biol. Med. *142*, 697–702

Vaughn, D.W., Green, S., Kalayanarooj, S., Innis, B.L., Nimmannitya, S., Suntayakorn, S., Endy, T.P., Raengsakulrach, B., Rothman, A.L., Ennis, F.A., and Nisalak, A. (2000). Dengue viremia titer, antibody response pattern and virus serotype correlate with disease severity. J. Infect. Dis. *181*, 2–9

Vazquez, S., Perez, A.B., Tuiz, R., Rodriguez, R., Pupo, M., Calzada, N., Gonzalez, L., Gonzalez, D., Castro, O., Serrrano, T., and Guzman, M.G. (2005). Serological markers during dengue 3 primary and secondary infections. J. Clin. Virol. 33, 132–137

Vazquez, S., Valdes, O., Pupo, M., Delgado, I., Alvarez, M., Pelegrino, J.L., and Guzman, M.G. (2003a). MAC-ELISA and ELISA inhibition methods for detection of antibodies after yellow fever vaccination. J. Virol. Methods *110*, 179–184

Vazquez, S. Lemus, G., Pupo, M., Ganzon, O., Palenzuela, D., Indart, A., and Guzman, M.G. (2003b). Diagnosis of dengue virus infection by the visual and simple AuBioDOT immunoglobulin M capture system. CDLI *10*, 1074–1077

Vazquez, S., Guzman, M.G, Guillen, G., Perez, A.B., Pupo, M., Rodriguez, R., Reyes, O., Glay, H.E., Delgado, I., Garcia, G., and Alvarez, M. (2002). Immune response to synthetic peptides of dengue prM protein. Vaccine *20*, 1823–1830

Vazquez, S., Saenz, E., Huelva, G., Gonzalez, A., Kouri, G., and Guzmán, M.G. (1998). Detection de IgM contra el virus del dengue en sangre entera absorbida en papel de filtro. Rev. Panam. Salud. Publica 3, 174–178

Vazquez, S., Bravo, J.R., Perez, A.B., and Guzman, M.G. (1997). ELISA de Inhibición. Su utilidad para clasificar un caso de dengue. Rev. Cubana Med. Trop. *49*, 108–112

Vazquez, S., Fernandez, R., and Llorente, C. (1991). Utilidad de sangre almacenada en papel de filtro para estudios serologicos por ELISA de Inhibicion. Rev. Inst. Med. Trop. Sao Paulo 33, 309–311

Vazquez, S., and Fernandez, R. (1989). Utilizacion de un metodo de inhibicion de ELISA en el diagnostico serologico del dengue. Reporte preliminar. Rev. Cubana Med. Trop. *41*, 18–26

Vazquez, S., de la Cruz, F., Guzman, M.G., and Fernandez, R. (1986). Comparacion de la tecnica de fijacion del complemento, la Inhibicion de la hemaglutinacion y el inmunoensayo enzimatico sobre fase solida para el diagnostico serologico del dengue. Rev. Cubana Med. Trop. 38, 7–14

Vorndam, V. and Beltran, M. (2002). Enzyme-linked immunosorbent assay-format microneutralization test for dengue viruses. Am. J. Trop. Med. Hyg. 66, 208–212

Vorndam, V., and Kuno, G. (1997). Laboratory diagnosis of dengue virus infections. In Gubler DJ, Kuno G (Eds), CAB International, New York, pp. 313–333

Wang, W.K., Chao, D.Y., Kao, C.L., Wu, H.C., Liu, Y.C., Li, C.M., Lin, S.C., Ho, S.T., Huang, J.H., and King, C.C. (2003). High levels of plasma dengue viral load during defervescence in patients with dengue hemorrhagic fever, implications for pathogenesis. Virology *305* 330–338.

Wang, W.K., Lee, C.N., Kao, C.L., Lin, Y.L., and King, C.C. (2000). Quantitative competitive reverse transcription-PCR for quantification of dengue virus RNA. J. Clin. Microbiol. *38*, 3306–10.

Waterman, S.H., Kuno, G., Gubler, D.J., and Sather, G.E. (1985). Low rates of antigen detection and virus isolation from peripheral blood leukocytes of dengue fever patients. Am. J. Trop. Med. Hyg. *34*, 380–384

Westaway, E.G. Brinton, M.A., Gaidamovich, S.Ya., Horzinek, M.C., Igarashi, A., Kaariainen, L., Lvov, D.K., Porterfield, J.S., Russell, P.K., and Trent, D.W. (1985). Flaviviridae. Intervirology *24*, 183–192

WHO/TDR. (2004). WHO/TDR and PDVI joint workshop on dengue diagnostics and dengue classification/case management, 4–6 October, 2004, Geneva, Switzerland.

Wilson, IG. (1997). Inhibition and facilitation of nucleic acid amplification. Appl. Environm. Microbiol. 63, 3741–3751.

Worobey, M., Rambaut, A., and Holmes, E.C. (1999). Widespread intra-serotype recombination in natural populations of dengue virus. Proc. Natl. Acad. Sci. USA *96*, 7352–7357

Wu, S.J., Lee, E.M., Putvatana, R., Shurtliff, R.N., Porter, K.R., Suharyono, W., Watts, D.M., King, C.C., Murphy, G.S., Hayes, C.G., and Romano, J.W. (2001). Detection of dengue viral RNA using a nucleic acid sequence-based amplification assay. J. Clin. Microbiol. *39*, 2794–2798

Wu, S.J.L., Hanson, B., Paxton, H., Nisalak, A., Vaughn, D.W., Rossi, C., Henchal., E.A., Porter, K.R., Watts, D.M., and Hayes, C.G. (1997). Evaluation of a dipstick enzyme-linked immunosorbent assay for detection of antibodies to dengue virus. Clin. Diagn. Lab. Immunol. *4*, 452–457

Yenchitsomanus, P.T., Sricharoen, P., Jaruthasana, I., Pattanakitsakul, S.N., Nitayaphan, S., Mongkolsapaya, J., and Malasit, P. (1996). Rapid detection and identification of dengue viruses by polymerase chain reaction (PCR). Southeast Asian Trop. Med. Public Health 27 228–36.

Young, P.R., Hilditch, P.A., Bletchly, C., and Halloran, W. (2000). An Antigen Capture Enzyme-Linked Immunosorbent Assay reveals high levels of the dengue virus protein NS1 in the sera of infected patients. J. Clin. Microbiol. *38*, 1053–1057

Young, P.R. (1989). An improved method for the detection of peroxidase-conjugated antibodies on immunoblots. J. Virol. Methods *24*, 227–236

Yuill, T.M., Sukhavachana, P., Nisalak, A., and Russell, P.K. (1968). Dengue virus recovery by direct and delayed plaques in LLCMK2 cells. Am. J. Trop. Med. Hyg. *17*, 441–448

Zhang, Y.M., Hayes, E.P., McCarty, T.C., Dubois, D. R., Summers, P. L., Eckels, K. H., Chanock, R. M. and Lai, C. J. (1988). Immunization of mice with dengue structural proteins and non-structural NS1 by baculovirus recombinant induces resistance to dengue virus encephalitis. J. Virol. *62*, 3027–3031

Japanese Encephalitis Virus: Molecular Biology and Vaccine Development

8

Sang-Im Yun and Young-Min Lee

Abstract

Japanese encephalitis virus (JEV), the leading cause of viral encephalitis, frequently causes mortality and morbidity, especially in children and young adults in Asia. The recent geographic expansion of JEV activity has become a worldwide public health threat. JEV is a mosquito-borne flavivirus with a single-stranded, positive-sense RNA genome approximately 11 000 nucleotides in length. The recent development of an efficient reverse genetics system for JEV, using a genetically stable infectious cDNA as a bacterial artificial chromosome, is enabling advancement of direct molecular genetic studies previously hampered by the difficulty of manipulating the viral genome. Although considerable information has been learned about JEV biology from comparison with other closely related flaviviruses, much more needs to be elucidated and demonstrated experimentally. This review discusses the molecular aspects of common features of conserved sequence elements and viral gene products involved in viral replication, as compared to other flaviviruses. This will assist current strategies for development of a safer, more effective, and less expensive antiviral vaccine against this pathogen as well as closely related flaviviruses.

Introduction

Outbreaks of encephalitis presumably caused by JEV infection were recognized in Japan as early as 1871 (Burke and Monath, 2001; Tiroumourougane *et al.*, 2002). In 1924, a severe epidemic with more than 6000 cases was recorded in Japan, and a filterable agent was determined to be responsible for these epidemics (Miyake, 1964). In 1934, a filterable agent from the brain of a human was transmitted to reproduce the encephalitis in intracerebrally inoculated monkeys. In 1935, the Nakayama strain of JEV was first isolated from the brain of a human and has been used continuously in development and production of the inactivated vaccine, which is still used in most affected areas worldwide (Solomon, 2000; Burke and Monath, 2001). Japanese encephalitis was originally termed Japanese B encephalitis (type B encephalitis) to distinguish it from von Economo's encephalitis lethargica (sleeping sickness), known as type A (Solomon *et al.*, 2000).

Approximately up to 50 000 sporadic and epidemic cases of Japanese encephalitis, 25–30% of which result in death, are reported annually and about half of the survivors have permanent neuropsychiatric sequelae (Okuno, 1978; Umenai *et al.*, 1985; Burke and Leake, 1988; Halstead, 1992;

Vaughn and Hoke, 1992; Tsai, 2000). The leading cause of viral encephalitis in Asia, JEV has occurred in many Asian countries, including the former Soviet Union, Korea, Japan, China, Taiwan, Nepal, Philippines, Malaysia, Vietnam, Cambodia, Thailand, India, Sri Lanka, and Pakistan. In recent years, however, JEV has expanded to Indonesia (Olson *et al.*, 1985; Wuryadi and Suroso, 1989) and even Australian territories (Hanna *et al.*, 1996; Shield *et al.*, 1996; Johansen *et al.*, 1997; Mackenzie *et al.*, 1997; Hanna *et al.*, 1999; Spicer *et al.*, 1999; Mackenzie *et al.*, 2002; Johansen *et al.*, 2004), posing a major public health concern worldwide.

The mode of transmission via mosquitoes was speculated in the early 1930s, and Mitamura and co-workers isolated the virus from *Culex tritaeniorhynchus* in 1938 (Burke and Monath, 2001). Classical ecological studies of JEV in Japan established that JEV is transmitted in an enzoonotic cycle between C. *tritaeniorhynchus* and vertebrate viremic-amplifying hosts, primarily pigs and birds (Buescher and Scherer, 1959; Buescher *et al.*, 1959a; Buescher *et al.*, 1959b; Scherer and Buescher, 1959; Scherer *et al.*, 1959a; Scherer *et al.*, 1959b; Scherer *et al.*, 1959c; Scherer *et al.*, 1959d). JEV grows by a complex cycle that involves pigs as amplifying hosts, ardeid birds as reservoirs, and mosquitoes as vectors. Although many animals can be infected with the virus, chiefly domestic pigs are the most important natural amplifying hosts for human transmission because of the prolonged and high viremia, the maintenance and amplification of the virus in the environment, the ability to produce uninfected offspring, and their close vicinity to humans (Innis, 1995; Gubler and Rochrig, 1998). As reservoirs, birds may also be responsible for disseminating the virus to new geographical regions. The virus does not typically cause clinical symptoms in these natural hosts.

The *Culex* species mosquito vector for JEV, of which C. *tritaeniorhynchus* is the most important vector for human infection, breeds in stagnant pools of water such as paddy fields for rice cultivation. Other *Culex* species appear to play a role in the transmission of JEV. The C. *vishnui* subgroup, namely C. *tritaeniorhynchus*, C. *vishnui*, and C. *pseudovishnui*, has long been implicated as capable of transmitting the virus and causing epidemics in India (Carey *et al.*, 1968; Dandawate *et al.*, 1969; Chakravarty *et al.*, 1975; Banerjee *et al.*, 1979; George *et al.*, 1987; Mourya *et al.*, 1989; Naik *et al.*, 1990; Dhanda *et al.*, 1997; Gajanana *et al.*, 1997). In this country, C. *gelidus* and C. *fuscocephala* may be important for the zoonotic cycle (Mourya *et al.*, 1989; Gajanana *et al.*, 1997). JEV has also been isolated from many other *Culex* species, including C. *bitaeniorhynchus* (Banerjee *et al.*, 1978; Banerjee *et al.*, 1979), C. *quinquefasciatus* (Mourya *et al.*, 1989), and C. *epidesmus* (Banerjee *et al.*, 1979). Recently, JEV has also been isolated from C. *sitiens* in Malaysia (Vythilingam *et al.*, 1994; Vythilingam *et al.*, 2002). In addition to the *Culex* species, JEV was also isolated from *Anopheles* species (Chakravarty *et al.*, 1975; Banerjee *et al.*, 1979; George *et al.*, 1987; Mourya *et al.*, 1989; Dhanda *et al.*, 1997) and *Mansonia* species (Dhanda *et al.*, 1997). However, the role of these mosquitoes in the transmission of JEV needs to be further investigated. Surprisingly, flavivirus-related integrated DNA sequences have recently been discovered in the genome of laboratory-bred and wild *Aedes albopictus* and *Aedes aegypti* mosquitoes, demonstrating an integration into a eukaryotic genome of a multigenic sequence from an RNA virus that replicates without a recognized DNA intermediate (Crochu *et al.*, 2004).

Different patterns of JEV transmission are observed within individual countries and from year to year (Inactivated JEV Vaccine, 1993). In temperate regions of Asia, JEV is transmitted seasonally, during the summer and early fall (Okuno, 1978; Umenai *et al.*, 1985; Burke and Leake, 1988; Halstead, 1992; Burke and Monath, 2001). Seasonal patterns of viral transmission are determined by the increased number of mosquito vectors and the abundance of vertebrate-amplifying hosts in the environment. These two factors could be affected by temperature and rainfall in the rainy season and may be correlated with the patterns of migrating avian-amplifying hosts. In tropical regions, however, the viral transmission takes places throughout the year as sporadic cases (Rosen, 1986; Thongcharoen, 1989; Sucharit *et al.*, 1989; Service, 1991), possibly due to the abundance of mosquito vectors thriving in irrigated rice paddy fields.

Virus classification

The family Flaviviridae currently consists of three genera, namely the flaviviruses, the pestiviruses, and the hepaciviruses (Heinz *et al.*, 2000; Lindenbach and Rice, 2001; Calisher and Gould, 2003). This family also contains a panel of unclassified viruses, the GB agents. Although members of this family share a similar genome structure and presumably a similar replication strategy, phylogenetic analyses indicate that they are distantly related (Simons *et al.*, 1995; Kuno *et al.*, 1998). About 70 viruses currently belong to the genus flavivirus, which is categorized into antigenic complexes of closely related viruses, clades, and clusters, based on serological and phylogenetic data (Calisher, 1988, Calisher *et al.*, 1989; Zanotto *et al.*, 1996; Kuno *et al.*, 1998; Heinz *et al.*, 2000). Most of these viruses are transmitted by mosquitoes or ticks to their vertebrate hosts. No arthropod vectors have been identified for some flaviviruses. Additionally, an insect virus known as Cell-Fusing Agent (CFA) has been categorized in the flaviviruses based on genome structural similarity and homology with other members of the flaviviruses (Cammisa-Parks *et al.*, 1992). JEV is a member of the mosquito-borne flaviviruses and is antigenically related to the flaviviruses of Cacipacore, Yaounde, Koutango, West Nile, Kunjin, Murray Valley encephalitis, St. Louis encephalitis, and Usutu viruses (Poidinger *et al.*, 1996; Heinz *et al.*, 2000).

Since the first isolation of JEV established as the prototype Nakayama strain, a number of geographically and temporally diverse JEV strains have been isolated from humans, mosquitoes, or pigs. Of these strains, partial nucleotide sequences of more than 500 JEV isolates are currently available from the GenBank database, while only 37 JEV isolates have been fully sequenced up to now. Phylogenetic analyses have mainly focused on partial sequence information from either the prM or E gene, which has important biological activities, such as hemagglutination, neutralization, virus binding to cellular receptors, and membrane fusion (Gritsun *et al.*, 1995). Data on the partial sequence information of the prM gene indicate that the JEV strains were divided into four genotypes (Chen *et al.*, 1990; Chen *et al.*, 1992; Huong *et al.*, 1993; Ali and Igarashi, 1997; Tsuchie *et al.*, 1997). Similar attempts were also made for the E gene (Ni and Barrett, 1995; Nam *et al.*, 1996; Paranjpe and Banerjee, 1996; Wu *et al.*, 1998; Mangada and Takegami, 1999). Several phylogenetic analyses using 6–27 fully sequenced JEV strains also revealed four distinct genotypes (Vrati *et al.*, 1999; Williams *et al.*, 2000; Nam *et al.*, 2001; Yun *et al.*, 2003a). However, several branch points dividing these genotypes were defined by poor bootstrap support (Williams *et al.*, 2000; Yun *et al.*, 2003a).

Thus, genetic characterization of additional full-length JEV isolates will need to be further examined to better elucidate the genetic relationships among JEV strains and to understand the clinical and functional significance of this divergence. Interestingly, an investigation of the geographical distribution of all known JEV isolates has recently suggested that JEV originated from its ancestral virus in the Indonesia–Malaysia tropical region, where all genotypes to date appear to have evolved from, before spreading across Asia (Solomon *et al.*, 2003).

Molecular biology

Our review focuses on comparative analysis with other flaviviruses to help define the conserved linear nucleotide sequence motifs and the predicted RNA secondary structures at the viral 5′ and 3′ noncoding regions of JEV genomic RNA, which might play an important role in the viral life cycle, such as translation, replication, and encapsidation of the genomic RNA into a virion. Furthermore, we summarize the functionally predictable important domains in the structural and nonstructural proteins of JEV involved in initiating viral infection and successful replication in a host cell, respectively. Direct investigations into functional significance of the *cis*-acting and *trans*-acting genetic elements of JEV conserved among the flaviviruses will undoubtedly be facilitated by recent development of the efficient reverse genetics system for JEV. This will advance the understanding of JEV biology and pathogenesis at the molecular level and will assist in the development of a genetically defined antiviral vaccine against this pathogen.

Virion morphology

Like other flaviviruses, JEV particles are spherical enveloped viruses, approximately 50 nm in diameter. A positive-sense, single-stranded genome RNA molecule of approximately 11 000 nucleotides is organized with a multiple copy of the capsid proteins to form an electron dense viral core which is surrounded by a lipid bilayer containing two viral surface proteins, prM/M and E (Murphy, 1980; Chambers *et al.*, 1990a; Heinz and Allison, 2003). During viral assembly and budding into the intracellular membrane compartment, the prM protein present on the surface of the intracellular immature virion is acquired with the E protein. During export of intracellular immature virions to the extracellular environment, this prM protein is cleaved by a cellular protease to generate the M protein, which is present on the surfaces of extracellular mature virions. Several features of each of the three structural proteins and their functions in the virus life cycle are summarized later in this chapter.

Significant progress toward understanding the structure and organization of flaviviral structural proteins has been made by solving the structure of the tryptic soluble form of the TBEV E protein by X-ray crystallography (Rey *et al.*, 1995) and by reconstructing the image of a recombinant TBEV subviral particle, produced by coexpression of the prM and E proteins by cryoelectron microscopy (Ferlenghi *et al.*, 2001). The structure of a dimeric ectodomain fragment of the TBEV E protein obtained after trypsin treatment of purified virions appears to have a long and elongated shape, consisting of three distinct domains—a central domain (I) that connects an Ig-like domain (III) to a dimerization domain (II) (Rey *et al.*, 1995). Based on this elongated and slightly curved shape (Rey *et al.*, 1995) and the binding sites for antibodies (Mandl *et al.*, 1989; Roehrig *et al.*, 1998) and heparin sulfate (Mandl *et al.*, 2001), the E protein appears to be laid along the surface of the virion.

The first structure of a flavivirus, DENV, was recently determined at a resolution of 24 Å by using a combination of cryoelectron microscopy, interpretation using the known structure of the E protein dimer X-ray structure, and a 3D image reconstruction (Kuhn *et al.*, 2002). This study showed the icosahedral scaffold consists of 90 E homodimers in such a way that a panel of three parallel dimers is arranged as a recognizable unit, and these 30 units are organized into a "herringbone" pattern on the surface of the virion (Kuhn *et al.*, 2002). The similarly structured WNV New York 99 strain, which is responsible for the outbreak in the US in 1999, has recently been determined at a resolution of 17 Å (Mukhopadhyay *et al.*, 2003). These structural characteristics have a significant impact on the nonsymmetric binding of cellular receptors and antibodies, and the majority of E dimers undergo a significant rearrangement into a trimer at low pH (Kuhn *et al.*, 2002). For the process of virus-cell membrane fusion, the distal β barrels of domain II of the E protein are suggested to be inserted into the cellular membrane. Similarly, the 9.5 Å resolution image of mature DENV has recently been determined by cryoelectron microscopy and image reconstruction (Zhang *et al.*, 2003a), which is consistent with the herringbone configuration of the E protein (Kuhn *et al.*, 2002). The most important and striking finding of this study is the structure and organization of the previously uncovered "stem" and "anchor" regions of the E and M protein. The amphipathic α-helical stem regions of the E protein and part of the N-terminal region of the M protein appear to be partially buried in the outer leaflet of the viral lipid bilayer (Zhang *et al.*, 2003a). The anchor regions of the E and M proteins have each been visualized to form antiparallel transmembrane α-helices, posing their respective C-termini toward the outside of the virion, which is consistent with the predicted topology of the unprocessed polyprotein (Zhang *et al.*, 2003a).

In addition, the structures of immature virions of DENV and YFV containing the uncleaved prM and E proteins have also been determined at a resolution of 16 Å and 9.5 Å, respectively (Zhang *et al.*, 2003b). The striking difference between the immature and mature virions has revealed that the surface of the immature virions contains 60 icosahedrally organized, asymmetric trimeric spikes with individual spikes containing three prM/E heterodimers (Zhang *et al.*, 2003b). In addition, the substantial pr portion of the prM protein in each spike appears to cover the fusion peptide of the E protein (Zhang *et al.*, 2003b) as a way of protecting the fusion peptide against premature fusion while moving through the secretory pathway to the extracellular compartment (Guirakhoo *et al.*, 1992). During the maturation process in the trans-Golgi network, the furin cleavage of the pr fragment from the prM protein induces a conformational rearrangement of the E proteins, resulting in a fusogenic virus (Stadler *et al.*, 1997).

Replication cycle

Importantly, JEV replication is believed to share characteristics with other flaviviruses (Figure 8.1). This scheme will undoubtedly be modified, as detailed molecular mechanisms of the viral replication

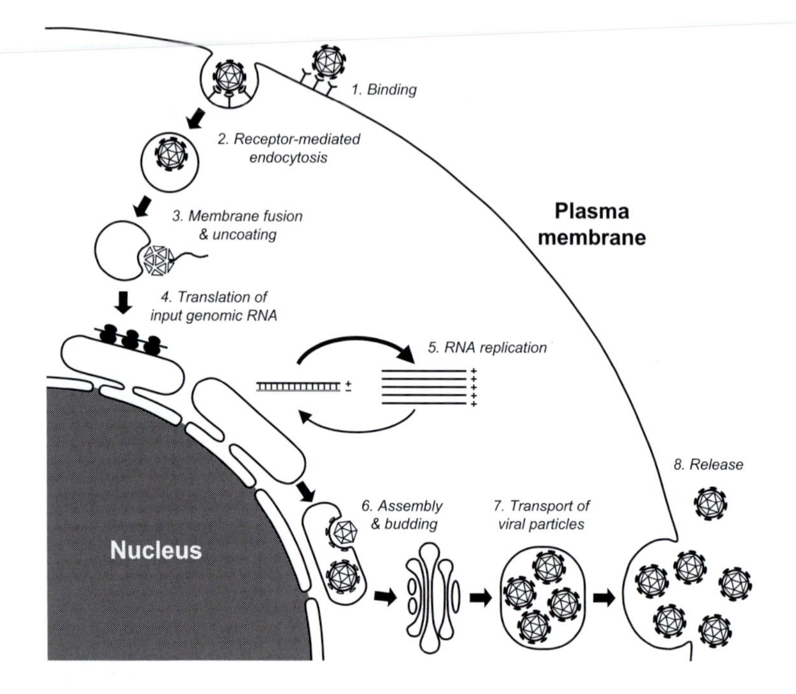

Figure 8.1 Overview of the JEV replication cycle. See text for the detailed description.

become understood in depth in the near future. According to the current working model for the flavivirus replication cycle, the first step is binding (attachment) to a cell surface molecule(s), which is necessary for viral entry [*Step 1*]. An important role of heparin sulfate has first been demonstrated for binding of DENV to vertebrate cells (Chen *et al.*, 1997), and subsequently in other members of flaviviruses (Hung *et al.*, 1999; Germi *et al.*, 2002; Lin *et al.*, 2002; Kroschewski *et al.*, 2003). Cell surface glycosaminoglycans have been shown to play a role in flavivirus entry, including JEV (Lee and Lobigs, 2000; Lee and Lobigs, 2002; Lee *et al.*, 2004). The highly sulfated form of glycosaminoglycans on cell surfaces appears to play an important role in the early stage of JEV infection (Su *et al.*, 2001; Liu *et al.*, 2004). Additionally, a 74 kDa molecule has been suggested to be a potential candidate for the cellular receptor for JEV (Kimura *et al.*, 1994). For DENV-4, two surface proteins from C6/36 cells have been shown to bind to the virus (Salas-Benito and del Angel, 1997), and potential molecules from Vero cells also may be involved in the infection (Martinez-Barragan and del Angel, 2001). Several additional molecules are suggested to bind to DENV-2 (Ramos-Castaneda *et al.*, 1997; Munoz *et al.*, 1998). A 105 kDa glycoprotein may bind to WNV and cause infection (Chu and Ng, 2003). For TBEV, several potential molecules appear to bind to the virus (Maldov *et al.*, 1992; Kopecky *et al.*, 1999). Considering the fact that the flaviviruses are able to infect insect and vertebrate-amplifying hosts in natural transmission cycles, flaviviruses may use a highly conserved cellular surface molecule(s). Alternatively, flaviviruses have been speculated to utilize multiple surface molecules as a cellular receptor(s) (Marianneau *et al.*, 1996).

Uptake of the flavivirus particles takes place via receptor-mediated endocytosis [*Step 2*]. The use of a cationic amphiphilic drug, chlorpromazine, has shown that JEV enters cells through the clathrin-dependent endocytic pathway (Nawa *et al.*, 2003). A recent study has also demonstrated that WNV enters through a clathrin-mediated endocytic pathway (Chu and Ng, 2004). The viral envelopes are fused with cellular membranes of endosomal vesicles in a low pH-dependent manner, which allows the viral core to be released into the cytoplasm. The RNA genome is then uncoated [*Step* 3] (Heinz and Allison, 2000; Heinz and Allison, 2003). JEV infection in C6/36 mosquito cells may proceed through the endocytic pathway involving intracellular acidic compartments, which are affected by bafilomycin A1 (Nawa, 1998). In addition, acidic pH may affect membrane fusion to play a role in WNV infection (Gollins and Porterfield, 1986; Kimura *et al.*, 1986; Kimura and Ohyama, 1988). Once the positive-sense viral RNA genome enters into the cytoplasm, it is translated to provide viral proteins essential for genome replication and mRNA synthesis [*Step 4*]. The single long open reading frame of the viral genome is translated to a large polyprotein. The polyprotein is co- or post-translationally processed into mature proteins by cellular proteases and the viral serine protease NS3 in association with NS2B (Falgout *et al.*, 1991; Chambers *et al.*, 1993).

Replication of the RNA genome occurs in cytoplasmic replication complexes which are localized at the perinuclear membranes [*Step* 5] (Lindenbach and Rice, 2003; Westaway *et al.*, 2003). These complexes contain the genomic RNA and viral replicases consisting of the viral nonstructural proteins in association with cellular proteins (Chu and Westaway, 1985; Chu and Westaway, 1987; Ng *et al.*, 1989; Westaway *et al.*, 1999). The first step in genome replication is copying of the input genomic RNA template to generate complementary genome-length minus-strand RNAs, which in turn serve as templates for the synthesis of multiple copies of genomic RNAs. The synthesis of plus-strand RNAs is about 10 times greater than that of minus-strand RNAs (Cleaves *et al.*, 1981; Chu and Westaway, 1985; Muylaert *et al.*, 1996). Previous studies on RNA synthesis have suggested that a nascent minus-strand RNA is copied from a plus-strand RNA template per cycle, which has been isolated exclusively in transiently stable double-stranded replicative forms (Wengler *et al.*, 1978; Cleaves *et al.*, 1981; Chu and Westaway, 1985). Free minus-strand RNAs have not been detected (Chu and Westaway, 1985; Khromykh and Westaway, 1997).

On the other hand, the minus-strand RNA serves as templates for reinitiation for the synthesis of multiple plus-strand RNAs per cycle, such that recently synthesized plus-strand RNAs are displaced by nascent plus-strands undergoing elongation (Cleaves *et al.*, 1981; Chu and Westaway, 1985). This RNA complex has been reported as replicative intermediates. Pulse-labeled replicative forms and replicative intermediates have been shown to chase into the genome-length plus-strand RNAs (Cleaves *et al.*, 1981; Chu and Westaway, 1985). Thus, flaviviral RNA synthesis appears to be semiconservative and asymmetric (Chu and Westaway, 1985). Lack of detection of virus-specific subgenomic RNAs in flavivirus-infected cells supports the notion that the genome-length plus-strand RNA is the only virus-specific mRNA.

All of the viral nonstructural proteins (NS1, NS2A, NS2B, NS3, NS4A, and NS5) appear to be involved in RNA replication. A temperature-sensitive mutation in NS1 has been shown to block RNA accumulation, suggesting an essential role in

RNA amplification (Muylaert *et al.*, 1997). Additionally, mutations introduced in NS1 have also been shown to affect the initiation of minus-strand synthesis (Lindenbach and Rice, 1997). NS1 and NS3 appear to be associated with double-stranded RNA within virus-induced membrane structures in flavivirus-infected cells, which are potential cytoplasmic sites of RNA replication (Westaway *et al.*, 1997). NS3 is a multifunctional protein shown to have protease, RNA helicase, and nucleotide triphosphatase activities (Wengler and Wengler, 1991; Warrener *et al.*, 1993; Utama *et al.*, 2000). The small hydrophobic proteins NS2A and NS4A also localize in the replication complex, which is located in virus-induced membrane structures (Mackenzie *et al.*, 1998). NS4A has been shown to genetically interact with NS1, suggesting its potential role as a replicase component in RNA replication (Lindenbach and Rice, 1999). NS5 is an RNA-dependent RNA polymerase (Tan *et al.*, 1996; Ranjith-Kumar *et al.*, 2001) and also has a potential methyltransferase activity, which may play a role in the capping of viral genomic RNAs (Koonin, 1993).

Progeny virions are assembled by encapsidating the genomic RNA into the core shell of capsid proteins, enveloped by two viral surface protein (prM and E)-embedded host-derived lipid membranes. Although the intracellular assembly of flaviviruses is not completely understood, the viral morphogenesis takes place by budding through intracellular membranes into cytoplasmic vesicles [*Step* 6] (Murphy, 1980; Heinz and Allison, 2003). Virus assembly intermediates or nucleocapsid particles during the replication process of the flaviviruses have rarely been visualized by electron microscopy, and viral particles first become visible at the intracellular membrane compartment in infected cells. A recent study using cryo-immunoelectron microscopy has described the nucleocapsid particles of WNV (Ng *et al.*, 2001). Intracellular viral particles appear to be present within the lumen of an intracellular membrane structure (Murphy *et al.*, 1968; Sriurairatna *et al.*, 1973; Matsumura *et al.*, 1977; Ohyama *et al.*, 1977; Sriurairatna and Bhamarapravati, 1977; Ko *et al.*, 1979; Leary and Blair, 1980; Deubel *et al.*, 1981; Hase *et al.*, 1987a; Hase *et al.*, 1987b; Ng, 1987; Ishak *et al.*, 1988; Mackenzie *et al.*, 1996; Wang *et al.*, 1998; Ng *et al.*, 2001), which are then transported to the plasma membrane via the cellular secretory pathway [*Step* 7].

During the transport to the plasma membrane, the intracellular viral particles undergo the maturation process. Glycans on prM and E protein are modified by trimming and terminal addition of sugar residues (Mason, 1989; Nowak *et al.*, 1989; Wengler and Wengler, 1989; Chambers *et al.*, 1990a; Chambers *et al.*, 1990b; Heinz *et al.*, 1994). The N-terminal portion of prM is cleaved by furin or a related protease in the *trans*-Golgi apparatus (Randolph and Stollar, 1990; Heinz *et al.*, 1994; Stadler *et al.*, 1997), and this cleavage is prevented by elevation of the pH in acidic intracellular compartments (Shapiro *et al.*, 1972; Randolph and Stollar, 1990; Heinz *et al.*, 1994). This prM cleavage is generally considered to distinguish extracellular viral particles from intracellular viral particles (Shapiro *et al.*, 1972) and is required for the generation of highly infectious viruses (Wengler and Wengler, 1989; Randolph *et al.*, 1990; Guirakhoo *et al.*, 1992; Heinz *et al.*, 1994; Allison *et al.*, 1995a; Stadler *et al.*, 1997; Elshuber *et al.*, 2003). The uncleaved prM interacts with E proteins, forms a fusion-inactive prM/E heterodimer, and keeps the E protein from undergoing an acid-induced conformational change during the transport to the plasma membrane (Wengler and Wengler, 1989; Randolph and Stollar, 1990; Guirakhoo *et al.*, 1991; Guirakhoo *et al.*, 1992; Heinz *et al.*, 1994; Heinz and Allison, 2000; Heinz and Allison, 2003).

Fusion of the vesicles with the plasma membrane releases the large amount of progeny virions into the extracellular compartment [*Step 8*] (Nowak *et al.*, 1989; Mason, 1989). Although it does not seem to be a dominant mode for viral morphogenesis, budding of viral particles directly at the plasma membrane has also been described (Matsumura *et al.*, 1977; Ohyama *et al.*, 1977; Sriurairatna and Bhamarapravati, 1977; Hase *et al.*, 1987a; Ng *et al.*, 2001). In addition to infectious viral particles, noninfectious virus-like particles known as slowly sedimenting hemagglutinin (SHA), are also produced from flavivirus-infected cells (Russell *et al.*, 1980). SHA appears to consist of E and M proteins with some prM. The expression of prM and E proteins in JEV has been shown to be sufficient for production of virus-like particles, which appear to elicit protective immunity (Mason *et al.*, 1991; Konishi *et al.*, 1992b; Konishi and Mason, 1993; Kojima *et al.*, 2003).

Genome structure and organization

The JEV genome is a single-stranded, positive-sense RNA molecule of approximately 11 000 nucleotides in length (Brinton, 1986; Rice *et al.*, 1986b; Westaway, 1987; Chambers *et al.*, 1990a; Lindenbach and Rice, 2001; Lindenbach and Rice, 2003). As schematically represented in Figure 8.2, the genomic RNA contains a single long open reading frame (ORF) flanked by 5′ and 3′ noncoding regions (NCRs). Like the other mosquito-borne flaviviruses (Wen-

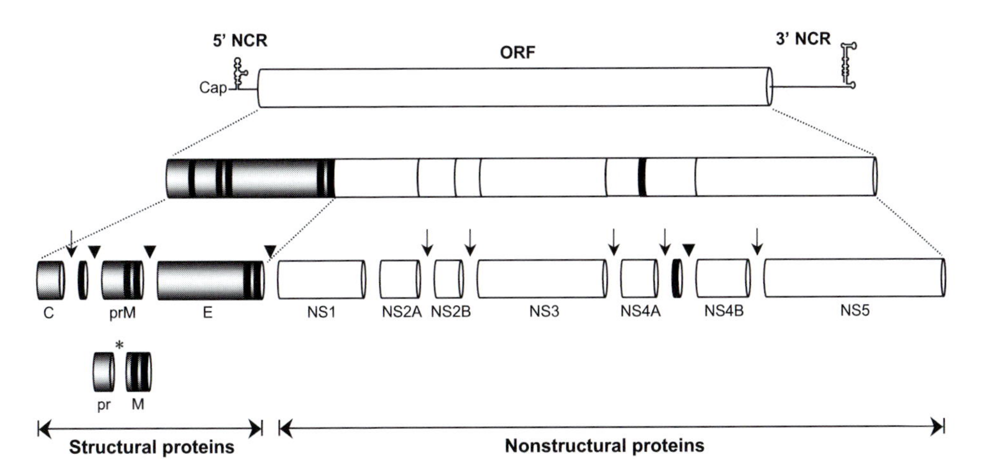

Figure 8.2 Schematic representation of the genome structure and polyprotein processing of JEV. The viral genome structure is illustrated at the top with a single open reading frame (a long open barrel) at both termini representing the 5′ and 3′ NCRs of the viral genome. Below the viral genome, the polyprotein precursor translated from the open reading frame is indicated. The structural proteins (C, prM, and E) are indicated in shaded barrels, whereas the nonstructural proteins (NS1-NS5) are in open barrels. Transmembrane domains are indicated by black bars. Arrow heads indicate the cleavage sites for host signalase and arrows indicate the cleavage sites for the viral NS2B-NS3 serine protease. The cleavage mediated by furin protease is indicated by an asterisk and the cleavage at the junction of NS1/NS2A is mediated by unknown proteases.

gler *et al.*, 1978; Cleaves and Dubin, 1979; Wengler and Wengler, 1981), the genomic RNA of JEV is believed to contain a type I cap at its 5′ end. The synthetic genome-length RNA transcripts capped with the $m^7G(5')ppp(5')A$ or $m^7G(5')ppp(5')G$ cap structure analog are infectious (Sumiyoshi *et al.*, 1992; Zhang *et al.*, 2001; Yun *et al.*, 2003b). At the 3′ end, the genomic RNA is reported to lack a poly(A) tail (Sumiyoshi *et al.*, 1987; Hashimoto *et al.*, 1988; Jan *et al.*, 1996; Vrati *et al.*, 1999; Williams *et al.*, 2000; Wu and Lee, 2001; Yun *et al.*, 2003a) and instead terminates with the highly conserved nucleotide sequences, which are predicted to form a strong stem–loop structure (Wengler and Wengler, 1981; Rice *et al.*, 1985; Brinton *et al.*, 1986), as described later.

The single long ORF of JEV encoding all the viral proteins is considered to be translated into a large polyprotein, which is the most striking process during replication of the flaviviruses (Brinton, 1986; Rice *et al.*, 1986b; Westaway, 1987; Chambers *et al.*, 1990a; Lindenbach and Rice, 2001; Lindenbach and Rice, 2003). As illustrated in Figure 8.2, the polyprotein is co- and post-translationally processed by cellular proteases and a viral serine protease into at least ten mature proteins, respectively. The three structural proteins designated as capsid (C), premembrane (prM, which is further processed into pr and M proteins), and envelope (E) proteins are encoded in the 5′ one-third of the ORF. The seven nonstructural proteins (NS1, NS2A, NS2B, NS3, NS4A, NS4B, and NS5, as arranged in the genome) are encoded in the remaining 3′ two-thirds.

The 5′ NCR of JEV is a highly conserved RNA sequence element among all different isolates and appears to be relatively short—invariably 95 nucleotides in length except for the Japanese Ishikawa strain (96 nucleotides) among the 37 fully sequenced JEV isolates currently available from the GenBank database (Sumiyoshi *et al.*, 1987; Hashimoto *et al.*, 1988; Jan *et al.*, 1996; Vrati *et al.*, 1999; Williams *et al.*, 2000; Wu and Lee, 2001; Yun *et al.*, 2003a). For translation of the JEV genomic RNA, this short 5′ NCR is likely to be sufficiently adapted to be facilitate ribosome binding to the cap structure at its 5′ end, which subsequently allows read down by ribosome scanning strategy from the 5′ cap structure. For YFV, the 5′ NCR functions as a potential enhancer for gene expression when linked to a heterologous mRNA derived from Germiston virus (Ruiz-Linares, 1989). In comparison, the 5′ NCR of hepatitis C virus is about 341 nucleotides in length (Fukushi *et al.*, 1994), and that of the poliovirus type 1, a member of the picornaviruses, is about 742 nucleotides in length (Rezapkin *et al.*, 1998). In these cases, the cap-independent translation is directed by an internal ribosome entry site, located in each 5′ NCR, which is a complex of well-organized secondary RNA structure (Brown *et al.*, 1992; Wang *et al.*, 1995; Bergamini *et al.*, 2000; Poyry *et al.*, 2001). The complement of the 5′ NCR, equivalent to the 3′ NCR of the minus-strand, must have *cis*-acting genetic elements required for directing the plus-strand RNA synthesis.

In comparison with the 37 fully sequenced JEV genomes currently available, the length of its 3′ NCR is reported to show a great deal of heterogeneity in length, ranging from 557 to 583 nucleotides (Sumiyoshi *et al.*, 1987; Hashimoto *et al.*, 1988; Jan *et al.*, 1996; Vrati *et al.*, 1999; Williams *et al.*, 2000; Wu and Lee, 2001; Yun *et al.*, 2003a). This discrepancy is largely explained by variable sizes of deletions that are uniformly found at the beginning of the 3′ NCR. The Taiwanese

Ling strain, isolated from the brain of a patient in 1965, was found to contain a 25-nucleotide deletion, 29 nucleotides apart from the ORF stop codon (Jan *et al.*, 1996). A deletion of 11–13 nucleotides immediately downstream of the ORF stop codon was also found in the first Australian FU (Williams *et al.*, 2000), the Korean K94P05 (Nam *et al.*, 2001), and the Japanese Ishikawa strains (GenBank accession number AB051292). Recently, the first Korean K87P39 isolate was also shown to contain a 9-nucleotide deletion 18 nucleotides apart from the ORF stop codon (Yun *et al.*, 2003a). Similarly, deletions of different lengths downstream of the stop codon were also observed in tick-borne encephalitis viruses (Gritsun *et al.*, 1997). Thus, the presence of these deletions strongly suggests that the variable region immediately downstream of the ORF stop codon in the JEV genome is dispensable for viral replication *per se*, although this should be addressed experimentally. In addition, computational analyses have suggested that this region does not exhibit secondary RNA structure (Proutski *et al.*, 1997; Rauscher *et al.*, 1997), suggesting that this variable region in the 3′ NCR may function in regulating the rate of viral RNA replication (Nam *et al.*, 2001).

For JEV, investigations to identify and characterize viral *cis*-acting and *trans*-acting genetic elements involved in the viral life cycle process at the molecular level have been extremely difficult due to the genetic instability of the cloned cDNA of the JEV genomic RNA, as described later. For this reason, little information is currently available at the molecular level. In this review, we will discuss common features of possible *cis*-acting and *trans*-acting genetic elements of JEV compared to those of other flaviviruses. This information may assist in directing future molecular genetic studies on JEV biology using the recently developed reverse genetics system (Yun *et al.*, 2003b).

Common features of conserved nucleotide sequence elements

In general, the short conserved RNA sequence motif and secondary RNA structure play important roles in various aspects of the viral life cycle, such as viral translation regulation, RNA replication, and encapsidation of a viral genomic RNA into a virion. The genomic RNA of flaviviruses including that of JEV competes for these processes in the viral life cycle, and certainly these 5′ and 3′ NCRs play an important regulatory role in these processes (Figure 8.3). Since the secondary structure of an RNA molecule can be significantly altered by protein binding (Puglisi *et al.*, 1992; Battiste *et al.*, 1994), and the same sequences can play a different role in various stages of viral replication (Goebel *et al.*, 2004), we cannot entirely rule out the possibility of an alternative form. The ability of RNA to fold alternatively could regulate these competing processes. Common features of 5′ and 3′ NCRs in both mosquito- and tick-borne flavivirus RNAs have been discussed extensively in a recent review (Markoff, 2003). Here, we have focused mainly on mosquito-borne flaviviruses including JEV.

The 5′-noncoding region

When Rice *et al.* determined the first complete nucleotide sequence of the entire RNA genome of the flavivirus YF 17D strain, they noticed two features in its 5′ NCR that are potentially important for RNA replication and/or encapsidation (Rice *et al.*, 1985). First, the extreme 5′-terminal dinucleotide AG and 3′-terminal dinucleotide CU of the YFV genome were

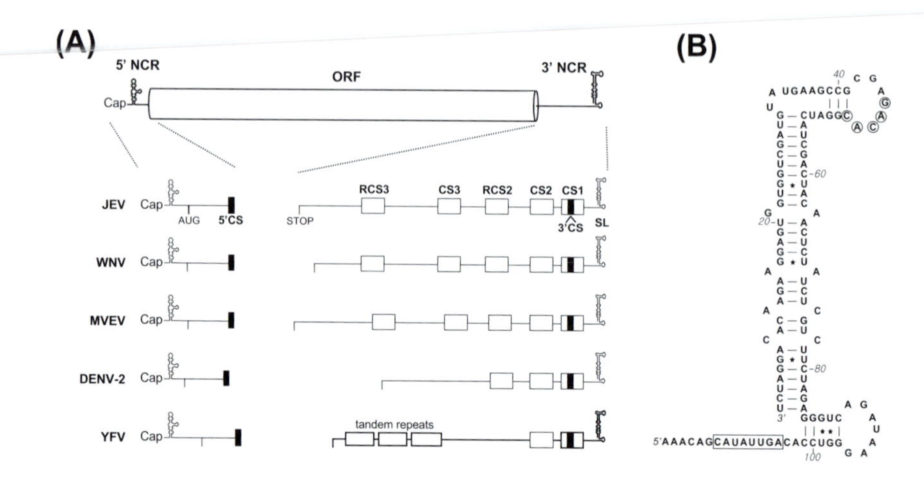

Figure 8.3 Conserved and repeated sequences. (A) Common features of the conserved and repeated sequence elements in the genome of JEV compared to other members of mosquito-borne flaviviruses including JEV, WNV, MVEV, DENV-2, and YFV (Chambers *et al.*, 1990; Markoff, 2003). At the top is the schematic of the flavivirus genome. Indicated is the 5′ NCR, the open reading frame encoding the structural and nonstructural proteins, and 3′ NCR. AUG indicates the location of the 5′ utmost start codon for synthesis of the single ORF in the genome of flaviviruses and STOP indicates the stop codon for the ORF. The highly conserved stem–loop structure at the 3′ end of the viral genome is indicated by SL. The perfectly conserved eight contiguous nucleotide sequences (5′ CS, solid bar) located in the 5′ coding region of capsid protein for several mosquito-borne flaviviruses complementary to the 3′-proximal conserved sequence element (3′ CS, solid bar), another conserved domain immediately adjacent to the potential 3′-terminal SL structure. Approximately 25 nucleotides of a highly conserved sequence region (CS1) were found just upstream of the SL structure. Additional conserved tandem repeats known as CS2 and R CS2 as well as CS3 and R CS3 have been highlighted in the corresponding regions of each flavivirus. In YFV, three tandem repeats at the upstream region of the 3′ NCR are also indicated. (B) Nucleotide sequences and model is from the K87P39 strain of JEV (GenBank accession number AY585242). Similar structures are also found in all JEV serocomplexes (Markoff, 2003). Nucleotides are numbered from the 3′ end of the viral genome. Circles indicate the conserved pentanucleotide sequence (5′-CACAG). Boxed is the cyclization sequence motif (5′-CAUAUUGA).

conserved at both termini of the WNV genome, which had been independently sequenced (Wengler and Wengler, 1981). The 3′-terminal dinucleotide of the plus-strand (CU-3′) is complementary to the 5′-terminal dinucleotide of the minus-strand (5′-AG), which is equivalent to the 5′-terminal dinucleotide AG of the plus-strand. Second, the complementary sequence of the short pentanucleotide (^{10}UGUGU^{14}) in the 5′ NCR of the YFV genome was present eight nucleotides from the end of the viral genome, which is also a feature of the WNV genome (Rice *et al.*, 1985). In that report, the investigators suggested that the viral replicase complex might have similar recognition sites for plus- and minus-strand synthesis (Rice *et al.*, 1985).

Importantly, the genomic RNA of all JEV isolates also begins with the dinucleotide AG and terminates with the dinucleotide CU. Moreover, these dinucleotide sequences at both termini of the genomes appear to be absolutely conserved among mosquito- and tick-borne flaviviruses (Cleaves and Dubin, 1979; Wengler and Wengler, 1981; Rice *et al.*, 1985; Brinton *et al.*, 1986; Wengler and Castle, 1986;

Brinton and Dispoto, 1988; Mandl *et al.*, 1993; Khromykh and Westaway, 1994; Shi *et al.*, 1996a; Mandl *et al.*, 1998). Recently, substitutions introduced in the conserved 3′-penultimate C at the 3′ terminus of KUNV RNA positive and negative strands completely blocked replication, and substitutions of the 3′-terminal U of either strand blocked or reduced replication (Khromykh *et al.*, 2003). One exception to this rule is the CFA, whose genome reportedly starts with 5′-GU and ends with GC-3′ (Cammisa-Parks *et al.*, 1992, NC 001564). In addition, the first noted pentanucleotide in YFV (^{10}UGUGU^{14}) and WNV (^{12}UGUGU^{16}) genomes is also well conserved at the predicted positions in the 5′ (^{12}UGUGU^{16}) and 3′ NCRs of JEV as well as in KUNV genomes, respectively. Among mosquito-borne flaviviruses, however, MVEV and DENV do not have such sequences in analogous positions in their genomes. Thus, the functional significance of these sequences needs to be demonstrated.

The linear primary nucleotide sequence of JEV 5′ NCR shows a high level of sequence homology among all JEV isolates but appears to be poorly conserved compared to other flaviviruses. Like other flaviviruses (Hahn *et al.*, 1987; Brinton and Dispoto, 1988; Cahour *et al.*, 1995; Shi *et al.*, 1996b), the 5′ NCR of JEV appears to contain a predicted secondary RNA structure that might regulate the translation of the viral genome. In addition, the 5′ NCR is the reverse complement to the 3′ NCR of the viral minus-strand, which is likely to function as the site for initiation of the plus-strand synthesis. Small deletions introduced in the 5′ NCR of DENV-4, which are predicted to alter or disrupt the local base-parings, affect the production of infectious viruses, but they do not correlate with the *in vitro* translation efficiency of the mutant RNAs (Cahour *et al.*, 1995). One of these deletion mutants (nt 82–87), which causes the greatest reduction in translation efficiency, produced small plaques on LLC-MK2 cells, but failed to form plaques on mosquito C6/36 cells and replicate in *Aedes aegypti* and *Aedes albopictus* following intrathoracic inoculation (Cahour *et al.*, 1995), suggesting potential interaction of this region with cellular factors in a cell type-specific manner. The minus-strand 3′ NCR of WNV, which is complementary to the plus-strand 5′ NCR, has been shown to form a corresponding predicted stem–loop structure and this region has been indicated to form three RNA–protein complexes resulting from the binding of four cellular proteins (Shi *et al.*, 1996b).

The 3′-noncoding region

The terminal nucleotides of the 3′ NCRs of flaviviruses can form highly conserved secondary and tertiary structures (Rice *et al.*, 1985; Brinton *et al.*, 1986; Brinton and Dispoto, 1988; Mandl *et al.*, 1993; Wallner *et al.*, 1995; Shi *et al.*, 1996a; Proutski *et al.*, 1997; Proutski *et al.*, 1999; Olsthoorn and Bol, 2001; Thurner *et al.*, 2004). Figure 8.3 summarizes a number of conserved nucleotide sequence elements (Figure 8.3A) and predicted RNA secondary structures of mosquito-borne flaviviruses (Figure 8.3B), which are likely to have important roles in a variety of life cycle processes, such as translation regulation and promoter activity for minus-strand synthesis.

The SL structure

The 3′-terminal 90 nucleotides are predicted to form the stem–loop (SL) structure, which is the best conserved distal part of its 3′ NCRs (Figure 8.3A). This SL structure is built with a long stem consisting of approximately 30 base pairings with several interrupting bulges and a loop of unpaired heptanucleotides at its top (Figure 8.3B). This SL structure was first noted at the 3′ terminus of YFV genomic RNA (Rice *et al.*, 1985). Since then, a large number of studies have indicated the existence of a stable and conserved RNA secondary structure in this region among divergent flaviviruses (Grange *et al.*, 1985; Brinton *et al.*, 1986; Takegami *et al.*, 1986; Wengler and Castle, 1986; Hahn *et al.*, 1987; Irie *et al.*, 1989; Mohan and Padmanabhan, 1991; Mandl *et al.*, 1993; Gultyaev *et al.*, 1995; Wallner *et al.*, 1995; Shi *et al.*, 1996a; Proutski *et al.*, 1997; Rauscher *et al.*, 1997; Zeng *et al.*, 1998). This SL structure seems to be undisturbed by base pairings with upstream sequences (Proutski *et al.*, 1997).

In addition, a conserved CACAG motif in the loop situated at the top of the SL structure has been noted among flaviviruses. Furthermore, Shi *et al.* (1996a) has indicated the presence of a pseudoknot tertiary structure resulting from base pairings between nucleotides contained in the lower portion of the long stem within the SL structure and nucleotides comprising the loop of an additional small stem–loop structure immediately upstream from the SL structure. The formation of this pseudoknot structure may occur by Watson-and-Crick as well as non-Watson-and-Crick base pairings such as G-U for most flaviviruses (Shi *et al.*, 1996a; Markoff, 2003). It will be interesting to examine the functional significance of this pseudoknot structure in the context of full-length infectious cDNAs from flaviviruses.

JEV appears to contain a similar SL structure at the 3′ end of its genomic RNA (Takegami *et al.*, 1986). Genetic mapping experiments have revealed that an approximately 200-nucleotide region, including the SL structure from the 3′ end of the viral genome, is absolutely required for replication (Choi, Y.J., and Lee, Y.M., manuscript in preparation). On the other hand, the approximately 300-nucleotide remaining portion of the JEV 3′ NCR appears to be dispensable for replication *per se* and infectious particle formation, but appears to affect the level of replication (Choi, Y.J., and Lee, Y.M., manuscript in preparation). We have also found this SL structure at the 3′ end of the JEV genome (Figure 8.3B), and suggest that it may function in viral replication by being involved in the base pairing of the long stem and formation of the linear nucleotide sequence of the loop (Choi, Y.J. and Lee, Y.M., manuscript in preparation). Screening of the neonatal mouse brain cDNA library has previously identified a 36-kDa Mov34 protein that interacts with JEV 3′ SL structural RNA (Ta and Vrati, 2000). However, further involvement of the linear primary nucleotide sequence in this SL structure remains to be demonstrated.

The cyclization sequence

Hahn *et al.* noted the existence of eight perfectly conserved contiguous nucleotide sequences (5′ CS; UCAAUAUG) located in the 5′ coding region of the capsid protein of several mosquito-borne flaviviruses, which are complementary to the 3′-proximal conserved sequence element (3′ CS) (another conserved domain immediately adjacent to the potential 3′-terminal SL structure) (Hahn *et al.*, 1987). The authors suggested that these 5′ CS and 3′ CS cyclization sequence motifs are likely to be involved in long-range base pairing between 5′ and 3′ end sequences, cyclizing the RNA to form a panhandle structure (Hahn *et al.*, 1987). The functionality of this cyclization sequence motif has been experimentally demonstrated *in vivo* for KUNV (Khromykh and Westaway, 1997; Khromykh *et al.*, 2001), YFV (Bredenbeek *et al.*, 2003;

Corver *et al.*, 2003), and WNV (Lo *et al.*, 2003) as well as in an *in vitro* replication system for DENV (You and Padmanabhan, 1999; You *et al.*, 2001). For several TBEV strains as well as Powassan and louping ill viruses, two pairs of potential cyclization sequence motifs have been suggested (Mandl *et al.*, 1993; Khromykh *et al.*, 2001).

For the JEV CNU/LP2 strain, both 5′ CS and 3′ CS cyclization sequence motifs residing in the capsid protein region (nt 136-nt 143) and in the 3′ NCR (nt 10858–10865), respectively, have been shown to be essential for viral replication and production of infectious viruses (Choi, Y.J. and Lee, Y.M., manuscript in preparation) (Figure 8.3A). Substitutions introduced into each cyclization sequence motif that compromise potential base pairings appear to be lethal, and compensatory mutations restore viral replication (Choi, Y.J. and Lee, Y.M., manuscript in preparation). Similar long-range interaction via base pairings between specific regions of the viral genome of RNA viruses has been suggested to be involved in the viral replication of rhinovirus (McKnight and Lemon, 1998), poliovirus (Goodfellow *et al.*, 2000), Qβ virus (Klovins *et al.*, 1998), and tobacco etch virus (Haldeman-Cahill, 1998).

The conserved sequence regions

Approximately 25 nucleotides of a highly conserved sequence region (CS1) common to mosquito-borne flavivirus genomes have been found just upstream of the SL structure (Figure 8.3A). This CS1 region includes the 3′ CS motif. The panhandle structure resulting from long-range base pairings of the 5′ CS and 3′ CS motif is suggested to be further stabilized by additional complementary nucleotides flanking the 5′ CS and 3′ CS cyclization motif (Hahn *et al.*, 1987). Thus, a total of 11–12 contiguous nucleotides appear to be involved in the long-range base pairings for cyclization of JEV, WNV, MVEV, DENV, and YFV, which may provide sufficient free energy for cyclization of flavivirus RNAs. Several studies have demonstrated the essential role of this CS1 region including the eight-nucleotide cyclization sequence motif for the viral replication in DENV (Men *et al.*, 1996; You and Padmanabhan, 1999; You *et al.*, 2001), KUNV (Khromykh and Westaway, 1997; Khromykh *et al.*, 2001), YFV (Bredenbeek *et al.*, 2003; Corver *et al.*, 2003), and WNV (Lo *et al.*, 2003).

Upstream of the CS1 region, a pair of conserved tandem repeats known as CS2 and R CS2 has been found among mosquito-borne flaviviruses, but not tick-borne flaviviruses (Hahn *et al.*, 1987). In YFV, CS2 is not repeated upstream in the YFV genome, whereas one to three unique tandem repeats were found in the 5′ terminus of its 3′ NCR. The repeat number is dependent on the YFV strain, which can be correlated with its geographical origin (Wang *et al.*, 1996). Upstream of the CS2 and R CS2 region, an additional pair of the conserved tandem repeat sequences (CS3 and R CS3) has been found in the JEV serocomplex viruses, such as JEV, WNV, and MVEV (Hahn *et al.*, 1987). The functional significance of these conserved sequences has been investigated by using a panel of eight different internal deletion mutants created in the 3′ NCR of DENV-4 (Men *et al.*, 1996). This study showed that the minimal region required for the replication of DENV-4 is 113 nucleotides from the 3′ end of the genome including CS1 and SL structure, and deletions in most of the upstream regions of the 3′ NCR appeared to slightly inhibit the replication (Men *et al.*, 1996). An investigation using KUNV subgenomic replicon showed that a deletion mutant, including two conserved regions CS3 and R CS3, slightly inhibited RNA replication (Khromykh and Westaway, 1997).

Common features of viral proteins

The structural proteins

The C protein

The C protein of approximately 11 kDa plays several roles in virus replication, including formation of a viral core, encapsidation/packaging of the RNA genome into a viral core, and interaction with cellular membranes. The nascent C proteins are first synthesized as membrane-anchored proteins, and the virion-associated form of C protein is generated by cleavage of the anchor domain by the viral serine protease NS3 in association with NS2B (Nowak *et al.*, 1989; Yamshchikov and Compans, 1993; Lobigs, 1993; Amberg *et al.*, 1994; Yamshchikov and Compans, 1994; Kummerer and Rice, 2002; Ma *et al.*, 2004). The C protein is highly basic (Trent, 1977; Boege *et al.*, 1983; Rice *et al.*, 1985), which is consistent with its role in binding to genomic RNA in the process of viral core formation (Kiermayr *et al.*, 2004). A conserved internal hydrophobic domain may play a role in association with the membrane (Ruiz-Linares *et al.*, 1989; Yamshchikov and Compans, 1994; Markoff *et al.*, 1997). Recently, Ma *et al.* determined the solution structure of the 200-residue homodimer of DENV-2 C protein (Ma *et al.*, 2004). This study proposed that one face of the dimer, which is rich in basic residues, interacts with the viral RNA and an extensive apolar region interacts with the viral membrane on its opposite side (Ma *et al.*, 2004). The cores from purified TBE virions have been isolated and could be disrupted into capsid dimers by increasing the salt concentration, and subsequently purified capsid dimers have been shown to assemble *in vitro* into capsid-like particles in the presence of *in vitro* transcribed viral RNA or single-stranded DNA oligonucleotides (Kiermayr *et al.*, 2004). The C-C interaction domain of the DENV-2 C protein has been shown to map to the internal hydrophobic domain (Wang *et al.*, 2004). Formation of virus particles has been suggested to be coordinated by interactions between the membrane-anchored C protein, presumably interacting with the genomic RNA, and the two membrane proteins prM and E (Lindenbach and Rice, 2001). This process is suggested to be regulated by proteolytic processing in the membrane-associated polyprotein precursors (Rice *et al.*, 1985; Kummerer and Rice, 2002).

The prM protein

The prM protein of about 25 kDa is a precursor of the virion-associated M protein. Host signal cleavages generate the N and C termini of prM protein. The generated C terminus, which contains the adjacent transmembrane domains, allows prM to anchor to the membrane and serves as the signal sequence for the translocation of E protein (Rice *et al.*, 1985). During the export of viral particles through the secretory pathway (Lindenbach and Rice, 2001), this protein is further cleaved by furin in the *trans*-Golgi network (Stadler *et al.*, 1997) into a membrane-anchored M protein of approximately 8 kDa, which is located in the C-terminus portion of prM and a soluble pr fragment, which is secreted into the extracellular compartment (Murray *et al.*, 1993). This cleavage leads to two different forms of virions: intracellular virions containing uncleaved prM and E proteins and extracellular virions containing exclusively cleaved M and E proteins with some uncleaved prM. The hydrophilic pr fragment contains six conserved cysteine residues, all of which are occupied in intramolecular disulfide bond formation (Nowak and Wengler, 1987). In addition to disulfide bridges, the pr fragment for members of flaviviruses also contains one to three potential N-linked glycosylation sites (Chambers *et al.*, 1990a; Chambers *et al.*, 1990b). In the case of JEV, a single potential N-linked glycosylation site has been found. The cleaved form of M protein found anchored on the surfaces of the extracellular virions contained a short ectodomain (≈ 41 residues) followed by two transmembrane domains (Lindenbach and Rice, 2001).

The precise role of prM/M protein in the early stage of viral infection remains to be investigated (Roehrig *et al.*, 1998; Burke and Monath, 2001; Lindenbach and Rice, 2001). For DENV, the prM/M protein has been shown to be able to actively induce protective immunity, and passive administration of anti-prM antibodies protects mice against a lethal challenge (Kaufman *et al.*, 1989; Bray and Lai, 1991; Falconar, 1999; Vazquez *et al.*, 2002). Some evidence has implied that monoclonal antibodies directed against prM provide protective immunity, perhaps because of their cross-reactivity with E protein (Kaufman *et al.*, 1989; Falconar, 1999), or possibly through their ability to neutralize uncleaved prM protein present on the surfaces of the extracellular virions (Kaufman *et al.*, 1989; Lindenbach and Rice, 2001). In the case of Langat virus, evidence for protection by a non-neutralizing antibody to the preM/M protein has been described (Iacono-Connors *et al.*, 1996).

During the viral assembly in the intracellular membrane, prM is noncovalently associated with the E protein as a heterodimer in the ER (Wengler and Wengler, 1989; Allison *et al.*, 1995b; Wang *et al.*, 1999; Courageot *et al.*, 2000; Lorenz *et al.*, 2002), which is subsequently incorporated into intracellular virions (Wengler and Wengler, 1989). PrM itself is able to form a completely folded structure in the absence of association with E (Lorenz *et al.*, 2002), whereas the proper folding, maturation, and transport of E requires the heterodimeric interaction with prM (Konishi and Mason, 1993; Allison *et al.*, 1995b). Thus, prM appears to function in a chaperone-like manner for the biosynthesis and efficient maturation of E. During the transport of intracellular virions through the secretory pathway, prM plays an important role in preventing E from undergoing the premature structural and oligomeric rearrangements induced by the low pH environment in the *trans*-Golgi network (Guirakhoo *et al.*, 1992; Heinz *et al.*, 1994; Allison *et al.*, 1995a; Stiasny *et al.*, 1996).

The cleavage of prM triggers the dissociation of prM/E heterodimers, followed by structural rearrangement of E, which is required for converting immature virions into mature virions, which are characterized by the ability to induce cell fusion, infectivity, and HA activity (Wengler and Wengler, 1989; Randolph *et al.*, 1990; Guirakhoo *et al.*, 1991; Heinz *et al.*, 1994; Stadler *et al.*, 1997). When the cleavage of prM is inhibited by raising the pH in intracellular compartments using acidotropic reagents or using specific inhibitors of furin (Randolph *et al.*, 1990; Guirakhoo *et al.*, 1991; Guirakhoo *et al.*, 1992; Heinz *et al.*, 1994; Stadler *et al.*, 1997), the resulting extracellular virions with higher proportions of prM appear to have impaired infectivity, and they block cell fusion activity at low pH (Randolph *et al.*, 1990; Guirakhoo *et al.*, 1992; Heinz *et al.*, 1994). Incomplete cleavage of prM has been described in some flaviviruses, for instance DENV (Randolph *et al.*, 1990; Murray *et al.*, 1993; Roehrig *et al.*, 1998; Wang *et al.*, 1999), JEV (Kimura-Kuroda and Yasui, 1983; Yun, S.I. and Lee, Y.M., unpublished), KUNV (Khromykh *et al.*, 1998), and Langat virus (Guirakhoo *et al.*, 1991; Iacono-Connors *et al.*, 1996). The significance of the relatively high proportion of prM in certain flaviviruses remains unclear.

Similar to many furin substrates (Nakayama, 1997; Steiner, 1998; Molloy *et al.*, 1999; Thomas, 2002), the prM cleavage site of tick- and mosquito-borne flaviviruses including JEV contains the consensus sequence RX(K/R)R, which is located proximal to the cleavage site (Chambers *et al.*, 1990a; Lindenbach and Rice, 2003; Keelapang *et al.*, 2004). In addition to these three basic residues, the diverse pr-M junction-proximal sequences of flaviviruses differentially influence pr-M cleavage when examined in a DENV prM background. Importantly, greatly enhanced prM cleavability adversely affects virus export while exerting a minimal effect on infectivity (Keelapang *et al.*, 2004).

The E protein

The E protein of approximately 50 kDa is the major surface protein of the virion. The mature E protein generated by signalase cleavages at its N and C termini contains two transmembrane domains in its C terminus; one for anchoring to the membrane and the other, a signal sequence, for translocation of NS1 protein (Rice *et al.*, 1985). This protein plays several important roles in the virus life cycle, for instance, in receptor binding and membrane fusion, and is the major target for the neutralizing antibody response. The E protein contains 12 absolutely conserved Cys residues that are involved in intramolecular disulfide bond formation (Nowak and Wengler, 1987). On the other hand, the N-linked glycosylation site is not absolutely conserved. The E protein of JEV has a single potential glycosylation site. In other flaviviruses, on the other hand, not only species-specific variations of glycosylation but also strain-specific variations have been described (Wengler *et al.*, 1985; Ballinger-Crabtree and Miller, 1990; Post *et al.*, 1992; Vorndam *et al.*, 1993; Johnson *et al.*, 1994; Adams *et al.*, 1995; Lee *et al.*, 1997; Chambers *et al.*, 1998). The functional significance of glycosylation of the E protein remains to be elucidated.

This protein has recently been implicated in flavivirus structure and membrane fusion. A tryptic, soluble, crystallizable dimeric form of the TBEV E proteins was isolated from purified virions (Heinz *et al.*, 1991), and its high resolution 3D structure was determined by X-ray crystallography (Rey *et al.*, 1995). These studies have shown that two monomers form a head-to-tail homodimer, which appears to be flat, elongated, and posited parallel to the surface of the virion. Individual monomers are recognized by three distinct domains designated as I, II, and III. Domain I, located in the N-terminus, is a central β-barrel containing two disulfide bridges and its external surface contains the single carbohydrate side chain. Domain II is an elongated dimerization region and three disulfide bridges in this domain stabilize a loop, which serves as an internal fusion peptide, which is highly conserved in all flaviviruses. Domain III, located in the C-terminus, is an Ig-like domain containing a single disulfide bridge. In some mosquito-borne flaviviruses, this domain contains a putative integrin-binding motif Arg-Gly-Asp. Thus, this domain has been proposed to play a critical role in receptor binding and this notion has been supported by a number of reports (Chen *et al.*, 1997; van der

Most *et al.*, 1999; Lee and Lobigs, 2000; Mandl *et al.*, 2000; Bhardwaj *et al.*, 2001; Crill and Roehrig, 2001; Hurrelbrink and McMinn, 2001; Thullier *et al.*, 2001). In addition, the entire external surface of the E protein is suggested to be involved in cell surface recognition (Mandl *et al.*, 2001), consistent with the observation that various locations in all three domains of the E protein have been mapped for binding sites for antibodies including neutralizing antibodies (Heinz and Allison, 2003).

The 3D structure of JEV E protein has been predicted by a homology modeling approach using X-ray structure data of TBEV E as a template (Kolaskar and Kulkarni-Kale, 1999). Recently, the solution structure of the major antigenic domain (domain III) of the JEV E protein has been determined, demonstrating that the domain III forms a β-barrel type structure composed of six antiparallel β-strands, similar to the immunoglobulin constant region (Wu *et al.*, 2003). The functional epitope determinants at Ser^{331} and Asp^{332} on domain III of the E protein appear to interact with a neutralizing monoclonal antibody (Lin *et al.*, 2003).

The nonstructural proteins

The NS1 protein

During protein synthesis, the 45 kDa NS1 glycoprotein is translocated into the lumen of the ER and its N-terminus is generated by cleavage at the E/NS1 junction by a signal peptidase (Rice *et al.*, 1986a; Chambers *et al.*, 1990a; Wengler *et al.*, 1990). The NS1 is released from the N-terminus of NS2A by a membrane-bound host protease in the ER (Falgout *et al.*, 1989; Falgout and Markoff, 1995). The nascent NS1 is a hydrophilic monomer, which subsequently forms hydrophobic homodimers associated with membranes (Smith and Wright, 1985; Winkler *et al.*, 1988; Winkler *et al.*, 1989). The NS1 protein exists within infected cells, presents at the plasma membrane of infected cells, and is secreted into the extracellular compartment (Smith and Wright, 1985; Winkler *et al.*, 1988; Mason, 1989; Winkler *et al.*, 1989). NS1 is secreted from mammalian cells, whereas it is not secreted from insect cells (Winkler *et al.*, 1988; Mason, 1989; Post *et al.*, 1991). Multiple species of NS1 exist presumably due to glycosylation variations, cleavage intermediates, or alternative cleavage sites for JEV (Mason *et al.*, 1987; Mason, 1989) as well as other flaviviruses (Nestorowicz *et al.*, 1994; Blitvich *et al.*, 1999). Interestingly, persistent infection with JEV has previously been established in murine neuroblastoma N18 cells, and these cells appear to produce truncated forms of NS1 due to an uncharacterized cellular adaptation/alteration rather than mutations within the viral genome, suggesting an association between abnormal NS1 expression and JEV persistency (Chen *et al.*, 1996).

The positions of the putative N-linked glycosylation sites of NS1 are highly conserved. All mosquito-borne flaviviruses contain two to three potential N-linked glycosylation sites and JEV appears to contain two potential Asn residues (Smith *et al.*, 1985; Sumiyoshi *et al.*, 1987; Lee *et al.*, 1989; Mason, 1989; Chambers *et al.*, 1990; Lindenbach and Rice, 2001). Additionally, the NS1 contains 12 cysteine residues that are highly conserved among all members of the flaviviruses except for DENV-4, suggesting their importance to the structure and function of this protein (Mackow *et al.*, 1987; Chambers *et al.*, 1990; Lindenbach and Rice, 2003). The intramolecular disulfide bond arrangement of the NS1 protein for DENV-2 has been determined (Wallis *et al.*, 2004) and that of the MVEV NS1 protein has been partially determined (Blitvich *et al.*, 2001). Site-directed mutagenesis study showed that the Ala substitution altered one of the last three Cys residues and blocked dimer formation, indicating that disulfide bridges play a critical role in dimerization of this protein (Pryor and Wright, 1993). Additionally, a single Pro→Leu substitution at residue 250 in the KUNV NS1 also prevents dimerization but does not affect the secretion of NS1 in monomeric form, and this mutant virus delayed virus replication and diminished the virulence in mice, indicating a functional role for NS1 dimerization (Hall *et al.*, 1999).

The secreted form of DENV-1 NS1 appeared to be a soluble noncovalently bound hexamer formed by three homodimers (Crooks *et al.*, 1994; Flamand *et al.*, 1999). Proper processing of N-glycans appears to be required for the NS1 protein to mature efficiently and be released from the infected cells, as suggested by the repressive effects of the N-glycan processing inhibitors in mammalian cells (Flamand *et al.*, 1999) and the absence of NS1 secretion in insect cells (Mason, 1989; Winkler, 1989; Flamand *et al.*, 1999) that produce glycoproteins with polymannose-rich structures rather than complex sugars (Kuroda *et al.*, 1990; Kulakosky *et al.*, 1998).

NS1 protein appears to play an essential role in RNA replication. NS1 has been demonstrated to co-localize with the viral dsRNA replicative form associated with intracellular membranes presumed to be sites for replication (Mackenzie *et al.*, 1996; Westaway *et al.*, 1997). Interestingly, a temperature-sensitive mutant in YFV, ts25, containing a single Ala substitution for Arg^{299} has a defect in RNA accumulation without any temperature-dependent differences in polyprotein processing, NS1 stability and secretion, or release of infectious virus (Muylaert *et al.*, 1997). Mutagenic analyses using infectious cDNAs of YFV and KUNV have demonstrated that viral RNA accumulation is impaired or blocked by specific mutations introduced in the NS1, and virus replication is restored when functional NS1 is provided *in trans* (Muylaert *et al.*, 1996; Lindenbach and Rice, 1997; Muylaert *et al.*, 1997; Pryor *et al.*, 1998; Khromykh *et al.*, 1999a). The DENV NS1 was unable to *trans*-complement the YFV RNA containing a deletion in the NS1 gene, suggesting that DENV NS1 is unable to productively interact with the YFV replicase (Lindenbach and Rice, 1999). A genetic screen to select YFV variants able to utilize DENV NS1 identified a single base mutation mapped to the NS4A gene, indicating that a genetic interaction between NS1 and NS4A is required for viral RNA replication (Lindenbach and Rice, 1999).

In DENV, the cellular glycosyl-phosphatidylinositol (GPI) linkage pathway is reportedly utilized for expression of a GPI-anchored form of NS1 on the surface of infected cells (Jacobs *et al.*, 2000). Addition of NS1-specific antibody resulted in signal transduction, as demonstrated by tyrosine phosphorylation of cellular proteins, suggesting that a mechanism of cellular activation may contribute to the pathogenesis of human disease (Jacobs *et al.*, 2000). In this regard, it will be very interesting to investigate other flaviviruses including JEV, since the mechanism of how NS1 protein is expressed on the surface of the infected cells is not known.

The NS2A and NS2B proteins

NS2A [≈ 22 kDa] and NS2B [≈ 14 kDa] are small, hydrophobic proteins (Lindenbach and Rice, 2001). Little is known about the function of NS2A and NS2B. The N-terminus of NS2A is generated by cleavage at the NS1/NS2A junction by a membrane-bound host protease within the ER (Falgout *et al.*, 1989; Falgout and Markoff, 1995), and its C-terminus is created by cleavage at the NS2A/NS2B junction by the viral serine protease on the cytoplasmic side, indicating that this protein spans the membrane (Lindenbach and Rice, 2001). The NS2A and NS4A appear to be localized in discrete foci in the cytoplasm of infected cells, associated with dsRNA, suggesting presumed sites for RNA replication; and NS2A fused to GST binds strongly to the 3′ NCR of Kunjin RNA and also to the NS3 and NS5 (Mackenzie *et al.*, 1998). Thus, these small hydrophobic proteins are thought to promote the assembly of viral replication complexes or the recruitment of RNA templates to the sites for replication.

Two forms of NS2A are found in YFV-infected cells. In addition to the full-length NS2A protein, an additional C-terminally truncated form of NS2A was also identified, which results from partial cleavage by the viral NS2B-NS3 serine protease (Chambers *et al.*, 1989; Nestorowicz *et al.*, 1994). Full-length RNA transcripts containing mutations at each or both of these sites which block cleavage appeared to be noninfectious (Nestorowicz *et al.*, 1994). A recent study showed that a substitution of serine for lysine at the alternative cleavage site (QK/T→QS/T) within NS2A blocks the cleavage and infectious virus production, but not at the level of RNA replication (Kummerer and Rice, 2002). The alternative cleavage site for NS2A could be suppressed by two classes of second-site mutations: insertions at the alternative cleavage site restore its basic charge character and cleavability, and substitutions occurring in the NS3 helicase domain at residue aspartate 343 restore infectious virus production in the absence of cleavage at the alternative cleavage site (Kummerer and Rice, 2002). This study indicates that NS2A in association with NS3 plays a role in the assembly and/or release of infectious virus particles (Kummerer and Rice, 2002). Using an inducible bacteria system and mammalian cells, the four small hydrophobic NS proteins of JEV such as NS2A, NS2B, NS4A, and NS4B appear to have various effects on host cell membrane permeability, contributing in part to virus-induced CPE in infected cells (Chang *et al.*, 1999). The mechanism by which modification of cell membrane permeability contributes to CPE in JEV-infected cells and the role of this modification for virus replication must be investigated further.

NS2B has been shown to form a complex with NS3 and function as a cofactor for the viral serine protease NS3 (Chambers *et al.*, 1991; Falgout *et al.*, 1991; Arias *et al.*, 1993; Chambers *et al.*, 1993; Falgout *et al.*, 1993; Jan *et al.*, 1995). Mutagenic studies demonstrated that a central conserved hydrophilic region of NS2B is a domain essential for the efficient processing function of the NS3 proteinase, and its hydrophobic regions are involved in membrane association of the serine proteinase precursor NS2B-NS3, which is required for efficient polyprotein processing (Chambers *et al.*, 1993; Falgout *et al.*, 1993; Jan *et al.*, 1995; Clum *et al.*, 1997; Brinkworth *et al.*, 1999; Droll *et al.*, 2000).

The NS3 protein

The NS3 [≈ 70 kDa] is a large multifunctional protein, with highly conserved sequence motifs for viral polyprotein processing and RNA replication (Lindenbach and Rice, 2001). About one-third of the N-terminal domain of the NS3 encodes a trypsin-like serine protease domain responsible for site-specific cleavages in the viral polyprotein (Bazan and Fletterick, 1989; Gorbalenya *et al.*, 1989; Bazan and

Fletterick, 1990; Wengler *et al.*, 1991). A number of mutagenic investigations into the proteinase activity of NS3 using various forms of purified proteins or their expression in host cells among various flaviviruses demonstrated an essential role in viral replication, catalytic active sites for the proteinase activity, and characteristics of the substrate specificity (Chambers *et al.*, 1990; Preugschat *et al.*, 1990; Falgout *et al.*, 1991; Wengler *et al.*, 1991; Zhang *et al.*, 1992; Pugachev *et al.*, 1993; Valle and Falgout, 1998).

In addition to homology modeling of DENV-2 NS3 proteinase (Brinkworth *et al.*, 1999), the high-resolution crystal structure of the protease domain of DENV-2 NS3 has been determined, and combination of its structure with modeling of peptide substrates into the active site suggests identities of residues involved in substrate recognition (Murthy *et al.*, 1999).

Membrane association of NS3 with NS2B is required for the efficient proteolytic processing of nonstructural proteins and virus replication (Chambers *et al.*, 1991; Falgout *et al.*, 1991; Wengler *et al.*, 1991; Arias *et al.*, 1993; Chambers *et al.*, 1993; Falgout *et al.*, 1993; Jan *et al.*, 1995; Clum *et al.*, 1997; Brinkworth *et al.*, 1999; Droll *et al.*, 2000). The N-terminal 184 residues of DENV NS3 are reportedly sufficient to interact with NS2B (Arias *et al.*, 1993). This NS2B–NS3 complex is responsible for proteolytic processing at multiple sites of nonstructural proteins (NS2A/2B, NS2B/3, NS3/4A, and NS4B/5) and also cleaves the anchor region or a transmembrane domain at the C-termini of C (Amberg *et al.*, 1994; Yamshchikov and Compans, 1994) and NS4A (Lin *et al.*, 1993), respectively. Additionally, internal processing within NS2A (Nestorowicz *et al.*, 1994) and NS3 (Arias *et al.*, 1993; Pugachev *et al.*, 1993; Teo and Wright, 1997) have also been reported, although these processings have not yet been reported in JEV. Consensus cleavage motifs consist of dibasic amino acids followed by an amino acid with a small side chain (Chambers *et al.*, 1993; Lin *et al.*, 1993; Nestorowicz *et al.*, 1994; Chambers *et al.*, 1995; Yusof *et al.*, 2000). The purified NS2B-NS3 complex may contribute to additional specific interactions with the P2 and P3 residues of the substrates (Yusof *et al.*, 2000), indicating that the flanking residues contribute to cleavage site selection (Lindenbach and Rice, 2001).

An overlapping region between amino acids 160 and 180 of DENV-2 NS3 protein appears to be the functional domain required for serine protease and RNA-stimulated NTPase activities (Li *et al.*, 1999). The remaining C-terminus of NS3 contains typical sequence motifs of an RNA helicase, an RNA-stimulated NTPase, and an RNA triphosphatase for RNA unwinding, NTP hydrolysis, and NTP binding (Wengler and Wengler, 1991; Wengler and Wengler, 1993; Gorbalenya *et al.*, 1989; Warrener *et al.*, 1993). The helicase/NTPase domain of the NS3 is categorized into the DExH protein superfamily of helicase superfamily II (Gorbalenya *et al.*, 1989; Koonin, 1991a; Kadare and Haenni, 1997). The RNA helicase activity has been shown for the NS3 proteins of various flaviviruses such as JEV (Utama *et al.*, 2000), WNV (Borowski *et al.*, 2001), and DENV (Li *et al.*, 1999). The RNA-stimulated NTPase activity has also been demonstrated in various flaviviruses such as JEV (Takegami *et al.*, 1995; Kuo *et al.*, 1996), WNV (Wengler and Wengler, 1991; Borowski *et al.*, 2001), and YFV (Warrener *et al.*, 1993), DENV (Li *et al.*, 1999). Since the 5′ cap structure is generally synthesized by an RNA triphosphatase, a guanylyltransferase and a methyltransferase, the RNA triphosphatase activity is likely to play a role in the 5′ capping modification of the 5′ end of the viral genome in the cytoplasm of infected cells (Wengler and Wengler, 1993).

These RNA helicase, RNA-stimulated NTPase, and RNA triphosphatase activities are believed to be involved in the replication of the viral genome, such as unwinding of RNA secondary structure of the viral genome using the energy of NTP hydrolysis for template recognition or initiation of RNA synthesis (Lindenbach and Rice, 2001; Brinton, 2002). This is consistent with the findings that the NS3 protein has been shown to interact with a region of the 3′ NCR containing the SL structure as well as the NS5 *in vivo* in a phosphorylation status-dependent manner (Kapoor *et al.*, 1995; Chen *et al.*, 1997; Cui *et al.*, 1998). The precise biological function of these activities in RNA replication remains to be further elucidated (Kadare and Haenni, 1997).

JEV NS3 appears to be associated with microtubules and tumor susceptibility gene 101, which are incorporated into the JEV-induced structure during JEV replication (Wang *et al.*, 1998; Chiou *et al.*, 2003). A similar association of the microtubule component of the cells has been observed in KUNV-infected cells (Ng and Hong, 1989). The involvement of microtubule components during virus replication and assembly requires further investigation.

The NS4A and NS4B proteins

In addition to NS2A and NS2B, the NS4A [≈ 16 kDa] and NS4B [≈ 28 kDa] are also small, hydrophobic proteins (Lindenbach and Rice, 2001). The N-terminus of NS4A is created by cleavage at NS3/NS4A junction by the NS2B-NS3 proteinase, and its C-terminus is generated by cleavage at NS4A/NS4B junction by host signalase. A transmembrane domain at the C-terminus of NS4A functions as a signal sequence to translocate the NS4B into the ER, and a novel cleavage site (NS4A/2K site) within NS4A just upstream of this signal sequence is processed by the NS2B-NS3 proteinase (Preugschat and Strauss, 1991; Lin *et al.*, 1993). This novel cleavage is required for the signalase cleavage generating the N-terminus of NS4B, which is conserved among flaviviruses (Lin *et al.*, 1993). Cleavage intermediates such as NS3-NS4A and NS4A-NS4B have been identified in flavivirus-infected cells (Chambers *et al.*, 1990; Preugschat and Strauss, 1991; Lobigs, 1992; Lindenbach and Rice, 2001). The functional significance of these cleavage intermediates in virus replication remains to be elucidated. NS4A appears to play a role in RNA replication, as suggested by its localization in KUNV-infected cells and interaction with the 3′ NCR of viral RNA as well as NS3 and NS5, which are major components of the replication complex (Mackenzie *et al.*, 1998). A genetic approach demonstrated that NS4A appears to interact with NS1, which is essential for RNA replication, as mentioned previously (Lindenbach and Rice, 1999).

The nascent protein of NS4B of approximately 30 kDa is then modified post-translationally to a 28 kDa protein by an unknown mechanism (Chambers *et al.*, 1990; Preugschat and Strauss, 1991). NS4B appears to be localized in the perinuclear membranes spreading outward in a reticular pattern, which presumably become sites for viral replication and reportedly enter the nucleus (Westaway *et al.*, 1997). The physiological significance of this subcellular localization of NS4B remains to be determined.

The NS5 protein

The NS5 protein [≈ 103 kDa] is the largest and most conserved protein, encoded at the C-terminus of the viral polyprotein (Lindenbach and Rice, 2001). Sequence comparisons showed that the C-terminus of NS5 protein contains sequence motifs characteristic of RNA-dependent RNA polymerases (RdRp) of other positive-sense RNA viruses, especially the highly conserved Gly-Asp-Asp (GDD) motif (Koonin, 1991b; Kamer and Argos, 1984; Rice *et al.*, 1985). The currently accepted model of the asymmetric and semiconservative replication strategy for flaviviruses is largely based on a number of investigations using KUNV as a model system (Westaway *et al.*, 2003), which agrees with the findings obtained with DENV (Cleaves *et al.*, 1981; Bartholomeusz and Wright, 1993). The NS5 appears to function as the viral RdRp in conjunction with other nonstructural proteins including NS3 and cellular factors to assemble efficient replication complexes, which are associated with membranes (Grun and Brinton, 1987; Chu and Westaway, 1992; Mackenzie *et al.*, 1999). This appears to be true for JEV NS5 (Takegami and Hotta, 1989; Edward and Takegami, 1993; Uchil and Satchidanandam, 2003).

The purified recombinant DENV NS5 expressed in *E. coli* possessed RdRp activity (Tan *et al.*, 1996; Guyatt *et al.*, 2001). In addition, the RdRp activity of JEV NS5 has also been demonstrated using the cell extracts obtained from the JEV-infected cells (Takegami and Hotta, 1989; Uchil and Satchidanandam, 2003). RdRp activities of other flaviviruses such as KUNV (Chu and Westaway, 1985; Chu and Westaway, 1987), WNV (Grun and Brinton, 1986; Grun and Brinton, 1987; Grun and Brinton, 1988), and DENV (Bartholomeusz and Wright, 1993; You and Padmanabhan, 1999) have been reported using infected cell lysates at various conditions. The essential role of KUNV NS3 in viral RNA replication has been implicated by the observation that antibodies against the NS3 and NS5 inhibited the conversion of the replicative form to the replicative intermediate (Bartholomeusz and Wright, 1993). Initiation of RNA synthesis at the 3′ end of exogenous viral RNA templates has been demonstrated to require 5′- and 3′-terminal complementary sequence motifs of the viral RNA including the complementary cyclization sequence motifs and stem–loop structures (You and Padmanabhan, 1999). Interestingly, RNA synthesis took place when the 5′-terminal region RNA was added *in trans* to the 3′ NCR RNA (You and Padmanabhan, 1999). These systems using infected cell lysates could be used to obtain information related to localization of the replication complexes, replication strategy in an *in vivo* environment, and metal ion requirements. On the other hand, these systems are limited for the investigation of chain elongation of RNA synthesis rather than *de novo* initiation or reinitiation.

The N-terminal portion of NS5 proteins appears to contain an additional conserved motif for a methyltransferase involved in viral RNA capping (Koonin, 1993). Site-directed mutagenic studies have shown that both polymerase and methyltransferase activities are required for virus replication, and NS5 could be *trans*-complemented (Khromykh *et al.*, 1998; Khromykh *et al.*, 1999a; Khromykh *et al.*, 1999b; Khromykh *et al.*, 2000). Flaviviral cap structure is likely to be synthesized by the RNA triphosphatase of NS3 and the methyltransferase of NS5, but the guanylyltransferase has not yet been found (Brinton, 2002).

NS5 protein appears to be phosphorylated on serine residues at multiple sites by unknown Ser/Thr kinases (Kapoor *et al.*, 1995; Morozova *et al.*, 1997; Reed *et al.*, 1998). NS5 has been shown to exist in differentially phosphorylated states in the nucleus and in the cytoplasm, and only the cytoplasmic form of NS5 has been indicated to interact with NS3, suggesting that the status of phosphorylation may regulate the interaction between these proteins (Buckley *et al.*, 1992; Kapoor *et al.*, 1995). In addition to the phosphorylation, a nuclear localization sequence of DENV NS5 (residues 369 to 405), which is capable of transport of β-galactosidase fusion protein to the nucleus, plays a role in subcellular localization of the NS5 protein (Forwood *et al.*, 1999). The N-terminal portion of NS5 appears to interact with both NS3 and importin-β, indicating that competition of NS3 and importin-β for the same binding site on NS5 may also involve regulation of the transport of NS5 to the nucleus (Johansson *et al.*, 2001).

Reverse genetics system for JEV

For positive-sense RNA viruses, major molecular advances have been made by a reverse genetics system involving the use of infectious cDNA clones of the viral genomes with the subsequent synthesis of infectious RNA for the generation of synthetic viruses. In the classical RNA-launched approach, synthetic viruses are recovered from the cells transfected with RNA transcripts made from the template of infectious cDNA clones (Liljestrom and Garoff, 1991; Rice *et al.*, 1987; Rice *et al.*, 1989; van Dinten *et al.*, 1997; Satyanarayana *et al.*, 1999). In the alternative DNA-launched approach first reported for poliovirus (Racaniello and Baltimore, 1981) and adapted for alphaviruses (Schlesinger and Dubensky, 1999), synthetic viruses are generated by direct transfection of infectious cDNA clones into susceptible cells. By using one of these two approaches, infectious cDNA clones have now been generated for members of many positive-sense RNA virus families, including coronaviruses which have the largest RNA genomes (Almazan *et al.*, 2000). Unlike other RNA viruses, however, the construction of a full-length infectious cDNA clone for JEV has long been hampered mainly by the genetic instability of the cloned cDNA (Sumiyoshi *et al.*, 1992; Sumiyoshi *et al.*, 1995; Mishin *et al.*, 2001; Zhang *et al.*, 2001).

It has previously been attempted to construct a full-length JEV cDNA by a variety of vector systems (Sumiyoshi *et al.*, 1992; Sumiyoshi *et al.*, 1995; Mishin *et al.*, 2001; Zhang *et al.*, 2001) including a cosmid vector in *E. coli* (Zhang *et al.*, 2001). In all cases, the cloned cDNA appeared to be genetically unstable during its propagation in a host. Similarly, we also have failed to clone stable full-length JEV cDNA in high-copy-number pUC-derived, medium-copy-number pBR322-derived, and low-copy-number pACYC184-derived vectors. Sumiyoshi and co-workers previously attempted to overcome this problem by designing a system where the template would be generated by *in vitro* ligation of two overlapping JEV cDNAs (Sumiyoshi *et al.*, 1992). This template was then used to synthesize infectious RNA transcripts *in vitro*. However, the specific infectivity of these transcripts was about 100 PFU/μg, which was too low to make this system useful for molecular and genetic studies of JEV biology (Sumiyoshi *et al.*, 1992). Recently, we have been able to overcome the genetic instability of JEV cDNA by cloning it into a bacterial artificial chromosome (BAC) that maintains one or two copies in *E. coli* (Yun *et al.*, 2003b).

Recently, an efficient reverse genetics system for JEV has been developed by employing a combination of three strategies (Yun *et al*., 2003b). (1) For the faithful synthesis of a full-length cDNA, the complete nucleotide sequence of an entire JEV RNA genome including the utmost 5′- and 3′-termini was determined by using RNase H(−) RT together with a low error rate *Pfu* DNA polymerase. (2) For the choice of a suitable vector that can stably accommodate and maintain a large cDNA insert, the full-length cDNA of the entire viral genome was cloned as a BAC based on *E. coli* and its single-copy plasmid F factor (Shizuya *et al*., 1992; Wang *et al*., 1997). Recently, BACs have been successfully used to clone and maintain large fragments of DNA from a variety of complex genomic sources into bacteria, including human (Shizuya *et al*., 1992) and herpes virus genomic DNA (Messerle *et al*., 1997). (3) For the production of synthetic RNA transcripts *in vitro*, promoters for bacteriophage SP6 or T7 RNA polymerase were engineered immediately upstream of the viral genome sequence and an artificial run-off site was introduced immediately downstream of the viral genome, such that the authentic 5′- and 3′-terminal nucleotide sequences of the viral genome are able to reproduce.

Yun *et al*. (2003b) showed that *in vitro* transcription of the full-length cDNA using the SP6 or T7 polymerase produces synthetic RNAs that, when transfected into susceptible BHK-21 cells, have a specific infectivity exceeding 10^6 PFU/μg. Synthetic JEVs recovered from the culture supernatant of the transfected cells 24 hr post-transfection exceeded 5×10^6 PFU/ml. Significantly, the infectious JEV BAC remained genetically stable even after 180 generations of serial growth in *E. coli* (Yun *et al*., 2003b). Thus, this reverse genetics system for JEV will greatly aid several basic and applied research areas. First, this system will help us to directly investigate the molecular mechanisms of JEV replication, transcription, and translation as well as to identify the JEV genetic elements involved in neurovirulence and pathogenesis. Second, in connection with the ongoing JEV epidemic in Asia and the recent expansion of the virus to Australia, the ability to generate recombinant JEV by targeted manipulation of the infectious JEV cDNA opens up new approaches in design of genetically modified anti-JEV vaccines. Finally, Yun *et al*. also demonstrated the possibility that JEV can be used as an attractive vector for the expression of heterologous genes in a wide variety of cell types for many applications in biological research (Yun *et al*., 2003b).

Vaccine development

There are two different kinds of vaccines against JEV, inactivated and live attenuated vaccines, currently available for use in humans. Formalin-inactivated vaccines have been used for over 30 years (Tsai, 2000). An inactivated vaccine available worldwide is prepared from mouse brains inoculated with the prototype Nakayama strain, which is manufactured by The Research Foundation for Microbial Diseases of Osaka University in Japan (Biken), designated as JE-VAX (Inactivated Japanese encephalitis virus vaccine, 1993). An additional inactivated mouse brain vaccine made from Beijing-1 strain is also used in Japan. A placebo-controlled, blinded, and randomized trial with two doses of monovalent (Nakayama strain) or bivalent (Nakayama plus Beijing strains) inactivated mouse brain vaccine showed that the combined efficacy in both vaccine groups was 91% (Hoke *et al*., 1988). However, three administrations of these vaccines are recommended to achieve a seroconversion rate of about 100%, whereas two doses produce only an 80% chance of seroconversion (Shlim and Solomon, 2002). After initial administration of the series, a booster vaccination is required to keep a neutralizing antibody titer high enough for protection (Shlim and Solomon, 2002). Formalin-inactivated mouse brain vaccines are relatively expensive and are associated with rare but clinically significant adverse effects (Monath, 2002). In addition, the complicated procedure for vaccine production using mouse brain led to development of an improved vaccine. An inactivated vaccine derived from primary hamster kidney cells inoculated with P3 strain has been used in China for over 40 years. The efficacy of this primary hamster kidney cell-derived vaccine appears to be only about 85% (Japanese encephalitis virus, 1998). A new inactivated tissue culture-derived vaccine manufactured in Vero cells is in advanced preclinical or early clinical development (Srivastava *et al.*, 2001; Sugawara *et al.*, 2002; Monath, 2002; Solomon, 2003).

In China, a live attenuated vaccine has been developed by empirically passaging the virulent SA14 strain in primary hamster kidney cells, designated as SA14-14-2 (Xin *et al.*, 1988). The genetic basis for attenuation of SA14-14-2 is partially understood (Ni *et al.*, 1994). The effectiveness of one dose was 80%, and that of two doses was 97.5% (Hennessy *et al.*, 1996). Thus, SA14-14-2 appears to be effective when administered in a two-dose regimen. Over 200 million doses of SA14-14-2 have been administered in China since 1988 with very few reported side effects (Solomon, 2003). Thus, the vaccine appears to be safe (Liu *et al.*, 1997). However, the regulatory concerns over manufacturing and control have restricted international distribution (Monath, 2002).

A new live attenuated vaccine (ChimeriVax-JE) in early clinical trials is a chimeric virus generated by replacing the genes encoding the prM and E of YF17D virus with the corresponding genes of an attenuated strain of JEV SA14-14-2 (Chambers *et al.*, 1999). This chimeric virus appears to grow efficiently in vertebrate and mosquito cells (Chambers *et al.*, 1999), but is restricted in its ability to infect and replicate in mosquito vectors (Bhatt *et al.*, 2000). The chimeric virus is less neurovirulent than the YFV 17D vaccine in mice and nonhuman primates (Guirakhoo *et al.*, 1999; Monath *et al.*, 1999). Since the prM and E proteins contain antigens conferring protective humoral and cellular immunity, the immune response to vaccination is principally directed at JEV (Monath *et al.*, 2000). Thus, ChimeriVax-JE is safe and immunogenic and provides protective efficacy (Monath *et al.*, 2000; Monath *et al.*, 2002; Monath *et al.*, 2003). Genetically, at least three distinct amino acid clusters in the E protein of the chimeric virus are responsible for its attenuation (Monath *et al.*, 2000; Arroyo *et al.*, 2001). Recently, using antisera raised against ChimeriVax-JE and the currently licensed vaccine JE-VAX, neutralization and passive transfer studies in mice demonstrated that greatest protection was provided against strains of genotypes II and III, although some protection was also observed against genotypes I and IV strains (Beasley *et al.*, 2004).

Highly attenuated vaccinia viruses such as NYVAC (Konishi *et al.*, 1991; Konishi *et al.*, 1992a; Konishi *et al.*, 1992b; Raengsakulrach *et al.*, 1999) and MVA (Nam *et al.*, 1999) are used as vectors to deliver the prM and E of JEV elicited neutralizing antibodies and to provide protection against a lethal challenge of JEV in animals. A controlled, randomized, double-blind clinical trial showed that NYVAC-JEV vaccine induced neutralizing antibody responses only in vaccinia-nonimmune recipients, while vaccinia-immune volunteers failed to develop protective antibodies, suggesting that preexisting immunity to poxvirus vector may suppress antibody responses to recombinant gene products of JEV (Kanesa-Thasan *et al.*, 2000). Recently, recombinant pseudorabies viruses expressing NS1 protein of JEV SA14-14-2 strain generated a good humoral and cellular immune response against JEV (Xu *et al.*, 2004). The extracellular subviral particles were produced by infecting mammalian cells with recombinant vaccinia virus designed to express the prM and E genes of JEV, and the purified subviral particles are highly immunogenic and provide long-lasting immunity in mice (Mason *et al.*, 1991; Konishi *et al.*, 1997). Using DNA vaccination techniques, the prM and E genes expressed from recombinant plasmid DNAs appeared to produce virus-like particles. This vaccination has been shown to be highly immunogenic and provided protective immunity from a lethal challenge of JEV (Konishi *et al.*, 1997; Chang *et al.*, 2000; Konishi *et al.*, 2000; Kojima *et al.*, 2003; Zhao *et al.*, 2003; Wu *et al.*, 2004). The expression of a variant form of E or NS1 protein has elicited a varied degree of protective immunity (Lin *et al.*, 1998; Ashok and Rangarajan, 1999; Chen *et al.*, 1999; Ashok and Rangarajan, 2002; Kaur *et al.*, 2002; Kaur *et al.*, 2004). Interestingly, oral immunization of mice with live JEV produces protective immunity, suggesting that oral immunization with JEV holds promise for the future (Ramakrishna *et al.*, 1999).

Conclusion and future directions

Since the isolation of JEV in 1935, research on the virus has generated a great deal of information to elucidate pathogenesis and biology of the virus. Yet, many questions are still unanswered about nearly every stage of viral life cycle, such as attachment, penetration, translation, RNA replication, assembly, and release. Several interesting questions are as follows. What is the cellular receptor(s) responsible for viral entry? How are the prM and E proteins involved and regulated in the process of membrane fusion between viral envelopes and intracellular endosomal vesicles? Why does RNA replication take place on the intracellular membrane compartment? How are the same genomic RNAs able to utilize templates for translation and RNA synthesis and how is this process coordinated to avoid collision between cellular translational machinery and viral replication machinery? Which cellular factors are involved in the different steps of viral replication and what are their roles? How do capsid proteins encapsidate the viral genomic RNA and bud into the intracellular compartment, where prM and E proteins are incorporated into the budding viral particles? At present, many key questions related to JEV biology could be properly addressed by use of the reverse genetics system, established by construction of an infectious cDNA molecular clone as a bacterial artificial chromosome (Yun *et al.*, 2003b).

JEV is the leading cause of acute encephalitis in most areas of Asia. In recent years, the viral activity has expanded to different geographic regions, where JEV had never before been isolated. Along with recent outbreaks of WNV infection in North America, the flaviviruses have become a potential threat to the global public health. The development of safer, less expensive, and more effective antiviral vaccines against these pathogens will be one of the top priorities in our research community over the next few years. Similarly, effective antiviral compounds against these pathogens are also highly desirable. These tasks will be facilitated by continuous efforts to understand the biology of these pathogens. As molecular virologists, we believe that many exciting discoveries lie ahead of us.

References

Adams, S.C., Broom, A.K., Sammels, L.M., Hartnett, A.C., Howard, M.J., Coelen, R.J., Mackenzie, J.S., and Hall, R.A. (1995). Glycosylation and antigenic variation among Kunjin virus isolates. Virology *206*, 49–56.

Ali, A., and Igarashi, A. (1997). Antigenic and genetic variations among Japanese encephalitis virus strains belonging to genotype 1. Microbiol. Immunol. *41*, 241–252.

Allison, S.L., Schalich, J., Stiasny, K., Mandl, C.W., Kunz, C., and Heinz, F.X. (1995a). Oligomeric rearrangement of tick-borne encephalitis virus envelope proteins induced by an acidic pH. J. Virol. *69*, 695–700.

Allison, S.L., Stadler, K., Mandl, C.W., Kunz, C., and Heinz, F.X. (1995b). Synthesis and secretion of recombinant tick-borne encephalitis virus protein E in soluble and particulate form. J. Virol. *69*, 5816–5820.

Almazan, F., Gonzalez, J.M., Penzes, Z., Izeta, A., Calvo, E., Plana-Duran, J., and Enjuanes, L. (2000). Engineering the largest RNA virus genome as an infectious bacterial artificial chromosome. Proc. Natl. Acad. Sci. USA *97*, 5516–5521.

Amberg, S.M., Nestorowicz, A., McCourt, D.W., and Rice, C.M. (1994). NS2B-3 proteinase-mediated processing in the yellow fever virus structural region: *in vitro* and *in vivo* studies. J. Virol. *68*, 3794–3802.

Arias, C.F., Preugschat, F., and Strauss, J.H. (1993). Dengue 2 virus NS2B and NS3 form a stable complex that can cleave NS3 within the helicase domain. Virology *193*, 888–899.

Arroyo, J., Guirakhoo, F., Fenner, S., Zhang, Z.X., Monath, T.P., and Chambers, T.J. (2001). Molecular basis for attenuation of neurovirulence of a yellow fever Virus/Japanese encephalitis virus chimera vaccine (ChimeriVax-JE). J. Virol. *75*, 934–942.

Ashok, M.S., and Rangarajan, P.N. (1999). Immunization with plasmid DNA encoding the envelope glycoprotein of Japanese Encephalitis virus confers significant protection against intracerebral viral challenge without inducing detectable antiviral antibodies. Vaccine *18*, 68–75.

Ashok, M.S., and Rangarajan, P.N. (2002). Protective efficacy of a plasmid DNA encoding Japanese encephalitis virus envelope protein fused to tissue plasminogen activator signal sequences: studies in a murine intracerebral virus challenge model. Vaccine *20*, 1563–1570.

Ballinger-Crabtree, M.E., and Miller, B.R. (1990). Partial nucleotide sequence of South American yellow fever virus strain 1899/81: structural proteins and NS1. J. Gen. Virol. *71*, 2115–2121.

Banerjee, K., Deshmukh, P.K., Ilkal, M.A., and Dhanda, V. (1978). Transmission of Japanese encephalitis virus by Culex bitaeniorhynchus Giles. Indian J. Med. Res. *67*, 889–893.

Banerjee, K., Mahadev, P.V., Ilkal, M.A., Mishra, A.C., Dhanda, V., Modi, G.B., Geevarghese, G., Kaul, H.N., Shetty, P.S., and George, P.J. (1979). Isolation of Japanese encephalitis virus from mosquitoes collected in Bankura district (West Bengal) during October 1974 to December (1975). Indian J. Med. Res. *69* 201–205.

Bartholomeusz, A.I., and Wright, P.J. (1993). Synthesis of dengue virus RNA *in vitro*: initiation and the involvement of proteins NS3 and NS5. Arch. Virol. *128*, 111–121.

Battiste, J.L., Tan, R., Frankel, A.D., and Williamson, J.R. (1994). Binding of an HIV Rev peptide to Rev responsive element RNA induces formation of purine-purine base pairs. Biochemistry *33*, 2741–2747.

Bazan, J.F., and Fletterick, R.J. (1989). Detection of a trypsin-like serine protease domain in flaviviruses and pestiviruses. Virology *171*, 637–639.

Bazan, J.F., and Fletterick, R.J. (1990). Structural and catalytic models of trypsin-like proteases. Semin. Virol. *1*, 311–322.

Beasley, D.W., Li, L., Suderman, M.T., Guirakhoo, F., Trent, D.W., Monath, T.P., Shope, R.E., and Barrett, A.D. (2004). Protection against Japanese encephalitis virus strains representing four genotypes by passive transfer of sera raised against ChimeriVax-JE experimental vaccine. Vaccine *22*, 3722–3726.

Bergamini, G., Preiss, T., and Hentze, M.W. (2000). Picornavirus IRESes and the poly(A) tail jointly promote cap-independent translation in a mammalian cell-free system. RNA *6*, 1781–1790.

Bhardwaj, S., Holbrook, M., Shope, R.E., Barrett, A.D., and Watowich, S.J. (2001). Biophysical characterization and vector-specific antagonist activity of domain III of the tick-borne flavivirus envelope protein. J. Virol. *75*, 4002–4007.

Bhatt, T.R., Crabtree, M.B., Guirakhoo, F., Monath, T.P., and Miller, B.R. (2000). Growth characteristics of the chimeric Japanese encephalitis virus vaccine candidate, ChimeriVax-JE (YF/JE SA14-14-2), in Culex tritaeniorhynchus, *Aedes albopictus*, and *Aedes aegypti* mosquitoes. Am. J. Trop. Med. Hyg. *62*, 480–484.

Blitvich, B.J., Scanlon, D., Shiell, B.J., Mackenzie, J.S., and Hall, R.A. (1999). Identification and analysis of truncated and elongated species of the flavivirus NS1 protein. Virus Res. *60*, 67–79.

Blitvich, B.J., Scanlon, D., Shiell, B.J., Mackenzie, J.S., Pham, K., and Hall, R.A. (2001). Determination of the intramolecular disulfide bond arrangement and biochemical identification of the glycosylation sites of the nonstructural protein NS1 of Murray Valley encephalitis virus. J. Gen. Virol. *82*, 2251–2256.

Boege, U., Heinz, F.X., Wengler, G., and Kunz, C. (1983). Amino acid compositions and amino-terminal sequences of the structural proteins of a flavivirus, European Tick-borne Encephalitis virus. Virology *126*, 651–657.

Borowski, P., Niebuhr, A., Mueller, O., Bretner, M., Felczak, K., Kulikowski, T., and Schmitz, H. (2001). Purification and characterization of West Nile virus nucleoside triphosphatase (NTPase)/helicase: evidence for dissociation of the NTPase and helicase activities of the enzyme. J. Virol. *75*, 3220–3229.

Bray, M., and Lai, C.J. (1991). Dengue virus premembrane and membrane proteins elicit a protective immune response. Virology *185*, 505–508.

Bredenbeek, P.J., Kooi, E.A., Lindenbach, B., Huijkman, N., Rice, C.M., and Spaan, W.J. (2003). A stable full-length yellow fever virus cDNA clone and the role of conserved RNA elements in flavivirus replication. J. Gen. Virol. *84*, 1261–1268.

Brinkworth, R.I., Fairlie, D.P., Leung, D., and Young, P.R. (1999). Homology model of the dengue 2 virus NS3 protease: putative interactions with both substrate and NS2B cofactor. J. Gen. Virol. *80*, 1167–1177.

Brinton, M.A. (1986). Replication of flaviviruses. In The Togaviridae and Flaviviridae, S. Schlesinger, and M.J. Schlesinger, eds. (New York: Plenum), pp. 327–365.

Brinton, M.A. (2002). The molecular biology of West Nile Virus: a new invader of the western hemisphere. Annu. Rev. Microbiol. 56 371–402.

Brinton, M.A., and Dispoto, J.H. (1988). Sequence and secondary structure analysis of the 5′-terminal region of flavivirus genome RNA. Virology *162*, 290–299.

Brinton, M.A., Fernandez, A.V., and Dispoto, J.H. (1986). The 3′-nucleotides of flavivirus genomic RNA form a conserved secondary structure. Virology *153*, 113–121.

Brown, E.A., Zhang, H., Ping, L.H., and Lemon, S.M. (1992). Secondary structure of the 5′ nontranslated regions of hepatitis C virus and pestivirus genomic RNAs. Nucleic Acids Res. 20, 5041–5045.

Buckley, A., Gaidamovich, S., Turchinskaya, A., and Gould, E.A. (1992). Monoclonal antibodies identify the NS5 yellow fever virus non-structural protein in the nuclei of infected cells. J. Gen. Virol. *73*, 1125–1130.

Buescher, E.L., and Scherer, W.F. (1959). Ecologic studies of Japanese encephalitis virus in Japan. IX. Epidemiologic correlations and conclusions. Am. J. Trop. Med. Hyg. *8*, 719–722.

Buescher, E.L., Scherer, W.F., Rosenberg, M.Z., Gresser, I., Hardy, J.L., and Bullock, H.R. (1959a). Ecologic studies of Japanese encephalitis virus in Japan. II. Mosquito infection. Am. J. Trop. Med. Hyg. *8*, 651–664.

Buescher, E.L., Scherer, W.F., McClure, H.E., Moyer, J.T., Rosenberg, M.Z., Yoshii, M., and Okada, Y. (1959b). Ecologic studies of Japanese encephalitis virus in Japan. IV. Avian infection. Am. J. Trop. Med. Hyg. *8*, 678–688.

Burke, D.S., and Leake, C.J. (1988). Japanese encephalitis. In The Arboviruses: Epidemiology and Ecology, Vol. III, T.P. Monath, ed. (Boca Raton, FL: CRC Press), pp. 63–92.

Burke, D.S., and Monath, T.P. (2001). Flaviviruses. In Fields Virology, 4th edition, D.M. Knipe, P.M. Howley, D.E. Griffin, R.A. Lamb, M.A. Martin, B. Roizman, and S.E. Straus, eds. (Philadelphia, PA: Lippincott Williams & Wilkins Publishers), pp. 1043–1125.

Cahour, A., Pletnev, A., Vazielle-Falcoz, M., Rosen, L., and Lai, C.J. (1995). Growth-restricted dengue virus mutants containing deletions in the 5′ noncoding region of the RNA genome. Virology *207*, 68–76.

Calisher, C.H. (1988). Antigenic classification and taxonomy of flaviviruses (family Flaviviridae) emphasizing a universal system for the taxonomy of viruses causing tick-borne encephalitis [see comment]. Acta Virol. *32*, 469–478.

Calisher, C.H., and Gould, E.A. (2003). Taxonomy of the virus family *Flaviviridae*. Adv. Virus. Res. 59, 1–19.

Calisher, C.H., Karabatsos, N., Dalrymple, J.M., Shope, R.E., Porterfield, J.S., Westaway, E.G.

and Brandt, W.E. (1989). Antigenic relationships between flaviviruses as determined by cross-neutralization tests with polyclonal antisera. J. Gen. Virol. *70*, 37–43.

Cammisa-Parks, H., Cisar, L.A., Kane, A., and Stollar, V. (1992). The complete nucleotide sequence of cell fusing agent (CFA): homology between the nonstructural proteins encoded by CFA and the nonstructural proteins encoded by arthropod-borne flaviviruses. Virology *189*, 511–524.

Carey, D.E., Reuben, R., and Myers, R.M. (1968). Japanese encephalitis studies in Vellore, South India. I. Virus isolation from mosquitoes. Indian J. Med. Res. *56*, 1309–1318.

Carey, D.E., Reuben, R., and Myers, R.M. (1969). Japanese encephalitis studies in Vellore, South India. V. Experimental infection and transmission. Indian J. Med. Res. *57*, 282–289.

Chakravarty, S.K., Sarkar, J.K., Chakravarty, M.S., Mukherjee, M.K., Mukherjee, K.K., Das, B.C., and Hati, A.K. (1975). The first epidemic of Japanese encephalitis studied in India-virological studies. Indian J. Med. Res. *63*, 77–82.

Chambers, T.J., Grakoui, A., and Rice, C.M. (1991). Processing of the yellow fever virus nonstructural polyprotein: a catalytically active NS3 proteinase domain and NS2B are required for cleavages at dibasic sites. J. Virol. *65*, 6042–6050.

Chambers, T.J., Hahn, C.S., Galler, R., and Rice, C.M. (1990a). Flavivirus genome organization, expression, and replication. Annu. Rev. Microbiol. *44*, 649–688.

Chambers, T.J., Halevy, M., Nestorowicz, A., Rice, C.M., and Lustig, S. (1998). West Nile virus envelope proteins: nucleotide sequence analysis of strains differing in mouse neuroinvasiveness. J. Gen. Virol. *79*, 2375–2380.

Chambers, T.J., McCourt, D.W., and Rice, C.M. (1989). Yellow fever virus proteins NS2A, NS2B, and NS4B: identification and partial N-terminal amino acid sequence analysis. Virology *169*, 100–109.

Chambers, T.J., McCourt, D.W., and Rice, C.M. (1990b). Production of yellow fever virus proteins in infected cells: identification of discrete polyprotein species and analysis of cleavage kinetics using region-specific polyclonal antisera. Virology *177*, 159–174.

Chambers, T.J., Nestorowicz, A., Amberg, S.M., and Rice, C.M. (1993). Mutagenesis of the yellow fever virus NS2B protein: effects on proteolytic processing, NS2B-NS3 complex formation, and viral replication. J. Virol. *67*, 6797–6807.

Chambers, T.J., Nestorowicz, A., Mason, P.W., and Rice, C.M. (1999). Yellow fever/Japanese encephalitis chimeric viruses: construction and biological properties. J. Virol. *73*, 3095–3101.

Chambers, T.J., Nestorowicz, A., and Rice, C.M. (1995). Mutagenesis of the yellow fever virus NS2B/3 cleavage site: determinants of cleavage site specificity and effects on polyprotein processing and viral replication. J. Virol. *69*, 1600–1605.

Chambers, T.J., Weir, R.C., Grakoui, A., McCourt, D.W., Bazan, J.F., Fletterick, R.J., and Rice, C.M. (1990). Evidence that the N-terminal domain of nonstructural protein NS3 from yellow fever virus is a serine protease responsible for site-specific cleavages in the viral polyprotein. Proc. Natl. Acad. Sci. USA *87*, 8898–8902.

Chang, G.J., Hunt, A.R., and Davis, B. (2000). A single intramuscular injection of recombinant plasmid DNA induces protective immunity and prevents Japanese encephalitis in mice. J. Virol. *74*, 4244–4252.

Chang, Y.S., Liao, C.L., Tsao, C.H., Chen, M.C., Liu, C.I., Chen, L.K., and Lin, Y.L. (1999). Membrane permeabilization by small hydrophobic nonstructural proteins of Japanese encephalitis virus. J. Virol. *73*, 6257–6264.

Chen, C.J., Kuo, M.D., Chien, L.J., Hsu, S.L., Wang, Y.M., and Lin, J.H. (1997). RNA-protein interactions: involvement of NS3, NS5, and 3′ noncoding regions of Japanese encephalitis virus genomic RNA. J. Virol. *71*, 3466–3473.

Chen, H.W., Pan, C.H., Liau, M.Y., Jou, R., Tsai, C.J., Wu, H.J., Lin, Y.L., and Tao, M.H. (1999). Screening of protective antigens of Japanese encephalitis virus by DNA immunization: a comparative study with conventional viral vaccines. J. Virol. *73*, 10137–10145.

Chen, L.K., Liao, C.L., Lin, C.G., Lai, S.C., Liu, C.I., Ma, S.H., Huang, Y.Y., and Lin, Y.L. (1996). Persistence of Japanese encephalitis virus is associated with abnormal expression of the nonstructural protein NS1 in host cells. Virology *217*, 220–229.

Chen, W.R., Tesh, R.B., and Rico-Hesse, R. (1990). Genetic variation of Japanese encephalitis virus in nature. J. Gen. Virol. *71*, 2915–2922.

Chen, W.R., Rico-Hesse, R., and Tesh, R.B. (1992). A new genotype of Japanese encephalitis virus from Indonesia. Am. J. Trop. Med. Hyg. *47*, 61–69.

Chen, Y., Maguire, T., Hileman, R.E., Fromm, J.R., Esko, J.D., Linhardt, R.J., and Marks, R.M. (1997). Dengue virus infectivity depends on envelope protein binding to target cell heparan sulfate. Nat. Med. *3*, 866–871.

Chiou, C.T., Hu, C.C., Chen, P.H., Liao, C.L., Lin, Y.L., and Wang, J.J. (2003). Association of Japanese encephalitis virus NS3 protein with microtubules and tumour susceptibility gene 101 (TSG101) protein. J. Gen. Virol. *84*, 2795–2805.

Chu, J.J., and Ng, M.L. (2003). Characterization of a 105-kDa plasma membrane associated glycoprotein that is involved in West Nile virus binding and infection. Virology *312*, 458–469.

Chu, J.J., and Ng, M.L. (2004). Infectious entry of West Nile virus occurs through a clathrin-mediated endocytic pathway. J. Virol. *78*, 10543–10555.

Chu, P.W., and Westaway, E.G. (1985). Replication strategy of Kunjin virus: evidence for recycling role of replicative form RNA as template in semiconservative and asymmetric replication. Virology *140*, 68–79.

Chu, P.W., and Westaway, E.G. (1987). Characterization of Kunjin virus RNA-dependent RNA polymerase: reinitiation of synthesis *in vitro*. Virology *157*, 330–337.

Chu, P.W., and Westaway, E.G. (1992). Molecular and ultrastructural analysis of heavy membrane fractions associated with the replication of Kunjin virus RNA. Arch. Virol. *125*, 177–191.

Cleaves, G.R., and Dubin, D,T. (1979). Methylation status of intracellular dengue type 2 40 S RNA. Virology *96*, 159–165.

Cleaves, G.R., Ryan, T.E., and Schlesinger, R.W. (1981). Identification and characterization of type 2 dengue virus replicative intermediate and replicative form RNAs. Virology *111*, 73–83.

Clum, S., Ebner, K.E., and Padmanabhan, R. (1997). Cotranslational membrane insertion of the serine proteinase precursor NS2B-NS3(Pro) of dengue virus type 2 is required for efficient *in vitro* processing and is mediated through the hydrophobic regions of NS2B. J. Biol. Chem. *272*, 30715–30723.

Corver, J., Lenches, E., Smith, K., Robison, R.A., Sando, T., Strauss, E.G. and Strauss, J.H. (2003). Fine mapping of a cis-acting sequence element in yellow fever virus RNA that is required for RNA replication and cyclization. J. Virol. *77*, 2265–2270.

Courageot, M.P., Frenkiel, M.P., Dos Santos, C.D., Deubel, V., and Despres, P. (2000). Alpha-glucosidase inhibitors reduce dengue virus production by affecting the initial steps of virion morphogenesis in the endoplasmic reticulum. J. Virol. *74*, 564–572.

Crill, W.D., and Roehrig, J.T. (2001). Monoclonal antibodies that bind to domain III of dengue virus E glycoprotein are the most efficient blockers of virus adsorption to Vero cells. J. Virol. *75*, 7769–7773.

Crochu, S., Cook, S., Attoui, H., Charrel, R.N., De Chesse, R., Belhouchet, M., Lemasson, J.J., de Micco, P., and de Lamballerie, X. (2004). Sequences of flavivirus-related RNA viruses persist in DNA form integrated in the genome of *Aedes* spp. mosquitoes. J. Gen. Virol. *85*, 1971–1980.

Crooks, A.J., Lee, J.M., Easterbrook, L.M., Timofeev, A.V., and Stephenson, J.R. (1994). The NS1 protein of tick-borne encephalitis virus forms multimeric species upon secretion from the host cell. J. Gen. Virol. *75*, 3453–3460.

Cui, T., Sugrue, R.J., Xu, Q., Lee, A.K., Chan, Y.C., and Fu, J. (1998). Recombinant dengue virus type 1 NS3 protein exhibits specific viral RNA binding and NTPase activity regulated by the NS5 protein. Virology *246*, 409–417.

Dandawate, C.N., Rajagopalan, P.K., Pavri, K.M., and Work, T.H. (1969). Virus isolations from mosquitoes collected in North Arcot district, Madras state, and Chittoor district, Andhra Pradesh between November 1955 and October (1957). Indian J. Med. Res. *57*, 1420–1426.

Deubel, V., Digoutte, J.P., Mattei, X., and Pandare, D. (1981). Morphogenesis of yellow fever virus in *Aedes aegypti* cultured cells. II. An ultrastructural study. Am. J. Trop. Med. Hyg. *30*, 1071–1077.

Dhanda, V., Thenmozhi, V., Kumar, N.P., Hiriyan, J., Arunachalam, N., Balasubramanian, A., Ilango, A., and Gajanana, A. (1997). Virus isolation from wild-caught mosquitoes during a Japanese encephalitis outbreak in Kerala in (1996). Indian J. Med. Res. *106*, 4–6.

Droll, D.A., Krishna Murthy, H.M., and Chambers, T.J. (2000). Yellow fever virus NS2B-NS3 protease: charged-to-alanine mutagenesis and deletion analysis define regions important for protease complex formation and function. Virology *275*, 335–347.

Edward, Z., and Takegami, T. (1993). Localization and functions of Japanese encephalitis virus nonstructural proteins NS3 and NS5 for viral RNA synthesis in the infected cells. Microbiol. Immunol. *37*, 239–243.

Elshuber, S., Allison, S.L., Heinz, F.X., and Mandl, C.W. (2003). Cleavage of protein prM is necessary for infection of BHK-21 cells by tick-borne encephalitis virus. J. Gen. Virol. *84*, 183–191.

Falconar, A.K. (1999). Identification of an epitope on the dengue virus membrane (M) protein defined by cross-protective monoclonal antibodies: design of an improved epitope sequence based on common determinants present in both envelope (E and M) proteins. Arch. Virol. *144*, 2313–2330.

Falgout, B., Chanock, R., and Lai, C.J. (1989). Proper processing of dengue virus nonstructural glycoprotein NS1 requires the N-terminal hydrophobic signal sequence and the downstream nonstructural protein NS2a. J. Virol. *63*, 1852–1860.

Falgout, B., and Markoff, L. (1995). Evidence that flavivirus NS1-NS2A cleavage is mediated by a membrane-bound host protease in the endoplasmic reticulum. J. Virol. *69*, 7232–7243.

Falgout, B., Miller, R.H., and Lai, C.J. (1993). Deletion analysis of dengue virus type 4 nonstructural protein NS2B: identification of a domain required for NS2B-NS3 protease activity. J. Virol. *67*, 2034–2042.

Falgout, B., Pethel, M., Zhang, Y.M., and Lai, C.J. (1991). Both nonstructural proteins NS2B and NS3 are required for the proteolytic processing of dengue virus nonstructural proteins. J. Virol. *65*, 2467–2475.

Ferlenghi, I., Clarke, M., Ruttan, T., Allison, S.L., Schalich, J., Heinz, F.X., Harrison, S.C., Rey, F.A., and Fuller, S.D. (2001). Molecular organization of a recombinant subviral particle from tick-borne encephalitis virus. Mol. Cell. *7*, 593–602.

Flamand, M., Megret, F., Mathieu, M., Lepault, J., Rey, F.A., and Deubel, V. (1999). Dengue virus type 1 nonstructural glycoprotein NS1 is secreted from mammalian cells as a soluble hexamer in a glycosylation-dependent fashion. J. Virol. *73*, 6104–6110.

Forwood, J.K., Brooks, A., Briggs, L.J., Xiao, C.Y., Jans, D.A., and Vasudevan, S.G. (1999). The 37-amino-acid interdomain of dengue virus NS5 protein contains a functional NLS and inhibitory CK2 site. Biochem. Biophys. Res. Commun. *257*, 731–737.

Fukushi, S., Katayama, K., Kurihara, C., Ishiyama, N., Hoshino, F.B., Ando, T., and Oya, A. (1994). Complete 5′ noncoding region is necessary for the efficient internal initiation of hepatitis C virus RNA. Biochem. Biophys. Res. Commun. *199*, 425–432.

Gajanana, A., Rajendran, R., Samuel, P.P., Thenmozhi, V., Tsai, T.F., Kimura-Kuroda, J., and Reuben, R. (1997). Japanese encephalitis in south Arcot district, Tamil Nadu, India: a three-year longitudinal study of vector abundance and infection frequency. J. Med. Entomol. *34*, 651–659.

George, S., Jacob, P.G., and Rao, J.A. (1987). Isolation of Japanese encephalitis and West Nile viruses from mosquitoes collected in Kolar district of Karnataka state during 1977–79. Indian J. Med. Res. *85*, 235–238.

Germi, R., Crance, J.M., Garin, D., Guimet, J., Lortat-Jacob, H., Ruigrok, R.W., Zarski, J.P., and Drouet, E. (2002). Heparan sulfate-mediated binding of infectious dengue virus type 2 and yellow fever virus. Virology *292*, 162–168.

Goebel, S.J., Hsue, B., Dombrowski, T.F., and Masters, P.S. (2004). Characterization of the RNA components of a putative molecular switch in the 3′ untranslated region of the murine coronavirus genome. J. Virol. *78*, 669–682.

Gollins, S.W., and Porterfield, J.S. (1986). The uncoating and infectivity of the flavivirus West Nile on interaction with cells: effects of pH and ammonium chloride. J. Gen. Virol. *67*, 1941–1950.

Goodfellow, I., Chaudhry, Y., Richardson, A., Meredith, J., Almond, J.W., Barclay, W., and Evans, D.J. (2000). Identification of a cis-acting replication element within the poliovirus coding region. J. Virol. *74*, 4590–4600.

Gorbalenya, A.E., Donchenko, A.P., Koonin, E.V., and Blinov, V.M. (1989). N-terminal domains of putative helicases of flavi- and pestiviruses may be serine proteases. Nucleic Acids Res. *17*, 3889–3897.

Gorbalenya, A.E., Koonin, E.V., Donchenko, A.P., and Blinov, V.M. (1989). Two related superfamilies of putative helicases involved in replication, recombination, repair and expression of DNA and RNA genomes. Nucleic Acids Res. *17*, 4713–4730.

Grange, T., Bouloy, M., and Girard, M. (1985). Stable secondary structures at the 3′-end of the genome of yellow fever virus (17D vaccine strain). FEBS Lett. *188*, 159–163.

Gritsun, T.S., Holmes, E.C., and Gould, E.A. (1995). Analysis of flavivirus envelope proteins reveals variable domains that reflect their antigenicity and may determine their pathogenesis. Virus Res. 35, 307–231.

Gritsun, T.S., Venugopal, K., Zanotto, P.M., Mikhailov, M.V., Sall, A.A., Holmes, E.C., Polkinghorne, I., Frolova, T.V., Pogodina, V.V., Lashkevich, V.A., and Gould, E.A. (1997). Complete sequence of two tick-borne flaviviruses isolated from Siberia and the UK:

Analysis and significance of the 5′- and 3′-UTRs. Virus Res. *49*, 27–39.

Grun, J.B., and Brinton, M.A. (1986). Characterization of West Nile virus RNA-dependent RNA polymerase and cellular terminal adenylyl and uridylyl transferases in cell-free extracts. J. Virol. *60*, 1113–1124.

Grun, J.B., and Brinton, M.A. (1987). Dissociation of NS5 from cell fractions containing West Nile virus-specific polymerase activity. J. Virol. *61*, 3641–3644.

Grun, J.B., and Brinton, M.A. (1988). Separation of functional West Nile virus replication complexes from intracellular membrane fragments. J. Gen. Virol. *69*, 3121–3127.

Gubler, D.J., and Rochrig, J.T. (1998). Arboviruses (togaviridae and flaviviridae). In Topley and Wilson's microbiology and microbial infection, 9th ed. L. Collier, A. Balows, and M. Sussman, eds. (London: Arnold), pp. 579–600.

Guirakhoo, F., Bolin, R.A., and Roehrig, J.T. (1992). The Murray Valley encephalitis virus prM protein confers acid resistance to virus particles and alters the expression of epitopes within the R2 domain of E glycoprotein. Virology *191*, 921–931.

Guirakhoo, F., Heinz, F.X., Mandl, C.W., Holzmann, H., and Kunz, C. (1991). Fusion activity of flaviviruses: comparison of mature and immature (prM-containing) tick-borne encephalitis virions. J. Gen. Virol. *72*, 1323–1329.

Guirakhoo, F., Zhang, Z.X., Chambers, T.J., Delagrave, S., Arroyo, J., Barrett, A.D., and Monath, T.P. (1999). Immunogenicity, genetic stability, and protective efficacy of a recombinant, chimeric yellow fever-Japanese encephalitis virus (ChimeriVax-JE) as a live, attenuated vaccine candidate against Japanese encephalitis. Virology *257*, 363–372.

Gultyaev, A.P., van Batenburg, F.H., and Pleij, C.W. (1995). The influence of a metastable structure in plasmid primer RNA on antisense RNA binding kinetics. Nucleic Acids Res. *23*, 3718–3725.

Guyatt, K.J., Westaway, E.G. and Khromykh, A.A. (2001). Expression and purification of enzymatically active recombinant RNA-dependent RNA polymerase (NS5) of the flavivirus Kunjin. J. Virol. Methods *92*, 37–44.

Hahn, C.S., Hahn, Y.S., Rice, C.M., Lee, E., Dalgarno, L., Strauss, E.G. and Strauss, J.H. (1987). Conserved elements in the 3′ untranslated region of flavivirus RNAs and potential cyclization sequences. J. Mol. Biol. *198*, 33–41.

Hall, R.A., Khromykh, A.A., Mackenzie, J.M., Scherret, J.H., Khromykh, T.I., and Mackenzie, J.S. (1999). Loss of dimerisation of the nonstructural protein NS1 of Kunjin virus delays viral replication and reduces virulence in mice, but still allows secretion of NS1. Virology *264*, 66–75.

Haldeman-Cahill, R., Daros, J.A., and Carrington, J.C. (1998). Secondary structures in the capsid protein coding sequence and 3′ nontranslated region involved in amplification of the tobacco etch virus genome. J. Virol. *72*, 4072–4079.

Halstead, S.B. (1992). Arboviruses of the Pacific and Southeast Asia. In Textbook of pediatric infectious diseases, 3rd edition, R.D. Feigin and J.D. Cherry, eds. (Philadelphia, PA: WB Saunders), pp. 1468–1475.

Hanna, J.N., Ritchie, S.A., Phillips, D.A., Shield, J., Bailey, M.C., Mackenzie, J.S., Poidinger, M., McCall, B.J., and Mills, P.J. (1996). An outbreak of Japanese encephalitis in the Torres Strait, Australia (1995). Med. J. Aust. *165*, 256–260.

Hanna, J.N., Ritchie, S.A., Phillips, D.A., Lee, J.M., Hills, S.L., van den Hurk, A.F., Pyke, A.T., Johansen, C.A., and Mackenzie, J.S. (1999). Japanese encephalitis in north Queensland, Australia (1998). Med. J. Aust. *170*, 533–536.

Hashimoto, H., Nomoto, A., Watanabe, K., Mori, T., Takezawa, T., Aizawa, C., Takegami, T., and Hiramatsu, K. (1988). Molecular cloning and complete nucleotide sequence of the genome of Japanese encephalitis virus Beijing-1 strain. Virus Genes *1*, 305–317.

Hase, T., Summers, P.L., Eckels, K.H., and Baze, W.B. (1987a). An electron and immunoelectron microscopic study of dengue-2 virus infection of cultured mosquito cells: maturation events. Arch. Virol. *92*, 273–291.

Hase, T., Summers, P.L., Eckels, K.H., and Baze, W.B. (1987b). Maturation process of Japanese encephalitis virus in cultured mosquito cells *in vitro* and mouse brain cells *in vivo*. Arch. Virol. *96*, 135–151.

Heinz, F.X., and Allison, S.L. (2000). Structures and mechanisms in flavivirus fusion. Adv. Virus Res. *55*, 231–269.

Heinz, F.X., and Allison, S.L. (2003). Flavivirus structure and membrane fusion. Adv. Virus Res. *59*, 63–97.

Heinz, F.X., Mandl, C.W., Holzmann, H., Kunz, C., Harris, B.A., Rey, F., and Harrison, S.C. (1991). The flavivirus envelope protein E: isolation of a soluble form from tick-borne encephalitis virus and its crystallization. J. Virol. *65*, 5579–5583.

Heinz, F.X., Collett, M.S., Purcell, R.H., Gould, E.A., Howard, C.R., Houghton, M., Moormann, R.J.M., Rice, C.M., and Thiel, H.J. (2000). Family Flaviviridae. In Virus Taxonomy: 7th International Committee for the Taxonomy of Viruses, M.H. van Regenmortel, C.M. Fauquet, D.H.L. Bishop, E.B. Carstens, M.K. Estes, S.M. Lemon, J. Maniloff, M.A. Mayo, D.J. McGeoch, C.R. Pringle, and R.B. Wickner, eds. (San Diego: Academic), pp. 859–878.

Heinz, F.X., Stiasny, K., Puschner-Auer, G., Holzmann, H., Allison, S.L., Mandl, C.W., and Kunz, C. (1994). Structural changes and functional control of the tick-borne encephalitis virus glycoprotein E by the heterodimeric association with protein prM. Virology *198*, 109–117.

Hennessy, S., Liu, Z., Tsai, T.F., Strom, B.L., Wan, C.M., Liu, H.L., Wu, T.X., Yu, H.J., Liu, Q.M., Karabatsos. N., Bilker, W.B., and Halstead, S.B. (1996). Effectiveness of live-attenuated Japanese encephalitis vaccine (SA14-14-2): a case-control study. Lancet *347*, 1583–1586.

Hoke, C.H., Nisalak, A., Sangawhipa, N., Jatanasen, S., Laorakapongse, T., Innis, B.L., Kotchasenee, S., Gingrich, J.B., Latendresse, J., Fukai, K., and Burke, D.S. (1988). Protection against Japanese encephalitis by inactivated vaccines. N. Engl. J. Med. *319*, 608–614.

Hung, S.L., Lee, P.L., Chen, H.W., Chen, L.K., Kao, C.L., and King, C.C. (1999). Analysis of the steps involved in Dengue virus entry into host cells. Virology *257*, 156–167.

Huong, V.T., Ha, D., and Deubel, V. (1993). Genetic study of Japanese encephalitis viruses from Vietnam. Am. J. Trop. Med. Hyg. *49*, 538–544.

Hurrelbrink, R.J., and McMinn, P.C. (2001). Attenuation of Murray Valley encephalitis virus by site-directed mutagenesis of the hinge and putative receptor-binding regions of the envelope protein. J. Virol. *75*, 7692–7702.

Iacono-Connors, L.C., Smith, J.F., Ksiazek, T.G., Kelley, C.L., and Schmaljohn, C.S. (1996). Characterization of Langat virus antigenic determinants defined by monoclonal antibodies to E, NS1 and preM and identification of a protective, non-neutralizing preM-specific monoclonal antibody. Virus Res. *43*, 125–136.

Inactivated Japanese encephalitis virus vaccine. (1993). Recommendations of the advisory committee on immunization practices (ACIP). Morb. Mortal. Wkly. Rep. *42*, 1–14.

Innis, B.L. (1995). Japanese encephalitis. In Exotic viral infections, J.S. Porterfield, ed. (London: Chapman and Hall), pp. 147–174.

Irie, K., Mohan, P.M., Sasaguri, Y., Putnak, R., and Padmanabhan, R. (1989). Sequence analysis of cloned dengue virus type 2 genome (New Guinea-C strain). Gene *75*, 197–211.

Ishak, R., Tovey, D.G., and Howard, C.R. (1988). Morphogenesis of yellow fever virus 17D in infected cell cultures. J. Gen. Virol. *69*, 325–335.

Jacobs, M.G., Robinson, P.J., Bletchly, C., Mackenzie, J.M., and Young, P.R. (2000). Dengue virus nonstructural protein 1 is expressed in a glycosyl-phosphatidylinositol-linked form that is capable of signal transduction. FASEB J. *14*, 1603–1610.

Jan, L.R., Chen, K.L., Lu, C.F., Wu, Y.C., and Horng, C.B. (1996). Complete nucleotide sequence of the genome of Japanese encephalitis virus ling strain: the presence of a 25-nucleotide deletion in the 3′-nontranslated region. Am. J. Trop. Med. Hyg. *55*, 603–609.

Jan, L.R., Yang, C.S., Trent, D.W., Falgout, B., and Lai, C.J. (1995). Processing of Japanese encephalitis virus non-structural proteins: NS2B-NS3 complex and heterologous proteases. J. Gen. Virol. *76*, 573–580.

Japanese encephalitis virus. (1998). WHO. Wkly. Epidemiol. Rec. *73*, 337–344.

Johansen, C.A., Nisbet, D.J., Foley, P.N., Van Den Hurk, A.F., Hall, R.A., Mackenzie, J.S., and Ritchie, S.A. (2004). Flavivirus isolations from mosquitoes collected from Saibai Island in the Torres Strait, Australia, during an incursion of Japanese encephalitis virus. Med. Vet. Entomol. *18*, 281–287.

Johansen, C., Ritchie, S., van den Hurk, A., Bockarie, M., Hanna, J., Phillips, D., Melrose, W., Poidinger, M., Scherret, J. Hall, R., and Mackenzie, J. (1997). The search for Japanese encephalitis virus in the Western province of Papua New Guinea. Arbovirus Research in Australia *7*, 131–136.

Johansson, M., Brooks, A.J., Jans, D.A., and Vasudevan, S.G. (2001). A small region of the dengue virus-encoded RNA-dependent RNA polymerase, NS5, confers interaction with both the nuclear transport receptor importin-beta and the viral helicase, NS3. J. Gen. Virol. *82*, 735–745.

Johnson, A.J., Guirakhoo, F., and Roehrig, J.T. (1994). The envelope glycoproteins of dengue 1 and dengue 2 viruses grown in mosquito cells differ in their utilization of potential glycosylation sites. Virology *203*, 241–249.

Kadare, G., and Haenni, A.L. (1997). Virus-encoded RNA helicases. J. Virol. *71*, 2583–2590.

Kamer, G., and Argos, P. (1984). Primary structural comparison of RNA-dependent polymerases from plant, animal and bacterial viruses. Nucleic Acids Res. *12*, 7269–7282.

Kanesa-thasan, N., Smucny, J.J., Hoke, C.H., Marks, D.H., Konishi, E., Kurane, I., Tang, D.B., Vaughn, D.W., Mason, P.W. and Shope, R.E. (2000). Safety and immunogenicity of NYVAC-JEV and ALVAC-JEV attenuated recombinant Japanese encephalitis virus—poxvirus vaccines in vaccinia-nonimmune and vaccinia-immune humans. Vaccine *19*, 483–491.

Kapoor, M., Zhang, L., Ramachandra, M., Kusukawa, J., Ebner, K.E., and Padmanabhan, R. (1995). Association between NS3 and NS5 proteins of dengue virus type 2 in the putative RNA replicase is linked to differential phosphorylation of NS5. J. Biol. Chem. *270*, 19100–19106.

Kaufman, B.M., Summers, P.L., Dubois, D.R., Cohen, W.H., Gentry, M.K., Timchak, R.L., Burke, D.S., and Eckels, K.H. (1989). Monoclonal antibodies for dengue virus prM glycoprotein protect mice against lethal dengue infection. Am. J. Trop. Med. Hyg. *41*, 576–580.

Kaur, R., Rauthan, M., and Vrati, S. (2004). Immunogenicity in mice of a cationic microparticle-adsorbed plasmid DNA encoding Japanese encephalitis virus envelope protein. Vaccine 22, 2776–2782.

Kaur, R., Sachdeva, G., and Vrati, S. (2002). Plasmid DNA immunization against Japanese encephalitis virus: immunogenicity of membrane-anchored and secretory envelope protein. J. Infect. Dis. *185*, 1–12.

Keelapang, P., Sriburi, R., Supasa, S., Panyadee, N., Songjaeng, A., Jairungsri, A., Puttikhunt, C., Kasinrerk, W., Malasit, P., and Sittisombut, N. (2004). Alterations of pr-M cleavage and virus export in pr-M junction chimeric dengue viruses. J. Virol. *78*, 2367–2381.

Kiermayr, S., Kofler, R.M., Mandl, C.W., Messner, P., and Heinz, F.X. (2004). Isolation of capsid protein dimers from the tick-borne encephalitis flavivirus and *in vitro* assembly of capsid-like particles. J. Virol. *78*, 8078–8084.

Kimura, T., and Ohyama, A. (1988). Association between the pH-dependent conformational change of West Nile flavivirus E protein and virus-mediated membrane fusion. J. Gen. Virol. *69*, 1247–1254.

Kimura, T., Gollins, S.W., and Porterfield, J.S. (1986). The effect of pH on the early interaction of West Nile virus with P388D1 cells. J. Gen. Virol. *67*, 2423–2433.

Kimura, T., Kimura-Kuroda, J., Nagashima, K., and Yasui, K. (1994). Analysis of virus-cell binding characteristics on the determination of Japanese encephalitis virus susceptibility. Arch. Virol. *139*, 239–251.

Kimura-Kuroda, J., and Yasui, K. (1983). Topographical analysis of antigenic determinants on envelope glycoprotein V3 (E) of Japanese encephalitis virus, using monoclonal antibodies. J. Virol. *45*, 124–32.

Khromykh, A.A., Kenney, M.T., and Westaway, E.G. (1998). *trans*-Complementation of flavivirus RNA polymerase gene NS5 by using Kunjin virus replicon-expressing BHK cells. J. Virol. *72*, 7270–7279.

Khromykh, A.A., Kondratieva, N., Sgro, J.Y., Palmenberg, A., and Westaway, E.G. (2003). Significance in replication of the terminal nucleotides of the flavivirus genome. J. Virol. *77*, 10623–10629.

Khromykh, A.A., Meka, H., Guyatt, K.J., and Westaway, E.G. (2001). Essential role of cyclization sequences in flavivirus RNA replication. J. Virol. *75*, 6719–6728.

Khromykh, A.A., Sedlak, P.L., Guyatt, K.J., Hall, R.A., and Westaway, E.G. (1999a). Efficient trans-complementation of the flavivirus kunjin NS5 protein but not of the NS1 protein requires its coexpression with other components of the viral replicase. J. Virol. *73*, 10272–10280.

Khromykh, A.A., Sedlak, P.L., and Westaway, E.G. (1999b). *trans*-Complementation analysis of the flavivirus Kunjin ns5 gene reveals an essential role for translation of its N-terminal half in RNA replication. J. Virol. *73*, 9247–9255.

Khromykh, A.A., Sedlak, P.L., and Westaway, E.G. (2000). *cis*- and *trans*-acting elements in flavivirus RNA replication. J. Virol. *74* 3253–3263.

Khromykh, A.A., Varnavski, A.N., and Westaway, E.G. (1998). Encapsidation of the flavivirus Kunjin replicon RNA by using a complementation system providing Kunjin virus structural proteins in *trans*. J. Virol. *72*, 5967–5977.

Khromykh, A.A., and Westaway, E.G. (1994). Completion of Kunjin virus RNA sequence and recovery of an infectious RNA transcribed from stably cloned full-length cDNA. J. Virol. *68*, 4580–4588.

Khromykh, A.A., and Westaway, E.G. (1997). Subgenomic replicons of the flavivirus Kunjin:

construction and applications. J. Virol. *71*, 1497–1505.

Klovins, J., Berzins, V., and van Duin, J. (1998). A long-range interaction in Qbeta RNA that bridges the thousand nucleotides between the M-site and the 3′ end is required for replication. RNA *4*, 948–957.

Ko, K.K., Igarashi, A., and Fukai, K. (1979). Electron microscopic observations on *Aedes albopictus* cells infected with dengue viruses. Arch. Virol. *62*, 41–52.

Kojima, A., Yasuda, A., Asanuma, H., Ishikawa, T., Takamizawa, A., Yasui, K., and Kurata, T. (2003). Stable high-producer cell clone expressing virus-like particles of the Japanese encephalitis virus e protein for a second-generation subunit vaccine. J. Virol. *77*, 8745–8755.

Kolaskar, A.S., and Kulkarni-Kale, U. (1999). Prediction of three-dimensional structure and mapping of conformational epitopes of envelope glycoprotein of Japanese encephalitis virus. Virology *261*, 31–42.

Konishi, E., and Mason, P.W. (1993). Proper maturation of the Japanese encephalitis virus envelope glycoprotein requires cosynthesis with the premembrane protein. J. Virol. *67*, 1672–1675.

Konishi, E., Pincus, S., Fonseca, B.A., Shope, R.E., Paoletti, E., Mason, P.W. (1991). Comparison of protective immunity elicited by recombinant vaccinia viruses that synthesize E or NS1 of Japanese encephalitis virus. Virology *185*, 401–410.

Konishi, E., Pincus, S., Paoletti, E., Laegreid, W.W., Shope, R.E., and Mason, P.W. (1992a). A highly attenuated host range-restricted vaccinia virus strain, NYVAC, encoding the prM, E, and NS1 genes of Japanese encephalitis virus prevents JEV viremia in swine. Virology *190*, 454–458.

Konishi, E., Pincus, S., Paoletti, E., Shope, R.E., Burrage, T., and Mason, P.W. (1992b). Mice immunized with a subviral particle containing the Japanese encephalitis virus prM/M and E proteins are protected from lethal JEV infection. Virology *188*, 714–720.

Konishi, E., Win, K.S., Kurane, I., Mason, P.W., Shope, R.E., and Ennis, F.A. (1997). Particulate vaccine candidate for Japanese encephalitis induces long-lasting virus-specific memory T lymphocytes in mice. Vaccine *15*, 281–286.

Konishi, E., Yamaoka, M., Kurane, I., and Mason, P.W. (2000). Japanese encephalitis DNA vaccine candidates expressing premembrane and envelope genes induce virus-specific memory B cells and long-lasting antibodies in swine. Virology *268*, 49–55.

Koonin, E.V. (1991a). Similarities in RNA helicases. Nature *352*, 290.

Koonin, E.V. (1991b). The phylogeny of RNA-dependent RNA polymerases of positive-strand RNA viruses. J. Gen. Virol. *72*, 2197–2206.

Koonin, E.V. (1993). Computer-assisted identification of a putative methyltransferase domain in NS5 protein of flaviviruses and lambda 2 protein of reovirus. J. Gen. Virol. *74*, 733–740.

Kopecky, J., Grubhoffer, L., Kovar, V., Jindrak, L., and Vokurkova, D. (1999). A putative host cell receptor for tick-borne encephalitis virus identified by anti-idiotypic antibodies and virus affinoblotting. Intervirology *42*, 9–16.

Kroschewski, H., Allison, S.L., Heinz, F.X., and Mandl, C.W. (2003). Role of heparan sulfate for attachment and entry of tick-borne encephalitis virus. Virology *308*, 92–100.

Kuhn, R.J., Zhang, W., Rossmann, M.G., Pletnev, S.V., Corver, J., Lenches, E., Jones, C.T., Mukhopadhyay, S., Chipman, P.R., Strauss, E.G. Baker, T.S., and Strauss, J.H. (2002). Structure of dengue virus: implications for flavivirus organization, maturation, and fusion. Cell *108*, 717–725.

Kulakosky, P.C., Shuler, M.L., and Wood, H.A. (1998). N-Glycosylation of a baculovirus-expressed recombinant glycoprotein in three insect cell lines. In Vitro Cell. Dev. Biol. *34*, 101–108.

Kuno, G., Chang, G.J., Tsuchiya, K.R., Karabatsos, N., and Cropp, C.B. (1998). Phylogeny of the genus Flavivirus. J. Virol. *72*, 73–83.

Kummerer, B.M., and Rice, C.M. (2002). Mutations in the yellow fever virus nonstructural protein NS2A selectively block production of infectious particles. J. Virol. *76*, 4773–4784.

Kuo, M.D., Chin, C., Hsu, S.L., Shiao, J.Y., Wang, T.M., and Lin, J.H. (1996). Characterization of the NTPase activity of Japanese encephalitis virus NS3 protein. J. Gen. Virol. *77*, 2077–2084.

Kuroda, K., Geyer, H., Geyer, R., Doerfler, W., and Klenk, H.D. (1990). The oligosaccharides of influenza virus hemagglutinin expressed in insect cells by a baculovirus vector. Virology *174*, 418–429.

Leary, K., and Blair, C.D. (1980). Sequential events in the morphogenesis of Japanese Encephalitis virus. J. Ultrastruct. Res. *72*, 123–129.

Lee, E., Hall, R.A., and Lobigs, M. (2004). Common E protein determinants for attenuation of glycosaminoglycan-binding variants of

Japanese encephalitis and West Nile viruses. J. Virol. *78*, 8271–8280.

Lee, E., and Lobigs, M. (2000). Substitutions at the putative receptor-binding site of an encephalitic flavivirus alter virulence and host cell tropism and reveal a role for glycosaminoglycans in entry. J. Virol. *74*, 8867–8875.

Lee, E., and Lobigs, M. (2002). Mechanism of virulence attenuation of glycosaminoglycan-binding variants of Japanese encephalitis virus and Murray Valley encephalitis virus. J. Virol. *76*, 4901–4911.

Lee, E., Weir, R.C., and Dalgarno, L. (1997). Changes in the dengue virus major envelope protein on passaging and their localization on the three-dimensional structure of the protein. Virology *232*, 281–290.

Lee, J.M., Crooks, A.J., and Stephenson, J.R. (1989). The synthesis and maturation of a non-structural extracellular antigen from tick-borne encephalitis virus and its relationship to the intracellular NS1 protein. J. Gen. Virol. *70*, 335–343.

Li, H., Clum, S., You, S., Ebner, K.E., and Padmanabhan, R. (1999). The serine protease and RNA-stimulated nucleoside triphosphatase and RNA helicase functional domains of dengue virus type 2 NS3 converge within a region of 20 amino acids. J. Virol. *73*, 3108–3116.

Liljestrom, P., and Garoff, H. (1991). A new generation of animal cell expression vectors based on the Semliki Forest virus replicon. Biotechnology *9*, 1356–1361.

Lin, C., Amberg, S.M., Chambers, T.J., and Rice, C.M. (1993). Cleavage at a novel site in the NS4A region by the yellow fever virus NS2B-3 proteinase is a prerequisite for processing at the downstream 4A/4B signalase site. J. Virol. *67*, 2327–2335.

Lin, C.W., and Wu, S.C. (2003). A functional epitope determinant on domain III of the Japanese encephalitis virus envelope protein interacted with neutralizing-antibody combining sites. J. Virol. *77*, 2600–2606.

Lin, Y.L., Chen, L.K., Liao, C.L., Yeh, C.T., Ma, S.H., Chen, J.L., Huang, Y.L., Chen, S.S., and Chiang, H.Y. (1998). DNA immunization with Japanese encephalitis virus nonstructural protein NS1 elicits protective immunity in mice. J. Virol. *72*, 191–200.

Lin, Y.L., Lei, H.Y., Lin, Y.S., Yeh, T.M., Chen, S.H., and Liu, H.S. (2002). Heparin inhibits dengue-2 virus infection of five human liver cell lines. Antiviral Res. *56*, 93–96.

Lindenbach, B.D., and Rice, C.M. (1997). *trans*-Complementation of yellow fever virus NS1 reveals a role in early RNA replication. J. Virol. *71*, 9608–9617.

Lindenbach, B.D., and Rice, C.M. (1999). Genetic interaction of flavivirus nonstructural proteins NS1 and NS4A as a determinant of replicase function. J. Virol. *73*, 4611–4621.

Lindenbach, B.D., and Rice, C.M. (2001). *Flaviviridae*: The viruses and their replication. In Fields Virology, 4th edition, D.M. Knipe, P.M. Howley, D.E. Griffin, R.A. Lamb, M.A. Martin, B. Roizman, and S.E. Straus, eds. (Philadelphia, PA: Lippincott Williams & Wilkins Publishers), pp. 991–1041.

Lindenbach, B.D., and Rice, C.M. (2003). Molecular biology of flaviviruses. Adv. Virus Res. *59*, 23–61.

Liu, H., Chiou, S.S., and Chen, W.J. (2004). Differential binding efficiency between the envelope protein of Japanese encephalitis virus variants and heparan sulfate on the cell surface. J. Med. Virol. *72*, 618–624.

Liu, Z.L., Hennessy, S., Strom, B.L., Tsai, T.F., Wan, C.M., Tang, S.C., Xiang, C.F., Bilker, W.B., Pan, X.P., Yao, Y.J., Xu, Z.W., and Halstead, S.B. (1997). Short-term safety of live attenuated Japanese encephalitis vaccine (SA14-14-2): results of a randomized trial with 26 239 subjects. J. Infect. Dis. *176*, 1366–1369.

Lo, M.K., Tilgner, M., Bernard, K.A., and Shi, P.Y. (2003). Functional analysis of mosquito-borne flavivirus conserved sequence elements within 3′ untranslated region of West Nile virus by use of a reporting replicon that differentiates between viral translation and RNA replication. J. Virol. *77*, 10004–10014.

Lobigs, M. (1992). Proteolytic processing of a Murray Valley encephalitis virus non-structural polyprotein segment containing the viral proteinase: accumulation of a NS3–4A precursor which requires mature NS3 for efficient processing. J. Gen. Virol. *73*, 2305–2312.

Lobigs, M. (1993). Flavivirus premembrane protein cleavage and spike heterodimer secretion require the function of the viral proteinase NS3. Proc. Natl. Acad. Sci. USA *90*, 6218–6222.

Lorenz, I.C., Allison, S.L., Heinz, F.X., and Helenius, A. (2002). Folding and dimerization of tick-borne encephalitis virus envelope proteins prM and E in the endoplasmic reticulum. J. Virol. *76*, 5480–5491.

Ma, L., Jones, C.T., Groesch, T.D., Kuhn, R.J., and Post, C.B. (2004). Solution structure of dengue virus capsid protein reveals another fold. Proc. Natl. Acad. Sci. USA *101*, 3414–3419.

Mackenzie, J.M., Jones, M.K., and Young, P.R. (1996). Immunolocalization of the dengue virus nonstructural glycoprotein NS1 suggests a role in viral RNA replication. Virology *220*, 232–240.

Mackenzie, J.M., Khromykh, A.A., Jones, M.K., and Westaway, E.G. (1998). Subcellular localization and some biochemical properties of the flavivirus Kunjin nonstructural proteins NS2A and NS4A. Virology *245*, 203–215.

Mackenzie, J.S., Johansen, C.A., Ritchie, S.A., van den Hurk, A.F., and Hall, R.A. (2002). Japanese encephalitis as an emerging virus: the emergence and spread of Japanese encephalitis virus in Australasia. Curr. Top. Microbiol. Immunol. *267*, 49–73.

Mackenzie, J.M., Jones, M.K., and Westaway, E.G. (1999). Markers for trans-Golgi membranes and the intermediate compartment localize to induced membranes with distinct replication functions in flavivirus-infected cells. J. Virol. *73*, 9555–9567.

Mackenzie, J.S., Poidinger, M., Phillips, D.A., Johansen, C.A., Hall, R.A., Hanna, J.N., Ritchie, S.A., Shield, J., and Graham, R. (1997). Emergence of Japanese encephalitis virus in the Australian region. In Factors in the Emergence of Arboviruses Diseases, J. Saluzzo, and B. Dodet, eds. (Paris: Elsevier), pp. 191–201.

Mackow, E., Makino, Y., Zhao, B.T., Zhang, Y.M., Markoff, L., Buckler-White, A., Guiler, M., Chanock, R., and Lai, C.J. (1987). The nucleotide sequence of dengue type 4 virus: analysis of genes coding for nonstructural proteins. Virology *159*, 217–228.

Maldov, D.G., Karganova, G.G., and Timofeev, A.V. (1992). Tick-borne encephalitis virus interaction with the target cells. Arch. Virol. *127*, 321–325.

Mandl, C.W., Allison, S.L., Holzmann, H., Meixner, T., and Heinz, F.X. (2000). Attenuation of tick-borne encephalitis virus by structure-based site-specific mutagenesis of a putative flavivirus receptor binding site. J. Virol. *74*, 9601–9609.

Mandl, C.W., Guirakhoo, F., Holzmann, H., Heinz, F.X., and Kunz, C. (1989). Antigenic structure of the flavivirus envelope protein E at the molecular level, using tick-borne encephalitis virus as a model. J. Virol. *63*, 564–571.

Mandl, C.W., Holzmann, H., Kunz, C., and Heinz, F.X. (1993). Complete genomic sequence of Powassan virus: evaluation of genetic elements in tick-borne versus mosquito-borne flaviviruses. Virology *194*, 173–184.

Mandl, C.W., Holzmann, H., Meixner, T., Rauscher, S., Stadler, P.F., Allison, S.L., and Heinz, F.X. (1998). Spontaneous and engineered deletions in the 3′ noncoding region of tick-borne encephalitis virus: construction of highly attenuated mutants of a flavivirus. J. Virol. *72*, 2132–2140.

Mandl, C.W., Kroschewski, H., Allison, S.L., Kofler, R., Holzmann, H., Meixner, T., and Heinz, F.X. (2001). Adaptation of tick-borne encephalitis virus to BHK-21 cells results in the formation of multiple heparan sulfate binding sites in the envelope protein and attenuation *in vivo*. J. Virol. *75*, 5627–5637.

Mangada, M.N.M., and Takegami, T. (1999). Molecular characterization of the Japanese encephalitis virus representative immunotype strain JaGAr01. Virus Res. *59*, 101–112.

Marianneau, P., Megret, F., Olivier, R., Morens, D.M., and Deubel, V. (1996). Dengue 1 virus binding to human hepatoma HepG2 and simian Vero cell surfaces differs. J. Gen. Virol. *77*, 2547–2554.

Markoff, L. (2003). 5′- and 3′-noncoding regions in flavivirus RNA. Adv. Virus Res. *59*, 177–228.

Markoff, L., Falgout, B., and Chang, A. (1997). A conserved internal hydrophobic domain mediates the stable membrane integration of the dengue virus capsid protein. Virology *233*, 105–117.

Martinez-Barragan, J.J., and del Angel, R.M. (2001). Identification of a putative coreceptor on Vero cells that participates in dengue 4 virus infection. J. Virol. *75*, 7818–7827.

Mason, P.W. (1989). Maturation of Japanese encephalitis virus glycoproteins produced by infected mammalian and mosquito cells. Virology *169*, 354–364.

Mason, P.W., McAda, P.C., Dalrymple, J.M., Fournier, M.J., and Mason, T.L. (1987). Expression of Japanese encephalitis virus antigens in Escherichia coli. Virology *158*, 361–372.

Mason, P.W., Pincus, S., Fournier, M.J., Mason, T.L., Shope, R.E., and Paoletti, E. (1991). Japanese encephalitis virus-vaccinia recombinants produce particulate forms of the structural membrane proteins and induce high levels of protection against lethal JEV infection. Virology *180*, 294–305.

Matsumura, T., Shiraki, K., Sashikata, T., and Hotta, S. (1977). Morphogenesis of dengue-1 virus in cultures of a human leukemic leukocyte line (J-111). Microbiol. Immunol. *21*, 329–334.

McKnight, K.L., and Lemon, S.M. (1998). The rhinovirus type 14 genome contains an internally located RNA structure that is required for viral replication. RNA *4*, 1569–1584.

Men, R., Bray, M., Clark, D., Chanock, R.M., and Lai, C.J. (1996). Dengue type 4 virus mutants containing deletions in the 3′ noncoding region of the RNA genome: analysis of growth restriction in cell culture and altered viremia pattern and immunogenicity in rhesus monkeys. J. Virol. *70*, 3930–3937.

Messerle, M., Crnkovic, I., Hammerschmidt, W., Ziegler, H., and Koszinowski, U.H. (1997). Cloning and mutagenesis of a herpesvirus genome as an infectious bacterial artificial chromosome. Proc. Natl. Acad. Sci. USA *94*, 14759–14763.

Mishin, V.P., Cominelli, F., and Yamshchikov, V.F. (2001). A 'minimal' approach in design of flavivirus infectious DNA. Virus Res. *81*, 113–123.

Miyake, M. (1964). The pathology of Japanese encephalitis. Bull. World Health Organ. *30*, 153–160.

Mohan, P.M., and Padmanabhan, R. (1991). Detection of stable secondary structure at the 3′ terminus of dengue virus type 2 RNA. Gene *108*, 185–191.

Molloy, S.S., Anderson, E.D., Jean, F., and Thomas, G. (1999). Bi-cycling the furin pathway: from TGN localization to pathogen activation and embryogenesis. Trends Cell. Biol. *9*, 28–35.

Monath, T.P. (2002). Japanese encephalitis vaccines: current vaccines and future prospects. Curr. Top. Microbiol. Immunol. *267*, 105–138.

Monath, T.P., Guirakhoo, F., Nichols, R., Yoksan, S., Schrader, R., Murphy, C., Blum, P., Woodward, S., McCarthy, K., Mathis, D., Johnson, C., and Bedford, P. (2003). Chimeric live, attenuated vaccine against Japanese encephalitis (ChimeriVax-JE): phase 2 clinical trials for safety and immunogenicity, effect of vaccine dose and schedule, and memory response to challenge with inactivated Japanese encephalitis antigen. J. Infect. Dis. *188*, 1213–1230.

Monath, T.P., Levenbook, I., Soike, K., Zhang, Z.X., Ratterree, M., Draper, K., Barrett, A.D., Nichols, R., Weltzin, R., Arroyo, J., and Guirakhoo, F. (2000). Chimeric yellow fever virus 17D-Japanese encephalitis virus vaccine: dose-response effectiveness and extended safety testing in rhesus monkeys. J. Virol. *74*, 1742–1751.

Monath, T.P., McCarthy, K., Bedford, P., Johnson, C.T., Nichols, R., Yoksan, S., Marchesani, R., Knauber, M., Wells, K.H., Arroyo, J., and Guirakhoo, F. (2002). Clinical proof of principle for ChimeriVax: recombinant live, attenuated vaccines against flavivirus infections. Vaccine *20*, 1004–1018.

Monath, T.P., Soike, K., Levenbook, I., Zhang, Z.X., Arroyo, J., Delagrave, S., Myers, G., Barrett, A.D., Shope, R.E., Ratterree, M., Chambers, T.J., and Guirakhoo, F. (1999). Recombinant, chimaeric live, attenuated vaccine (ChimeriVax) incorporating the envelope genes of Japanese encephalitis (SA14-14-2) virus and the capsid and nonstructural genes of yellow fever (17D) virus is safe, immunogenic and protective in non-human primates. Vaccine *17*, 1869–1882.

Morozova, O.V., Tsekhanovskaya, N.A., Maksimova, T.G., Bachvalova, V.N., Matveeva, V.A., Kit, Y.Y. (1997). Phosphorylation of tick-borne encephalitis virus NS5 protein. Virus Res. *49*, 9–15.

Mourya, D.T., Ilkal, M.A., Mishra, A.C., Jacob, P.G., Pant, U., Ramanujam, S., Mavale, M.S., Bhat, H.R., and Dhanda, V. (1989). Isolation of Japanese encephalitis virus from mosquitoes collected in Karnataka state, India from 1985 to (1987). Trans. R. Soc. Trop. Med. Hyg. 83, 550–552.

Mukhopadhyay, S., Kim, B.S., Chipman, P.R., Rossmann, M.G., and Kuhn, R.J. (2003). Structure of West Nile virus. Science *302*, 248.

Munoz, M.L., Cisneros, A., Cruz, J., Das, P., Tovar, R., and Ortega, A. (1998). Putative dengue virus receptors from mosquito cells. FEMS Microbiol. Lett. *168*, 251–258.

Murphy, F.A. (1980). Togavirus morphology and morphogenesis. In The Togaviruses: Biology, Structure, Replication, R.W. Schlesinger, ed. (New York: Academic Press), pp. 241–316.

Murphy, F.A., Harrison, A.K., Gary, G.W. Jr, Whitfield, S.G., and Forrester, F.T. (1968). St. Louis encephalitis virus infection in mice. Electron microscopic studies of central nervous system. Lab. Invest. *19*, 652–662.

Murthy, H.M., Clum, S., and Padmanabhan, R. (1999). Dengue virus NS3 serine protease. Crystal structure and insights into interaction of the active site with substrates by molecular modeling and structural analysis of mutational effects. J. Biol. Chem. *274*, 5573–5580.

Murray, J.M., Aaskov, J.G., and Wright, P.J. (1993). Processing of the dengue virus type 2 proteins prM and C-prM. J. Gen. Virol. *74*, 175–182.

Muylaert, I.R., Chambers, T.J., Galler, R., and Rice, C.M. (1996). Mutagenesis of the N-linked glycosylation sites of the yellow fever virus NS1 protein: effects on virus replication and mouse neurovirulence. Virology *222*, 159–168.

Muylaert, I.R., Galler, R., and Rice, C.M. (1997). Genetic analysis of the yellow fever virus NS1 protein: identification of a temperature-sensitive mutation which blocks RNA accumulation. J. Virol. *71*, 291–298.

Naik, P.S., Ilkal, M.A., Pant, U., Kulkarni, S.M., and Dhanda, V. (1990). Isolation of Japanese encephalitis virus from Culex pseudovishnui Colless, 1957 (Diptera: Culicidae) in Goa. Indian J. Med. Res. *91*, 331–333.

Nakayama, K. (1997). Furin: a mammalian subtilisin/Kex2p-like endoprotease involved in processing of a wide variety of precursor proteins. Biochem. J. *327*, 625–635.

Nam, J.H., Chae, S.L., Won, S.Y., Kim, E.J., Yoon, K.S., Kim, B.I., Jeong, Y.S., and Cho, H.W. (2001). Short report: genetic heterogeneity of Japanese encephalitis virus assessed via analysis of the full-length genome sequence of a Korean isolate. Am. J. Trop. Med. Hyg. *65*, 388–392.

Nam, J.H., Chung, Y.J., Ban, S.J., Kim, E.J., Park, Y.K., and Cho, H.W. (1996). Envelope gene sequence variation among Japanese encephalitis viruses isolated in Korea. Acta Virol. *40*, 303–309.

Nam, J.H., Wyatt, L.S., Chae, S.L., Cho, H.W., Park, Y.K., and Moss, B. (1999). Protection against lethal Japanese encephalitis virus infection of mice by immunization with the highly attenuated MVA strain of vaccinia virus expressing JEV prM and E genes. Vaccine *17*, 261–268.

Nawa, M. (1998). Effects of bafilomycin A1 on Japanese encephalitis virus in C6/36 mosquito cells. Arch. Virol. *143*, 1555–1568.

Nawa, M., Takasaki, T., Yamada, K., Kurane, I., and Akatsuka, T. (2003). Interference in Japanese encephalitis virus infection of Vero cells by a cationic amphiphilic drug, chlorpromazine. J. Gen. Virol. *84*, 1737–1741.

Nestorowicz, A., Chambers, T.J., and Rice, C.M. (1994). Mutagenesis of the yellow fever virus NS2A/2B cleavage site: effects on proteolytic processing, viral replication, and evidence for alternative processing of the NS2A protein. Virology *199*, 114–123.

Ng, M.L., and Hong, S.S. (1989). Flavivirus infection: essential ultrastructural changes and association of Kunjin virus NS3 protein with microtubules. Arch Virol. *106*, 103–120.

Ng, M.L. (1987). Ultrastructural studies of Kunjin virus-infected *Aedes albopictus* cells. J. Gen. Virol. *68*, 577–582.

Ng, M.L., Tan, S.H., and Chu, J.J. (2001). Transport and budding at two distinct sites of visible nucleocapsids of West Nile (Sarafend) virus. J. Med. Virol. *65*, 758–764.

Ni, H., and Barret, A.D.T. (1995). Nucleotide and deduced amino acid sequence of the structural protein genes of Japanese encephalitis viruses from different geographical locations. J. Gen. Virol. *76*, 401–407.

Ni, H., Burns, N.J., Chang, G.J., Zhang, M.J., Wills, M.R., Trent, D.W., Sanders, P.G., and Barrett, A.D. (1994). Comparison of nucleotide and deduced amino acid sequence of the 5′ non-coding region and structural protein genes of the wild-type Japanese encephalitis virus strain SA14 and its attenuated vaccine derivatives. J. Gen. Virol. *75*, 1505–1510.

Nowak, T., and Wengler, G. (1987). Analysis of disulfides present in the membrane proteins of the West Nile flavivirus. Virology *156*, 127–137.

Nowak, T., Farber, P.M., Wengler, G., and Wengler, G. (1989). Analyses of the terminal sequences of West Nile virus structural proteins and of the *in vitro* translation of these proteins allow the proposal of a complete scheme of the proteolytic cleavages involved in their synthesis. Virology *169*, 365–376.

Ohyama, A., Ito, T., Tanimura, E., Huang, S.C., and Hsue, J. (1977). Electron microscopic observation of the budding maturation of group B arboviruses. Microbiol. Immunol. *21*, 535–538.

Okuno, T. (1978). An epidemiological review of Japanese encephalitis. World Health Stat. Q. *31*, 120–133.

Olson, J.G., Ksiazek, T.G., Lee, V.H., Tan, R., and Shope, R.E. (1985). Isolation of Japanese encephalitis virus from Anopheles annularis and Anopheles vagus in Lombok, Indonesia. Trans. R. Soc. Trop. Med. Hyg. *79*, 845–847.

Olsthoorn, R.C., and Bol, J.F. (2001). Sequence comparison and secondary structure analysis of the 3′ noncoding region of flavivirus genomes reveals multiple pseudoknots. RNA *7*, 1370–1377.

Paranjpe, S., and Banerjee, K. (1996). Phylogenetic analysis of the envelope gene of Japanese encephalitis virus. Virus Res. *42*, 107–117.

Poidinger, M., Hall, R.A., and Mackenzie, J.S. (1996). Molecular characterization of the Japanese encephalitis serocomplex of the flavivirus genus. Virology *218*, 417–421.

Post, P.R., Carvalho, R., and Galler, R. (1991). Glycosylation and secretion of yellow fever virus nonstructural protein NS1. Virus Res. *18*, 291–302.

Post, P.R., Santos, C.N., Carvalho, R., Cruz, A.C., Rice, C.M., and Galler, R. (1992). Heterogeneity in envelope protein sequence and N-linked glycosylation among yellow fever virus vaccine strains. Virology *188*, 160–167.

Poyry, T.A., Hentze, M.W., and Jackson, R.J. (2001). Construction of regulatable picornavirus IRESes as a test of current models of the mechanism of internal translation initiation. RNA *7*, 647–660.

Preugschat, F., and Strauss, J.H. (1991). Processing of nonstructural proteins NS4A and NS4B of dengue 2 virus *in vitro* and *in vivo*. Virology *185*, 689–697.

Preugschat, F., Yao, C.W., and Strauss, J.H. (1990). In vitro processing of dengue virus type 2 nonstructural proteins NS2A, NS2B, and NS3. J. Virol. *64*, 4364–4374.

Proutski, V., Gritsun, T.S., Gould, E.A., and Holmes, E.C. (1999). Biological consequences of deletions within the 3′-untranslated region of flaviviruses may be due to rearrangements of RNA secondary structure. Virus Res. *64*, 107–123.

Proutski, V., Gould, E.A., and Holmes, E.C. (1997). Secondary structure of the 3′ untranslated region of flaviviruses: similarities and differences. Nucleic Acids Res. *25*, 1194–1202.

Pryor, M.J., Gualano, R.C., Lin, B., Davidson, A.D., and Wright, P.J. (1998). Growth restriction of dengue virus type 2 by site-specific mutagenesis of virus-encoded glycoproteins. J. Gen. Virol. *79*, 2631–2639.

Pryor, M.J., and Wright, P.J. (1993). The effects of site-directed mutagenesis on the dimerization and secretion of the NS1 protein specified by dengue virus. Virology *194*, 769–780.

Pugachev, K.V., Nomokonova, N.Y., Dobrikova, E.Y., and Wolf, Y.I. (1993). Site-directed mutagenesis of the tick-borne encephalitis virus NS3 gene reveals the putative serine protease domain of the NS3 protein. FEBS Lett. *328*, 115–118.

Puglisi, J.D., Tan, R., Calnan, B.J., Frankel, A.D., and Williamson, J.R. (1992). Conformation of the TAR RNA-arginine complex by NMR spectroscopy. Science *257*, 76–80.

Racaniello, V.R., and Baltimore, D. (1981). Cloned poliovirus complementary DNA is infectious in mammalian cells. Science *214*, 916–919.

Raengsakulrach, B., Nisalak, A., Gettayacamin, M., Thirawuth, V., Young, G.D., Myint, K.S., Ferguson, L.M., Hoke, C.H. Jr., Innis, B.L., and Vaughn, D.W. (1999). Safety, immunogenicity, and protective efficacy of NYVAC-JEV and ALVAC-JEV recombinant Japanese encephalitis vaccines in rhesus monkeys. Am. J. Trop. Med. Hyg. *60*, 343–349.

Ramakrishna, C., Desai, A., Shankar, S.K., Chandramuki, A., and Ravi, V. (1999). Oral immunisation of mice with live Japanese encephalitis virus induces a protective immune response. Vaccine *17*, 3102–3108.

Ramos-Castaneda, J., Imbert, J.L., Barron, B.L., and Ramos, C. (1997). A 65-kDa trypsin-sensible membrane cell protein as a possible receptor for dengue virus in cultured neuroblastoma cells. J. Neurovirol. *3*, 435–440.

Randolph, V.B., and Stollar, V. (1990). Low pH-induced cell fusion in flavivirus-infected *Aedes albopictus* cell cultures. J. Gen. Virol. *71*, 1845–1850.

Randolph, V.B., Winkler, G., and Stollar, V. (1990). Acidotropic amines inhibit proteolytic processing of flavivirus prM protein. Virology *174*, 450–458.

Ranjith-Kumar, C.T., Gajewski, J., Gutshall, L., Maley, D., Sarisky, R.T., and Kao, C.C. (2001). Terminal nucleotidyl transferase activity of recombinant Flaviviridae RNA-dependent RNA polymerases: implication for viral RNA synthesis. J. Virol. *75*, 8615–8623.

Rauscher, S., Flamm, C., Mandl, C.W., Heinz, F.X., and Stadler, P.F. (1997). Secondary structure of the 3′-noncoding region of flavivirus genomes: Comparative analysis of base pairing probabilities. RNA *3*, 779–791.

Reed, K.E., Gorbalenya, A.E., and Rice, C.M. (1998). The NS5A/NS5 proteins of viruses from three genera of the family flaviviridae are phosphorylated by associated serine/threonine kinases. J. Virol. *72*, 6199–6206.

Rezapkin, G.V., Alexander, W., Dragunsky, E., Parker, M., Pomeroy, K., Asher, D.M., and Chumakov, K.M. (1998). Genetic stability of Sabin 1 strain of poliovirus: implications for quality control of oral poliovirus vaccine. Virology *245*, 183–187.

Rey, F.A., Heinz, F.X., Mandl, C., Kunz, C., and Harrison, S.C. (1995). The envelope glycoprotein from tick-borne encephalitis virus at 2 Å resolution. Nature *375*, 291–298.

Rice, C.M., Aebersold, R., Teplow, D.B., Pata, J., Bell, J.R., Vorndam, A.V., Trent, D.W., Brandriss, M.W., Schlesinger, J.J., and Strauss, J.H. (1986a). Partial N-terminal amino acid sequences of three nonstructural proteins of two flaviviruses. Virology *151*, 1–9.

Rice, C.M., Grakoui, A., Galler, R., and Chambers, T.J. (1989). Transcription of infectious yellow fever virus RNA from full-length cDNA templates produced by *in vitro* ligation. New Biol. *1*, 285–296.

Rice, C.M., Lenches, E.M., Eddy, S.R., Shin, S.J., Sheets, R.L., and Strauss, J.H. (1985). Nucleotide sequence of yellow fever virus: implications for flavivirus gene expression and evolution. Science *229*, 726–733.

Rice, C.M., Levis, R., Strauss, J.H., and Huang, H.V. (1987). Production of infectious RNA transcripts from Sindbis virus cDNA clones: mapping of lethal mutations, rescue of a temperature-sensitive marker, and *in vitro* mutagenesis to generate defined mutants. J. Virol. *61*, 3809–3819.

Rice, C.M., Strauss, E.G. and Strauss, J.H. (1986b). Structure of the flavivirus genome. In The Togaviridae and Flaviviridae, S. Schlesinger, and M.J. Schlesinger, eds. (New York: Plenum), pp. 279–326.

Roehrig, J.T., Bolin, R.A., and Kelly, R.G. (1998). Monoclonal antibody mapping of the envelope glycoprotein of the dengue 2 virus, Jamaica. Virology *246*, 317–328.

Rosen, L. (1986). The natural history of Japanese encephalitis virus. Annu. Rev. Microbiol. *40*, 395–414.

Ruiz-Linares, A., Bouloy, M., Girard, M., and Cahour, A. (1989). Modulations of the *in vitro* translational efficiencies of Yellow Fever virus mRNAs: interactions between coding and noncoding regions. Nucleic Acids Res. *17*, 2463–2476.

Ruiz-Linares, A., Cahour, A., Despres, P., Girard, M., and Bouloy, M. (1989). Processing of yellow fever virus polyprotein: role of cellular proteases in maturation of the structural proteins. J. Virol. 63, 4199–4209.

Russell, P.K., Brandt, W.E., and Dalrymple, J.M. (1980). Chemical and antigenic structure of flaviviruses. In The Togaviruses: Biology, Structure, Replication, R.W. Schlesinger, ed. (New York: Academic Press), pp. 503–529.

Salas-Benito, J.S., and del Angel, R.M. (1997). Identification of two surface proteins from C6/36 cells that bind dengue type 4 virus. J. Virol. *71*, 7246–7252.

Satyanarayana, T., Gowda, S., Boyko, V.P., Albiach-Marti, M.R., Mawassi, M., Navas-Castillo, J., Karasev, A.V., Dolja, V., Hilf, M.E., Lewandowski, D.J., Moreno, P., Bar-Joseph, M., Garnsey, S.M., and Dawson, W.O. (1999). An engineered closterovirus RNA replicon and analysis of heterologous terminal sequences for replication. Proc. Natl. Acad. Sci. USA *96*, 7433–7438.

Scherer, W.F., and Buescher, E.L. (1959). Ecologic studies of Japanese encephalitis virus in Japan. I. Introduction. Am. J. Trop. Med. Hyg. *8*, 644–650.

Scherer, W.F., Buescher, E.L., Flemings, M.B., Noguchi, A., and Scanlon, J. (1959a). Ecologic studies of Japanese encephalitis virus in Japan. III. Mosquito factors. Zootropism and vertical flight of Culex tritaeniorhynchus with observations on variations in collections from animal-baited traps in different habitats. Am. J. Trop. Med. Hyg. *8*, 665–677.

Scherer, W.F., Buescher, E.L., Southam, C.M., Flemings, M.B., and Noguchi, A. (1959b). Ecologic studies of Japanese encephalitis virus in Japan. VIII. Survey for infection of wild rodents. Am. J. Trop. Med. Hyg. *8*, 716–718.

Scherer, W.F., Kitaoka, M., Okuno, T., and Ogata, T. (1959c). Ecologic studies of Japanese encephalitis virus in Japan. VII. Human infection. Am. J. Trop. Med. Hyg. *8*, 707–715.

Scherer, W.F., Moyer, J.T., Izumi, T., Gresser, I., and McCown, J. (1959d). Ecologic studies of Japanese encephalitis virus in Japan. VI. Swine infection. Am. J. Trop. Med. Hyg. *8*, 698–706.

Schlesinger, S., and Dubensky, T.W. (1999). Alphaviruses vectors for gene expression and vaccines. Curr. Opin. Biotechnol. *10*, 434–439.

Service, M.W. (1991). Agricultural development and arthropod-borne disease-a review. Revista Saude Publica 25, 165–178.

Shapiro, D., Brandt, W.E., and Russell, P.K. (1972). Change involving a viral membrane glycoprotein during morphogenesis of group B arboviruses. Virology *50*, 906–911.

Shi, P.Y., Brinton, M.A., Veal, J.M., Zhong, Y.Y., and Wilson, W.D. (1996a). Evidence for the existence of a pseudoknot structure at the 3′ terminus of the flavivirus genomic RNA. Biochemistry 35, 4222–4230.

Shi, P.Y., Li, W., and Brinton, M.A. (1996b). Cell proteins bind specifically to West Nile virus minus-strand 3′ stem–loop RNA. J. Virol. *70*, 6278–6287.

Shield, J., Hanna, J., and Phillips, D. (1996). Reappearance of the Japanese encephalitis virus in the Torres Strait (1996). Commun. Dis. Intell. *20*, 191.

Shizuya, H., Birren, B., Kim, U.J., Mancino, V., Slepak, T., Tachiiri, Y., and Simon, M. (1992). Cloning and stable maintenance of 300-kilobase-pair fragments of human DNA in *Escherichia coli* with an F-factor-based vector. Proc. Natl. Acad. Sci. USA *89*, 8794–8797.

Shlim, D.R., and Solomon, T. (2002). Japanese encephalitis vaccine for travelers: exploring the limits of risk. Clin. Infect. Dis. 35, 183–188.

Simons, J.N., Leary, T.P., Dawson, G.J., Pilot-Matias, T.J., Muerhoff, A.S., Schlauder, G.G.,

Desai, S.M., and Mushahwar, I.K. (1995). Isolation of novel virus-like sequences associated with human hepatitis. Nat. Med. *1*, 564–569.

Smith, G.W., and Wright, P.J. (1985). Synthesis of proteins and glycoproteins in dengue type 2 virus-infected vero and *Aedes albopictus* cells. J. Gen. Virol. *66*, 559–571.

Solomon, T. (2000). Japanese encephalitis. In Neurobase, S. Gilman, G.W. Goldstein, S.G. Waxman, eds. (San Diego: Medlink Publishing), Available on CD-ROM.

Solomon, T. (2003). Recent advances in Japanese encephalitis. J. Neurovirol. *9*, 274–283.

Solomon, T., Dung, N.M., Kneen, R., Gainsborough, M., Vaughn, D.W., and Khanh, V.T. (2000). Japanese encephalitis. J. Neurol. Neurosurg. Psychiatry *68*, 405–415.

Solomon, T., Ni, H., Beasley, D.W., Ekkelenkamp, M., Cardosa, M.J., and Barrett, A.D. (2003). Origin and evolution of Japanese encephalitis virus in southeast Asia. J. Virol. *77*, 3091–3098.

Spicer, P.E., Phillips, D., Pike, A., Johansen, C., Melrose, W., and Hall, R.A. (1999). Antibodies to Japanese encephalitis virus in human sera collected from Irian Jaya. Follow-up of a previously reported case of Japanese encephalitis in that region. Trans. R. Soc. Trop. Med. Hyg. *93*, 511–514.

Sriurairatna, S., and Bhamarapravati, N. (1977). Replication of dengue-2 virus in *Aedes albopictus* mosquitoes. An electron microscopic study. Am. J. Trop. Med. Hyg. *26*, 1199–1205.

Sriurairatna, S., Bhamarapravati, N., and Phalavadhtana, O. (1973). Dengue virus infection of mice: morphology and morphogenesis of dengue type-2 virus in suckling mouse neurones. Infect. Immun. *8*, 1017–1028.

Srivastava, A.K., Putnak, J.R., Lee, S.H., Hong, S.P., Moon, S.B., Barvir, D.A., Zhao, B., Olson, R.A., Kim, S.O., Yoo, W.D., Towle, A.C., Vaughn, D.W., Innis, B.L., and Eckels, K.H. (2001). A purified inactivated Japanese encephalitis virus vaccine made in Vero cells. Vaccine *19*, 4557–4565.

Stadler, K., Allison, S.L., Schalich, J., and Heinz, F.X. (1997). Proteolytic activation of tick-borne encephalitis virus by furin. J. Virol. *71*, 8475–8481.

Steiner, D.F. (1998). The proprotein convertases. Curr. Opin. Chem. Biol. *2*, 31–39.

Stiasny, K., Allison, S.L., Marchler-Bauer, A., Kunz, C., and Heinz, F. X. (1996). Structural requirements for low-pH-induced rearrangements in the envelope glycoprotein of tick-borne encephalitis virus. J. Virol. *70*, 8142–8147.

Su, C.M., Liao, C.L., Lee, Y.L., and Lin, Y.L. (2001). Highly sulfated forms of heparin sulfate are involved in Japanese encephalitis virus infection. Virology *286*, 206–215.

Sucharit, S., Surathin, K., and Shrestha, S.R. (1989). Vectors of Japanese encephalitis virus (JEV): species complexes of the vectors. Southeast Asian J. Trop. Med. Public Health. *20*, 611–621.

Sugawara, K., Nishiyama, K., Ishikawa, Y., Abe, M., Sonoda, K., Komatsu, K., Horikawa, Y., Takeda, K., Honda, T., Kuzuhara, S., Kino, Y., Mizokami, H., Mizuno, K., Oka, T., and Honda, K. (2002). Development of Vero cell-derived inactivated Japanese encephalitis vaccine. Biologicals *30*, 303–314.

Sumiyoshi, H., Hoke, C.H., and Trent, D.W. (1992). Infectious Japanese encephalitis virus RNA can be synthesized from *in vitro*-ligated cDNA templates. J. Virol. *66*, 5425–5431.

Sumiyoshi, H., Mori, C., Fuke, I., Morita, K., Kuhara, S., Kondou, J., Kikuchi, Y., Nagamatu, H., and Igarashi, A. (1987). Complete nucleotide sequence of the Japanese encephalitis virus genome RNA. Virology *161*, 497–510.

Sumiyoshi, H., Tignor, G.H., and Shope, R.E. (1995). Characterization of a highly attenuated Japanese encephalitis virus generated from molecularly cloned cDNA. J. Infect. Dis. *171*, 1144–1151.

Ta, M., and Vrati, S. (2000). Mov34 protein from mouse brain interacts with the 3′ noncoding region of Japanese encephalitis virus. J. Virol. *74*, 5108–5115.

Takegami, T., and Hotta, S. (1989). In vitro synthesis of Japanese encephalitis virus (JEV) RNA: membrane and nuclear fractions of JEV-infected cells possess high levels of virus-specific RNA polymerase activity. Virus Res. *13*, 337–350.

Takegami, T., Sakamuro, D., and Furukawa, T. (1995). Japanese encephalitis virus nonstructural protein NS3 has RNA binding and ATPase activities. Virus Genes *9*, 105–112.

Takegami, T., Washizu, M., and Yasui, K. (1986). Nucleotide sequence at the 3′ end of Japanese encephalitis virus genomic RNA. Virology *152*, 483–486.

Tan, B.H., Fu, J., Sugrue, R.J., Yap, E.H., Chan, Y.C., and Tan, Y.H. (1996). Recombinant dengue type 1 virus NS5 protein expressed in Escherichia coli exhibits RNA-dependent RNA polymerase activity. Virology *216*, 317–325.

Teo, K.F., and Wright, P.J. (1997). Internal proteolysis of the NS3 protein specified by dengue virus 2. J. Gen. Virol. *78*, 337–341.

Thomas, G. (2002). Furin at the cutting edge: from protein traffic to embryogenesis and disease. Nat. Rev. Mol. Cell. Biol. *3*, 753–766.

Thongcharoen, P. (1989). Japanese encephalitis virus encephalitis: an overview. Southeast Asian J. Trop. Med. Public Health. *20*, 559–573.

Thullier, P., Demangel, C., Bedouelle, H., Megret, F., Jouan, A., Deubel, V., Mazie, J.C., and Lafaye, P. (2001). Mapping of a dengue virus neutralizing epitope critical for the infectivity of all serotypes: insight into the neutralization mechanism. J. Gen. Virol. *82*, 1885–1892.

Thurner, C., Witwer, C., Hofacker, I.L., and Stadler, P.F. (2004). Conserved RNA secondary structures in Flaviviridae genomes. J. Gen. Virol. *85*, 1113–1124.

Tiroumourougane, S.V., Raghava, P., and Srinivasan, S. (2002). Japanese viral encephalitis. Postgrad. Med. J. *78*, 205–215.

Trent, D.W. (1977). Antigenic characterization of flavivirus structural proteins separated by isoelectric focusing. J. Virol. *22*, 608–618.

Tsai, T.F. (2000). New initiatives for the control of Japanese encephalitis by vaccination: minutes of a WHO/CVI meeting, Bangkok, Thailand, 13–15 October (1998). Vaccine *18(Suppl 2)*, 1–25.

Tsuchie, H., Oda, K., Vythilingam, I., Thayan, R., Vijayamalar, B., Sinniah, M., Singh, J., Wada, T., Tanaka, H., Kurimura, T., and Igarashi, A. (1997). Genotypes of Japanese encephalitis virus isolated in three states in Malaysia. Am. J. Trop. Med. Hyg. *56*, 153–158.

Uchil, P.D., and Satchidanandam, V. (2003). Characterization of RNA synthesis, replication mechanism, and *in vitro* RNA-dependent RNA polymerase activity of Japanese encephalitis virus. Virology *307*, 358–371.

Umenai, T., Krzysko, R., Bektimirov, T.A., and Assaad, F.A. (1985). Japanese encephalitis: current worldwide status. Bull. World Health Organ. *63*, 625–631.

Utama, A., Shimizu, H., Morikawa, S., Hasebe, F., Morita, K., Igarashi, A., Hatsu, M., Takamizawa, K., and Miyamura, T. (2000). Identification and characterization of the RNA helicase activity of Japanese encephalitis virus NS3 protein. FEBS Lett. *465*, 74–78.

Valle, R.P., and Falgout, B. (1998). Mutagenesis of the NS3 protease of dengue virus type 2. J. Virol. *72*, 624–632.

van der Most, R.G., Corver, J., and Strauss, J.H. (1999). Mutagenesis of the RGD motif in the yellow fever virus 17D envelope protein. Virology *265*, 83–95.

van Dinten, L.C., den Boon, J.A., Wassenaar, A.L., Spaan, W.J., and Snijder, E.J. (1997). An infectious arteriivirus cDNA clone: identification of a replicase point mutation that abolishes discontinuous mRNA transcription. Proc. Natl. Acad. Sci. USA *94*, 991–996.

Vaughn, D.W., and Hoke, C.H. Jr. (1992). The epidemiology of Japanese encephalitis: prospects for prevention. Epidemiol. Rev. *14*, 197–221.

Vazquez, S., Guzman, M.G., Guillen, G., Chinea, G., Perez, A.B., Pupo, M., Rodriguez, R., Reyes, O., Garay, H.E., Delgado, I., Garcia, G., and Alvarez, M. (2002). Immune response to synthetic peptides of dengue prM protein. Vaccine *20*, 1823–1830.

Vorndam, V., Mathews, J.H., Barrett, A.D., Roehrig, J.T., and Trent, D.W. (1993). Molecular and biological characterization of a non-glycosylated isolate of St Louis encephalitis virus. J. Gen. Virol. *74*, 2653–2660.

Vrati, S., Giri, R.K., Razdan, A., and Malik, P. (1999). Complete nucleotide sequence of an Indian strain of Japanese encephalitis virus: sequence comparison with other strains and phylogenetic analysis. Am. J. Trop. Med. Hyg. *61*, 677–680.

Vythilingam, I., Oda, K., Tsuchie, H., Mahadevan, S., and Vijayamalar, B. (1994). Isolation of Japanese encephalitis virus from Culex sitiens mosquitoes in Selangor, Malaysia. J. Am. Mosq. Control Assoc. *10*, 228–229.

Vythilingam, I., Tan, S.B., and Krishnasamy, M. (2002). Susceptibility of Culex sitiens to Japanese encephalitis virus in peninsular Malaysia. Trop. Med. Int. Health. *7*, 539–540.

Wallis, T.P., Huang, C.Y., Nimkar, S.B., Young, P.R., and Gorman, J.J. (2004). Determination of the disulfide bond arrangement of dengue virus NS1 protein. J. Biol. Chem. *279*, 20729–20741.

Wallner, G., Mandl, C.W., Kunz, C., and Heinz, F.X. (1995). The flavivirus 3′-noncoding region: extensive size heterogeneity independent of evolutionary relationships among strains of tick-borne encephalitis virus. Virology *213*, 169–178.

Wang, C., Le, S.Y., Ali, N., and Siddiqui, A. (1995). An RNA pseudoknot is an essential structural element of the internal ribosome entry site located within the hepatitis C virus 5′ noncoding region. RNA *1*, 526–537.

Wang, E., Weaver, S.C., Shope, R.E., Tesh, R.B., Watts, D.M., and Barrett, A.D. (1996). Genetic variation in yellow fever virus: duplication in the 3′ noncoding region of strains from Africa. Virology *225*, 274–281.

Wang, J.J., Liao, C.L., Yang, C.I., Lin, Y.L., Chiou, C.T., and Chen, L.K. (1998). Localizations of NS3 and E proteins in mouse brain infected with mutant strain of Japanese encephalitis virus. Arch. Virol. *143*, 2353–2369.

Wang, K., Boysen, C., Schizuya, H., Simon, M.I., and Hood, L. (1997). Complete nucleotide sequence of two generations of a bacterial artificial chromosome cloning vector. Biotechniques *23*, 992–994.

Wang, S., He, R., and Anderson, R. (1999). PrM- and cell-binding domains of the dengue virus E protein. J. Virol. *73*, 2547–2551.

Wang, S.H., Syu, W.J., and Hu, S.T. (2004). Identification of the homotypic interaction domain of the core protein of dengue virus type 2. J. Gen. Virol. *85*, 2307–2314.

Warrener, P., Tamura, J.K., and Collett, M.S. (1993). RNA-stimulated NTPase activity associated with yellow fever virus NS3 protein expressed in bacteria. J. Virol. *67*, 989–996.

Wengler, G., and Castle, E. (1986). Analysis of structural properties which possibly are characteristic for the 3′-terminal sequence of the genome RNA of flaviviruses. J. Gen. Virol. *67*, 1183–1188.

Wengler, G., Castle, E., Leidner, U., Nowak, T., and Wengler, G. (1985). Sequence analysis of the membrane protein V3 of the flavivirus West Nile virus and of its gene. Virology *147*, 264–274.

Wengler, G., Czaya, G., Farber, P.M., and Hegemann, J.H. (1991). In vitro synthesis of West Nile virus proteins indicates that the amino-terminal segment of the NS3 protein contains the active centre of the protease which cleaves the viral polyprotein after multiple basic amino acids. J. Gen. Virol. *72*, 851–858.

Wengler, G., and Wengler, G. (1981). Terminal sequences of the genome and replicative-from RNA of the flavivirus West Nile virus: absence of poly(A) and possible role in RNA replication. Virology. *113*, 544–555.

Wengler, G., and Wengler, G. (1989). Cell-associated West Nile flavivirus is covered with E+pre-M protein heterodimers which are destroyed and reorganized by proteolytic cleavage during virus release. J. Virol. *63*, 2521–2526.

Wengler, G., and Wengler, G. (1991). The carboxy-terminal part of the NS 3 protein of the West Nile flavivirus can be isolated as a soluble protein after proteolytic cleavage and represents an RNA-stimulated NTPase. Virology *184*, 707–715.

Wengler, G., and Wengler, G. (1993). The NS 3 nonstructural protein of flaviviruses contains an RNA triphosphatase activity. Virology *197*, 265–273.

Wengler, G., Wengler, G., and Gross, H.J. (1978). Studies on virus-specific nucleic acids synthesized in vertebrate and mosquito cells infected with flaviviruses. Virology *89*, 423–437.

Wengler, G., Wengler, G., Nowak, T., and Castle, E. (1990). Description of a procedure which allows isolation of viral nonstructural proteins from BHK vertebrate cells infected with the West Nile flavivirus in a state which allows their direct chemical characterization. Virology *177*, 795–801.

Westaway, E.G. (1987). Flavivirus replication strategy. Adv. Virus Res. *33*, 45–90.

Westaway, E.G. Khromykh, A.A., and Mackenzie, J.M. (1999). Nascent flavivirus RNA colocalized in situ with double-stranded RNA in stable replication complexes. Virology *258*, 108–117.

Westaway, E.G. Khromykh, A.A., Kenney, M.T., Mackenzie, J.M., and Jones, M.K. (1997). Proteins C and NS4B of the flavivirus Kunjin translocate independently into the nucleus. Virology *234*, 31–41.

Westaway, E.G. Mackenzie, J.M., Kenney, M.T., Jones, M.K., and Khromykh, A.A. (1997). Ultrastructure of Kunjin virus-infected cells: colocalization of NS1 and NS3 with double-stranded RNA, and of NS2B with NS3, in virus-induced membrane structures. J. Virol. *71*, 6650–6661.

Westaway, E.G. Mackenzie, J.M., and Khromykh, A.A. (2003). Kunjin RNA replication and applications of Kunjin replicons. Adv. Virus Res. *59*, 99–140.

Winkler, G., Maxwell, S.E., Ruemmler, C., and Stollar, V. (1989). Newly synthesized dengue-2 virus nonstructural protein NS1 is a soluble protein but becomes partially hydrophobic and membrane-associated after dimerization. Virology *171*, 302–305.

Winkler, G., Randolph, V.B., Cleaves, G.R., Ryan, T.E., and Stollar, V. (1988). Evidence that the mature form of the flavivirus nonstructural protein NS1 is a dimer. Virology *162*, 187–196.

Williams, D.T., Wang, L.F., Daniels, P.W., and Mackenzie, J.S. (2000). Molecular characterization of the first Australian isolate of Japanese encephalitis virus, the FU strain. J. Gen. Virol. *81*, 2471–2480.

Wu, S., Lian, W., Hsu, L., Wu, Y., and Liau, M. (1998). Antigenic characterization of nine wild-type Taiwanese isolates of Japanese encephalitis virus as compared with two vaccine strains. Virus Res. 55, 83–91.

Wuryadi, S., and Suroso, T. (1989). Japanese encephalitis in Indonesia. Southeast Asian J. Trop. Med. Public Health *20*, 575–580.

Wu, K.P., Wu, C.W., Tsao, Y.P., Kuo, T.W., Lou, Y.C., Lin, C.W., Wu, S.C., and Cheng, J.W. (2003). Structural basis of a flavivirus recognized by its neutralizing antibody: solution structure of the domain III of the Japanese encephalitis virus envelope protein. J. Biol. Chem. *278*, 46007–46013.

Wu, S.C., and Lee, S.C. (2001). Complete nucleotide sequence and cell-line multiplication pattern of the attenuated variant CH2195LA of Japanese encephalitis virus. Virus Res. *73*, 91–102.

Wu, Y., Zhang, F., Ma, W., Song, J., Huang, Q., and Zhang, H. (2004). A plasmid encoding Japanese encephalitis virus prM and E proteins elicits protective immunity in suckling mice. Microbiol. Immunol. *48*, 585–590.

Xin, Y.Y., Ming, Z.G., Peng, G.Y., Jian, A., and Min, L.H. (1988). Safety of a live-attenuated Japanese encephalitis virus vaccine (SA14-14-2) for children. Am. J. Trop. Med. Hyg. *39*, 214–217.

Xu, G., Xu, X., Li, Z., He, Q., Wu, B., Sun, S., and Chen, H. (2004). Construction of recombinant pseudorabies virus expressing NS1 protein of Japanese encephalitis (SA14-14-2) virus and its safety and immunogenicity. Vaccine *22*, 1846–1853.

Yamshchikov, V.F., and Compans, R.W. (1993). Regulation of the late events in flavivirus protein processing and maturation. Virology *192*, 38–51.

Yamshchikov, V.F., and Compans, R.W. (1994). Processing of the intracellular form of the west Nile virus capsid protein by the viral NS2B-NS3 protease: an *in vitro* study. J. Virol. *68*, 5765–5771.

You, S., and Padmanabhan, R. (1999). A novel *in vitro* replication system for Dengue virus. Initiation of RNA synthesis at the 3′-end of exogenous viral RNA templates requires 5′- and 3′-terminal complementary sequence motifs of the viral RNA. J. Biol. Chem. *274*, 33714–33722.

You, S., Falgout, B., Markoff, L., and Padmanabhan, R. (2001). In vitro RNA synthesis from exogenous dengue viral RNA templates requires long range interactions between 5′- and 3′-terminal regions that influence RNA structure. J. Biol. Chem. *276*, 15581–15591.

Yun, S.I., Kim, S.Y., Choi, W.Y., Nam, J.H., Ju, Y.R., Park, K.Y., Cho, H.W., and Lee, Y.M. (2003a). Molecular characterization of the full-length genome of the Japanese encephalitis viral strain K87P39. Virus Res. *96*, 129–140.

Yun, S.I., Kim, S.Y., Rice, C.M., and Lee, Y.M. (2003b). Development and application of a reverse genetics system for Japanese encephalitis virus. J. Virol. *77*, 6450–6465.

Yusof, R., Clum, S., Wetzel, M., Murthy, H.M., and Padmanabhan, R. (2000). Purified NS2B/NS3 serine protease of dengue virus type 2 exhibits cofactor NS2B dependence for cleavage of substrates with dibasic amino acids *in vitro*. J. Biol. Chem. *275*, 9963–9969.

Zanotto, P.M., Gould, E.A., Gao, G.F., Harvey, P.H., and Holmes, E.C. (1996). Population dynamics of flaviviruses revealed by molecular phylogenies [see comment]. Proc. Natl. Acad. Sci. USA. *93*, 548–553.

Zeng, L., Falgout, B., and Markoff, L. (1998). Identification of specific nucleotide sequences within the conserved 3′- SL in the dengue type 2 virus genome required for replication. J. Virol. *72*, 7510–7522.

Zhang, F., Huang, Q., Ma, W., Jiang, S., Fan, Y., and Zhang, H. (2001). Amplification and cloning of the full-length genome of Japanese encephalitis virus by a novel long RT-PCR protocol in a cosmid vector. J. Virol. Methods *96*, 171–182.

Zhang, L., Mohan, P.M., and Padmanabhan, R. (1992). Processing and localization of Dengue virus type 2 polyprotein precursor NS3-NS4A-NS4B-NS5. J. Virol. *66*, 7549–7554.

Zhang, W., Chipman, P.R., Corver, J., Johnson, P.R., Zhang, Y., Mukhopadhyay, S., Baker, T.S., Strauss, J.H., Rossmann, M.G., and Kuhn, R.J. (2003a). Visualization of membrane protein domains by cryo-electron microscopy of dengue virus. Nat. Struct. Biol. *10*, 907–912.

Zhang, Y., Corver, J., Chipman, P.R., Zhang, W., Pletnev, S.V., Sedlak, D., Baker, T.S., Strauss, J.H., Kuhn, R.J., and Rossmann, M.G. (2003b). Structures of immature flavivirus particles. EMBO J. *22*, 2604–2613.

Zhao, Z., Wakita, T., and Yasui, K. (2003). Inoculation of plasmids encoding Japanese encephalitis virus PrM-E proteins with colloidal gold elicits a protective immune response in BALB/c mice. J. Virol. 77, 4248–4260. [Erratum in: J. Virol. (2003). 77, 9107.]

Evolution, Epidemiology, and Control of the Japanese Encephalitis Virus

9

Ma Shao-Ping and Makino Yoshihiro

Abstract

Research on the classification of the Japanese encephalitis (JE) virus in Japan has been related to the control of JE with vaccine. To develop a more suitable vaccine, epidemiological studies have been carried out in Japan. Until the mid-1980s, research to classify field isolates based on their antigenicity has been limited.

The molecular epidemiological methods (finger printing and phylogenetic analyses) have drastically revealed that (1) JE viruses are composed of five geographically related but distinct genotypes and (2) new genotypes have occasionally been introduced into new geographical areas.

Japan has successfully reduced the numbers of JE cases. The reduction of JE cases was mainly due to vaccination. Water control of rice fields also resulted in the reduction of the number of vector mosquitoes.

Lastly, we discuss the recent vaccine developments (cell line derived JE vaccine and some new developing vaccines) and the antigenicity of genotype 1 strains recently introduced into Japan.

Introduction

Japanese encephalitis (JE) is a serious acute viral encephalitis with high incidence of fatality; approximately half of the survivors suffer from grave sequelae accompanied by neuropsychiatric disorders. Its causative agent, the JE virus, is a member of the genus *Flavivirus* of the family of Flaviviridae, and is a typical mosquito-borne arbovirus according to its mode of transmission. Its principal vector is *Culex tritaeniorhynchus* and related species, and its major amplifier is the swine. Humans are relatively insensitive to the JE virus and are considered as dead-end hosts in the transmission cycle (Igarashi, 2002). The JE virus has a single strand positive RNA as a genome. The genome encodes a single open reading frame (ORF), three structural proteins and seven nonstructural (NS) proteins which are required for viral replication are coded on its ORF. In Japan annually 1000–5000 cases were reported in the 1960s epidemic era. Though annual swine surveys show that the JE virus still exists and NS1 antibody prevalence, which is a marker of natural infection, occurs in 10–20% of the residents of western Japan (Konishi *et al.*, 2002), less than 10 cases have been reported in the recent decade (Figure 9.1). Although the true factors are still unclear, the reason for this drastic reduction in reported case has been speculated to be due to three factors: immunization, vector control, and a change in the virulence of circulating virus (Takegami *et al.*, 2000). In this chapter, the viral factor

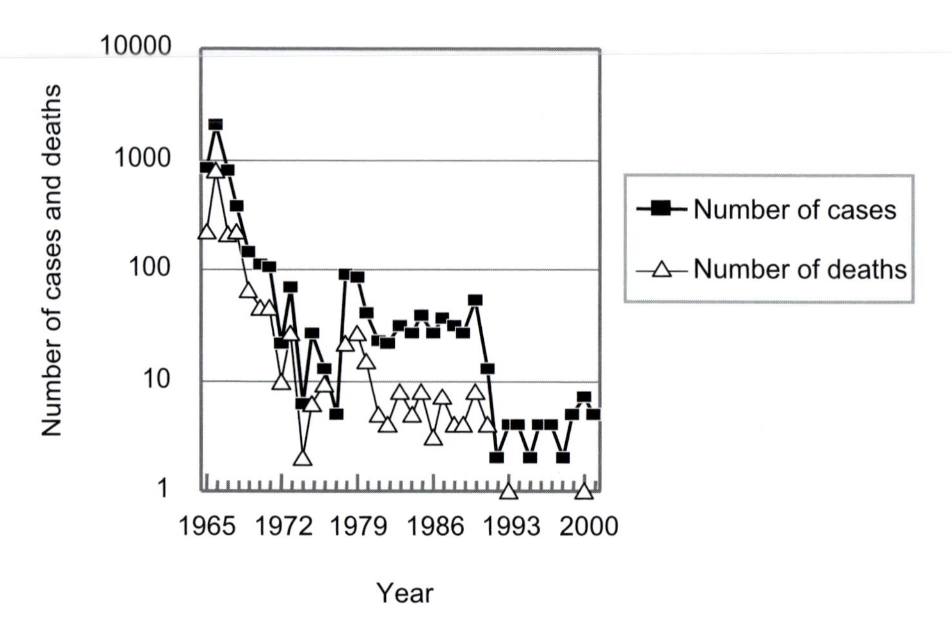

Figure 9.1 The reported JE cases and deaths in Japan between 1965 and 2001.

is approached by molecular epidemiological methods. The immunization and vector control factors are also described.

Molecular epidemiology of Japanese encephalitis virus

Epidemiological surveys before phylogenetic analysis

Immunotyping

The JE virus had been thought to have no antigenic variation since it was first isolated from human brain and cerebrospinal fluid in Tokyo in 1935 (Mitamura *et al.*, 1936). For example, an epidemic in Japan in 1951 was thought to be caused by a "variant" virus because (1) the epidemic occurred later than the average years and (2) many patients' sera did not react to known virus antigens by complement fixation (CF) test. But the JE isolates from the epidemic in 1951 showed similar reactivity to "known" JE strains by both the neutralization test and CF test. Thus, it was concluded that no "variant" viruses existed in the epidemic in 1951 (Kitaoka *et al.*, 1952).

The existence of antigenic variants of the JE virus was first reported by Hale and Lee. They analyzed six Malaya strains by the cross neutralization test and found that Muar and Tengha were different from that of Nakayama (Hale *et al.*, 1954).

In Japan, the earliest report of the difference among JE virus strains was made by Kobayashi. He characterized a newly isolated strain, G-1, which was isolated from human brain in 1949, and compared it with the Nakayama. He concluded that at least two immunotype viruses were in existence in Japan (Kobayashi, 1959). Okuno and others analyzed 22 Japanese strains and four Malaya strains isolated between 1935 and 1966 mainly by the antibody abortion hemagglutination-inhibition (HI) test. They demonstrated that two immunotypes could be distinguished.

One was the I-58 immunotype which was composed two strains, Nakayama and I-58. The other was the JaGAr01 immunotype (Okuno *et al.*, 1968). It is interesting that they considered the Muar strain to be a member of the JaGAr01 immunotype. Further studies demonstrated that the immuotype of "recent" field strains was different from that of Nakayama and JaGAr01 (Ozaki *et al.*, 1967: Oda *et al.*, 1976).

Advances in the monoclonal antibody technique led to more detailed immunotyping. Kimura-Kuroda and others obtained monoclonal antibodies directed against envelope glycoprotein V3 (E) of the JE virus (Kimura-Kuroda *et al.*, 1983) and revealed that the main antigenic structure of the three strains of the JE virus, which represents the above three immunotypes, is homologous (Kimura-Kuroda *et al.*, 1986). Kobayashi and others carried out the characterization of 27 strains of the JE virus, 24 isolated from various parts of Japan during 1935 and 1979 and three strains were from Southeast Asia, using five JE species-specific monoclonal antibodies. The 27 strains divided into three immunologically different groups: Group I contained only the Nakayama. Group II contained the strains isolated during 1935 to 1959 and indicated that at least two immunologically distinct strains existed in Japan in 1935. Group III-1 contained the strains frequently isolated in various areas of Japan since 1952. They concluded that at least three antigenically distinct strains of the JE virus had existed in Japan, and that the recently isolated strains were immunologically different from the Nakayama and JaGAr01 strains (Kobayashi *et al.*, 1984, 1985). The immunotype of the field strains of the 1980s in Japan was different from the Nakayama and JaGAr01 (Ochiai *et al.*, 1989: Tadano *et al.*, 1994).

Subsequently, immunotyping of JE virus strains isolated in Southeast Asia was carried out.

Yoshida produced monoclonal antibodies against the Muar and 691004 strains, which showed different reactivities from the three groups reported previously. He analyzed 25 JE virus strains, 21 from Japan and four from Southeast Asia, and these strains were classified into four serotypes: Nakayama, Beijing 1, Kamiyama, and Muar. The 691004 strain was classified into a subtype of the Nakayama serotypes (Yoshida, 1994a). He also analyzed 22 JE virus strains, which were isolated from Taiwan, Singapore, Thailand and India between 1935 and 1984, and revealed that the currently prevalent JE virus strains in those areas belonged to the Kamiyama serotypes and that one Thailand strain isolated in 1983 (KE093) showed an outstanding difference from the other stains. This result revealed the immunological diversity of the JE virus (Yoshida, 1994b). These results were also confirmed by Hasegawa and others (Hasegawa *et al.*, 1995).

Ali and others characterized three JE virus Thailand strains isolated in 1992 and 1993 (ThCMAr4492, ThCMAr10592, and ThCMAr6793) by the HI test and the neutralization test. The results were that these isolates could not be assigned to either the Nakayama and JaGAr01 subtype. They analyzed the antigenic relationships among 12 Thai strains belonging to genotype 1 with hyperimmune antisera by neutralization tests. The result was that there are three subtypes: (1) the majority of Thailand strains belonged ThCMAr6793 subtype, (2) ThCMAr4492 was antigenically different from all other strains, and (3) two Thai strains (Sunbin and KE-093) were significantly less neutralized by all the antisera tested. However, the relation between the subtype ThCMAr6793 and the Kamiyama serotype was still unclear (Ali *et al.*, 1995).

Oligonucleotide fingerprint

The early molecular epidemiological analyses of the JE virus based on the viral genome were performed by RNase-T1 resistant oligonucleotide fingerprinting.

Hori and others analyzed 12 JE strains: seven were isolated from Japan and five from Thailand. All Japanese strains were relatively similar to each other and the similarity was generally greater depending on the year of isolation. Recent strains in Thailand were very similar. Those recent Thailand strains generally showed a low similarity to recent Japanese strains isolated after 1982. From these results they concluded that mutations and selections of the JE virus genome would have progressed independently in these geographically distant areas (Hori *et al.*, 1986).

Banerjee and Ranadive analyzed 14 JE viruses; 2 were isolated from Japan, 1 from China, 1 from Sri Lanka and 10 from India. The Nakayama strain bore a greater similarity to one West Bengal strain whereas the other West Bengal strain showed a marked difference. The geographical proximity of the place of isolation did not contribute much to the similarity of the fingerprint of the strains, nor did temporal proximity. They discussed that the mutations and selections of virus strains are indeed taking place rapidly at different locations (Banerjee *et al.*, 1989).

Phylogenetic analyses and genotyping of JE viruses

Establishment of JE virus genotypes based on the core premembrane gene

In 1987 Sumiyoshi and others first determined the complete nucleotide sequence of JE virus strain JaOArS982, isolated in Osaka, Japan in 1982 (Sumiyoshi *et al.*, 1987). It has been possible to compare the nucleotide sequences among JE virus strains since then. In 1990, Chen and others analyzed 46 JE viruses from a variety of geographic areas in Asia representing a 52-year time span by primer extension sequencing based on 240 nucleotides (240-nt) of the core premembrane (prM) junction (C/prM) region. Using 12% divergence as a cutoff point, the 46 isolates fell into three distinct genotypes. One genotype consisted of JE virus isolates from northern Thailand and Cambodia. A second genotype was composed of those from southern Thailand, Malaysia, Sarawak and Indonesia. The remainder of the isolates was from Japan, China, Taiwan, the Philippines, Sri Lanka, India and Nepal, and made up the third group. They confirmed that JE virus isolates from the same geographic region and time period were very similar, but that genetic variation occurs among strains from diverse regions or from different time periods in the same region. This result confirmed that the JE virus has been continuously evolving in nature. They speculated the possibility of the regional difference of JE virus virulence based on isolates from endemic areas where reported cases were relatively low; southern Thailand, Malaysia, Sarawak and Indonesia, formed one genotype. They also speculated that JE viruses are maintained locally in temperate countries, because JE virus strains from different years in the same country are genetically similar to each other. Because the authors did not designate each genotype in the original report, there are two nomenclatures: one is the nomenclature proposed by Monath and Heintz. They proposed a nomenclature as follows: genotype I consists of virus strains from a broad region encompassing northern Asia

(Japan, China, Taiwan), Vietnam, Nepal, India, and Sri Lanka; genotype II consists of those from northern Thailand and Cambodia; the third genotype contains the strains from Indonesia (Sarawak and Java), Malaysia, and southern Thailand; the fourth genotype is composed of the strains from Indonesia (Java, Bali, and Flores) (Monath *et al.*, 1996). The other is the nomenclature after the original order which Chen and others described, i.e. genotype 1 (G1), genotype 2 (G2), and genotype 3 (G3) (Chen *et al.*, 1990, 1992; Ali *et al.*, 1997). In this chapter we take the nomenclature proposed by Chen and Ali.

In 1992 Chen and others also analyzed 240-nt of the C/PrM junction region of 19 JE virus isolates representing various geographic regions of Asia and a 50-year time span; 12 of the JE strains from the Indonesian Archipelago and the Philippines had not been previously examined; the remainders were the representatives of the three genotypes reported previously and Murray Valley encephalitis, West Nile, and Kunjin virus. Using 12% divergence as a cut-off point, the 19 JE strains fell into the four distinct genotypic groups. The newly recognized, fourth genotype (G4), was composed of five Indonesian strains. It is generally accepted that the eastern limit of JE virus activity in tropical Asia corresponds roughly to the ecological boundary defined by Wallace's Line, a hypothetical demarcation that separates the Oriental zoogeographic zone from that of Australia and New Guinea. East and south of Wallace's Line is the boundary of Malay valley encephalitis virus activity. One JE virus was isolated from a mosquito pool in Flores. Flores is east of Wallace's Line, suggesting that JE virus was circulating in a broader geographic region than recognized earlier in the 1990. It was speculated that the JE virus could extend into the new region, when susceptible hosts are available, but genotype displacement seemed not to occur as commonly as with the JE virus (Chen *et al.*, 1992).

This speculation was confirmed by analyzing Vietnamese strains. Huong and others analyzed 285-nt of the C/prM region of 10 JE virus strains from northern Vietnam isolated between 1964 and 1976 and 6 JE virus strains from southern Vietnam isolated during 1978 and 1988. A low temporal evolution was observed between strains isolated in northern Vietnam from 1964 to 1976 and the southern Vietnamese JE virus strains, which showed a closer relationship with the earlier isolated Nakayama and Beijing-1 strains. It was also reported that despite the relative geographic proximity of Thailand and Cambodia, no new genotype has been introduced in Vietnam (Huong *et al.*, 1993). In 1996, Chung and others analyzed five Korean strains of the JE virus between 1982 and 1994. The 198-nt sequence in the C/prM gene region of the five Korean strains indicated that they were classified into G3. Four of five Korean isolates formed a unique phylogenetic tree within the G3, although the last one (K93P05) was genetically highly related to the Nakayama-NIH strain. The result was that the 251-nt of the E gene of the 5 Korean strains was more divergent than the C/prM region. These data suggested that at least two subgenotypes existed in Korea and that one of them was unique to the Korean strains. The result of immunotyping of the Korean strains showed that at least two antigenically different JE virus strains existed in Korea (Chung *et al.*, 1996). Ma and others analyzed 23 JE virus strains isolated in Okinawa, Japan, between 1968 and 1992. The phylogenetic analyses were carried out on both the 240-nt of the C/prM gene region and the 111-nt of the E gene region. The result of the phylogenetic analysis was that all the Okinawan isolates showed more than 94% homology in the nucleotide sequence in the each region and that they were chronologically divided into two groups: one group contained nine strains before 1979 and the other group contained 14 strains after 1985. In comparison with the reference strains of mainland Japan, the Okinawan strains belonged to the same genotype (G3) as that of mainland Japan. Although the island of Okinawa is surrounded by the sea and is isolated from mainland Japan, an Okinawan strain isolated in 1992 showed a close genetic relation to the strain isolated in 1992 in Osaka, 1000 km from Okinawa. An Okinawan strain isolated in 1985 also showed a close genetic relation to the strain in Hokkaido, which is the northernmost region of Japan, isolated in 1985 (Ma *et al.*, 1996). Jan and others analyzed the 280-nt of C/prM region of 44 JE virus strains from mosquitoes from 1983 and 1994 in Taiwan, and three strains isolated from human brain. The result was that all isolates fell into G3. A high nucleotide homology was observed among isolates from different regions of Taiwan and different time periods, whereas high variation existed among isolates from the same region and time cluster. Phylogenetic analysis showed that 47 Taiwan isolates fell into three clusters: clusters 1, 2, and 3. The majority of isolates from northern and central Taiwan belonged to cluster 1, whereas most isolates from southern Taiwan belonged to cluster 2. They concluded that isolates of cluster 1 were more specific to Taiwan than isolates of cluster 2 and cluster 3, compared with other Asian JE virus strains. They concluded that Taiwan JE virus strains progressively changed from cluster 1 to cluster 2 (Jan *et al.*, 2000). Until the 1990s, JE virus G3 strains had been dominant in the east Asian countries.

Phylogenetic analysis based on the envelope gene region

The envelope (E) protein is believed to play key roles in a number of important processes of receptor binding, and is the principal target for neutralizing antibodies (Rice, 1996). The E gene region was also chosen for molecular epidemiological analyses. Hasegawa and others, who had carried out the immunotyping of JE viruses using monoclonal antibodies, determined the structural gene region of five JE viruses; Nakayama, Beijing-1, Kamiyama, 691004 and Muar, which represent their five antigenic groups. They observed that the Muar strain was the most structurally different from the other four strains, and concluded that it would be possible to compare the antigenic differences among the JE virus strains using monoclonal antibodies with different specificities (Hasegawa *et al.*, 1994). Ni and others analyzed 13 JE virus strains isolated from different Asian countries; Vietnam, Nepal, Indonesia, Thailand, India, Japan, and China. A dendrogram was constructed based on the amino acid sequences of the E protein of 13 strains. Strain variation could not be related to geographical location on the basis of the wild-JE virus strains studied. They speculated that the most probable explanation for the strain variation of the JE virus was the variation in particular amino acid in the structural proteins (Ni *et al.*, 1995).

Paranjpe and others determined four E gene sequences of JE virus India strains and compared these with the published E gene sequences of 16 JE strains. Phylogenetic analyses of the E gene sequences by neighbor joining, maximum parsimony, and maximum likelihood methods identified four clusters. Viral groupings did not correspond to geographic origin, isolation host or virulence (Paranjpe *et al.*, 1996). Mangada and Takegami determined the full genomic sequence of the JaGAr01 strain, which represented one immunotype. Phylogenetic analyses were also carried out including the 34 published JE virus E genes both for the nucleotides and the deduced amino acids of the E gene region. The result showed no temporal, geographical or virulence correlation which agreed with the result of Paranjpe and others. The phylogenetic analyses also revealed that the Beijing-1 (a member of the JaGAr01 immunotype) and Nakayama grouping were different from JaGAr01 (Mangada *et al.*, 1999). This may reflect no direct relation between immunotyping and genotyping. Tsarev and others analyzed for 92 complete E sequences of JE isolates available in the GenBank. Two distant genetic groups were identified; genotype I, which included only the Muar strain, and genotype II, which include the remainder. These groupings were identified both on an amino acid and nucleotide level. They stated that no association between genotypes and geographical, temporal, or source (human, mosquito) factors were found. According to the minimum amino acid differences between poliovirus serotypes (18%), they concluded that all known JE virus isolates comprised a single serotype (Tsarev *et al.*, 2000). On the other hand, Williams and others analyzed 64 E gene sequences from temporally and geographically diverse JE virus strains. For comparison with the E gene phylogeny, phylogenetic analysis using cognate prM gene sequences was also carried out. The E gene analysis corresponded well with the prM gene analysis and with previous genotyping studies using the prM gene (Williams *et al.*, 2000). The cluster which corresponded to the C/prM based G3 could be further delineated into four subgroups. These subgroups did not indicate a clear differentiation among the strains based on geographic or temporal boundaries (Holbrook *et al.*, 2002). The correspondence of phylogenetic analysis based on the C/prM region and the E region was confirmed by the analysis of an Indian strain. Uchil and others analyzed 107 strains of the E sequence available from different geographic locations worldwide. The phylogenetic tree classified the JE virus strains into five distinct genotypic groups differing by a minimum nucleotide divergence of 7% with high bootstrap support values. The result suggested that the JE virus was introduced into India twice on two occasions separated by at least 17 years. They stated that except for the newly found fifth genotype, which includes only the Muar strain, all four of the groups broadly matched the four genotypes that were proposed earlier using the prM gene region (Uchil, *et al.*, 2001).

Expansion of JE virus distribution

The discrepancy between the genotype and its geographic distribution was first reported based on an analysis of JE virus strains isolated in Malaysia. Vythilingam and others sequenced the 240-nt of C/prM region of a Malaysia JE strain; MaSAr39692; isolated in Peninsula Malaysia in 1992 and found a close homology to the strains from epidemic region areas (Vythilingam *et al.*, 1994). Tsuchie and others analyzed 10 JE virus strains isolated in Malaysia in 1992 (Tsuchie *et al.*, 1994). The result was that 10 strains belonged to the largest genotype (G3) and differed in 32 nucleotides (13.3%) from the WTP/70/22 strain isolated in from Malaysia in 1970 (G2). In 1997, Tsuchie and others analyzed 12 JE strains isolated from three different regions of Malaysia from 1993 and 1994. The four JE virus strains isolated from Sarawak in 1994 and four strains isolated from Sepang, Selangor in 1993 were classified into G2. The four JE virus strains from

Ipoh, Perak in 1994 were classified into G1. Taken together with the earlier report on 10 strains from Selanger in 1992 (Tsuchie *et al.*, 1994), they concluded that at least three genotypes of the JE virus have been circulating in Malaysia. They speculated that the strains classified into G1 and G3 might have been introduced into Malaysia recently by migrating birds or by the international transportation system because in tropical countries, susceptible animals and mosquitoes are available in all seasons, and the JE virus can be expanded into new regions (Tsuchie *et al.*, 1997).

It was believed that the east limit of distribution of the JE virus in the south was somewhere in the Indonesian Archipelago, because the closely related Murray Valley encephalitis virus has been isolated on the island of New Guinea, and it is unlikely that two viruses occupying a similar ecologic niche would coexist there (Rosen, 1986). Over a two-week period in March–April 1995, three cases of JE occurred among residents of the island of Badu in the Torres Strait, Australia. The Torres Strait outbreaks were the first time that it had been recognized in Australia. Two strains (FU and NO) were isolated from the sera of the Badu residents. A detailed sequence comparison of the "prM region" (amplified 381-nt of the 440–821 region, but no detail information about the sequencing region was available) showed that the Badu strains were most closely related to other viruses of G2. They suggested that migratory birds and/or wind-blown mosquitoes could have imported the virus into the Torres Strait from a focus of viral activity, possibly in Papua New Guinea (Hanna *et al.*, 1996). To confirm that the JE virus strains isolated in the Torres Strait in 1995 were of Papua New Guinean (PNG) origin, almost 400 000 adult mosquitoes from 26 locations in the Western Province of PNG between February 1996 and February 1998 were processed for viral isolation. Two JE virus strains were recovered. Nucleic acid sequences of the PNG JE strains were determined in the 219-nt of the prM region and the 424-nt of the fifth nonstructural protein and the 3′ untranslated region (NS5–3′ NCR). The PNG strains were most closely related to G2 of Malaysia and Indonesia. The high level of nucleotide sequence similarity between the PNG and Torres Strait strains suggested a common source. These evidences provided solid evidence that PNG is the source of incursions of the JE virus into Australia. The mechanism by which the JE virus is being introduced into Australia was not defined (Johansen *et al.*, 2000). In mid-January 2000, the reappearance of JE virus activity in the Australasian region was first demonstrated by the isolation of the JE virus from three sentinel pigs on Badu Island in the Torres Strait. Further evidence of JE virus activity was revealed through the isolation of the JE virus from *Culex gelidus* mosquitoes collected on Badu Island and the detection of specific JE virus neutralizing antibodies in three pigs on Moa Island. Nucleotide sequencing and phylogenetic analyses were performed using both the prM gene and 432-nt of the E gene regions. Both trees were virtually identical in topology with four distinct phylogenetic groupings that corresponded to four genotypes. With the exception of two 2000 Torres Strait strains, all Australasian strains clustered into G2. The 2000 Torres Strait strains isolated from pig (TS00) and from mosquito (TS4152) were separately clustered into G1. All previous Australasian JE virus strains belong to genotype 2. Therefore, for the first time, the appearance and transmission of a second genotype of the JE virus in the Australasian region has been demonstrated. They reconfirmed the hypothesis that the high degree of sequence identity between Australian and PNG isolates support the hypothesis that the New Guinea landmass is the source of the JE virus. The emergence of G2 strains in the Australasian region was conceivable because this genotype contains strains from the adjacent geographic region of Indonesia. But the relatively low sequence identity of the two Australian G1 strains with PNG and Australian G2 strains suggested a separate source. G1 strains have not been reported in PNG or neighboring Indonesia, so a potential source of the Australian G1 strains may therefore be Thailand or Malaysia (Pyke *et al*, 2001).

Based on the interest of evolution, not from the epidemiological interest of the JE virus, the origin and evolution of the JE virus are proposed. The JE virus may have diverged from an ancestral African virus, which died out in Asia as recently as a few centuries ago (Gould, 1997; 2002). The JE virus originated from its ancestral virus in the Indonesia-Malaysia region and evolved there into different genotypes which then spread across Asia (Solomon *et al.*, 2003).

As described above, the genotype of the JE virus in northeast Asia has been considered to belong to G3. During September and December in 1994, a ecological survey was carried out in Ishikawa prefecture, Japan. Virus isolation was performed in the survey and a JE virus strain, Ishikawa, was isolated from swine mononuclear cells. The nucleotide sequence of the prM gene of Ishikawa strain was remarkably different from those of strains isolated previously in Japan, and showed a close relation to the Thailand strains (Miyamoto *et al.*, 1997). Takegami and others reported that the nucleotide sequences of two JE virus strains isolated from mosquito in Ishikawa prefecture in 1998, were similar to Japanese isolates belonging to G3 and that the nucleotide sequence of JE virus-specific amplicon from mosquitoes in 1999 showed a similarity to the Ishikawa strain (i.e. belonged to G1). They speculated that the change in the distribution of JE virus strains in Japan might have occurred in around 1993–1994, but the origin of the Ishikawa strain remained unclear (Takegami *et al.*, 2000). Nam and others determined the full-length genomic sequence of the JE virus K94P05 strain, which was isolated from mosquitoes in Wando, Korea. The phylogenetic topology constructed from full-length genome nucleotide sequences showed that K94P05 belonged in a clearly different genetic group. They speculated that the Korean JE virus strain may have spread to Korea from another Asian country many years ago and may have varied independently from the other Asian JE virus strains (Nam *et al.*, 2001). Ma and others analyzed the 240-nt C/prM gene region of 23 JE virus strains isolated in Tokyo and Oita, Japan. The strains included 14 from Tokyo and nine from Oita, isolated mainly from swine sera between 1965 and 2001. All the strains were clustered into two distinct genotypes; G1 and G3. Except for one strain isolated in 1991 (JaOArK36–91, isolated from mosquitoes in Osaka, Japan in 1991), all the strains isolated before 1991 belonged to G3, whereas those isolated after 1994 and JaOArK36–91 belonged to G1. To evaluate the genetic relation among the G1 strains, eleven Japanese G1 strains (nine from both Tokyo and Oita, Chiba, near Tokyo isolate 10–173, and Ishikawa) were compared with the published sequence data of G1 strains including four G1 strains from Thailand, Malaysia, and Korea in the 1990s. All 11 Japanese G1 strains, the Korean strain, and the Malaysian strain showed a close genetic relation with the Thailand strains, while the G1 strains isolated in Australia were found to belong to those isolated in Thailand before 1985. The result clearly showed that G3 was the major genotype of the JE virus, circulating in Tokyo and Oita, Japan until 1991, but it has since changed to G1. The origin of the G1 strains isolated in Japan is unknown. Taken together with the Ishikawa strain, the swift change of the genotype during the mid-1990s appears to be a common phenomenon in a wide area of Japan. From this result it was speculated that the origin of Australian G1 strains and north Asian strains might be different (Ma *et al.*, 2003). Nga, Parquet and others analyzed the evolutionary relatedness of 16 JE virus strains (nine from northern Vietnam and seven from Nagasaki prefecture, Japan) to previously published JE virus strains using E gene sequence data. Vietnamese and Japanese strains isolated between 1986 and 1990 were found to cluster into G3. However, more recent Vietnamese and Japanese strains isolated between 1995 and 2002 were grouped within G1, now a dominant though previously unreported genotype in Vietnam. In addition, strains isolated between 1995 and 2002 were more closely related to those isolated in the 1990s than to the older G1 strains. How and why JE virus G1 expanded its area of distribution is unknown. A contribution of wind-blown mosquitoes to the dispersal of JE virus G1 would be expected in the area composed of East Coastal China-Korea-Japan. They speculated that it was possible that the JE virus G1 strains acquired increased infectivity for birds. In the case of the JE virus, migrating birds are thought to be important in its dispersion to new geographical areas. They speculated that the spread by viremic migratory birds may provide a sound explanation to the widening distribution of JE virus genotypes which was observed synchronously in Korea, Japan, and Vietnam (Nga and Parquet *et al.*, 2004). Recently, 77 JE strains from various part of Japan between 1935 and 2002, most of which are field strains mainly isolated after 1965, were analyzed and a phylogenetic tree based on 196-nt of the C/prM region showed that these 77 strains were classified into at least six chronological clusters. These clusters were arbitrarily designated cluster A through GI (lacks cluster F). Figure 9.2 shows the chronological shift of clusters to which circulating field strains belonged in Japan. Cluster A contained the Nakayama strain and was thought to represent strains from the 1930s in Japan. Cluster B contained JaGAr01 and had the strains from the 1950s. Cluster C contained strains from the 1960s and late the 1980s. Cluster D contained a few strains from the 1980s. Cluster E contained most strains from the 1980s. This cluster may be further divided into subcluster E1 and E2. Cluster GI contains the Japanese G1 strains after the

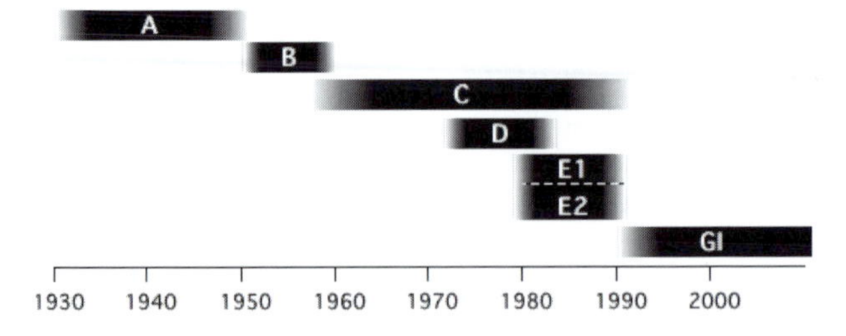

Figure 9.2 Chronological order of the putative dominant clusters of JE virus, A-E and GI, from Japan between 1935 and 2002.

1990s. The dominant strains were unique in Japan in the 1960s, and three clusters coexisted in Japan in 1980. Although G3 stains have shifted to G1 strains in Japan and Korea since the early 1990s, G3 strains are still dominant in Taiwan. From this result it was speculated that Okinawa Island and Taiwan belonged to genetically and ecologically separate groups, though both islands have a geographical proximity and are located under the pathway of migration (Nga and Parquet *et al.*, 2004). They speculated that these clusters of the JE virus have been introduced into Japan from the west of Japan.

Recently, the sequences of 11 Chinese JE virus strains were deposited in the DDBJ. Figure 9.3 shows the 99 JE strains isolated from Japan, Korea, China, Taiwan, Vietnam, and Malaysia. Five clusters were also confirmed. Additionally, a new cluster appeared that contained one Vietnamese strain isolated in 1976 and Chinese G3 strains isolated in 2002 appeared. This cluster was arbitrarily designated cluster F. The strains belonging to cluster E were found only in Japan and Korea in our previous analysis. Two Vietnamese strains (HT-59 and BT-CT-548, isolated in 1964 and 1976, respectively) seemed to share the ancestor of the cluster E strains. It is noteworthy that a Malaysian G3 strain MakAr32292 and a Taiwanese strain CH2195 showed a very close relationship. It is also noteworthy that a Vietnamese strain BT-CT-549 was located among the Okinawan strains. These findings might support the speculation that these strains belong to a cluster widely distributed in east Asia. Table 9.1 shows the distribution of seven clusters in east Asian countries. A plus sign in the column means that the cluster was observed, and an empty column means that the cluster was not observed in that country in the period. The number in each parenthesis is the nomenclature of Taiwanese isolates reported of Japan and others (Jan *et al.*, 2000). Our analysis does not support that "cluster 1 is more specific to Taiwan." And the coexistence of cluster A, B, and C in Taiwan may explain the observation that "high nucleotide homology was observed among isolates from different regions of Taiwan and different time periods. On the other hand, a high variation existed among isolates from the same region and time period" without any discrepant (Jan *et al.*, 2000). These results may propose that the ecology of JE in Taiwan is specific to Taiwan.

We previously speculated that the strains introduced into Japan came from the west of Japan (Ma *et al.*, in submission). Though the sequence information of the recent Chinese field strains is limited, the genetic relation between Vietnamese strains and cluster E strains shown in Figure 9.2 may support the speculation that the JE viruses came from Southeast Asia proposed by Nga and Parquet and others (Nga and Parquet *et al.*, 2004).

Control of Japanese encephalitis

Control of JE with vaccine immunization

JE vaccine currently used in Japan

The JE virus is maintained in the cycles of the amplifier swine and vector mosquitoes. Humans are the terminal host. Swine vaccination is one of the effective ways to control JE. For example, a mass vaccination for swine was carried out in Kumamoto prefecture, one of the relatively endemic locations in Kyushu, Japan 1985. In the Kumamoto prefecture, except for 1982 and 1983, 10–25 human JE cases were reported each year between

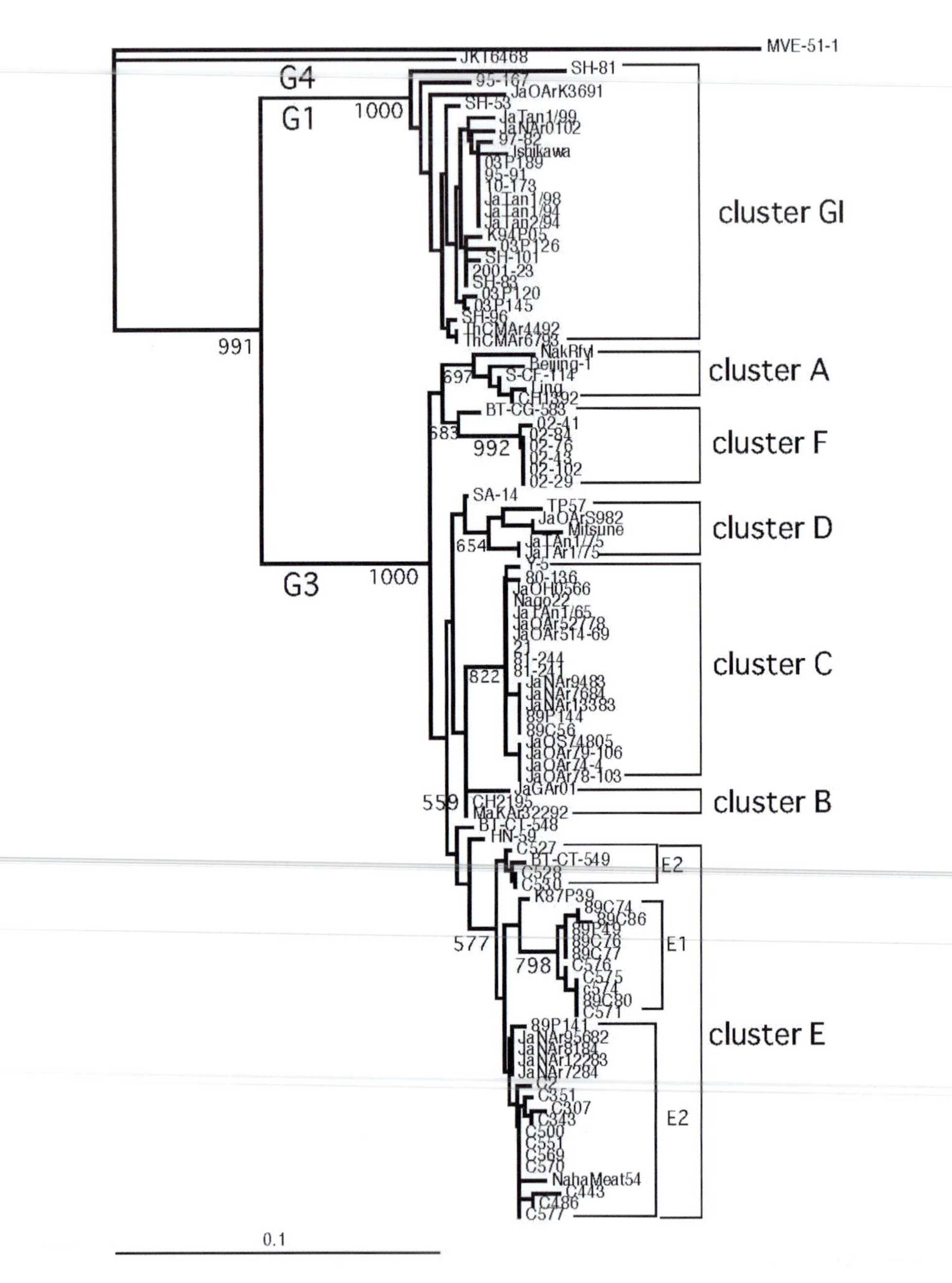

Figure 9.3 Phylogenetic relationship of the 99 JE virus strains predicted from capsid/premembrane protein gene sequences. Phylogenetic groupings corresponding to the genotyping classification of Chen and Ali and Igarashi are indicated (G1 and G3). The tree was constructed by the neighbor-joining method in ClustalW. The scale indicates the number of nucleotide substitutions per site. The horizontal branch lengths are proportional to genetic distance, and the vertical branch lengths have no significance. The numbers at each branch represent the percentage of 1000 bootstrap replicate supports. The tree was rooted using the sequence information of the Murray Valley encephalitis (MVE) virus, a member of the JEV serocomplex.

Table 9.1 Distribution of clusters A-F and GI in Japan, Korea, China, Taiwan, Vietnam, and Malaysia.

Country	Period	Cluster						
		G3						G1
		A	B	C	D	E	F	GI
Japan	1935–2002	+	+	+	+	+		+
Korea	1982–1994	+				+		+
China	2002					+	+	
Taiwan	1958–1994	+(3)	+(2)		+(1)			
Vietnam	1964–1988	+				+	+	
Malaysia	1992–1994	+					+	

1987 and 1984. Almost 25 000 pigs were inoculated with m strain derived live attenuate vaccine in Kumamoto city and 10 vicinity towns. The result appeared to be significant, because the reported human cases were reduced from 10 in 1984 to one in 1985 in Kumamoto city. But because of the following reasons, swine immunization has not generally been used: (1) the high turn-over rate (usually 6 months) of swine requires immunization of large number of newborn swine each year, and (2) the period for effective swine immunization using live-attenuated JE vaccine is limited by the maternal antibodies (Igarashi, 2002). So, human vaccination is considered to be the only certain way to prevent human JE cases in Japan.

Crude, formalinized mouse brain vaccines, were tested in humans in Japan in the later 1930s (Mitamura *et al.*, 1940) and were licensed in 1954. Partial purification with protamine sulfate was reported in 1958 and became a routine commercial procedure in 1962. Since a large double-blind, placebo controlled trial of protamine-sulfate purified vaccine was carried out in Taiwan and demonstrated an efficacy of 81 percent in 1965 (Hsu, 1971), the vaccine has been routinely administrated to millions of Japanese schoolchildren, without postvaccinal neurological accidents. This study also showed that two doses of the vaccine were more effective than a single dose in preventing human JE. The protamine sulfate-precipitate vaccine was further purified by ultracentrifugation in 1965, but the efficacy of this highly purified vaccine was not evaluated until 1988 (Monath, 1988). Another placebo-controlled, blinded, randomized trial in a northern Thai province, with two doses of monovalent (Nakayama strain) and bivalent (Nakayama and Beijing strains) inactivated JE vaccine derived from mouse brain in order to cover different strains of JE virus that might be circulating in the study area. Between November 1984 and March 1985, 65 224 children received two doses of JE vaccine (including 21 516 children who received a tetanus toxoid placebo) with only minor side effects. The efficacy of both vaccine groups combined was 91 percent. They concluded that two doses of inactivated JE vaccine, either monovalent or bivalent, could protect against JE (Hoke *et al.*, 1988). Inactivated mouse brain-derived Nakayama strain vaccine is produced in India, Thailand, Vietnam, Taiwan, and Korea with the technical assistance from Japan (Tsai, 2000).

In Japan, the vaccine strain was changed from the Nakayama strain to the Beijing-1 strain, which belongs to the JaGAr01 immunotype in 1989. Beijing-1 strain was considered a wide immunological spectrum and immunogenicity (Kitano *et al.*, 1986). Following two inoculations of the Beijing-1 vaccine with a 7- to 10-day interval, the seroconversion rate of the neutralizing antibodies was 97.6% against the homologous Beijing-1 strain with a geometric mean neutralizing antibody titer (GMT) of 2.58, whereas the seroconversion rate against the heterologous Nakayama strain was 89.5% with a GMT of 2.40. These values were significantly elevated following a third immunization given 4–6 weeks after the second immunization, which elicited a 100% seroconversion rate with a GMT of 3.46 against the homologous Beijing-1 strain, and a 97.2% seroconversion rate with a GMT of 2.56 against the heterologous Nakayama strain (Oya *et al.*, 1987). A field study to compare the immune response in children aged 1–6 years to the Nakayama and Beijing strains JE vaccine was carried out in northwest Thailand, where there was a low incidence of JE virus infection. Ninety-nine children received the Nakayama vaccine and 100 children received the Beijing-1 vaccine. The seroconversion rate after two doses dropped from 94.38% immediately after the second dose to 77.5% in the Nakayama, 93.98% to 87.5% in the Beijing-1 vaccine. After the third dose of vaccine, the seroconversion rate rose to 100% in both groups. The GMT in the Beijing vaccine was slightly higher than the Nakayama. Therefore, they concluded that either the Nakayama or the Beijing vaccine could be used in Thailand (Nimmannitya *et al.*, 1995). The standard vaccination schedule in Japan is two doses within a 1-week interval, followed by a booster after a year (Tsai, 2000).

Though the current mouse derived JE vaccine is highly purified and its safety has been established, some theoretical risks remain, i.e. adventitious infectious agents and traces of impurities derived from mouse brains may lead to adverse neurological events, such as acute disseminated encephalomyelitis (Sugawara *et al.*, 2002). Formalin inactivated vaccine prepared in PHK cells infected with the P3 strain has been widely used in China since the late 1960s. The vaccine was not purified; it was stabilized with 0.1% human serum albumin and presented as a liquid formulation. The seroconversion rate of the inactivated PHK P3 strain vaccine appeared as low as 60–70% and had relatively low antibody titers (Monath, 2002).

Today, two vaccine manufactures succeeded in producing the Vero cell-derived Beijing-1 vaccine in Japan. The Research Foundation for Microbial Disease (Biken) has developed a Vero cell-derived vaccine. Residual DNA was less than 1 ng/ml. The vaccine was more potent than the current mouse brain vaccine, being 10-fold higher than the current vaccine at equivalent vaccine doses of 0.3 μg. The vaccine underwent extensive toxicological evaluation, prior to entering clinical trials in 2001 (Monath, 2002). Another manufacturer: The Chemo-Sero Therapeutic Research Institute (Kaketsuken), Kumamoto, Japan, has also developed a similar Beijing-1 vaccine in Vero cells (Monath, 2002). Sugawara and others reported that to circumvent the possible risks associated with the current vaccine, a production system for JE vaccine derived from Vero cells was established. Except for an additional purification step, the downstream purification processes were similar to those used for the current mouse brain derived vaccine. The Vero cell-derived vaccine has a similar immunological potency, a broad spectrum of immunity, and immunogenicity to those of the current mouse derived vaccine (Sugawara *et al.*, 2002). Subsequently, the safety and effectiveness of Vero cell-derived inactivated JE vaccine were compared with those of a current JE vaccine in nonclinical studies and phase I clinical trials. The seroconversion rate after three doses of the Vero cell-derived vaccine and the current vaccine was 100% and the GMTs were 2.35 and 2.03, respectively. These data suggest that the safety and effectiveness of the Vero cell-derived vaccine are equal to those of the current mouse derived vaccine, and that the Vero cell-derived vaccine could be a useful second-generation JE vaccine (Kuzuhara *et al.*, 2003).

Another JE vaccine currently in use

In developing countries, wide-scale immunization with purified mouse-brain vaccines is not likely because of its high cost (mouse-brain derived vaccine, Biken vaccine costs approximately $2.30 per dose). Less expensive inactivated and live vaccines produced in primary hamster kidney (PHK) cells have been used successfully in China (Monath, 1988). The SA-14-14-2 strain has been attenuated through an empirical process of serial passage principally in PHK cells and had fine balance of safety and immunogenicity. SA-14-14-2 was stably neuroattenuated and was immunogenic in 85–100% of vaccinated children (Tsai, 2000). Before the SA-14-14-2 strain was licensed in 1988, a field trial was carried out in a low JE epidemic area in 1985. A total of 1026 children received a SA-14-14-2 vaccine. Seroconversion rates were 100% (GMT 35.3, n=11), 100% (GMT 31.7, n=12), and 83.3% (GMT 23, n=10) in groups receiving vaccine diluted 1:3, 1:5, and 1:50, respectively. The results showed that the SA-14-14-2 live attenuated vaccine was immunogenic and safe (Yu *et al.*, 1988). A matched case-control study to measure the efficacy of a single dose of SA-14-14-2 vaccine was carried out in Nepal in 1999 and a total of 160 000 doses of the SA-14-14-2 vaccine were given to children aged 1–15 years. The efficacy of a single dose of SA-14-14-2 vaccine was 99.3%. They concluded that a single dose of the SA-14-14-2 vaccine was highly efficacious in preventing JE when administered only days or weeks before exposure to infection (Bista *et al.*, 2001).

Attempts for developing the next generation JE vaccine

One attempt for developing a "second generation" JE vaccine, which is most promising, is the ChimeriVax-JE. ChimeriVax-JE is a live attenuated JE vaccine. ChimeriVax-JE is a hybrid virus that carries the prM and E gene of the JE vaccine SA-14-14-2 strain within the backbone as a molecular clone of the yellow fever (YF) 17D vaccine strain (Chambers *et al.*, 1999). ChimeriVax-JE has an improved safety profile when compared with commercial YF 17D vaccine, in mice and monkeys (Chambers *et al.*, 1999: Guirakhoo *et al.*, 1999). The safety and immunogenicity profile induced by a single dose of ChimeriVax-JE vaccine in six volunteers appeared to be similar to that of YF 17D (Monath *et al.*, 2002). Phase 2 clinical trials revealed that ChimeriVax-JE was well tolerated, with a safety profile and viremia pattern consistent with those of the YF vaccine. ChimeriVax-JE rapidly elicited high titers of neutralizing antibodies after a single inoculation at very low doses, an advantage over existing inactivated vaccines that require multiple doses (Monath *et al.*, 2003).

Previously, Konish and others reported on the evaluation of both replication deficient canary pox and highly attenuated NYVAC vaccinia viruses as vectors for prM-E or prM-E-NS1 genes of JE virus (Konishi *et al.*, 1992; 1994). Another development of second-generation vaccines has been carried out by Konishi and others. They constructed two JE DNA vaccine candidates based on commercial plasmids encoding the JE virus signal-prM-E gene cassette and evaluated them for their immunogenicity in swine. The results showed the possibility that DNA vaccines are able to induce virus-specific memory B cells and long-lasting antibodies in swine, which were of higher levels than those of obtained with a commercial formalin-inactivated JE vaccine. Also, the levels of antibody responses induced by these DNA vaccine plasmid were comparable or higher than those induced in swine by a recombinant vaccinia virus encoding the JE virus signal-prM-E gene cassette (Konishi, *et al.*, 2000). The combined immunization with protein vaccines is a useful strategy to increase the ability of DNA vaccines to induce neutralizing antibodies. A series of works were reported that: (1) the synergistic increase in the neutralizing antibody titer by simultaneous immunization with DNA and protein vaccines was observed and this phenomenon would reduce the amount of DNA needed for immunization (Konishi *et al.*, 2003a), (2) the safety of the DNA vaccine in cynomolgus monkey (Tanabayashi *et al.*, 2003), (3) the neutralizing antibody induced by DNA vaccine prevented virus dissemination from the peripheral site to the brain, and (4) this antibody mediated mechanism of protection is more efficient than the immunity induced by plasmids that generate cytotoxic T lymphocytes (Konishi *et al.*, 2003b).

Heterologous protection against G1 stains in Japan

The difference of immunotype between vaccine strains and filed strains has been speculated. Based on a survey of neutralizing antibodies of Taiwan residents, Shyu and others conclude that neutral antibodies induced by the Nakayama vaccine strain might not be protected efficiently from infection by wild strain JE virus (Shyu *et al.*, 1997). This report is notable because the G1 strains, whose origin is considered to be Southeast Asia, are circulating in Japan. In 2004, Kondo and others reported that a case of encephalitis in a horse occurred in Tottori, Japan in August 2003, and that a G1 strain was isolated from the horse brain (T. Kondo *et al.*, unpublished). And, as described above, the dominant immunotype of G1 strains is thought to be different from that of G3 and even that an unique immunotype existed (Ali *et al.*, 1997). Since very few cases have been reported in Japan in recent decades, there is the thought that the recent low incidence of reported cases has been due to the virulence of circulating viruses shifting to a lesser or avirulence. But this case clearly demonstrated that G1 strains circulating in Japan had the same virulence as those of the G3 strains. To evaluate the efficacy of current the Beijing-1 vaccine strain against G1 strains, sero-reactivity against both G1 and G3 was analyzed (Y. Makino, unpublished). Figure 9.4 shows the relationship of the neutralizing antibody titers of 20 healthy adults among JaGAr01 and two Japanese G1 strains (95-91 and 95-167). The clear correlations were observed in the both inter- and intra-genotypic comparisons. This result also raises the suspicion of the existence of a minor antigenic difference within G1. Based on our result, we speculate that the current Beijing-1

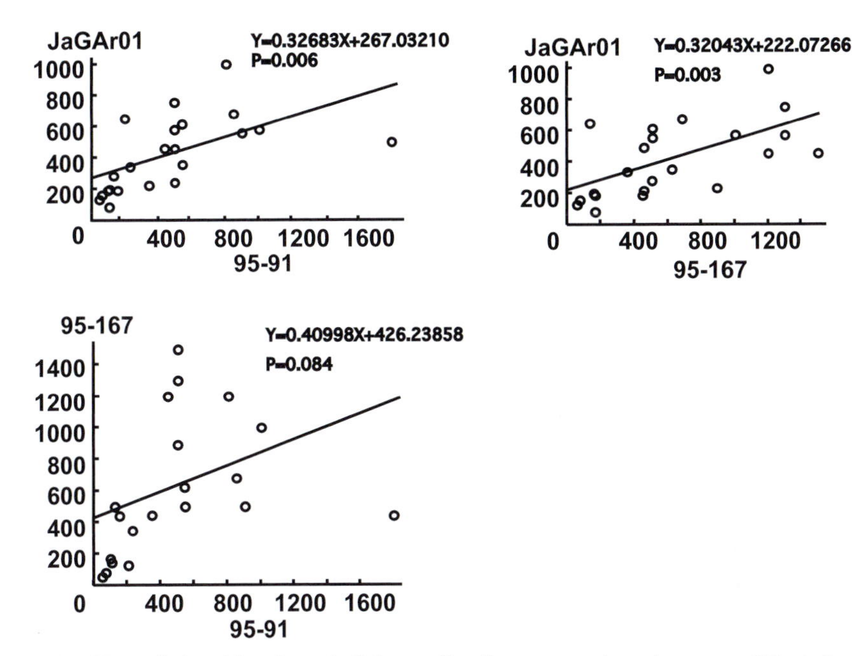

Figure 9.4 The relationship of neutralizing antibodies among two Japanese G1 strains and JaGAr01 from 20 adult residents in Oita, Japan.

vaccine might protect against the Japanese G1 strains that are dominantly circulating in Japan. It was reported that (1) there is no consistent pattern of antigenic variation across strains from different geographic regions, (2) no evidence that such differences influence vaccine efficacy, and (3) these phenomena are due to the breadth of the cross-reactive antibody response to multiple inoculations of vaccine antigen and post-exposure to wild-type virus (Monath, 2002). These findings might support our serological survey and the speculation.

Control of JE by vector control

In east Asia, rice is the staple. The rice field is a good breeding site of vector mosquito *Culex tritaeniorhynchus*, and are spread all over Japan. The larval control project of vector mosquitoes of JE was carried out. A 5% aerial application of granules of fenthion and/or 1.5% granules of EPN (ethylparanitrophenylthiobenzene phosphonate, a type of organophosphorous insecticide) was applied for larval control by aircraft on rice fields in Kyoto city area, Japan from 1967 to 1977 (Maeda *et al.*, 1981). The project was evaluated as being effective for controlling mosquito larvae over a period of 3–4 weeks after application of the insecticides (Igarashi, 2002). JE vector populations were significantly reduced all over Japan during the 1970s (Kamimura *et al.*, 1972; Wada *et al.*, 1975; Maeda, *et al.*, 1978; Mogi, 1978; Ogata *et al.*, 1991), even in areas where the JE vector control project was not carried out. There is a general consensus that the reduction in the JE vector population was caused by the widespread spraying of pesticide in rice fields by farmers to protect rice crops from insect pests (Igarashi, 2002). It had been considered that *Culex tritaeniorhynchus* is sensitive against organophosphorous insecticides. However, the *Culex tritaeniorhynchus* mosquitoes collected in 1982 was highly resistant to organophosphorous insecticides (Kamimura *et al.*, 1983). Sixteen samples of *Culex tritaeniorhynchus* collected in 14 prefectures in Japan in 1984 were assessed for their resistance to insecticides. All of the samples were highly resistant to organophosphorous insecticides. Insensitivity to acetylcholinesterases (AChEs), which are the principle factors regulating the resistance to organophosphorous insecticides, appeared in 1978 and rapidly replaced most of the susceptible mosquitoes within 2 or 3 years (Yasutomi *et al.*, 1987). In Toyama prefecture, where organophosphorous resistant mosquitoes were first found, the number of *Culex tritaeniorhynchus* at that time was lowest in 1977, then began to increase in 1978. The number of organophosphorous insecticide resistant mosquitoes was reduced by one tenth when compared with its peak in 1982 (Kamimura *et al.*, 1997). Because of (1) the appearance of insecticide-resistant mosquitoes, (2) environmental pollution, and (3) its transient effect, vector control by insecticide application has not been used routinely to control human JE in Japan (Igarashi, 2002).

One hypothesis has been made to explain the reduction of JE vector mosquitoes. Mogi considered that the answer to reduce the vector mosquitoes could easily be obtained, theoretically. He stated that an appropriate management of water of the rice field could greatly reduce the effective vector mosquitoes breeding area and that this seemed to be rather impractical at that time. He considered that farmers would not accept the water control, which was an additional procedure and did not improve the rice yield directly, if the water control would be effective for JE vector control (Mogi, 1978). Kamimura and others considered that the reduction of vector mosquitoes was due to the advance of the water management of rice fields, i.e. water-logged rice fields will harm both the rice and the rice field (Kamimura *et al.*, 1972). The water management of rice fields was principally water-logged in the 1970s. As the advance of agricultural techniques, the current water management of the rice fields in Japan is the cyclic irrigation. Except for the following two weeks after implantation, rice field is kept wet. The draining of the rice field is carried out for two weeks during the early June (Figure 9.5). Though there has been no direct demonstration, the change of the water management of rice fields might have reduced the larval breeding site of vector mosquitoes.

References

Ali, A., Igarahi, A., Peneru, N.R., Hasebe, F., Morita, K., Tagami, M., Suwonkerd, W., Tsuda, Y., and Wada, Y. (1995). Characterization of two Japanese encephalitis virus strains isolated in Thailand. Arch. Virol. *140*, 1557–1575.

Ali, A., and Igarashi A,. (1997). Antigenic and genetic variations among Japanese encephalitis virus strains belonging to genotype 1. Microbiol. Immunol. *41*, 241–252.

Banerjee K., and Ranadive, S.N. (1989). Oligonucleotide finger print analysis of Japanese encephalitis virus strains of different geographical origin. Indian J. Med. Res. *89*, 201–216.

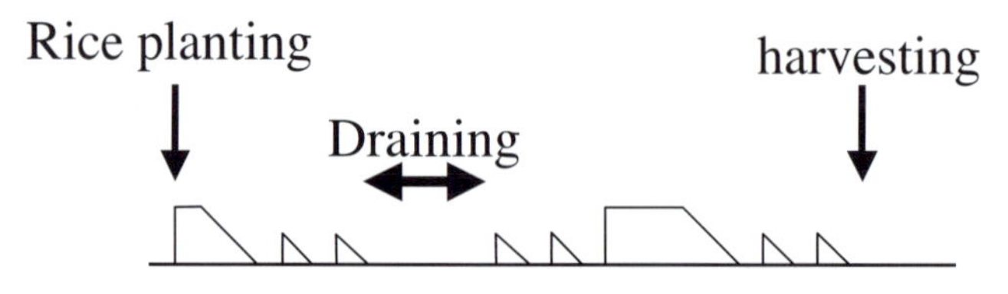

Figure 9.5 The current water management schedule of rice fields in Japan. The shaded areas mean the periods rice fields are water-logged. White triangles mean cyclic irrigation.

Bista, M.B., Banerjee, M.K., Shin, S.H., Tandan, J.B., Kim, M.H., Sohn, Y.M., Ohrr, H.C., Tang, J.L., and Halstead, S.B. (2001). Efficacy of single-dose SA 14-14-2 vaccine against Japanese encephalitis: a case control study. Lancet *358*, 791–795.

Chambers, T.J., Nestorowicz, A., Mason, P.W., and Rice, C.M. (1999). Yellow fever/Japanese encephalitis chimeric viruses: construction and biological properties. J. Virol. *73*, 3095–3101.

Chen, W.-R., Tesh, R.B., and Rico-Hesse, R. (1990). Genetic variation of Japanese encephalitis virus in nature. J. Gen. Virol. *71*, 2915–2922.

Chen, W.-R., Rico-Hesse, R., and Tesh, R.B. (1992). A new genotype of Japanese encephalitis virus from Indonesia. Am. J. Trop. Med. Hyg. *47*, 61–69.

Chung, Y.-J., Nam, J.-H., Ban, S.-J., and Cho, H.-W. (1996). Antigenic and genetic analysis of Japanese encephalitis viruses isolated from Korea. Am. J. Trop. Med. Hyg. *55*, 91–97.

Gould, E.A., Zanotto, P.M., and Holmes, E.C. (1997). The genetic evolution of flaviviruses. In Factors the emergence of arbovirus disease, Mackenzie, A.D., Barrett, A.D., and Deubel, V, ed. (Berlin, Germany: Springer-Verlag), pp. 51–63.

Gould, E.A. (2002). Evolution of Japanese encephalitis virus serocomplex viruses. Curr. Top. Microbiol. Immunol. *267*, 391–404.

Guirakhoo, F., Zhang, Z.-X., Chambers, T.J., Delagrave, S., Arroyo, J., Barrett, A.D.T, and Monath, P.T. (1999). Immunogenicity, genetic stability, and Protective efficacy of a recombinant, chimeric yellow fever-Japanese encephalitis virus (ChimeriVac-JE) as a live, attenuated vaccine candidate against Japanese encephalitis. Virology *257*, 363–372.

Hale, J.H., and Lee, L.H. (1954). A serological investigation of six encephalitis viruses isolated in Malaya. Br. J. Exp. Pathol. *35*, 426–33.

Hanna, J.N., Richie, S.A., Phillips, D.A., Shield, J., Bailey, M., Mackenzie, J.S., Poidinger, M., McCall, B.J., and Mills, P.J. (1996). An outbreak of Japanese encephalitis in the Torres Strait, Australia 1995. Med. J. Aust. *165*, 256–260.

Hasegawa, H., Yoshida, M., Fujita, S., and Kobayashi, Y. (1994). Comparison of structural proteins among antigenically different Japanese encephalitis virus strains. Vaccine. *12*, 841–844.

Hasegawa, H., Yoshida, M., Kobayashi, Y., and Fujita, S. (1995). Antigenic analysis of Japanese encephalitis viruses in Asia by using monoclonal antibodies. Vaccine *13*, 1713–1721.

Hoke, C.H., Nisalak, A., Sangawhiph, N., Jatanasen, S., Laorakapongse, T., Innis, B.L., Kotchasenee, S., Gingrich, J.B., Latendresse, J., Fukai, K., and Burke, D.S. (1988). Protection against Japanese encephalitis by inactivated vaccines. N. Engl. J. Med. *319*, 608–614.

Holbrook, M.R., and Barrett, A.D.T. (2002). Molecular epidemiology of Japanese encephalitis virus. Curr. Top. Microbiol. Immunol. *267*, 75–90.

Hori, H., Morita, K., and Igarashi, A. (1986). Oligonucleotide fingerprint analysis on Japanese encephalitis virus strains isolated in Japan and Thailand. Acta Virol. *30*, 353–359.

Houng, V.T.Q., Ha, D.Q., and Deubel, V. (1993). Genetic study of Japanese encephalitis viruses from Vietnam. Am. J. Trop. Med. Hyg. *49*, 538–544.

Hsu, T.C., Chow L.P., Wei, H.Y., *et al.* (1971). A completed field trial for an evaluation of the effectiveness of mouse-brain Japanese encephalitis vaccine. In Immunization for Japanese encephalitis: Hammon W.M., Kitaoka, M., Downs W.G. ed. (Tokyo, Japan: Igaku-Shoin), pp. 258–265.

Igarashi, A. (2002). Control of Japanese encephalitis in Japan: Immunization of Human and Animals, and vector control. Curr. Top. Microbiol. Immunol. *267*, 139–195.

Jan, L.-R., Yuhe, Y.-Y., Wu, Y.-C., Horng, C.-B., and Wang, G.-R. (2000). Genetic variation of Japanese encephalitis virus in Taiwan. Am. J. Trop. Med. Hyg. *62*, 446–452.

Johansen, C.A., Van Den Hurk, A.F., Ritch, S.A., Zbiriwski, P., Nisbet, D.J., Paru, R., Bockarie, M.J., Macdonald J., Drew, A.C., Khromykh, T.I., and Mackenzie, J.S. (2000). Isolation of Japanese encephalitis virus from mosquitoes (diptera: culicidae) collected in the western province of Papua New Guinea, 1997–1998. Am. J. Trop. Med. Hyg. *62*, 631–638.

Kamimura K., and Matsuda M. (1972). Consideration n the decrease of Japanese encephalitis cases (in Japanese). J. Toyama Soc. Rural Med. 3, 66–86.

Kamimura, K., and Maruyama, Y. (1983). Appearance of highly resistant strain of *Culex tritaeniorhynchus* to organophosphorus insecticides. Jap. J. Sanit. Zool. *34*, 33–37.

Kamimura, K., Kusumoto, I.T., Endo, Y., Watanabe, M., and Arakawa, R. (1997). Recent seasonal prevalence and insecticide resistance of *Culex tritaeniorhynchus* in Toyama. J. Ecol. for J. E. V. 28, 1 (in Japanese).

Kimura-Kroda, J., and Yasui, K. (1983). Topographical analysis of antigenic determinants on envelope glycoprotein V3 (E) of

Japanese encephalitis virus, using monoclonal antibodies. J. Virol. *45*, 124–132.
Kimura-Kuroda, J., and Yasui, K. (1986). Antigenic comparison of envelope protein E between Japanese encephalitis virus and some other flaviviruses using monoclonal antibodies. J. Gen. Virol. *67*, 2663–2672.
Kitano, T., Yabe, S., Kobayashi, M., Oya, A., and Ogata, T. (1986). Immunogenicity of JE Nakayama and Beijing-1 vaccines. JE HFRS Bull. *1*, 37–41.
Kitaoka, M., Miura, T., and Ogata, T. (1952). The virus isolation from Japanese encephalitis patients in 1950 and (1951). Jap. Encephalitis. *1950–1951*, 17–19 (in Japanese).
Kobayashi, I. (1959). On the immunological differences recognized between two strains of Japanese B encephalitis virus. Virus. *9*, 475–482 (in Japanese).
Kobayashi, Y., Hasegawa, H., Oyama, T., Tamai, T., and Kusaba, T. (1984). Antigenic analysis of Japanese encephalitis virus by using monoclonal antibodies. Infect. Immun. *44*, 117–123.
Kobayashi, Y., Hasegawa, H., and Yamaguchi, T. (1985). Studies on the antigenic structure of Japanese encephalitis virus using monoclonal antibodies. Microbiol. Immunol. *29*, 1069–1082.
Konishi, E., Pincus, S., Paoletti, E., Laegreid, W.W., Shope, R.E., and Mason, P.W. (1992). A highly attenuated host range-restricted vaccinia virus strain, NYVAC, encoding the prM, E, and NS1 genes of Japanese encephalitis virus prevents JEV viremia in swine. Virology. *190*, 454–458.
Konishi, E., Pincus, S., Paoletti, E., Shope, R.E., and Wason, P.W. (1994). Avipox virus-vectored Japanese encephalitis virus vaccines: use as vaccine candidates in combination with purified subunit immunogens. Vaccine. *12*, 633–8.
Konishi, E., Yamaoka, M., Kurane, I., and Mason, P.W. (2000). Japanese encephalitis DNA vaccine candidates expressing premembrane and envelope genes induce virus-specific memory B cells and long-lasting antibodies in swine. Virology *268*, 49–55.
Konishi, E., and Suzuki, T. (2002). Ratios of subclinical to clinical Japanese encephalitis (JE) virus infections in vaccinated populations: evaluation of an inactivated JE vaccine by comparing the ratios those in unvaccinated populations. Vaccine *21*, 98–107.
Konishi, E., Terazawa, A., and Imoto, J. (2003a). Simultaneous immunization with DNA and protein vaccines against Japanese encephalitis or dengue synergistically increases their own ability to induce neutralizing antibody in mice. Vaccine *21*, 1826–1832.
Konishi, E., Ajiro, N., Mukuzuma, C., Mason, P.M., and Kurane, I. (2003b). Comparison of protective efficacies of plasmid DNAs encoding Japanese encephalitis virus proteins that induce neutralizing antibody or cytotoxic T lymphocytes in mice. Vaccine *21*, 3675–3683.
Kuzuhara, S., Nakamura, H., Hayashida, K., Obata, J., Abe, M., Sonoda, K., Mishiyama, K., Sugawara, K., Takeda, K., Honda, T., Matsui, H., Shigaki, T., Kino, Y., Mizokami, H., Tanaka, M., Mizuno, K., and Ueba, K. (2003). Non-clinical and phase I clinical trials of a Vero cell-derived inactivated Japanese encephalitis vaccine. Vaccine *21*, 4519–4526.
Ma, S.-P., Arakaki, S., Makino, Y., and Fukunaga, T. (1996). Molecular epidemiology of Japanese encephalitis virus in Okinawa. Microbiol. Immunol. *40*, 847–855.
Ma, S.-P., Yoshida, Y., Makino, Y., Tadano, M., Ono, T., and Ogawa, M. (2003). Short report: A major genotype of Japanese encephalitis virus currently circulating in Japan. Am. J. Trop. Med. Hyg. *69*, 151–154.
Maeda, O., Takenokuma, K., Karoji, Y., and Matsuyama, Y. (1978). Epidemiological studies on Japanese encephalitis in Kyoto city area, Japan. Japan. J. Med. Sci. Biol. *31*, 27–37.
Maeda, O., Uemoto, K., Nakawaza, T., and Matsuyama. (1981). Mosquito control project by aircraft application of larvicidal granules on rice fields in Kyoto City area. Jap. J. Sanit. Zool. 32, 193–202.
Mangada, M.N.M., and Takegami, T. (1999). Molecular characterization of the Japanese encephalitis virus representative immunotype JaGAr01. Virus Res. *59*, 101–112.
Mitamura, T., Kitaoka, M., Watanabe, S., Okubo, K., Tenjin, S., Yamada, S., and Mori, K. (1936). Uber das Virus der epidemischen Enzephalitis (Sommerenzephalitis) mit besonderer Berucksichtigung seiner immunologishchen Eigenschften und seiner Beziehung zu einem Virusstamm der Winterenzephalitis. Trans. Soc. Path. Jpn. *26*, 429–452 (in German).
Mitamura, T., Kitaoka, M., Watanabe, M., Iwasaki, T., and Tenjin, S. (1940). A research of Japanese epidemic encephalitis virus vaccine. Nihonigaku oyobi Kenkouhoken. *3208*, 5–9.
Miyamoto, C., Xu, K.-S., and Takegemi, T. (1997). Japanese encephalitis virus inn Ishikawa. J. Ecol. for JEV. *28*, 7 (in Japanese).
Mogi, M. (1978). Population studies on mosquitoes in the rice field area of Nagasaki, Japan,

especially on *Culex tritaeniorhynchus*. Trop. Med. 20, 173–263.

Monath, T.P. and Heinz, F.X. (1996). Flaviviruses. In Fields Virology, third edition: Fields, B.N., Knipe, P.N., Howley, P.M., *et al.*, ed. (Philadelphia: Lippincott-Raven Publishers), pp. 961–1034.

Monath, T.P. (1988). Japanese encephalitis—a plague of the orient. N. Engl. J. Med. *319*, 641–643.

Monath, T.P. (2002). Japanese encephalitis virus vaccines: current vaccines and future prospects. Curr. Top. Microbiol. Immunol. *267*, 105–138.

Monath, P.T., McCarthy, K., Bedford, P., Johnson, C.T., Nichols, R., Yoksan, S., Marchesani, R., Knauber, M., Wells, K.H., Arroyo, J., and Guirakhoo F. (2002). Clinical proof of principle for ChimeriVaxTM: recombinant live, attenuated vaccines against flavivirus infections. Vaccine *20*, 1004–1018.

Monath, P.T., Guirakhoo F., Nichols, R., Yoksan, S., Schrader, R., Murphy, C., Blum, P., Woodward, S., McCarthy, K., Mathis, D., Johnson, C., and Bedford, P. (2003). Chimeric live, attenuated vaccine against Japanese encephalitis (ChimeriVac-JE): Phase 2 clinical trials for safety and immunogenicity, effect of vaccine dose and schedule, and memory response to challenge with inactivated Japanese encephalitis antigen. J. Infect. Dis. *188*, 1213–1230.

Nam, J.-H., Chae, S.-L., Won, S.-Y., Kim, E.-J., Yoon, K.-S., Kim, B.-I., Jeong, Y.-S., and Cho, H.-E. (2001). Genetic heterogeneity of Japanese encephalitis virus assessed via analysis of the full-length genome sequence of a Korean isolate. Am. J. Trop. Med. Hyg. *65*, 388–392.

Nga, P.T., Parquet, M.d.C., Cuong, V.D., Ma, S.-P., Hasebe, F., Inoue, S., Makino, Y., Takagi, M., Nam, V.S., and Morita, K. (2004). Shift in Japanese encephalitis virus (JEV) genotype circulating in northern Vietnam: implications for frequent introductions of JEV from Southeast Asia to East Asia. J. Gen. Virol. *85*, 1625–1631.

Ni, H., and Barrett, A.D.T. (1995). Nucleotide and deduced amino acid sequence of the structural protein genes of Japanese encephalitis viruses from different geographical locations. J. Gen. Virol. *76*, 401–407.

Nimmannitya, S., Hutamai, S., Kalayanarooj, S., and Rojanasuphot, S. (1995). A filed study on Nakayama and Beijing strains of Japanese encephalitis vaccines. Southeast Asian J. Trop. Med. Public Health. *26*, 689–693.

Ochiai, K., Takashima, I., and Hashimoto, N. (1989). Brief communication: Antigenic analysis of Japanese encephalitis virus isolated in Hokkaido with monoclonal antibodies. Jpn. J. Vet. Res. *37*, 21–26.

Oda K. (1976). Antigenic characterization among strains of Japanese encephalitis virus isolated in Hyogo Prefecture by the antibody-absorption test. Kobe J. Med. Sci. *22*, 123–137.

Ogata, M., and Meguro, T. (1991). Variation in incidence rate and casual factors of Japanese encephalitis in Okayama Prefecture from 1950 to 1989 (Epidemiological study on Japanese encephalitis, 101). Acta Medica Okayama *103: suppl. Japanese encephalitis issue XXXII*, 27–33.

Okuno, T., Okada, A., Kondo, A., Suzuki, M., Kobayashi, M., and Oya, A. (1968). Immunotyping of different strains of Japanese encephalitis virus by antibody absorption, haemagglutination-inhibition and complement-fixation tests. Bull. WHO. *38*, 547–563.

Oya, A. (1987). New development of criteria on Japanese encephalitis vaccine requirements in Japan. JE HFRS Bull. *2*, 11–13.

Ozaki, Y., and Tabeyi, K. (1967). Studies on the neutralization of Japanese encephalitis virus. I. Application of kinetic neutralization to the measurement of the neutralizing potency of antiserum. J. Immunol. *98*, 1218–1223.

Paranjpe, S., and Banerjee, K. (1996). Phylogenetic analysis of the gene of Japanese encephalitis virus. Virus Res. *42*, 107–117.

Pyke, A.T., Williams, D.T., Nisbet, D.J., Van Den Hurk, A.F., Taylor, C.T., Johansen, C.A., Macdonald J., Hall, R.A., Simmons, R.J., Mason, R.J.V., Lee, J.M., Ritchie, S.A., Smith, G.A., and Mackenzie, J.S. (2001). The appearance of a second genotype of Japanese encephalitis virus in the Australasian region. Am. J. Trop. Med. Hyg. *65*, 747–753.

Rice, M.C. (1996). Flaviviridae: the viruses and their replication. In Fields Virology, third edition: Fields, B.N., Knipe, P.N., Howley, P.M., *et al.*, ed. (Philadelphia: Lippincott-Raven Publishers), pp. 951–960.

Rosen, L. (1986). The natural history of Japanese encephalitis virus. Annu. Rev. Microbiol. *40*, 395–414.

Shyu, W.R.-H., Wang, Y.-C., Chin, C., and Chen, W.-J. (1997). Assessment of neutralizing antibodies elicited by a vaccine (Nakayama) strain of Japanese encephalitis virus in Taiwan. Epidemiol. Infect. *119*, 79–83.

Solomon, T., Ni, H., Beasley, D.W.C., Ekkekenkamp, M., Cardosa, M.J., and Barrett, A.D.T. (2003). Origin and evolution of

Japanese encephalitis virus in southeast Asia. J. Virol. *77*, 3091–3098.

Sugawara, K., Nishiyama, K., Ishikawa, Y., Abe, M., Sonoda, K., Komatsu, K., Horikawa, Y., Takeda, K., Honda, T., Kuzuhara, S., Kino, Y., Mizokami, H., Mizuno, K., Oka, T., and Honda, K. (2002). Development of Vero cell-derived inactivated Japanese encephalitis vaccine. Biologicals *30*, 303–314.

Sumiyoshi, H., Mori, C., Fuke, C., Morita, K., Kuhara, S., Kondou, J., Kikuchi, Y., Nagamatu, H., and Igarashi, A. (1987). Complete nucleotide sequence of the Japanese encephalitis virus genome RNA. Virology, *161*, 497–510.

Tadano, M., Kanemura, K., Hasegawa, H., Makino, Y., and Fukunaga, T. (1994). Epidemiological and ecological study of Japanese encephalitis in Okinawa, subtropical area in Japan. I. Investigations on antibody levels to Japanese encephalitis virus in swine sera and vector mosquito in Okinawa, Miyako and Ishigaki islands. Microbiol. Immunol. *38*, 117–122.

Takegami, T., Ishak, H., Miyamato, C., Hirai, Y., and Kamimura, K. (2000). Isolation and molecular comparison of Japanese encephalitis virus in Ishikawa, Japan. Jpn. J. Infect. Dis. *53*, 178–179.

Tanabayashi, K., Mukai, R., Yamada, A., Takasaki, T., Kurane, I., Yamada, M., Terazawa, A., and Konishi, E. (2003). Immunogenicity of a Japanese encephalitis DNA vaccine candidate in cynomolgus monkeys. Vaccine *21*, 2338–2345.

Tsai, T.F. (2000). New initial for the control of Japanese encephalitis by vaccination: minutes of a WHO/CVI meeting, Bangkok, Thailand, 13–15 October (1998). Vaccine *18 supple 2*, 1–25.

Tsarev, S.A., Sanders, M.L., Vaughn, D.W., and Innis, B.L. (2000). Phylogenetic analysis suggests only one serotype of Japanese encephalitis virus. Vaccine. *18*, 36–43.

Tsuchie, H., Oda, K., Vythilingam, I., Thayan, R., Vijayamalar, B., Sinniah, M., Hossain, M., Kurimura T., and Igarashi, A. (1994). Genetic study of Japanese encephalitis viruses isolated in Malaysia. Jpn. J. Med. Sci. Biol., *47*, 101–107.

Tsuchie, H., Oda, K., Vythilingam, I., Thayan, R., Vijayamalar, B., Sinniah, M., Singh, J., Wada, T., Tanaka, H., Kurimura T., and Igarashi, A. (1997). Genotypes of Japanese encephalitis virus isolated in three states in Malaysia. Am. J. Trop. Med. Hyg. *56*, 153–158.

Uchil, P., and Satchidanandam, V. (2001). Phylogenetic analysis of Japanese encephalitis virus: envelope gene based analysis reveals a fifth genotype, geographic clustering, and multiple introductions of the virus into the Indian subcontinent. Am. J. Trop. Med. Hyg. *65*, 242–251.

Vythilingam, I., Morita, K., and Igarashi, A. (1994). Nucleotide and amino acid sequences in the preM region of a Japanese encephalitis virus strain isolated from a pool of *Aedes albopictus* and *Ae. butleri* mosquitoes captured in Peninsula Malaysia in 1992. Trop. Med. *36*, 51–56.

Wada, Y., Oda, T., Mogi, M., Mori, A., and Omori, N. (1975). Ecology of Japanese encephalitis virus in Japan. II. The population of vector mosquitoes and the epidemic of Japanese encephalitis. Trop. Med. *17*, 111–127.

Williams, D.T., Wang, L.-F., Daniel, P.W., and Mackenzie, J.S. (2000). Molecular characterization of the first Australian isolate of Japanese encephalitis virus, the FU strain. J. Gen. Virol. *81*, 2471–2480.

Yasutomi, K., and Takahashi, M. (1987). Insecticidal resistance of *Culex tritaeniorhynchus* (Diptera: Culicidae) in Japan: a country-wide survey of resistance to insecticides. J. Med. Entomol. *24*, 604–608.

Yoshida, M. (1994a). Immunological study of Japanese encephalitis virus—Characterization of monoclonal antibodies against Muar and 691004 strains. Kansensyougaku Zasshi. *68*, 520–528.

Yoshida, M. (1994b). Immunological study of Japanese encephalitis virus—Serological analysis of strains isolated from Thailand, India, Singapore, and Taiwan. Kansensyougaku Zasshi. *68*, 529–535.

Yu, Y.X., Zhang, G.M., Guo, Y.P., Ao, J., and Li, H.M. (1988). Safety of a live-attenuated Japanese encephalitis virus vaccine (SA14-14-2) for children. Am. J. Trop. Med. Hyg. *39*, 214–217.

Maturation and Assembly of Hepatitis C Virus Core Protein

10

Tetsuro Suzuki and Ryosuke Suzuki

Abstract

Hepatitis C virus (HCV) core protein, which is derived from the N-terminus of the viral polyprotein, forms the viral nucleocapsid. The amino acid sequence of this protein is well conserved among different HCV strains, compared to other HCV proteins. The N-terminal domain of the core protein is highly basic, while its C-terminus is hydrophobic. The core protein is primarily detected in the cytoplasm in associating with endoplasmic reticulum (ER), lipid droplets, and mitochondria. However, it has also been detected in the nucleus. In fact, a nuclear localization signal has been identified in the N-terminal region of the protein. Its C-terminal hydrophobic region is thought to act as a membrane anchor for the core protein and as a signal sequence for E1 protein.

Nucleocapsid formation presumably involves interactions between the core protein and viral RNA and envelope protein(s), as well as self-interaction of the core protein. Interactions between the HCV core protein and specific regions of viral genomic RNA have been reported. In addition, the core protein has been observed to form homo-multimeric, as well as hetero-dimeric, complexes with E1 protein. These interactions might be important for nucleocapsid formation, and may activate the biological activity of various HCV structural proteins.

Introduction

Hepatitis C virus (HCV) is the major etiologic agent of posttransfusion- and sporadic non-A, non-B hepatitis (Kuo *et al.*, 1989) and presently infects approximately 170 million people worldwide (Grakoui *et al.*, 2001; Lauer *et al.*, 2001). Persistent infection with HCV is associated with the development of chronic hepatitis, hepatic steatosis, cirrhosis and hepatocellular carcinoma (HCC) (Saito *et al.*, 1990; Alter, 1995; Di Bisceglie, 1997; Lauer *et al.*, 2001; Poynard *et al.*, 2003; Pawlotsky, 2004). Although HCV research has long been hampered by the lack of a cell culture system by which to propagate the virus, molecular cloning and expression of HCV gene products in cell cultures has yielded a lot of information. HCV is classified as belonging to the *Hepacivirus* genus of the Flaviviridae family (Houghton *et al.*, 1991; Robertson *et al.*, 1998). Its genome consists of a single-stranded positive-sense RNA of approximately 9.6 kb, which contains an open reading frame coding for a polyprotein precursor of approximately 3000 residues (Choo *et al.*, 1989). This precursor is co- and post-translationally processed into structural and nonstructural proteins by cellular and viral proteases (Houghton *et al.*, 1991; Hijikata *et al.*, 1993; Grakoui *et al.*, 1993). The structural proteins are located in the N-terminal one-fourth of the polyprotein and are cleaved by cellular membrane proteinases.

The core protein, which is derived from the N-terminus of the polyprotein, most likely forms the viral nucleocapsid given similarities between its position and that of sequences encoding viral nucleocapsids in the genomes of other flaviviruses. The amino acid sequence of the core protein is highly conserved among different HCV strains, compared with other HCV proteins. This protein has been extensively used in serologic assays since anti-core antibodies are highly prevalent among HCV-infected individuals. The N-terminal domain of the core protein is highly basic, while its C-terminus is hydrophobic (Figure 10.1). Although several core proteins of varying molecular weights have been reported (Harada *et al.*, 1991; Liu *et al.*, 1997; Lo *et al.*, 1994; Lo *et al.*, 1995; Suzuki *et al.*, 2001; Yasui *et al.*, 1998), two processing events result in the predominant production of a 21 kDa core protein. The core protein is primarily detected in the cytoplasm, by associating with the endoplasmic reticulum (ER), lipid droplets, and mitochondria (Harada *et al.*, 1991; Selby *et al.*, 1993; Lo *et al.*, 1995; Suzuki *et al.*, 1995; Suzuki *et al.*, 1996; Moradpour *et al.*, 1996; Barba *et al.*, 1997; Yasui *et al.*, 1998; Moriya *et al.*, 1998; Sabile *et al.*, 1999; Hope *et al.*, 2000; McLauchlan *et al.*, 2002; Okuda *et al.*, 2002; Schwer *et al.*, 2004; Suzuki *et al.*, 2005). In some studies, a fraction of the core protein has also been found in the nucleus (Chang *et al.*, 1994; Suzuki *et al.*, 1995; Lo *et al.*, 1995; Liu *et*

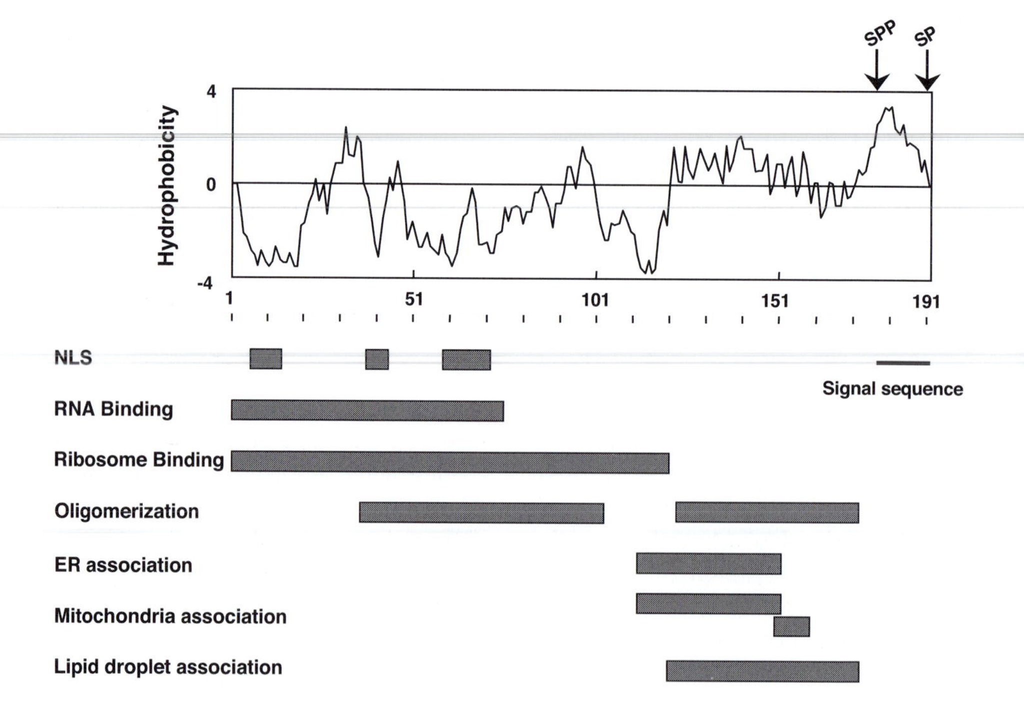

Figure 10.1 Hydrophobicity profile and regions within HCV core protein involved in subcellular distribution and nucleocapsid assembly. NLS, nuclear localization signal; SPP, signal peptide peptidase; SP, signal peptidase.

al., 1997; Moriya *et al.*, 1998; Yasui *et al.*, 1998, Falcon *et al.*, 2003).

Although the functions of the core protein within various subcellular compartments are unclear to date, the core protein is likely multifunctional and essential for viral replication, maturation, and pathogenesis. It is involved not only in formation of the HCV virion but also has a number of regulatory functions, such as influencing signaling pathways, cellular and viral gene expression, cell transformation, apoptosis, and immune presentation.

Processing by membrane-associated proteases

A precursor polyprotein, which is translated from an open reading frame of approximately 9 kb, is processed by both host and viral proteases. The HCV structural proteins include the nucleocapsid or core protein and two envelope glycoproteins, E1 and E2. Secondary structure analysis of the core protein has revealed that all major alpha helices are located in the C-terminal half of the protein. A predicted alpha helix encoded by amino acids (aa) 174–191 is extremely hydrophobic and resembles other signal peptide sequences. Membrane and secretory proteins are generally targeted to the appropriate intracellular membrane by their signal peptides. In eukaryotes, signal peptides are 15–50 amino acids long and a typical signal peptide is composed of three distinct regions: a polar N-terminal region that may have a net positive charge, a central hydrophobic sequence that consists of 6–15 hydrophobic residues, and a polar carboxyl-terminal region that contains the cleavage site for the signal peptidase (von Heijne, 1983; von Heijne, 1985; Martoglio *et al.*, 1998).

The N-terminus of E1 protein has been mapped to a signal-peptidase-like cleavage site at aa 192 of the precursor polyprotein (Hijikata *et al.*, 1991). In addition, cleavage occurs on microsomal membranes and depends on a signal recognition particle. Furthermore, inhibition of cleavage by mutations at the possible recognition sites by the signal peptidase has been shown (Hijikata *et al.*, 1991; Santolini *et al.*, 1994).

Cleavage of the core/E1 junction between aa 191 and 192 by the signal peptidase results in anchoring of the core protein within the ER membrane by the C-terminal signal peptide. Further processing within or at the N-terminus of the signal sequence mediated by signal peptide peptidase (SPP) is thought to be involved in maturation of the core protein (Hussy *et al.*, 1996; McLauchlan *et al.*, 2002; Lemberg *et al.*, 2002; Okamoto *et at.*, 2004).

SPP has recently been identified (Welhofen *et al.*, 2002) and exhibits protease activity within cellular membranes, resulting in cleavage of peptide bonds in the plane of lipid bilayers. The peptidase is an aspartic protease and catalyses intramembrane proteolysis of signal sequences and possibly membrane proteins, within the ER (Lemberg *et al.*, 2004). SPP has been reported to cleave human lymphocytic antigen (HLA) molecules, thereby promoting the release of HLA-E epitope-containing peptides from the ER membrane into the cytosol, resulting in recognition by the immune system (Lemberg *et al.*, 2001). The signal sequence at the C-terminus of the core protein is also a substrate for SPP (McLauchlan *et al.*, 2002). It has been shown that (1) intramembrane cleavage by SPP is abolished when helix-breaking and -bending residues in the C-terminal signal sequence are replaced by basic residues, (2) the signal sequence itself and three hydrophobic amino acids Leu-139, Val-140, and Leu-144 of the core protein are required for SPP cleavage, and (3) none of these residues are essential for cleavage at the core-E1 junction by the signal peptidase, or for translocation of E1 into the ER (Okamoto *et al.*, 2004). Experiments using cDNA clones of HCV isolated from patients with different clinical phenotypes suggest that certain amino acid residues within the C-terminus of the core protein influence processing by SPP (Kato *et al.*, 2003). More work is required to identify the SPP cleavage site within the core protein, although Leu-179 (Hussy *et al.*, 1996), Leu-182 (Hussy *et al.*, 1996), and Ser-173 (Santolini *et al.*, 1994), have all been cited as potential P1 sites of cleavage.

Various studies of HCV cDNA expression *in vitro* and in cultured cells have generally demonstrated two forms of core protein with 21- and 23-kDa (p21 and p23). The p21 form of the core protein predominates in cultured cells, and also in viral particles isolated from the sera of hepatitis C patients (Yasui *et al.*, 1998). Thus, this form of the core protein, which presumably results from two consecutive membrane-dependent cleavages as described above, is thought to be the mature form of the core protein and to constitute the viral capsid. The other form of the core protein, p23, is a 191-residue product, which contains a signal sequence for directing E1 protein to the ER. In addition, production of a 16 kDa form of the core protein (p16) has been reported by HCV genotype 1a (Lo *et al.*, 1994; Lo *et al.*, 1995). However, later studies have identified this protein as F protein, which is encoded by an alternative reading frame from the core region and is expressed by translational ribosomal frameshift (Xu *et al.*, 2001; Walewski *et al.*, 2001; Varakliotl *et al.*, 2002; Vassllaki *et al.*, 2003).

Mechanisms of subcellular localization

Analysis of the subcellular localization of HCV within the hepatic tissue of HCV-infected individuals has proven difficult due to low levels of viral replication. Nevertheless, immunostaining of liver biopsy specimens has revealed a predominance of core protein within the cytoplasm of infected hepatocytes, and often shows a punctate granular distribution within cells (Yap *et al.*, 1994; Gonzalez-Peralta *et al.*, 1994; Gowans, 2000; Sansonno *et al.*, 2004). As shown in Figure 10.2, in mammalian cells, when core protein alone or the entire viral polyprotein has been expressed, the core protein has primarily been observed within the cytoplasm. Data from a number of studies have identified its co-localization at the ER (Harada *et al.*, 1991; Selby *et al.*, 1993; Suzuki *et al.*, 1995; Lo *et al.*, 1995; Moradpour *et al.*, 1996; Moriya *et al.*, 1997; Yasui *et al.*, 1998), lipid droplets (Barba *et al.*, 1997; Sabile *et al.*, 1999; Hope *et al.*, 2000; McLauchlan *et al.*, 2002), and mitochondria (Moriya *et al.*, 1998; Okuda *et al.*, 2002; Schwer *et al.*, 2004; Suzuki *et al.*, 2005). In addition, a fraction of the core protein has been detected in the nucleus (Chang *et al.*, 1994; Suzuki *et al.*, 1995; Lo *et al.*, 1995; Liu *et al.*, 1997; Moriya *et al.*, 1998; Yasui *et al.*, 1998; Yamanaka *et al.*, 2002).

It has been proposed that, after SPP cleavage, a large part of the core protein remains within the cytoplasmic leaflets of the ER membrane due to preservation of the original transmembrane domain (McLauchlan *et al.*, 2002). The cytoplasmic leaflets become distended with accumulated lipid between two membrane leaflets. The core protein diffuses freely and is transferred along with part of the ER membrane to the surface of a nascent lipid droplet before the droplet buds off the ER.

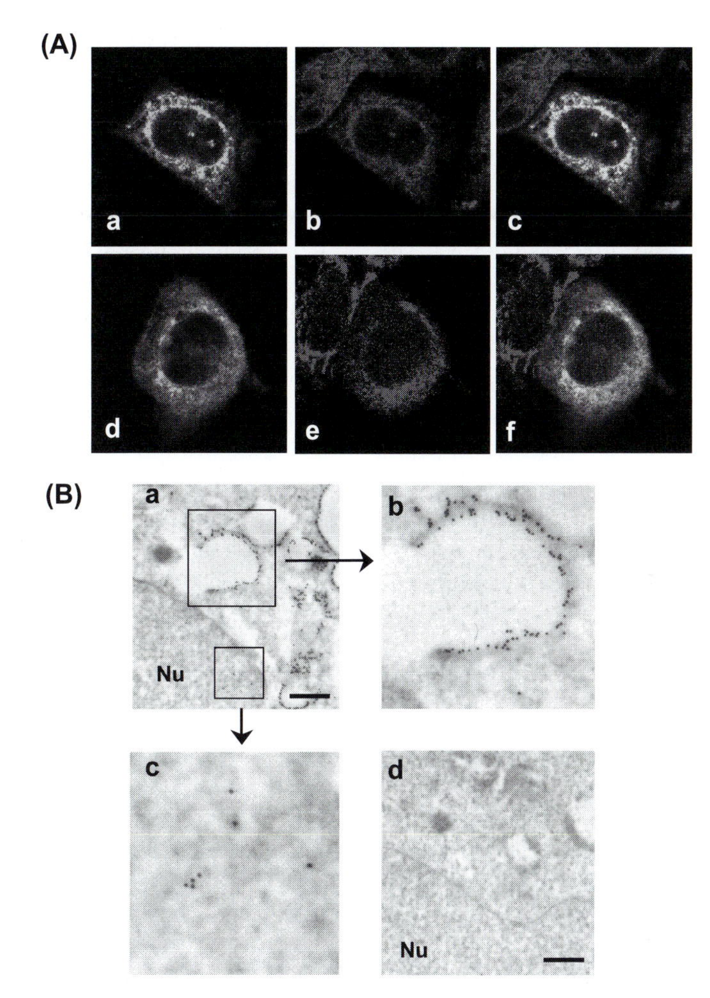

Figure 10.2 (A) Confocal microscopic images of HepG2 cells transfected with the HCV core (aa 1–191) cDNA. Cells were double stained with anti-core and anti-calregulin, ER marker, or Mitotracker, mitochondrial marker. a, d: core protein, b: calregulin, e: Mitotracker, c, f: overlay. (B) Immunoelectron microscopy. Cells expressing the core protein (a, b, c) or nonexpressing cells (d) were fixed and immunogold labeled with anti-core. Gold particles were found at cytoplasmic membranes (a, b) and in the nucleus (a, c). Bars, 0.5 μm. This figure is reproduced in color in the color section at the end of the book.

This model has recently been modified to indicate that other membranes within the ER network, such as mitochondrion-associated membranes, might be targeted by the mature core protein (Schwer *et al.*, 2004).

Hydrophobic profiling (Figure 10.1) and amino acid sequencing has identified three domains within the core protein (Hope *et al.*, 2000). An N-terminal half domain of 118 residues contains clusters of Lys and Arg residues, which are required for nuclear localization of the core protein and for binding to viral RNA as described below. The 56-residue domain spanning aa 119–174 has a few basic residues and is more hydrophobic compared than the N-terminal domain. The C-terminal 17-residue domain (aa 175–191) is highly hydrophobic and is predicted to have an alpha-helix structure. This feature is consistent with its role as a signal peptide for E1 protein. Interestingly, the HCV core protein (p21) is much larger than flavi- and pesti-virus capsid proteins. Kunjin virus and classical swine fever virus, for example, are approximately 100 residues in length (Speight *et al.*, 1989; Stark *et al.*, 1993; Rumenapf *et al.*, 1993). Although significant amino acid identity has not been observed between the core protein of HCV and the capsid proteins of other flaviviruses, their N-terminal regions are all rich in basic residues and their C-terminal hydrophobic domains act as signal sequences for the translocation of prM or E1 proteins. A 56-residue domain-like region has also been observed within the core sequence of GB virus-B, which is closely related to HCV, but not among other flavi- and pestiviruses (Hope *et al.*, 2002).

Mutational analyses and immunostaining have shown that a considerable length of this domain is indispensable for the associations between the core protein and ER membrane and lipid droplets to occur (Hope *et al.*, 2000; Okamoto *et al.*, 2004; Suzuki *et al.*, 2005). Sequences in the 56-residue domain required for lipid droplet association also facilitate maturation of the core protein (Hope *et al.*, 2000). Results from our laboratory have demonstrated that a region spanning aa 112–152 of the core protein plays a key role in ER retention of the mature core protein (Suzuki *et al.*, 2005). From examination of the secondary structure of the core protein, a long helical segment (aa 116–134) and two short α-helices (aa 146–152, aa 155–159) are predicted. It has been suggested that an amphipathic alpha-helix spanning aa 116–134 may be required for association of the core protein with the ER membrane. A helical wheel plot of this region shows an amphipathic structure with hydrophobic residues on one side and polar residues on the other side of the α-helix, which are often observed in membrane-associated proteins. This helical wheel conformation might be important in directing the core protein to the ER membrane.

Localization of a fraction of the core protein in the mitochondria of cultured cells has been reported (Okuda *et al.*, 2002; Schwer *et al.*, 2004; Suzuki *et al.*, 2005). In addition, mitochondrial localization has been observed in the transgenic mice, following disruption of the bilayer structure of the mitochondrial membrane by expression of the core protein (Moriya *et al.*, 1998; Moriya *et al.*, 2001). Subcellular fractionation and protease protection assays, as well as immunoelectron microscopy, have recently demonstrated localization of the core protein to the mitochondrial outer membrane (Schwer *et al.*, 2004; Suzuki *et al.*, 2005).

Translocation of nuclear-encoded mitochondrial proteins usually depends on N-terminal sequences, known as mitochondrial targeting sequences (Neupert, 1997). However, a significant proportion of mitochondrial proteins lack these N-terminal mitochondrial-targeting sequences. Specifically, a number of outer membrane proteins do not have cleavable sequences at their N-terminus, rather, they are targeted to mitochondria by means of internal or C-terminal signals (Mihara, 2000). There is some controversy regarding the identity of which sequence is responsible for targeting the core protein to the mitochondria, based on results from two groups of researchers. Through fusion experiments with green fluorescent protein, a short stretch extending from aa 149–158 within the C-terminal hydrophobic region of the core protein, has been observed to play a role in mitochondrial targeting (Schwer *et al.*, 2004). However, another study based on a similar approach has identified a 41-residue region extending from aa 112–152 as the sequence responsible for association between the core protein and mitochondria (Suzuki *et al.*, 2005). This 41-residue region is same as that required for association of the core protein with the ER membrane, as mentioned above. This might suggest that the mature core protein moves freely within the ER network, which might include mitochondria enveloped by cytoplasmic extensions of the ER membrane, after which the core protein might translocate to the mitochondrial surface.

Although eukaryotic proteins which target the ER and mitochondria generally have different signal sequences and follow distinct transport routes, recent evidence suggests that chimeric signals for bimodal targeting might exist, as suggested by the behavior of the core protein. For example, cytochrome P4502E1 (CYP2E1), which has been implicated in cellular pathology and toxicity related to oxidative stress, is able to localize to both the ER and mitochondria. The last 29 residues of the CYP2E1 N-terminal encode a putative transmembrane anchor domain responsible for targeting the ER, while a sequence extending from aa 21–31 might represent a mitochondrial targeting signal activated by cAMP-dependent phosphorylation at Ser-129 of the protein (Robin *et al.*, 2002). Activation of the mitochondrial signal has been associated with increased efficiency of association between the target protein and cytoplasmic chaperones and/or mitochondrial translocases. HCV core protein can be phosphorylated, probably by cellular protein kinase A and C, as described below (Lu *et al.*, 2002). It has been suggested that phosphorylation of the core protein might be required for some of its biological activity. Thus, it would be useful to investigate whether post-translational modification, such as protein phosphorylation, might mediate ER/mitochondria localization of the core protein.

The known association between HCV core protein and the mitochondrial membrane suggests that the core protein has the ability to modulate mitochondrial function, presumably, at least in part, by altering permeability of the mitochondrial membrane. The core protein induces the production of cellular reactive oxygen species (ROS) in the livers of transgenic mice and in cell lines expressing the protein (Moriya *et al.*, 2001). ROS, predominantly generated in the mitochondria, induce genetic mutations and act as secondary messengers to regulate a variety of cellular functions, including gene expression and proliferation (Adler *et al.*, 1999). Although the exact molecular mechanism by which the core protein induces ROS production has not been determined, HCV core protein is known to impair the mitochondrial electron transfer system (Moriya *et al.*, 2001). The core protein may also modulate apoptosis, since mitochondria play a major role in regulating programmed cell death. Expression of HCV proteins, including the core protein, suppresses the release of cytochrome c from mitochondria to the cytoplasm in HCV-transgenic mice, thus inhibiting Fas-mediated apoptosis (Machida *et al.*, 2001).

Finally, in addition to its association with the cytoplasmic membrane, nuclear localization of the core protein has been observed in cultured cells over-expressing the protein, as well as in liver tissue isolated from patients with hepatitis C (Falcon *et al.*, 2003) and in transgenic mice (Moriya *et al.*, 1998). Although three-dimensional structural data regarding the core protein is lacking, studies using a series of monoclonal antibodies have shown that cytoplasmic and nuclear forms of the core protein are recognized by distinct antibodies, suggesting that cytoplasmic and nuclear forms of the core protein have different tertiary structures (Yasui *et al.*, 1998).

The N-terminal half of the core protein contains three nuclear localization signal sequences, which are composed of three stretches of sequences rich in basic residues (Suzuki *et al.*, 1995; Suzuki *et al.*, 1996; Figure 10.1). C-terminal truncated versions of the core protein, such as that encoded by aa 1–152, are known to localize exclusively to the nucleus, suggesting that the hydrophobic region of the C-terminal might determine whether the core protein localizes to the nucleus or cytoplasm.

In general, the NLS sequences fall into one of two distinct classes termed monopartite NLSs, containing a single cluster of basic residues, and bipartite NLSs, comprising two basic clusters separated by an unconserved spacer sequence of variable length. The conventional NLSs are recognized by the same receptor protein termed importin or karyopherin (reviewed in Damelin *et al.*, 2002; Weis, 2002). Importin α contains the NLS-binding site, and importin β docks importin-substrate complexes to the cytoplasmic filaments of a nuclear pore complex. Thus, importin α functions as an adaptor between the *bona fide* import receptor and the NLS-carrying protein.

All three NLS motifs of the core protein are able to bind to importin α, and at least two NLS motifs are required for efficient nuclear distribution of the core protein. It appears that mutations of two of the three NLS motifs (double mutant proteins) decrease the ability of the core protein to bind importin α. These observations suggest that NLS motifs within the core protein have a bipartite function and binding between double mutants and importin α leads to little, or no, active translocation of the core protein into the nucleus. Double mutants may also block subsequent interactions with importin β_1, GTPase Ran, and/or NTF2/p10, all of which are required for translocation through nuclear pore complexes.

Crystallographic studies of the structural basis of the NLS recognition by importin α have shown that the two basic residue clusters of bipartite NLSs occupy separate binding sites on importin α. In contrast, monopartite NLSs can bind to the same sites but primarily use the binding site for the C-terminal basic cluster of the bipartite NLSs, which is known as the major NLS binding site on importin α (Conti *et al.*, 1998; Fontes *et al.*, 2000). An importin α variant containing a mutation in its major binding site has shown a decreased ability to bind both monopartite and bipartite NLSs. Another importin α variant with a mutation in its minor binding site, has shown a decreased ability to bind bipartite NLS-containing proteins, making importin α nonfunctional *in vivo* (Leung *et al.*, 2003). Thus, we favor a model in which the core protein bipartite NLS, composed of any two of the three basic clusters, occupies both major and minor binding sites on importin α, resulting in efficient nuclear translocation. Importin α may be equally accessible to all clusters given their close proximity to one another, as well as the distinct conformational flexibility of the approximately 70-residue N-terminal region of the core protein.

A proteasome activator, PA28γ, has been implicated in nuclear localization of the core protein. Yeast two-hybrid screening has identified PA28γ as the core-binding protein. PA28γ specifically interacts with the core protein in cultured cells, as well as in the livers of transgenic mice and hepatitis C patients. Interaction of the core protein with PA28γ plays an important role in retention of the core protein, especially the C-terminal truncated form, in the nucleus (Moriishi *et al.*, 2003). A yeast model system has shown a requirement for small GTPase Ran/Gsp1p activity mediated by Kap123p, for nuclear transport of the core protein, however, neither importin α nor importin β are (Isoyama *et al.*, 2002).

Thus, multiple functional domains within the core protein appear to play a role in its subcellular localization, which might ultimately depend on the balance achieved between competing signals.

Possible post-translational modification

The core protein can be phosphorylated in insect cells (Lanford *et al.*, 1993), reticulocyte lysates (Shih *et al.*, 1995), and mammalian cells (Lu *et al.*, 2002). Possible sites of phosphorylation include Ser-53, -93, -96, and -116 of the core protein (Shih *et al.*, 1995), and the basal phosphorylated residues have been identified as Ser-53 and -116 (Lu *et al.*, 2002). Cellular protein kinase A (PKA) and C (PKC) are presumably responsible for their phosphorylation since activation of PKA and PKC enhances phosphorylation, while inhibition of PKA and PKC negates this effect (Shih *et al.*, 1995; Lu *et al.*, 2002). Results from mutational analyses suggest that phosphorylation at Ser-116 may regulate nuclear localization of the core protein (Lu *et al.*, 2002). Phosphorylation of the core protein might also be required for its biological activity, including inhibition of replication and gene expression of hepatitis B virus (Shih *et al.*, 1995).

The core protein is a substrate of tissue transglutaminase (Lu *et al.*, 2001), which catalyzes calcium-dependent acyl transfer reactions between the γ-carboxamide groups of Gln residues and the ε-amino groups of Lys residues within peptides, resulting in the formation of a gamma-glutamyl-ε-lysine isopeptide bond (Lorand *et al.*, 1984; Greenberg *et al.*, 1991). Presumably, a primary function of the core protein is formation of the viral nucleocapsid. Indeed, the core protein has been observed to form dimers and multimers (Baumert *et al.*, 1998; Nolandt *et al.*, 1997; Kunkel *et al.*, 2001; Matsumoto *et al.*, 1996). A small fraction of the core dimer is highly stable and resistant to denaturation and reduction by SDS and β-mercaptoethanol (Lu *et al.*, 2001). A potential role for tissue transglutaminase in core protein dimer formation has been proposed, based on the results of induction and inhibition experiments. Also, post-translational modification of the core protein by tissue transglutaminase has been observed to generate multimers *in vitro* (Lu *et al.*, 2001).

The ubiquitin-proteasome pathway is the major route by which selective protein degradation occurs in eukaryotic cells and is now emerging as an essential mechanism of cellular regulation (reviewed in Hershko *et al.*, 1998; Finley *et al.*, 2004). This pathway is also involved in the post-translational regulation of the core protein. The core protein is unstable in cells when expressed as the C-terminal truncated forms such as aa 1–173 (21 kDa) and 1–152 (17 kDa) (Suzuki *et al.*, 2001; Moriishi *et al.*, 2003). Specific inhibitors of the 20S proteasome stabilize these short-lived forms of the core protein, suggesting that the proteasome machinery is responsible for their degradation. By contrast, the full-length form of the core protein (aa 1–191) is long-lived and its life is only minimally prolonged by treatment with proteasome inhibitors. Although both C-terminal truncated and full-length forms of the core protein can be ubiquitylated, only conjugation of C-terminal truncated forms to form multiubiquitin chains has been observed. The predominant stable form of the core protein links to a single or only a few ubiquitin moieties.

The homopolymeric multiubiquitin chain, which links multiple ubiquitin molecules through isopeptide bonds between Lys-48 and Gly-76, is a signal for targeting various substrates to the proteasome. Studies of structurally defined multiubiquitylated substrates have revealed that a chain of four or longer is required for efficient proteasomal targeting (Thrower *et al.*, 2000). The same scenario has been suggested for destabilization of the core protein. Differences in the stability of various forms of the core protein depend on how the ubiquitin chain is assembled, such as the number of ubiquitin molecules conjugated to the core protein. Diversity in ubiquitin-chain assembly might result from variations in the conformation of the core protein. The conformation of C-terminal truncated versions of the core protein might favor multiubiquitylation. Although more stable forms of the core protein are recognized by the ubiquitin-conjugating system, ubiquitylation ceases after only one or a few cycles. This may be due to steric constraints. It has been hypothesized that the conformation of stable forms of the core protein might interfere with elongation of the ubiquitin chain.

Recently, a role for the core-binding protein PA28γ in degradation of the core protein has been suggested (Moriishi *et al.*, 2003). Overexpression of PA28γ promotes proteolysis of the core protein. PA28γ predominates in the nucleus and forms a homopolymer, which associates with the 20S proteasome (Tanahashi *et al.*, 1997), thereby enhancing proteasomal activity (Realini *et al.*, 1997). Both nuclear retention and core protein stability are regulated via a PA28γ-dependent pathway.

The biological significance of ubiquitylation of the core protein is not well understood. In eukaryotic cells, targeted protein degradation is increasingly understood to be an important mechanism by which cells regulate levels of specific proteins, and thereby regulate their function. Presumably, the core protein plays a key role in viral replication and pathogenesis since it forms the viral particle and regulates a number of host cell functions. Degradation of the core protein by the ubiquitin-proteasome pathway might result in downregulation of viral assembly and/or inhibition of a number of intracellular processes mediated by the core protein.

Viral assembly

A crucial function of the core protein is assembly of the viral nucleocapsid. However, the molecular mechanism by which this occurs is still uncertain. In addition to mature, enveloped HCV virions (Kaito *et al.*, 1994; Kanto *et al.*, 1994), nonenveloped nucleocapsid-like particles and viral structures expressing the core protein on their surface are present in the plasma of HCV-infected patients (Takahashi *et al.*, 1992; Ishida *et al.*, 2001; Maillard *et al.*, 2001). Thus, in spite of an internal component of infectious particles, the HCV nucleocapsid may be a feature of the viral morphogenesis and be released from the cells. Nucleocapsid-like particles obtained from patient plasma are spherical particles, 33–40 nm in diameter, with a buoyant density of 1.22–1.25 g/ml in sucrose density gradients (Ishida *et al.*, 2001) or those with 38–43 or 54–62 nm appeared at 1.32–1.34 g/ml in CsCl (Malliad *et al.*, 2001). They have similar morphology and physicochemical properties to HCV nucleocapsids isolated by detergent treatment of putative virions.

Several expression systems have been used to investigate HCV capsid assembly using mammalian, insect, yeast, bacteria, and reticulocyte lysates, as well as purified recombinant proteins (Baumert *et al.*, 1998; Falcon *et al.*, 1999; Shimizu *et al.*, 1996; Kunkel *et al.*, 2001; Lorenzo *et al.*, 2001; Acosta-Rivero *et al.*, 2001; Kunkel *et al.*, 2002; Acosta-Rivero *et al.*, 2003; Blanchard *et al.*, 2003; Majeau *et al.*, 2004; Klein *et al.*, 2004). The results suggest that immunogenic nucleocapsid-like particles are heterologous in size and range from 30 to 80 nm in diameter. The N-terminal half of the core protein is important for nucleocapsid formation (Kunkel *et al.*, 2001; Majeau *et al.*, 2004; Klein *et al.*, 2004). Assembly of the nucleocapsid does not require presence of the viral envelope or nonstructural proteins. In addition, HCV capsid formation occurs in the presence or absence of ER-derived membrane, which supports cleavage of the signal peptide at the C-terminus (Klein *et al.*, 2004).

Nucleocapsid assembly generally involves oligomerization of the capsid protein and encapsidation of genomic RNA. In fact, study of a recombinant mature core protein has shown it to exist as a large multimer in solution under physiological conditions, within which stable secondary structures have been observed (Kunkel *et al.*, 2004). There is conflicting data regarding the sequence responsible for self-interaction of the core protein. Studies using yeast two-hybrid systems have identified a potential homotypic interaction domain within the N-terminal region of the core protein (aa 1–115 or–122), with particular emphasis on the region encompassing aa 82–102 (Matsumoto *et al.*, 1996; Nolandt *et al.*, 1997). However, more recent studies have identified two C-terminal regions, extending from aa 123–191 and 125–179, as responsible for self-interaction. Furthermore, Pro substitution within these C-terminal regions has been observed to abolish core protein self-interaction (Yan *et al.*, 1998, Kunkel *et al.*, 2004). Circular dichroism spectroscopy has further shown that a Trp-rich region spanning aa 76–113 is largely solvent-exposed and unlikely to play a role in multimerization (Kunkel *et al.*, 2004).

Once a HCV nucleocapsid is formed in the cytoplasm, it acquires an envelope as it buds through intracellular membranes. Interactions between the core protein and E1/E2 envelope proteins are thought to determine viral morphology. Expression of HCV structural proteins using recombinant virus vectors has succeeded in generating virus-like particles with similar ultrastructural properties to HCV virions. Packaging of these HCV-like particles into intracellular vesicles as a result of budding from the ER has been noted (Baumert *et al.*, 1998; Ezelle *et al.*, 2002; Clayton *et al.*, 2002). Mapping studies to determine the nature of interaction between core and E1 proteins have demonstrated the importance of C-terminal regions in this interaction (Lo *et al.*, 1996; Ma *et al.*, 2002). Since corresponding sequences are not well conserved among various HCV isolates, interactions between core and E1 proteins might depend more on hydrophobicity than specific sequences.

The core protein is highly basic, especially its N-terminal half. Therefore, the protein may encapsulate the viral genome within the nucleocapsid and/or associate with RNA in infected cells. Interactions between the core protein and homologous and heterologous RNA have been analyzed. The results of an *in vitro* North-Western analysis suggests that the core protein binds to ribosomes and the 5′ untranslated region (5′ UTR) of viral genomic RNA, regardless of the specific RNA sequences involved (Santolini *et al.*, 1994; Hwang *et al.*, 1995; Fan *et al.*, 1999). By contrast, an *in vivo* system has been used to show preferential interaction between the core protein and positive-sense viral RNA containing the 5′ UTR and part of the structural protein coding region (Shimoike *et al.*, 1999). A study employing surface plasmon resonance technology has demonstrated selective binding between the core protein and synthetic oligonucleotides corresponding to the 5′ UTR, as well as short homopolymeric oligomers. In addition, it has been shown that the core protein has strong affinity for the stem–loop IIId domain of the 5′ UTR and for (G)-rich sequences (Tanaka *et al.*, 2000). Formation of HCV nucleocapsid-like particles by purified core protein occurs in the presence of full-length or partial 5′ UTR RNA, but not in the absence of nucleic acids, indicating that interaction between the core protein and RNA is essential for HCV assembly (Kunkel *et al.*, 2001). Recently, the core protein has been reported to have nucleic acid chaperone activity (Cristofari *et al.*, 2004). The core protein enhances hybridization of complementary DNA and RNA sequences, and allows the formation of stable structures by strand exchange. The core protein also mediates dimerization of viral positive-stranded 3′ UTR RNA

In addition to the importance of core-RNA interactions for virion formation, these interactions are also thought to regulate 5′ UTR internal ribosome entry site (IRES)-mediated viral translation, which is required for the initiation of cap-independent translation. It has been shown that the core protein down-regulates HCV translation through interactions with viral IRES RNA (Shimoike *et al.*, 1999). Although a conflicting report has suggested that inhibition of HCV translation is due to an RNA–RNA interaction, rather than an interaction between RNA and the core protein (Wang *et al.*, 2000), later studies support the role of a core protein sequence spanning aa 34–44 in inhibition of viral translation through its interaction with IRES (Zhang *et al.*, 2002). Furthermore, the N-terminal 20 residues of the core protein have been shown to selectively inhibit translation mediated by HCV IRES in a cell type-specific manner (Li *et al.*, 2003). This implies that the core protein might contribute to virus persistence by maintaining a low level of HCV replication and expression.

Future perspectives

Although progress is being made in understanding biosynthesis and biochemical properties of the HCV core protein, a number of questions remain regarding its maturation and morphogenesis into the viral nucleocapsid. The development of subgenomic- and full-length replicons has greatly contributed to our knowledge regarding the molecular biology of HCV. However, cells expressing HCV replicons do not support virion formation, including the nucleocapsid assembly. There is particular need for the development of an improved cell culture system, by which HCV virions can be efficiently produced and released.

References

Acosta-Rivero, N., Aguilar, J.C., Musacchio, A., Falcon, V., Vina, A., de la Rosa, M.C., and Morales, J. (2001). Characterization of the HCV core virus-like particles produced in the methylotrophic yeast Pichia pastoris. Biochem. Biophys. Res. Commun. *287*, 122–125.

Acosta-Rivero, N., Falcon, V., Alvarez, C., Musacchio, A., Chinea, G., Cristina de la Rosa, M., Rodriguez, A., Duenas-Carrera, S., Tsutsumi, V., Shibayama, M., *et al.* (2003). Structured HCV nucleocapsids composed of P21 core protein assemble primary in the nucleus of Pichia pastoris yeast. Biochem. Biophys. Res. Commun. *310*, 48–53.

Adler, V., Yin, Z., Tew, K.D., and Ronai, Z. (1999). Role of redox potential and reactive oxygen species in stress signaling. Oncogene *18*, 6104–6111.

Alter, M.J. (1995). Epidemiology of hepatitis C in the West. Semin Liver Dis. *15*, 5–14.

Barba, G., Harper, F., Harada, T., Kohara, M., Goulinet, S., Matsuura, Y., Eder, G., Schaff, Z., Chapman, M.J., Miyamura, T., and Brechot, C. (1997). Hepatitis C virus core protein shows a cytoplasmic localization and associates to cellular lipid storage droplets. Proc. Natl. Acad. Sci. USA *94*, 1200–1205.

Baumert, T.F., Ito, S., Wong, D.T., and Liang, T.J. (1998). Hepatitis C virus structural proteins assemble into virus like particles in insect cells. J. Virol. *72*, 3827–3836.

Blanchard, E., Hourioux, C., Brand, D., Ait-Goughoulte, M., Moreau, A., Trassard, S., Sizaret, P.Y., Dubois, F., and Roingeard, P. (2003). Hepatitis C virus-like particle budding: role of the core protein and importance of its Asp111. J. Virol. *77*, 10131–10138.

Chang, S.C., Yen, J.H., Kang, H.Y., Jang, M.H., and Chang, M.F. (1994). Nuclear localization signals in the core protein of hepatitis C virus. Biochem. Biophys. Res. Commun. *205*, 1284–1290.

Choo, Q.L., Kuo, G., Weiner, A.J., Overby, L.R., Bradley, D.W., and Houghton, M. (1989). Isolation of a cDNA clone derived from a blood-borne non-A, non-B viral hepatitis genome. Science *244*, 359–362.

Clayton, R.F., Owsianka, A., Aitken, J., Graham, S., Bhella, D., and Patel, A.H. (2002). Analysis of antigenicity and topology of E2 glycoprotein present on recombinant hepatitis C virus-like particles. J. Virol. *76*, 7672–7682.

Conti, E., Uy, M., Leighton, L., Blobel, G., and Kuriyan, J. (1998). Crystallographic analysis of the recognition of a nuclear localization signal by the nuclear import factor karyopherin alpha. Cell *94*, 193–204.

Cristofari, G., Ivanyi-Nagy, R., Gabus, C., Boulant, S., Lavergne, J.P., Penin, F., and Darlix, J.L. (2004). The hepatitis C virus Core protein is a potent nucleic acid chaperone that directs dimerization of the viral (+) strand RNA *in vitro*. Nucleic. Acids Res. *32*, 2623–2631.

Damelin, M., Silver, P.A., and Corbett, A.H. (2002). Nuclear protein transport. Methods Enzymol. *351*, 587–607.

Di Bisceglie, A.M. (1997). Hepatitis C and hepatocellular carcinoma. Hepatology *26*, 34S-38S.

Ezelle, H.J., Markovic, D., and Barber, G.N. (2002). Generation of hepatitis C virus-like particles by use of a recombinant vesicular stomatitis virus vector. J. Virol. *76*, 12325–12334.

Falcon, V., Acosta-Rivero, N., Chinea, G., de la Rosa, M.C., Menendez, I., Duenas-Carrera, S., Gra, B., Rodriguez, A., Tsutsumi, V., Shibayama, M., *et al.* (2003). Nuclear localization of nucleocapsid-like particles and HCV core protein in hepatocytes of a chronically HCV-infected patient. Biochem. Biophys. Res. Commun. *310*, 54–58.

Falcon, V., Garcia, C., de la Rosa, M.C., Menendez, I., Seoane, J., and Grillo, J. M. (1999). Ultrastructural and immunocytochemical evidences of core-particle formation in the methylotrophic Pichia pastoris yeast when expressing HCV structural proteins (core-E1). Tissue Cell *31*, 117–125.

Fan, Z., Yang, Q.R., Twu, J.S., and Sherker, A.H. (1999). Specific *in vitro* association between the hepatitis C viral genome and core protein. J. Med. Virol. *59*, 131–134.

Finley, D., Ciechanover, A., and Varshavsky, A. (2004). Ubiquitin as a central cellular regulator. Cell *116*, S29-S32.

Fontes, M.R., Teh, T., and Kobe, B. (2000). Structural basis of recognition of monopartite and bipartite nuclear localization sequences by mammalian importin-alpha. J. Mol. Biol. *297*, 1183–1194.

Gonzalez-Peralta, R.P., Fang, J.W., Davis, G.L., Gish, R., Tsukiyama-Kohara, K., Kohara, M., Mondelli, M.U., Lesniewski, R., Phillips, M.I., Mizokami, M., and *et al.* (1994). Optimization for the detection of hepatitis C virus antigens in the liver. J. Hepatol. *20*, 143–147.

Gowans, E.J. (2000). Distribution of markers of hepatitis C virus infection throughout the body. Semin. Liver Dis. *20*, 85–102.

Grakoui, A., Hanson, H.L., and Rice, C.M. (2001). Bad time for Bonzo? Experimental models of hepatitis C virus infection, replication, and pathogenesis. Hepatology *33*, 489–495.

Grakoui, A., Wychowski, C., Lin, C., Feinstone, S.M., and Rice, C.M. (1993). Expression and identification of hepatitis C virus polyprotein cleavage products. J. Virol. *67*, 1385–1395.

Greenberg, C.S., Birckbichler, P.J., and Rice, R.H. (1991). Transglutaminases: multifunctional cross-linking enzymes that stabilize tissues. FASEB J. *5*, 3071–3077.

Harada, S., Watanabe, Y., Takeuchi, K., Suzuki, T., Katayama, T., Takebe, Y., Saito, I., and Miyamura, T. (1991). Expression of processed core protein of hepatitis C virus in mammalian cells. J. Virol. *65*, 3015–3021.

Hershko, A., and Ciechanover, A. (1998). The ubiquitin system. Annu. Rev. Biochem. *67*, 425–479.

Hijikata, M., Kato, N., Ootsuyama, Y., Nakagawa, M., and Shimotohno, K. (1991). Gene mapping of the putative structural region of the hepatitis C virus genome by *in vitro* processing analysis. Proc. Natl. Acad. Sci. USA *88*, 5547–5551.

Hijikata, M., Mizushima, H., Akagi, T., Mori, S., Kakiuchi, N., Kato, N., Tanaka, T., Kimura, K., and Shimotohno, K. (1993). Two distinct proteinase activities required for the processing of a putative nonstructural precursor protein of hepatitis C virus. J. Virol. *67*, 4665–4675.

Hope, R.G., and McLauchlan, J. (2000). Sequence motifs required for lipid droplet association and protein stability are unique to the hepatitis C virus core protein. J. Gen. Virol. *81*, 1913–1925.

Hope, R.G., Murphy, D.J., and McLauchlan, J. (2002). The domains required to direct core proteins of hepatitis C virus and GB virus-B to lipid droplets share common features with plant oleosin proteins. J. Biol. Chem. *277*, 4261–4270.

Houghton, M., Weiner, A., Han, J., Kuo, G., and Choo, Q.L. (1991). Molecular biology of the hepatitis C viruses: implications for diagnosis, development and control of viral disease. Hepatology *14*, 381–388.

Hussy, P., Langen, H., Mous, J., and Jacobsen, H. (1996). Hepatitis C virus core protein: carboxy-terminal boundaries of two processed species suggest cleavage by a signal peptide peptidase. Virology *224*, 93–104.

Hwang, S.B., Lo, S.Y., Ou, J.H., and Lai, M.M. (1995). Detection of Cellular Proteins and Viral Core Protein Interacting with the 5′ Untranslated Region of Hepatitis C Virus RNA. J. Biomed. Sci. *2*, 227–236.

Ishida, S., Kaito, M., Kohara, M., Tsukiyama-Kohora, K., Fujita, N., Ikoma, J., Adachi, Y., and Watanabe, S. (2001). Hepatitis C virus core particle detected by immunoelectron microscopy and optical rotation technique. Hepatol. Res. *20*, 335–347.

Isoyama, T., Kuge, S., and Nomoto, A. (2002). The core protein of hepatitis C virus is imported into the nucleus by transport receptor Kap123p but inhibits Kap121p-dependent nuclear import of yeast AP.-like transcription factor in yeast cells. J. Biol. Chem. *277*, 39634–39641.

Kaito, M., Watanabe, S., Tsukiyama-Kohara, K., Yamaguchi, K., Kobayashi, Y., Konishi, M., Yokoi, M., Ishida, S., Suzuki, S., and Kohara, M. (1994). Hepatitis C virus particle detected by immunoelectron microscopic study. J. Gen. Virol. *75*, 1755–1760.

Kanto, T., Hayashi, N., Takehara, T., Hagiwara, H., Mita, E., Oshita, M., Katayama, K., Kasahara, A., Fusamoto, H., and Kamada, T. (1995). Serial density analysis of hepatitis C virus particle populations in chronic hepatitis C patients treated with interferon-alpha. J. Med. Virol. *46*, 230–237.

Kato, T., Miyamoto, M., Furusaka, A., Date, T., Yasui, K., Kato, J., Matsushima, S., Komatsu, T., and Wakita, T. (2003). Processing of hepatitis C virus core protein is regulated by its C-terminal sequence. J. Med. Virol. *69*, 357–366.

Klein, K.C., Polyak, S.J., and Lingappa, J.R. (2004). Unique features of hepatitis C virus capsid formation revealed by de novo cell-free assembly. J. Virol. *78*, 9257–9269.

Kunkel, M., Lorinczi, M., Rijnbrand, R., Lemon, S.M., and Watowich, S.J. (2001). Self-assembly of nucleocapsid-like particles from recombinant hepatitis C virus core protein. J. Virol. *75*, 2119–2129.

Kunkel, M., and Watowich, S.J. (2002). Conformational changes accompanying self-assembly of the hepatitis C virus core protein. Virology *294*, 239–245.

Kunkel, M., and Watowich, S.J. (2004). Biophysical characterization of hepatitis C virus core protein: implications for interactions within the virus and host. FEBS Lett. *557*, 174–180.

Kuo, G., Choo, Q.L., Alter, H.J., Gitnick, G.L., Redeker, A.G., Purcell, R.H., Miyamura, T., Dienstag, J.L., Alter, M.J., Stevens, C.E., *et al.* (1989). An assay for circulating antibodies to a major etiologic virus of human non-A, non-B hepatitis. Science *244*, 362–4.

Lanford, R.E., Notvall, L., Chavez, D., White, R., Frenzel, G., Simonsen, C., and Kim, J. (1993). Analysis of hepatitis C virus capsid, E1, and E2/NS1 proteins expressed in insect cells. Virology *197*, 225–235.

Lauer, G.M., and Walker, B.D. (2001). Hepatitis C virus infection. N. Engl. J. Med. *345*, 41–52.

Lemberg, M.K., Bland, F.A., Weihofen, A., Braud, V.M., and Martoglio, B. (2001). Intramembrane proteolysis of signal peptides: an essential step in the generation of HLA-E epitopes. J. Immunol. *167*, 6441–6446.

Lemberg, M.K., and Martoglio, B. (2002). Requirements for signal peptide peptidase-catalyzed intramembrane proteolysis. Mol. Cell *10*, 735–744.

Lemberg, M.K., and Martoglio, B. (2004). On the mechanism of SPP-catalysed intramembrane proteolysis; conformational control of peptide bond hydrolysis in the plane of the membrane. FEBS Lett. *564*, 213–218.

Leung, S.W., Harreman, M.T., Hodel, M.R., Hodel, A.H., and Corbett, A.H. (2003). Dissection of the karyopherin alpha nuclear localization signal (NLS)-binding groove: functional requirements for NLS binding. J. Biol. Chem. *278*, 41947–41953.

Li, D., Takyar, S.T., Lott, W.B., and Gowans, E.J. (2003). Amino acids 1–20 of the hepatitis C virus (HCV) core protein specifically inhibit HCV IRES-dependent translation in HepG2 cells, and inhibit both HCV IRES- and cap-dependent translation in HuH7 and CV-1 cells. J. Gen. Virol. *84*, 815–825.

Liu, Q., Tackney, C., Bhat, R.A., Prince, A.M., and Zhang, P. (1997). Regulated processing of hepatitis C virus core protein is linked to subcellular localization. J. Virol. *71*, 657–662.

Lo, S. Y., Masiarz, F., Hwang, S.B., Lai, M.M., and Ou, J.H. (1995). Differential subcellular localization of hepatitis C virus core gene products. Virology *213*, 455–461.

Lo, S. Y., Selby, M., Tong, M., and Ou, J.H. (1994). Comparative studies of the core gene products of two different hepatitis C virus isolates: two alternative forms determined by a single amino acid substitution. Virology *199*, 124–131.

Lo, S. Y., Selby, M.J., and Ou, J.H. (1996). Interaction between hepatitis C virus core protein and E1 envelope protein. J. Virol. *70*, 5177–5182.

Lorand, L., and Conrad, S.M. (1984). Transglutaminases. Mol. Cell. Biochem. *58*, 9–35.

Lorenzo, L. J., Duenas-Carrera, S., Falcon, V., Acosta-Rivero, N., Gonzalez, E., de la Rosa, M.C., Menendez, I., and Morales, J. (2001). Assembly of truncated HCV core antigen into virus-like particles in Escherichia coli. Biochem. Biophys. Res. Commun. *281*, 962–965.

Lu, W., and Ou, J.H. (2002). Phosphorylation of hepatitis C virus core protein by protein kinase A and protein kinase C. Virology *300*, 20–30.

Lu, W., Strohecker, A., and Ou J.H. (2001). Post-translational modification of the hepatitis C virus core protein by tissue transglutaminase. J. Biol. Chem. *276*, 47993–47999.

Ma, H.C., Ke, C.H., Hsieh, T.Y., and Lo, S.Y. (2002). The first hydrophobic domain of the hepatitis C virus E1 protein is important for interaction with the capsid protein. J. Gen. Virol. *83*, 3085–3092.

Machida, K., Tsukiyama-Kohara, K., Seike, E., Tone, S., Shibasaki, F., Shimizu, M., Takahashi, H., Hayashi, Y., Funata, N., Taya, C., *et al.* (2001). Inhibition of cytochrome c release in Fas-mediated signaling pathway in transgenic mice induced to express hepatitis C viral proteins. J. Biol. Chem. *276*, 12140–12146.

Maillard, P., Krawczynski, K., Nitkiewicz, J., Bronnert, C., Sidorkiewicz, M., Gounon, P., Dubuisson, J., Faure, G., Crainic, R., and Budkowska, A. (2001). Nonenveloped nucleocapsids of hepatitis C virus in the serum of infected patients. J. Virol. *75*, 8240–8250.

Majeau, N., Gagne, V., Boivin, A., Bolduc, M., Majeau, J. A., Ouellet, D., and Leclerc, D. (2004). The N-terminal half of the core protein of hepatitis C virus is sufficient for nucleocapsid formation. J. Gen. Virol. *85*, 971–981.

Martoglio, B., and Dobberstein, B. (1998). Signal sequences: more than just greasy peptides. Trends Cell. Biol. *8*, 410–415.

Matsumoto, M., Hwang, S.B., Jeng, K.S., Zhu, N., and Lai, M.M. (1996). Homotypic interaction and multimerization of hepatitis C virus core protein. Virology *218*, 43–51.

McLauchlan, J., Lemberg, M.K., Hope, G., and Martoglio, B. (2002). Intramembrane proteolysis promotes trafficking of hepatitis C virus core protein to lipid droplets. EMBO J. *21*, 3980–3988.

Mihara, K. (2000). Targeting and insertion of nuclear-encoded preproteins into the mitochondrial outer membrane. Bioessays *22*, 364–371.

Moradpour, D., Wakita, T., Tokushige, K., Carlson, R.I., Krawczynski, K., and Wands, J.R. (1996) Characterization of three novel monoclonal antibodies against hepatitis C virus core protein. J. Med. Virol. *48*, 234–41.

Moriishi, K., Okabayashi, T., Nakai, K., Moriya, K., Koike, K., Murata, S., Chiba, T., Tanaka, K., Suzuki, R., Suzuki, T., *et al.* (2003). Proteasome activator PA28gamma-dependent nuclear retention and degradation of hepatitis C virus core protein. J. Virol. *77*, 10237–10249.

Moriya, K., Fujie, H., Shintani, Y., Yotsuyanagi, H., Tsutsumi, T., Ishibashi, K., Matsuura, Y., Kimura, S., Miyamura, T., and Koike, K. (1998). The core protein of hepatitis C virus induces hepatocellular carcinoma in transgenic mice. Nat. Med. *4*, 1065–1067.

Moriya, K., Fujie, H., Yotsuyanagi, H., Shintani, Y., Tsutsumi, T., Matsuura, Y., Miyamura, T., Kimura, S., and Koike, K. (1997). Subcellular localization of hepatitis C virus structural proteins in the liver of transgenic mice. Jpn. J. Med. Sci. Biol. *50*, 169–177.

Moriya, K., Nakagawa, K., Santa, T., Shintani, Y., Fujie, H., Miyoshi, H., Tsutsumi, T., Miyazawa, T., Ishibashi, K., Horie, T., *et al.* (2001). Oxidative stress in the absence of inflammation in a mouse model for hepatitis C virus-associated hepatocarcinogenesis. Cancer Res. *61*, 4365–4370.

Neupert, W. (1997). Protein import into mitochondria. Annu. Rev. Biochem. *66*, 863–917.

Nolandt, O., Kern, V., Muller, H., Pfaff, E., Theilmann, L., Welker, R., and Krausslich, H.G. (1997). Analysis of hepatitis C virus core protein interaction domains. J. Gen. Virol. *78*, 1331–1340.

Okamoto, K., Moriishi, K., Miyamura, T., and Matsuura, Y. (2004). Intramembrane proteolysis and endoplasmic reticulum retention of hepatitis C virus core protein. J. Virol. *78*, 6370–6380.

Okuda, M., Li, K., Beard, M.R., Showalter, L.A., Scholle, F., Lemon, S.M., and Weinman, S.A. (2002). Mitochondrial injury, oxidative stress, and antioxidant gene expression are induced by hepatitis C virus core protein. Gastroenterology *122*, 366–375.

Pawlotsky, J.M. (2004). Pathophysiology of hepatitis C virus infection and related liver disease. Trends Microbiol. *12*, 96–102.

Poynard, T., Yuen, M. F., Ratziu, V., and Lai, C.L. (2003). Viral hepatitis C. Lancet *362*, 2095–2100.

Realini, C., Jensen, C.C., Zhang, Z., Johnston, S.C., Knowlton, J.R., Hill, C.P., and Rechsteiner, M. (1997). Characterization of recombinant REGalpha, REGbeta, and REGgamma proteasome activators. J. Biol. Chem. *272*, 25483–25492.

Robertson, B., Myers, G., Howard, C., Brettin, T., Bukh, J., Gaschen, B., Gojobori, T., Maertens, G., Mizokami, M., Nainan, O., *et al.* (1998). Classification, nomenclature, and database development for hepatitis C virus (HCV) and related viruses: proposals for standardization. International Committee on Virus Taxonomy. Arch. Virol. *143*, 2493–2503.

Robin, M. A., Anandatheerthavarada, H.K., Biswas, G., Sepuri, N.B., Gordon, D.M., Pain, D., and Avadhani, N.G. (2002). Bimodal targeting of microsomal CYP2E1 to mitochondria through activation of an N-terminal chimeric signal by cAMP-mediated phosphorylation. J. Biol. Chem. *277*, 40583–40593.

Rumenapf, T., Unger, G., Strauss, J.H., and Thiel, H.J. (1993). Processing of the envelope glycoproteins of pestiviruses. J. Virol. *67*, 3288–3294.

Sabile, A., Perlemuter, G., Bono, F., Kohara, K., Demaugre, F., Kohara, M., Matsuura, Y., Miyamura, T., Brechot, C., and Barba G. (1999) Hepatitis C virus core protein binds to apolipoprotein AII and its secretion is modulated by fibrates. Hepatology *30*, 1064–76.

Saito, I., Miyamura, T., Ohbayashi, A., Harada, H., Katayama, T., Kikuchi, S., Watanabe, Y., Koi, S., Onji, M., Ohta, Y., *et al.* (1990). Hepatitis C virus infection is associated with the development of hepatocellular carcinoma. Proc Natl. Acad. Sci. USA *87*, 6547–6549.

Sansonno, D., Lauletta, G., and Dammacco, F. (2004). Detection and quantitation of HCV core protein in single hepatocytes by means of laser capture microdissection and enzyme-linked immunosorbent assay. J. Viral. Hepatol. *11*, 27–32.

Santolini, E., Migliaccio, G., and La Monica, N. (1994). Biosynthesis and biochemical properties of the hepatitis C virus core protein. J. Virol. *68*, 3631–3641.

Schwer, B., Ren, S., Pietschmann, T., Kartenbeck, J., Kaehlcke, K., Bartenschlager, R., Yen, T.S., and Ott, M. (2004). Targeting of hepatitis C virus core protein to mitochondria through a novel C-terminal localization motif. J. Virol. *78*, 7958–7968.

Selby, M.J., Choo, Q.L., Berger, K., Kuo, G., Glazer, E., Eckart, M., Lee, C., Chien, D., Kuo, C., and Houghton, M. (1993). Expression, identification and subcellular localization of the proteins encoded by the hepatitis C viral genome. J. Gen. Virol. *74*, 1103–1113.

Shih, C.M., Chen, C.M., Chen, S.Y., and Lee, Y.H. (1995). Modulation of the trans-suppression activity of hepatitis C virus core protein by phosphorylation. J. Virol. *69*, 1160–1171.

Shimizu, Y.K., Feinstone, S.M., Kohara, M., Purcell, R.H., and Yoshikura, H. (1996). Hepatitis C virus: detection of intracellular virus particles by electron microscopy. Hepatology *23*, 205–209.

Shimoike, T., Mimori, S., Tani, H., Matsuura, Y., and Miyamura, T. (1999). Interaction of hepatitis C virus core protein with viral sense RNA and suppression of its translation. J. Virol. *73*, 9718–9725.

Speight, G., and Westaway, E.G. (1989). Carboxy-terminal analysis of nine proteins specified by the flavivirus Kunjin: evidence that only the intracellular core protein is truncated. J. Gen. Virol. *70*, 2209–2214.

Stark, R., Meyers, G., Rumenapf, T., and Thiel, H.J. (1993). Processing of pestivirus polyprotein: cleavage site between autoprotease and nucleocapsid protein of classical swine fever virus. J. Virol. *67*, 7088–7095.

Suzuki, R., Matsuura, Y., Suzuki, T., Ando, A., Chiba, J., Harada, S., Saito, I., and Miyamura, T. (1995). Nuclear localization of the truncated hepatitis C virus core protein with its hydrophobic C terminus deleted. J. Gen. Virol. *76*, 53–61.

Suzuki, R., Sakamoto, S., Tsutsumi, T., Rikimaru, A., Tanaka, K., Shimoike, T., Moriishi, K., Iwasaki, T., Mizumoto, K., Matsuura, Y., Miyamura, T., and Suzuki, T. (2005). Molecular determinants for subcellular localization of hepatitis C virus core protein. J. Virol. *79*, 1270–1281.

Suzuki, R., Tamura, K., Li, J., Ishii, K., Matsuura, Y., Miyamura, T., and Suzuki, T. (2001). Ubiquitin-mediated degradation of hepatitis C virus core protein is regulated by processing at its carboxyl terminus. Virology *280*, 301–309.

Suzuki, T., Matsuura, Y., Harada, T., Suzuki, R., Saito, I., and Miyamura, T. (1996). Molecular basis of subcellular localization of HCV core protein. Liver *16*, 221–224.

Takahashi, K., Kishimoto, S., Yoshizawa, H., Okamoto, H., Yoshikawa, A., and Mishiro, S. (1992). p26 protein and 33-nm particle associated with nucleocapsid of hepatitis C virus recovered from the circulation of infected hosts. Virology *191*, 431–434.

Tanahashi, N., Yokota, K., Ahn, J. Y., Chung, C.H., Fujiwara, T., Takahashi, E., DeMartino, G.N., Slaughter, C.A., Toyonaga, T., Yamamura, K., *et al.* (1997). Molecular properties of the proteasome activator PA28 family proteins and gamma-interferon regulation. Genes Cells *2*, 195–211.

Tanaka, Y., Shimoike, T., Ishii, K., Suzuki, R., Suzuki, T., Ushijima, H., Matsuura, Y., and Miyamura, T. (2000). Selective binding of hepatitis C virus core protein to synthetic oligonucleotides corresponding to the 5′ untranslated region of the viral genome. Virology *270*, 229–236.

Thrower, J.S., Hoffman, L., Rechsteiner, M., and Pickart, C.M. (2000). Recognition of the polyubiquitin proteolytic signal. EMBO J. *19*, 94–102.

Varaklioti, A., Vassilaki, N., Georgopoulou, U., and Mavromara, P. (2002). Alternate translation occurs within the core coding region of the hepatitis C viral genome. J. Biol. Chem. *277*, 17713–17721.

Vassilaki, N., and Mavromara, P. (2003). Two alternative translation mechanisms are responsible for the expression of the HCV ARFP/F/core+1 coding open reading frame. J. Biol. Chem. *278*, 40503–40513.

von Heijne, G. (1983). Patterns of amino acids near signal-sequence cleavage sites. Eur. J. Biochem. *133*, 17–21.

von Heijne, G. (1985). Signal sequences. The limits of variation. J. Mol. Biol. *184*, 99–105.

Walewski, J.L., Keller, T.R., Stump, D.D., and Branch, A.D. (2001). Evidence for a new hepatitis C virus antigen encoded in an overlapping reading frame. RNA *7*, 710–721.

Wang, T.H., Rijnbrand, R.C., and Lemon, S.M. (2000). Core protein-coding sequence, but not core protein, modulates the efficiency of cap-independent translation directed by the internal ribosome entry site of hepatitis C virus. J. Virol. *74*, 11347–11358.

Weihofen, A., Binns, K., Lemberg, M.K., Ashman, K., and Martoglio, B. (2002). Identification of signal peptide peptidase, a presenilin-type aspartic protease. Science *296*, 2215–2218.

Weis, K. (2002). Nucleocytoplasmic transport: cargo trafficking across the border. Curr. Opin. Cell. Biol. *14*, 328–335.

Xu, Z., Choi, J., Yen, T. S., Lu, W., Strohecker, A., Govindarajan, S., Chien, D., Selby, M. J., and Ou, J. (2001). Synthesis of a novel hepatitis C virus protein by ribosomal frameshift. EMBO J. *20*, 3840–3848.

Yamanaka, T., Uchida, M., and Doi, T. (2002). Innate form of HCV core protein plays an important role in the localization and the function of HCV core protein. Biochem. Biophys. Res. Commun. *294*, 521–527.

Yan, B.S., Tam, M.H., and Syu, W.J. (1998). Self-association of the C-terminal domain of the hepatitis-C virus core protein. Eur. J. Biochem. *258*, 100–106.

Yap, S.H., Willems, M., Van den Oord, J., Habets, W., Middeldorp, J.M., Hellings, J.A., Nevens, F., Moshage, H., Desmet, V., and Fevery, J. (1994). Detection of hepatitis C virus antigen by immuno-histochemical staining: a histological marker of hepatitis C virus infection. J. Hepatol. *20*, 275–281.

Yasui, K., Wakita, T., Tsukiyama-Kohara, K., Funahashi, S. I., Ichikawa, M., Kajita, T., Moradpour, D., Wands, J. R., and Kohara, M. (1998). The native form and maturation process of hepatitis C virus core protein. J. Virol. *72*, 6048–6055.

Zhang, J., Yamada, O., Yoshida, H., Iwai, T., and Araki, H. (2002). Autogenous translational inhibition of core protein: implication for switch from translation to RNA replication in hepatitis C virus. Virology *293*, 141–150.

Molecular Virology and Pathogenesis of Hepatitis C

11

Darius Moradpour and Hubert E. Blum

Abstract

Hepatitis C virus (HCV) infection is a major cause of chronic hepatitis, liver cirrhosis and hepatocellular carcinoma worldwide. Exciting advances have recently been made in the understanding of the molecular virology and pathogenesis of hepatitis C. These advances have translated into the identification of novel antiviral targets and the development of innovative therapeutic and preventive strategies, some of which are already in early phase clinical evaluation. Much work remains to be done with respect to the virion structure, the early and late steps of the HCV life cycle, the mechanism and regulation of RNA replication, and the pathogenesis of HCV-induced liver disease. However, given the current pace of HCV research, progress in these areas may be expected in the near future.

Introduction

Hepatitis C virus (HCV) infection is a major cause of chronic hepatitis, liver cirrhosis, and hepatocellular carcinoma (HCC) worldwide (NIH, 2002). A protective vaccine is not available to date, and therapeutic options are still limited. Current standard therapy, pegylated interferon-α (PEG-IFN-α) combined with ribavirin, results in a sustained virologic response in 20–80% of patients, depending on the HCV genotype and other factors (Manns *et al.*, 2001; Fried *et al.*, 2002; Hadziyannis *et al.*, 2004; Muir *et al.*, 2004). However, in clinical practice many patients do not qualify for or do not tolerate IFN-based therapy (Falck-Ytter *et al.*, 2002). As a consequence, the number of patients presenting with long-term sequelae of chronic hepatitis C, including HCC, is expected to further increase for the next 20–30 years (Davis *et al.*, 2003).

HCV was identified in 1989 as the most common etiologic agent of posttransfusion and sporadic non-A, non-B hepatitis (Choo *et al.*, 1989). Investigation of the viral life cycle has been limited by the low viral titers found in sera and livers of infected individuals and the lack of an efficient cell culture system permissive for HCV infection and replication. Despite these obstacles, considerable progress has been made using heterologous expression systems, infectious cDNA clones (Kolykhalov *et al.*, 1997), the replicon system (Lohmann *et al.*, 1999), functional HCV pseudoparticles (Bartosch *et al.*, 2003a; Hsu *et al.*, 2003), and, most recently, systems that allow the production of recombinant infectious virus in tissue culture (Lindenbach *et al.*, 2005a; Wakita *et al.*, 2005; Zhong *et al.*, 2005) (see Bartenschlager *et al.*, 2004; Penin *et al.*, 2004b; Lindenbach *et al.*,

2005b for recent reviews). These and other milestones in HCV research are listed in Table 11.1.

Taxonomy

HCV has been classified in the *Hepacivirus* genus within the Flaviviridae family which includes the classical flaviviruses, such as yellow fever (YFV) and dengue viruses, the animal pestiviruses, such as bovine viral diarrhea virus (BVDV), and the as yet unassigned GB viruses A (GBV-A), GBV-B and GBV-C (van Regenmortel *et al.*, 2000). GBV-C was also designated as hepatitis G virus (HGV). However, it was found subsequently that GBV-C/HGV is not a common agent of viral hepatitis and its pathogenic relevance, if any, remains unknown.

An important feature of HCV is its high genetic variability (Simmonds, 2004). The E1 and E2 regions are particularly variable, whereas the core and some of the nonstructural protein sequences are more conserved. The highest degree of sequence conservation is found in the 5′ and 3′ noncoding regions (NCRs).

HCV isolates fall into three major categories, depending on the degree of sequence divergence: genotypes, subtypes, and isolates. There are six major genotypes that differ in their nucleotide sequence by 30–35%. Within an HCV genotype, several subtypes (designated a, b, c, etc.) can be defined that differ in their nucleotide sequence by 20–25%. The term quasispecies refers to the genetic heterogeneity of the population of HCV genomes coexisting in an infected individual.

The genetic variability of HCV may have important implications for the pathogenesis, natural course and prevention of hepatitis C. However, the clinical significance of HCV genotypes is mainly in the response to antiviral therapy in that patients infected with genotype 1 have a poorer response to IFN-α therapy than those infected with genotype 2 or 3.

Genetic organization

HCV contains a 9.6-kb positive-strand RNA genome composed of a 5′ NCR, a long open reading frame encoding a polyprotein precursor of about 3000 amino acids, and a 3′ NCR (Figure 11.1).

Since in the absence of a robust tissue culture system the only read-out for infectivity was the direct inoculation of *in vitro* transcribed, synthetic RNA into the liver of a chimpanzee, it took 8 years from the discovery of HCV to the establishment of the first infectious cDNA clone (Kolykhalov *et al.*, 1997). In addition, due to the variation present in the quasispecies and errors in-

Table 11.1 Milestones in HCV research.

1989	Identification of HCV
1993	Polyprotein processing
1996	Three-dimensional structure of the NS3 serine protease
1997	Infectious clone of HCV
1998	Interferon-alpha and ribavirin combination therapy
1999	Replicon system
2003	Functional HCV pseudoparticles
2003	Proof-of-concept clinical studies of an HCV protease inhibitor
2005	Production of recombinant infectious HCV in tissue culture

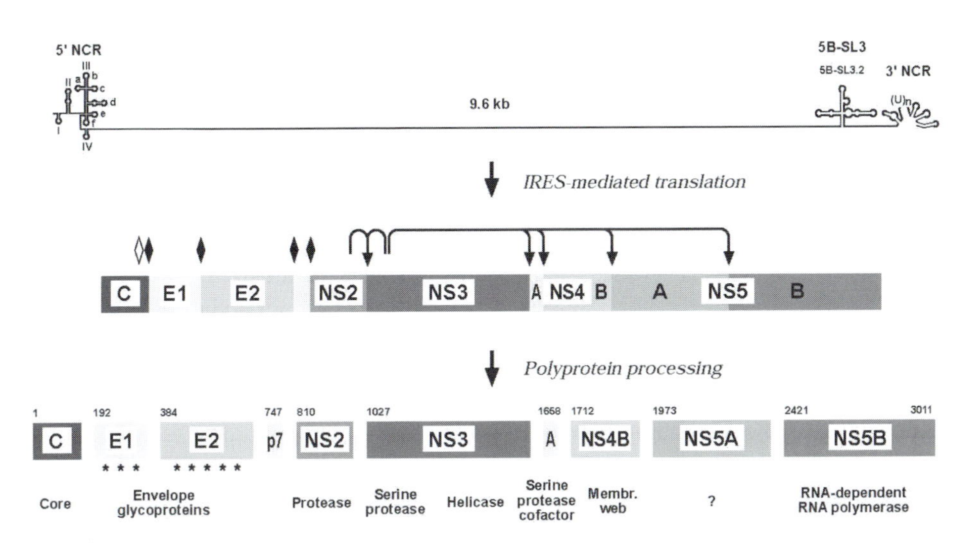

Figure 11.1 Genetic organization and polyprotein processing of HCV. The 9.6-kb positive-strand RNA genome is schematically depicted at the top. Simplified RNA secondary structures in the 5′ and 3′ noncoding regions (NCRs) as well as in the NS5B stem–loop 3 *cis*-acting replication element (5B- SL3) are shown. Internal ribosome entry site (IRES)-mediated translation yields a polyprotein precursor of about 3000 amino acids that is processed into the mature structural and nonstructural proteins. Amino acid positions are shown above each protein (HCV H strain; genotype 1a; GenBank accession number AF009606). Solid diamonds denote cleavage sites of the HCV polyprotein precursor by the endoplasmic reticulum signal peptidase. The open diamond indicates further C-terminal processing of the core protein by signal peptide peptidase. Arrows indicate cleavages by the HCV NS2–3 and NS3 proteases. Asterisks in the E1 and E2 region indicate glycosylation of the envelope proteins. Note that polyprotein processing, illustrated here as a separate step for simplicity, occurs co- and posttranslationally.

troduced by PCR amplification, construction of infectious cDNA clones required preparation of a consensus sequence from a number of clones. Functional cDNA clones now exist for genotypes 1a, 1b and 2a. Genetic studies using infectious clones have shown the essential nature of the HCV enzymes, the conserved elements of the 3′ NCR and difficult-to-study proteins such as p7 (Yanagi *et al.*, 1999; Kolykhalov *et al.*, 2000; Sakai *et al.*, 2003).

The 5′ NCR is highly conserved among different HCV isolates and contains an internal ribosome entry site (IRES) essential for cap-independent translation of the viral RNA (reviewed in Rijnbrand and Lemon, 2000; Sarnow, 2003). It contains four highly ordered domains numbered I to IV (Figure 11.1). Domain I is not required for IRES activity, but is essential for HCV RNA replication (Friebe *et al.*, 2001). Domains II and III include two large stem–loops. Subdomain IIIf forms a pseudoknot with domain IV which contains the translation initiation codon. Domains II, III and IV, together with the first 24–40 nucleotides of the core coding region, constitute the IRES. The key element is domain III that permits direct binding of the 40 S ribosomal subunit to subdomains IIIa, IIIc, IIId and IIIe as well as of eukaryotic translation initiation factor 3 (eIF3) to subdomain IIIb. The three-dimensional structure of the HCV IRES bound to the 40 S ribosomal subunit was resolved at 20 Å resolution by cryoelectron microscopy (Spahn *et al.*, 2001). In addition, high-resolution structures are now available for critical elements of the IRES, including stem–loops II, IIIb, IIId and IIIe as well as the IIIabc four-way junction, facilitating the design of small molecule inhibitors of HCV translation initiation (Lukavsky *et al.*, 2000; Collier *et al.*, 2002; Kieft *et al.*, 2002; Lukavsky *et al.*, 2003).

A current model of HCV translation initiation includes formation of a binary complex between the IRES and the 40 S ribosomal subunit, followed by the assembly of a 48 S-like complex at the AUG initiation codon upon association of eIF3 and ternary complex (eIF2•Met-$tRNA_i^{Met}$•GTP) and, in a rate limiting step, GTP-dependent association of the 60 S subunit to form the 80 S complex (Otto and Puglisi, 2004).

The 3′ NCR is composed of a short variable region, a poly(U/UC) tract with an average length of 80 nucleotides, and an almost invariant 98 nucleotide RNA element, designated the X-tail (Kolykhalov *et al.*, 1996; Tanaka *et al.*, 1996; Friebe and Bartenschlager, 2002; Yi and Lemon, 2003a and 2003b). A new *cis*-acting replication elements (CRE) was recently discovered in the sequence encoding the C-terminal domain of the nonstructural protein 5B (NS5B) (You *et al.*, 2004). An essential stem–loop, designated 5B-SL3.2, was identified within a larger cruciform RNA element, designated 5B-SL3 (Figure 11.1). More recently, it was shown that the upper loop of 5B-SL3.2 is engaged in a kissing interaction with a stem–loop in the X-tail, suggesting that a pseudoknot structure essential for RNA replication is formed at the 3′ end of the HCV genome (Friebe *et al.*, 2005).

Polyprotein processing

IRES-mediated translation of the HCV open reading frame yields a polyprotein precursor that is co- and posttranslationally processed by cellular and viral proteases into the mature structural and nonstructural proteins (Figure 11.1). The structural proteins include the core protein and the envelope glycoproteins E1 and E2. These are released from the polyprotein precursor by the endoplasmic reticulum (ER) signal peptidase. The structural proteins are separated from the nonstructural proteins by the p7 polypeptide. The nonstructural proteins include the NS2–3 protease and the NS3 serine protease, an RNA helicase/NTPase located in the C-terminal two-thirds of NS3, the NS4A polypeptide, the NS4B and NS5A proteins, and the NS5B RNA-dependent RNA polymerase (RdRp). The NS2–3 protease cleaves at the NS2/NS3 site while the NS3 serine protease is responsible for processing of the downstream nonstructural proteins (Figure 11.1).

Structural proteins

Core

The first structural protein encoded by the HCV open reading frame is the core protein. During translation of the HCV polyprotein, the nascent polypeptide is targeted to the ER membrane for translocation of the E1 ectodomain into the ER lumen, a process mediated by an internal signal sequence located between the core and E1 sequences. Cleavage of the signal sequence by signal peptidase yields an immature 191-amino-acid core protein which contains the E1 signal peptide at its C-terminus. Further processing of this signal peptide by the signal peptide peptidase (SPP) yields the mature 21-kDa core protein of approximately 179 amino acids (McLauchlan *et al.*, 2002).

The N-terminal hydrophilic domain of core contains a high proportion of basic amino acid residues and has been implicated both in RNA binding and homooligomerization. When expressed in mammalian cells, core is found on membranes of the ER, in seemingly ER-derived membranous webs (see below), on the surface of lipid droplets and, to a limited extent, in the nucleus (Moradpour *et al.*, 1996; Barba *et al.*, 1997; Egger *et al.*, 2002b; Hope *et al.*, 2002). It is presently unclear if the association with lipid droplets, which is mediated by the central, relatively hydrophobic domain of core and was detected in different heterologous expression systems, in transgenic mice and in liver specimens from HCV-infected chimpanzees, plays a role during viral replication or virion morphogenesis. It has been speculated that the interaction of core with lipid droplets may affect lipid metabolism, which in turn may contribute to the development of liver steatosis. The observation that certain HCV core-transgenic mice develop steatosis and HCC has lent further support to this hypothesis (Moriya *et al.*, 1998; Lerat *et al.*, 2002).

Little is known about the assembly of core into nucleocapsids. *In vitro* studies with recombinant HCV core proteins demonstrated that the N-terminal 124 amino acid residues are sufficient for the assembly of nucleocapsid-like structures and that the presence of structured RNA is required for this process (Kunkel *et al.*, 2001). However, under these experimental conditions RNA encapsidation is not specific, and the signals and processes that mediate RNA packaging and nucleocapsid assembly during HCV replication are unknown.

Intriguingly, the core protein has been reported to interact with a variety of cellular proteins and to influence numerous host cell functions, including apoptosis, cell cycle control, gene expression and many others (reviewed in Tellinghuisen and Rice, 2002). However, the relevance of these observations, derived mainly from heterologous overexpression experiments, for the natural course and pathogenesis of hepatitis C is presently unknown.

ARFP/F protein

An alternative reading frame (ARF) was recently identified in the HCV core region that, as a result of a -2/+1 ribosomal frameshift, has the potential to encode a protein of up to 160 amino acids, designated ARFP (alternative reading frame protein) or F (frameshift) protein (Walewski *et al.*, 2001; Xu *et al.*, 2001; Varaklioti *et al.*, 2002) (reviewed in Branch *et al.*, 2005). Expression of the ARFP/F protein of HCV genotype 1a *in vitro* or in mammalian cells yields a 17-kDa protein. Amino acid sequencing indicated that the frameshift likely occurs at or near codon 11 of the core protein sequence (Xu *et al.*, 2001). However, multiple frameshifting events have been reported in this region, and a 1.5-kDa protein could also be produced by −1/+2 frameshifting (Choi *et al.*, 2003). In addition, the frameshift position seems to be genotype-dependent, as a +1 frameshift at codon 42 was reported for genotype 1b (Boulant *et al.*, 2003). Detection of antibodies (Komurian-Pradel *et al.*, 2004) and T cells (Bain *et al.*, 2004) specific for the ARFP/F protein in patients with hepatitis C suggests that this protein is expressed during HCV infection. However, given that the ARF is not present in subgenomic HCV replicons, the ARFP/F protein is not required for HCV RNA replication *in vitro*. Thus, the functions, if any, of the ARFP/F protein in the life cycle and pathogenesis of HCV remain to be elucidated.

Envelope glycoproteins

The envelope proteins E1 and E2 are extensively glycosylated and have an apparent molecular weight of 30–35 and 70–

72 kDa, respectively. They form a noncovalent complex, which is believed to represent the building block for the viral envelope (Op De Beeck *et al.*, 2001). E2 seems to make contact with the cellular receptor(s) for HCV while E1 has been predicted to possess fusion activity.

E1 and E2 are type I transmembrane glycoproteins. Interestingly, the transmembrane domains, located at their C-termini, are involved in heterodimerization and have ER retention properties (Cocquerel *et al.*, 2002). The ectodomains of E1 and E2 contain numerous highly conserved cysteine residues that may form 4 and 9 intramolecular disulfide bonds, respectively. In addition, E1 and E2 contain up to 5 and 11 glycosylation sites, respectively. Thus, HCV glycoprotein maturation and folding is a complex process that involves the ER chaperone machinery and depends on disulfide bond formation as well as glycosylation.

A structural model for E2 based on the structure of the envelope protein from tick-borne encephalitis virus (TBE; a member of the flavivirus genus) (Rey *et al.*, 1995) was proposed (Yagnik *et al.*, 2000). According to this model, E2 forms an elongated and flat head-to-tail homodimer. The fact that the envelope protein of Semliki Forest virus (Lescar *et al.*, 2001), a more distantly related alphavirus, has a similar structure as the envelope proteins of TBE (Rey *et al.*, 1995) and dengue virus (Modis *et al.*, 2003), suggests that HCV may have a similar surface architecture. However, virtually nothing is known about the actual structure of the HCV E1-E2 complex, and the processes that mediate viral attachment, entry, and fusion have only recently become amenable to systematic study (see below).

As discussed above, the genes encoding the envelope glycoproteins E1 and E2 are particularly variable. A hypervariable region (HVR) of approximately 28 amino acids in the N-terminal domain of E2 has been termed HVR1 (Hijikata *et al.*, 1991; Weiner *et al.*, 1991). The HVR1 amino acid sequence differs by up to 80% among HCV isolates. However, despite high variability at the sequence level the structure of this domain was found to be quite conserved (Penin *et al.*, 2001). HVR1 appears to contain a neutralization epitope (Farci *et al.*, 1996) and variability, therefore, may be driven by antibody selection.

p7

p7 is a 63-amino-acid polypeptide that is often incompletely cleaved from E2. It has two transmembrane domains connected by a short hydrophilic segment which forms a cytoplasmic loop, while the N- and C-termini are oriented toward the ER lumen (Carrère-Kremer *et al.*, 2002). p7 of the related pestivirus BVDV is essential for the production of infectious progeny, but not for RNA replication (Harada *et al.*, 2000). Similarly, HCV p7 is not required for HCV RNA replication because it is not present in subgenomic replicons. However, it is essential for virus infectivity *in vivo*, as shown by genetic studies using infectious HCV cDNA clones (Sakai *et al.*, 2003). p7 has recently been reported to form hexamers and to possess ion channel activity (Griffin *et al.*, 2003; Pavlovic *et al.*, 2003). These properties suggest that p7 belongs to the viroporin family and may represent an attractive target for antiviral intervention.

Virion structure

While exciting progress has been made with respect to related flavi- (Ferlenghi *et al.*, 2001; Kuhn *et al.*, 2002; Mukhopadhyay *et al.*, 2003) and alphaviral virion structures (Lescar *et al.*, 2001), the structure of

HCV remains largely unknown. By analogy to these related viruses, one can assume that the core protein and the envelope glycoproteins E1 and E2 represent the principal structural components of the virion. E1 and E2 are presumably anchored to a host cell-derived lipid envelope that surrounds a nucleocapsid composed of multiple copies of the core protein and encapsidating the genomic RNA. HCV particles are believed to have a diameter of 40–70 nm (Kaito *et al.*, 1994; Shimizu *et al.*, 1996a). These observations were recently confirmed by immunoelectron microscopy of infectious viral particles produced in tissue culture (Wakita *et al.*, 2005).

HCV circulates in various forms in the infected host, including virions bound to low-density (LDL) and very-low-density lipoproteins (VLDL) (André *et al.*, 2002), which appear to represent the infectious fraction, virions bound to immunoglobulins, and free virions.

Nonstructural proteins

NS2–3 protease

Cleavage of the polyprotein precursor at the NS2/NS3 junction is accomplished by a protease encoded by NS2 and the N-terminal one-third of NS3. NS2 is dispensable for replication of subgenomic replicons *in vitro* (see below) and is thus not essential for the formation of a functional replication complex. However, the NS2–3 protease activity is essential for the replication of full-length HCV genomes *in vivo*. Site-directed mutagenesis has shown that amino acids His 143 (corresponding to His 952 of the HCV polyprotein), Glu 163 (Glu 972) and Cys 184 (Cys 993) are essential for catalytic activity (Grakoui *et al.*, 1993; Hijikata *et al.*, 1993). Recombinant proteins lacking the N-terminal membrane domain of NS2 have been found to retain cleavage activity, allowing further characterization of this unique enzymatic activity (Pallaoro *et al.*, 2001; Thibeault *et al.*, 2001).

NS3–4A complex

A distinct serine protease located in the N-terminal one-third of NS3 is responsible for the downstream cleavage events in the nonstructural region (reviewed in De Francesco and Steinkühler, 2000). In addition, an RNA helicase/NTPase domain is found in the C-terminal two-thirds of NS3 (reviewed in Kwong *et al.*, 2000). The catalytic triad of the NS3 serine protease is formed by His 57 (corresponding to His 1083 of the HCV polyprotein), Asp 81 (Asp 1107) and Ser 139 (Ser 1165). The NS4A polypeptide functions as a cofactor for the NS3 serine protease and is incorporated as an integral component into the enzyme core. Complex formation occurs via a tight interaction of the N-terminal residues of NS3 with 12 amino acid residues in the center of NS4A. NS3 by itself has no membrane anchor. The N-terminal domain of NS4A is strongly predicted to form a transmembrane α-helix responsible for membrane anchorage of the NS3–4A complex (Wölk *et al.*, 2000).

The crystal structures of the serine protease (Kim *et al.*, 1996; Love *et al.*, 1996; Yan *et al.*, 1998) and RNA helicase domains of NS3 (Yao *et al.*, 1997; Kim *et al.*, 1998) as well as the entire NS3 protein (Yao *et al.*, 1999) have been elucidated. These enzymes are essential for viral replication and have emerged as prime targets for the design of specific inhibitors as antiviral agents (Lamarre *et al.*, 2003; Hinrichsen *et al.*, 2004; reviewed in De Francesco and Migliaccio, 2005).

The NS3 helicase is a member of the so-called helicase superfamily 2. These are also called DEXH/D helicases according to a characteristic signature sequence in one of the essential enzyme motifs. It was recently shown that NS3 unwinds RNA through a highly coordinated cycle of fast ripping and local pausing that occurs with regular spacing along the duplex substrate, suggesting that nucleic acid motors can function in a manner analogous to cytoskeletal motor proteins (Serebrov and Pyle, 2004).

NS4B

NS4B is a 27-kDa integral membrane protein with a multispanning topology (Hügle *et al.*, 2001; Lundin *et al.*, 2003). The NS4B proteins of HCV, pesti- and flaviviruses are similar in size, amino acid composition, and hydrophobic properties. No function, however, has yet been ascribed to NS4B in any of these related viruses. More recently, it was found by electron microscopy that expression of HCV NS4B induces the formation of a seemingly ER-derived specific cellular membrane alteration, designated the membranous web, that harbors the viral replication complex (Egger *et al.*, 2002b; Gosert *et al.*, 2003) (see below). Thus, a function of NS4B may be to induce the specific membrane alteration that serves as a scaffold for the HCV replication complex.

NS5A

NS5A is a phosphoprotein of unknown structure and function. It is found in a basally phosphorylated form of 56 kDa and in a hyperphosphorylated form of 58 kDa. Basal phosphorylation requires domains in the center and C-terminus of NS5A. The centrally located serine residues 225, 229 and 232 (corresponding to Ser 2197, Ser 2201 and Ser 2204 of the HCV polyprotein) are important for NS5A hyperphosphorylation (Figure 11.2). However, it is unknown if these serine residues are actually phosphorylated or if they affect phosphorylation indirectly. The only phosphoacceptor sites that have been mapped experimentally are serine residues 349 (Ser 2321) (Reed and Rice, 1999) and 222 (Ser 2194) (Katze *et al.*, 2000) in genotypes 1a and 1b, respectively. A number of kinases capable of phosphorylating NS5A have been identified. However, it is unknown which cellular kinase generates the different phosphoforms of NS5A during the viral life cycle.

Interestingly, adaptive mutations have been found to cluster in the central region of NS5A in the context of selectable subgenomic HCV replicons, suggesting that NS5A is involved—either directly or by interaction with cellular proteins and pathways—in the viral replication process. The detection of adaptive mutations in NS5A, evidence for physical and functional interactions of NS5A with other HCV nonstructural proteins (Shirota *et al.*, 2002; Dimitrova *et al.*, 2003), and the results of site-directed mutagenesis (Elazar *et al.*, 2003; Penin *et al.*, 2004a) point towards an essential role of NS5A in HCV RNA replication. However, its function remains elusive.

HCV NS5A has attracted considerable interest because of its potential role in modulating the IFN response (reviewed in Tan and Katze, 2001). Studies performed in Japan first described a correlation between mutations within a discrete region of NS5A, termed interferon sensitivity determining region (ISDR) (Figure 11.2), and a favorable response to IFN-α therapy (Enomoto *et al.*, 1996). However, this controversial issue has thus far not translated into clinically applicable predictors. The same applies to other regions of NS5A that have been associated with the response to IFN therapy, such as a variable region in the C-terminal domain of NS5A

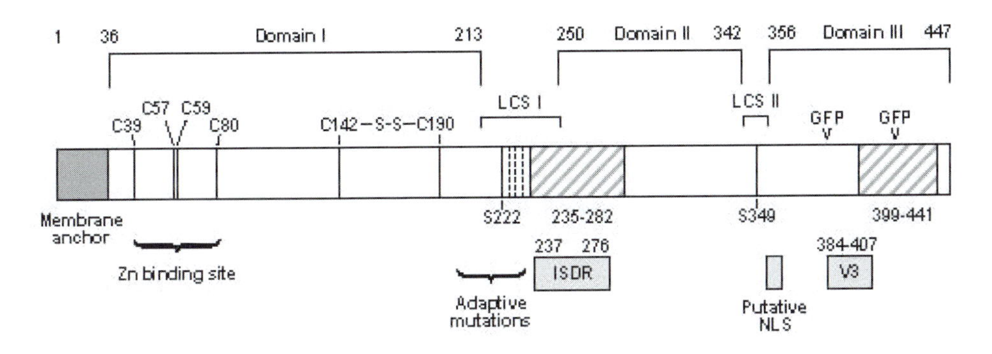

Figure 11.2 Overview of the HCV NS5A protein. NS5A is drawn to scale as a box. Amino acid positions relate to the HCV Con1 sequence (genotype 1b; GenBank accession number AJ238799; add 1972 amino acids to obtain positions relative to the HCV polyprotein). The domain organization proposed by Tellinghuisen *et al.* (2004) is shown. Domains I–III are connected by low complexity sequences (L CS) I and II. Cysteine residues 39, 57, 59 and 80 coordinate one zinc atom per NS5A protein. Cysteine residues 142 and 190 form a disulfide bond. The N-terminal amphipathic α-helix, which mediates in-plane membrane association of NS5A, is highlighted by a gray box. Phosphoacceptor sites mapped for genotype 1a (Ser 349) (Reed and Rice, 1999) and 1b (Ser 222) (Katze *et al.*, 2000) isolates and are highlighted. Dashed lines denote serine residues 225, 229 and 232 that affect hyperphosphorylation of NS5A. The region where cell culture-adaptive mutations have been found to cluster in the replicon system, the so called interferon sensitivity determining region (ISDR), a putative nuclear localization signal (NLS), and variable region 3 (V3) are highlighted. Deletions found to be tolerated in the replicon system are indicated by shaded boxes (Blight *et al.*, 2000; Zhu *et al.*, 2003; Appel *et al.*, 2005). GFP relates to sites that were found to tolerate the insertion of green fluorescent protein (Moradpour *et al.*, 2004b). From Moradpour *et al.*, 2005, with permission.

termed V3 (Figure 11.2). Numerous other functions and a plethora of interaction partners have been postulated for NS5A (reviewed in Tellinghuisen and Rice, 2002; Macdonald and Harris, 2004). However, only few of these potential interactions have been validated in a meaningful experimental context involving active HCV RNA replication and surprisingly little effort has been devoted to the basic biochemical characterization of this protein.

An N-terminal amphipathic α-helix mediates membrane association of NS5A (Brass *et al.*, 2002; Elazar *et al.*, 2003; Penin *et al.*, 2004a). This helix exhibits a hydrophobic, tryptophan-rich side embedded in the cytosolic leaflet of the membrane bilayer, while the polar, charged side is exposed to the cytosol. Thus, NS5A is a monotopic protein with an in-plane amphipathic α-helix as membrane anchor. Comparative sequence analyses and limited proteolysis of recombinant NS5A protein have recently led to a proposed domain organization of NS5A (Tellinghuisen *et al.*, 2004) (Figure 11.2). The relatively highly conserved domain I immediately following the membrane-anchoring α-helix has been shown to contain four absolutely conserved cysteine residues that coordinate one zinc atom per NS5A protein (Tellinghuisen *et al.*, 2004). Mutation of these residues abolishes HCV RNA replication, indicating that the zinc is essential for NS5A structure and/or function. The structure of NS5A domain I has recently been solved, revealing a novel protein fold, a new zinc coordination motif, a dimerization interface, and a putative protein–protein interaction surface. Furthermore, analyses of the molecular surface together with the recently reported structure of the N-terminal membrane anchor domain of NS5A (Penin *et al.*, 2004a) allowed the authors to tentatively position the molecule with respect to the membrane and to propose a potential RNA-binding groove exposed towards the cytosol (Tellinghuisen *et al.*, 2005).

NS5B

HCV replication proceeds *via* synthesis of a complementary negative-strand RNA using the genome as a template and the subsequent synthesis of genomic positive-strand RNA from this negative-strand RNA template. The key enzyme responsible for both of these steps is the NS5B RdRp. This essential viral enzyme has been extensively characterized at the biochemical (Behrens *et al.*, 1996; Lohmann *et al.*, 1997; Yamashita *et al.*, 1998; Ferrari *et al.*, 1999) and structural level (Ago *et al.*, 1999; Bressanelli *et al.*, 1999; Lesburg *et al.*, 1999; Bressanelli *et al.*, 2002) and has emerged as a major target for antiviral intervention. The HCV NS5B protein contains motifs shared by all RdRps, including the hallmark GDD sequence within motif C, and possesses, based on the similarity of the enzyme structure with the shape of a right hand, the classical fingers, palm, and thumb subdomains. A special feature of the HCV RdRp is that extensive interactions between the fingers and thumb subdomains result in a completely encircled active site (Figure 11.3). More recently, it was found that this feature is shared by

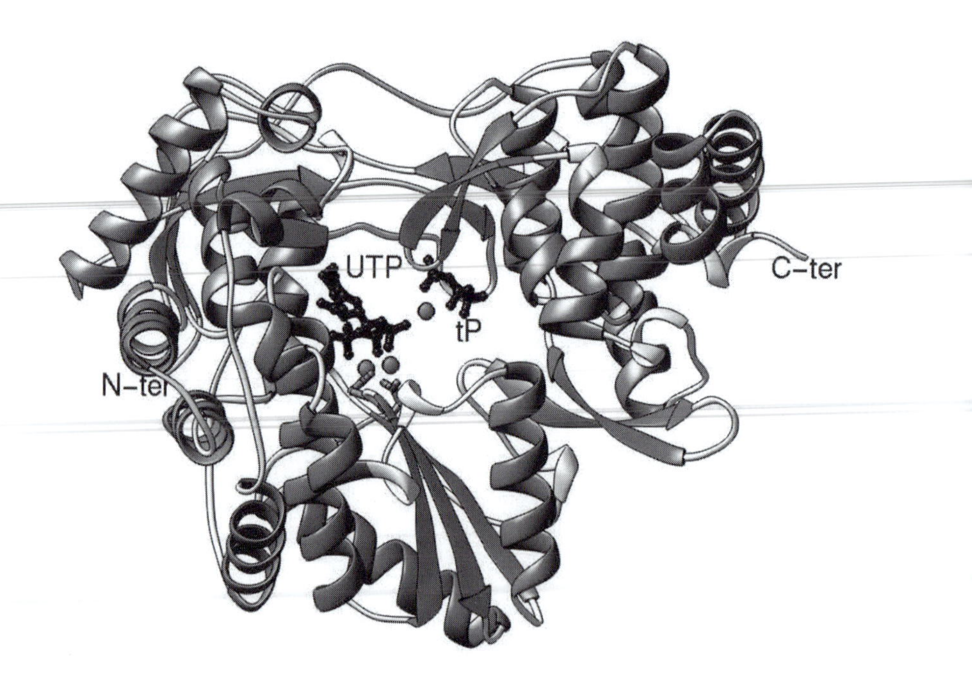

Figure 11.3 Crystal structure of the catalytic domain of the HCV RNA-dependent RNA polymerase (Bressanelli *et al.*, 1999; PDB accession code 1GX6). Ribbon diagram of C-terminally truncated NS5B protein complexed with UTP and Mn^{2+}. The bound nucleotide and the side chains of the catalytic aspartic acids (Asp 220 and Asp 318) in the center of the structure are represented as ball-and-stick. Mn^{2+} ions are shown as spheres. Also labeled is the triphosphate (tP) moiety of a nucleotide bound to the "priming" site. This figure was kindly provided by Dr. Felix A. Rey, Department of Virology, Pasteur Institute, Paris, France.

other RdRps, including those of the bacteriophage Φ6 and of BVDV (Butcher *et al.*, 2001; Choi *et al.*, 2004).

Membrane association of the HCV RdRp is mediated by the C-terminal 21 amino acid residues which are dispensable for polymerase activity *in vitro*. Membrane targeting occurs by a post-translational mechanism and results in integral membrane association of NS5B (Schmidt-Mende *et al.*, 2001). These features define the HCV RdRp as a member of the so called tail-anchored proteins. The HCV RdRp insertion sequence crosses the membrane bilayer as a transmembrane segment (Ivashkina *et al.*, 2002), is essential for HCV RNA replication in cells, and is likely to possess additional functions apart from its role as a membrane anchor (Moradpour *et al.*, 2004a).

Model systems

Given the lack of a robust cell culture system allowing natural infection, replication, and release of viral progeny, various *in vitro* and *in vivo* models have been used to study HCV (Table 11.2) (reviewed in Lanford and Bigger, 2002).

In vitro models

Infection of primary hepatocytes and established cell lines *in vitro* yielded only low-level replication and often poorly reproducible results. In their present format, some of these systems may be useful for neutralization assays, but not for a systematic investigation of the viral life cycle (Shimizu *et al.*, 1996b; Castet *et al.*, 2002). An alternative approach involves the generation of cell lines constitutively or inducibly expressing viral sequences from chromosomally integrated cDNA (reviewed in Moradpour *et al.*, 2003b). Moreover, viable chimeras of certain positive-strand RNA viruses, such as polio and Sindbis virus, with HCV genetic elements, such as the IRES (Zhao *et al.*, 1999) or NS3 (Filocamo *et al.*, 1999), have been constructed

Table 11.2 *In vitro* and *in vivo* models to study HCV.

In vitro models
In vitro transcription–translation
Transient cellular expression systems
Stably transfected cell lines (constitutive/inducible expression)
Infection of primary hepatocytes and established cell lines
Retroviral pseudoparticles displaying functional HCV glycoproteins
Replicons (subgenomic/full-length; selectable/transient)
Production of infectious recombinant HCV in tissue culture
Chimeric viruses (e.g. poliovirus—HCV)
Related viruses (e.g. BVDV)
In vivo models
Transgenic mice
Immunodeficient mice/hepatocellular reconstitution models
Chimpanzee (Pan troglodytes)
Tree shrew (Tupaia belangeri chinensis)?
Related viruses (e.g. GBV-B in tamarins)

and may facilitate the screening of selected antiviral compounds.

The replicon system

The development of a replicon system for HCV was a particularly important breakthrough which revolutionized the investigation of HCV RNA replication (Lohmann *et al.*, 1999; reviewed in Bartenschlager, 2005). The prototype subgenomic replicon was a bicistronic RNA where the structural region and part of the nonstructural region of HCV were replaced by the neomycin phosphotransferase gene and translation of the nonstructural proteins 3–5B was driven by second, heterologous IRES from encephalomyocarditis virus (Figure 11.4). By this approach, it became possible, for the first time, to study efficient and genuine HCV RNA replication in Huh-7 human HCC cells *in vitro*. Interestingly, certain amino acid substitutions, i.e. cell culture adaptive mutations, were found to increase the efficiency of replication by several orders of magnitude (Blight *et al.*, 2000; Lohmann *et al.*, 2001). Such adaptive mutations cluster in the central region of NS5A (Figure 11.2), the C-terminal portion of the NS3 serine protease and the N-terminal portion of the NS3 RNA helicase domains as well as two positions in NS4B (Lohmann *et al.*, 2003). In NS5A,

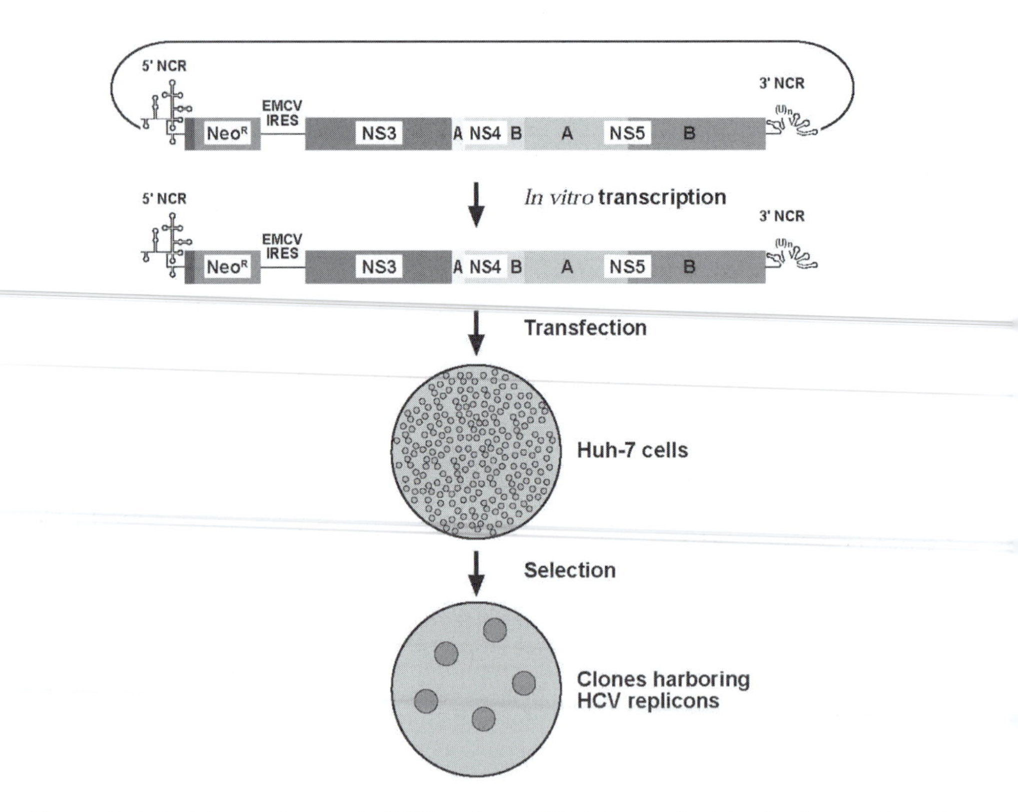

Figure 11.4 Prototype subgenomic HCV replicon. RNA is transcribed *in vitro* from a plasmid containing the HCV internal ribosome entry site (IRES) followed by a neomycin resistance cassette, a second heterologous IRES from encephalomyocarditis virus (EMCV), the HCV nonstructural region (NS3 to NS5B), and the HCV 3′ NCR. RNA is subsequently transfected into Huh-7 human hepatoma cells, followed by selection with G418 of clones harboring autonomously replicating subgenomic HCV RNA.

these changes often affect serine residues required for hyperphosphorylation, suggesting that hyperphosphorylation of NS5A reduces HCV RNA replication (Evans *et al.*, 2004; Neddermann *et al.*, 2004; Appel *et al.*, 2005). In addition, the efficiency of replicon RNA amplification was found to be determined by selection for particularly permissive cells within a given population of Huh-7 cells (Blight *et al.*, 2002; Lohmann *et al.*, 2003). Accordingly, removal of a replicon from a given cell clone by treatment with IFN-α or a selective drug often results in cell lines that support higher levels of HCV RNA replication as compared to naive Huh-7 cells. In one prominent example, an IFN-α-"cured," Huh-7-derived cell line designated Huh-7.5 (Blight *et al.*, 2002), a single point mutation in the double-strand RNA sensor retinoic acid-inducible gene I (RIG-I) appears to determine higher permissiveness for HCV RNA replication (Sumpter *et al.*, 2005). In most other cases, however, the molecular basis for increased permissiveness is unknown.

The replicon system has allowed genetic dissection of HCV RNA elements and proteins, provided material for biochemical and ultrastructural characterization of the viral replication complex, and facilitated discovery efforts as well as the investigation of antiviral resistance. Moreover, the replicon system has been exploited for analyses of the effect of cytokines on HCV RNA replication (Frese *et al.*, 2001 and 2002), and was instrumental in a recent series of elegant studies looking at other aspects of the interaction between HCV and the host cell (Foy *et al.*, 2003; Evans *et al.*, 2004; Foy *et al.*, 2005; Li *et al.*, 2005; Wang *et al.*, 2005; Watashi *et al.*, 2005).

Since the original reports of functional genotype 1b replicons, replicons for genotype 1a (Blight *et al.*, 2003) and 2a (Kato *et al.*, 2003) as well as derivatives expressing easily quantifiable marker enzymes in a separate cistron have been made to facilitate genetic studies as well as drug screening and evaluation (Krieger *et al.*, 2001; Yi *et al.*, 2002; Murray *et al.*, 2003).

In addition, full-length replicons and HCV genomes efficiently replicating in tissue culture have been developed (Blight *et al.*, 2002; Ikeda *et al.*, 2002; Pietschmann *et al.*, 2002), and the spectrum of permissive host cells has been expanded (Zhu *et al.*, 2003; Kato *et al.*, 2005). Finally, replicons have been established that allow tracking of functional HCV replication complexes in living cells (Moradpour *et al.*, 2004b).

Production of recombinant infectious HCV in tissue culture

A disappointing observation was that full-length HCV genomes with adaptive mutations were incapable of producing infectious virus. It was thought that either the host cell, i.e. the Huh-7 cell line, lacked some factor(s) critical for particle formation and release or that the adaptive mutations required for efficient replication in tissue culture interfered with packaging, assembly or release of virus. In this context, an inverse correlation was found between mutations that permit efficient replication of HCV RNA in HuH-7 cells *in vitro* and productive replication after intrahepatic inoculation into chimpanzees *in vivo* (Bukh *et al.*, 2002). Thus, one possibility is that cell culture adaptive mutations that promote efficient RNA replication *in vitro* may be deleterious *in vivo* because they compromise particle assembly and release.

The foundation for a long-awaited breakthrough in HCV research was laid when a HCV genotype 2a clone isolated from a patient with fulminant hepatitis C, designated as JFH-1 (for *J*apanese *f*ulminant *h*epatitis 1), was found to replicate in

Huh-7 cells and also in other liver-derived (Date *et al.*, 2004) as well as nonhepatic cell lines (Kato *et al.*, 2005) without the requirement for adaptive mutations. Indeed, it was recently reported that cloned JFH-1 genomes transfected into Huh-7 cells not only replicate efficiently but produce virus that is infectious for naive Huh-7 cells, allowing, for the first time, to study the complete life cycle of HCV *in vitro* (Lindenbach *et al.*, 2005a; Wakita *et al.*, 2005; Zhong *et al.*, 2005) (Figure 11.5).

In vivo models

The restricted host range of HCV has hampered the development of a suitable small animal model of viral replication and pathogenesis. Apart from a single report on the transmission of HCV to tree shrews (*Tupaia belangeri chinensis*) (Xie *et al.*, 1998), the chimpanzee (*Pan troglodytes*) is the only animal known to be susceptible to HCV infection (Bukh, 2004). Indeed, the chimpanzee was essential in the early characterization of the agent of non-A, non-B hepatitis and has allowed to determine important aspects of HCV replication, pathogenesis and prevention. Recent studies in chimpanzees have provided new insight into the host immune response against hepatitis C (Cooper *et al.*, 1999; Grakoui *et al.*, 2003; Shoukry *et al.*, 2003), and the chimpanzee remains the only faithful model to test the immunogenicity and efficacy of vaccine candidates (reviewed in Forns *et al.*, 2002; Houghton and Abrignani, 2005). However, ethical and financial restrictions limit the use of primates to highly selected experimental questions.

Expression of HCV proteins in transgenic mice provided some insights into the pathogenesis of HCV-induced liver disease (Moriya *et al.*, 1998; Lerat *et al.*, 2002). However, expression of HCV proteins from chromosomally integrated cDNA does not appropriately reflect the viral life cycle and studies on viral entry and replication are hardly conceivable in this system.

GBV-B, the closest relative of HCV within the Flaviviridae family, can be transmitted to tamarins (*Saguinus* sp.) and may represent a valuable surrogate model for HCV. Remarkably, GBV-B can be cultured in tamarin hepatocytes *in vitro* (Beames *et al.*, 2000). In addition, infectious cDNA clones (Bukh *et al.*, 1999) and replicons (De Tomassi *et al.*, 2002) have been established for GBV-B.

Progress in the development of a small animal model of HCV replication was achieved with the successful HCV infection of immunodeficient mice reconstituted with human hepatocytes (Mercer *et al.*, 2001). The properties of two different mouse strains, the *Alb-uPA*-transgenic and the immunodeficient SCID mouse, were combined to develop a model system that allows orthotopic engraftment of human hepatocytes. Inoculation of such mice with serum from patients with hepatitis C resulted in persistent HCV viremia with viral titers similar to those found in infected humans. However, the handling of these fragile animals represents a challenge and requires specialized expertise. Moreover, access to fresh human hepatocytes remains limited for many investigators.

Replication cycle

The life cycle of HCV includes (1) binding to an as yet unidentified cell surface receptor and internalization into the host cell, (2) cytoplasmic release and uncoating of the viral RNA genome, (3) IRES-mediated translation and polyprotein processing by cellular and viral proteases, (4) RNA replication, (5) packaging and assembly, and (6) virion maturation and release from the host cell (Figure 11.5).

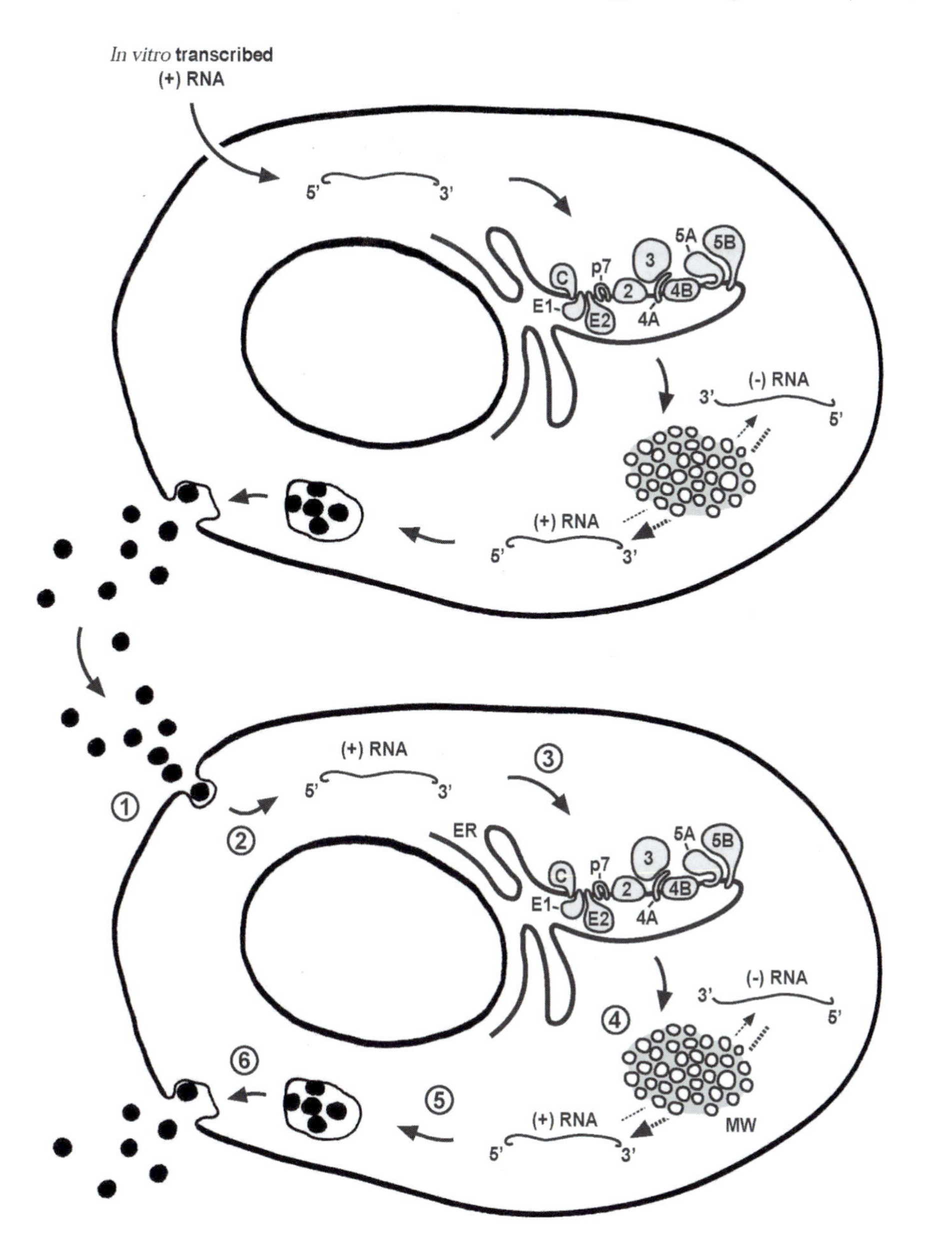

Figure 11.5 Production of recombinant infectious HCV in tissue culture. Transfection of Huh-7 cells with *in vitro* transcribed JFH-1 clone RNA leads to the production of viral particles that are infectious for naive Huh-7 cells and can be serially passaged, thereby allowing, for the first time, to study the entire life cycle of HCV in tissue culture. Steps of the viral life cycle include (1) virus binding and internalization, (2) cytoplasmic release and uncoating, (3) IRES-mediated translation and polyprotein processing, (4) RNA replication in a membrane associated replication complex, (5) packaging and assembly, (6) virion maturation and release. The topology of HCV structural and nonstructural proteins at the endoplasmic reticulum (ER) membrane is shown schematically. HCV RNA replication occurs in a specific membrane alteration, the membranous web (MW). Note that IRES-mediated translation and polyprotein processing as well as membranous web formation and RNA replication, illustrated here as separate steps for simplicity, may occur in a tightly coupled fashion. From Berke and Moradpour, 2005, with permission.

Receptor candidates

CD81, a tetraspanin molecule found on the surface of many cell types including hepatocytes (Pileri *et al.*, 1998), the low density lipoprotein receptor (LDLR) (Agnello *et al.*, 1999) and scavenger receptor class B type I (SR-BI) (Scarselli *et al.*, 2002) have, among others, been proposed as HCV receptors or components of a receptor complex.

Both CD81 and SR-BI bind E2 and are currently viewed as necessary, but not sufficient for HCV entry (Bartosch *et al.*, 2003a; Bartosch *et al.*, 2003b; Hsu *et al.*, 2003; Cormier *et al.*, 2004b; Zhang *et al.*, 2004). Expression of CD81 in CD81-negative liver-derived cell lines confers susceptibility to HCV pseudoparticles (see below) as well as recombinant virus produced in tissue culture, and blocking antibodies against CD81 or SR-BI, recombinant CD81, or siRNA-mediated downregulation of CD81 expression reduce infectivity. However, additional, yet unidentified hepatocyte-specific factors are required for HCV entry.

The LDLR has been attractive as a candidate receptor since infectious HCV has been reported to be associated with LDL or VLDL (see above). Presently, however, it is unclear whether interaction of HCV with the LDLR can lead to productive infection. HCV E2 also binds to DC-SIGN (dendritic cell-specific intercellular adhesion molecule-3-grabbing nonintegrin) and L-SIGN (liver/lymph node-specific intercellular adhesion molecule-3-grabbing integrin). The latter is a calcium-dependent lectin expressed on liver sinusoidal endothelial cells that may facilitate the infection process by trapping the virus for subsequent interaction with the receptor (Gardner *et al.*, 2003; Pöhlmann *et al.*, 2003; Cormier *et al.*, 2004a; Lozach *et al.*, 2004).

Identification and validation of HCV receptor candidates has until recently been limited by the paucity of systems for analyses of the early steps of the viral life cycle. Given the lack of native HCV particles and efficient cell culture systems, various alternatives have been explored to study the early steps of HCV infection. Soluble C-terminal truncated versions of HCV envelope glycoprotein E2 (Pileri *et al.*, 1998; Flint *et al.*, 1999; Scarselli *et al.*, 2002; Pöhlmann *et al.*, 2003), liposomes reconstituted with HCV E1 and E2 (Lambot *et al.*, 2002), and virus-like particles expressed in insect cells (Triyatni *et al.*, 2002; Wellnitz *et al.*, 2002) have been used to study HCV glycoprotein interactions with the cell surface. The production of virus-like particles has also been described in mammalian cells (Blanchard *et al.*, 2002). However, it is unclear how virus-like particles produced in insect or mammalian cells will compare to authentic HCV virions. Pseudotyped vesicular stomatitis virus (VSV) or influenza virus particles have been reported incorporating chimeric E1 and/or E2 glycoproteins whose C-terminal transmembrane domains were modified to allow transport to the cell surface (Lagging *et al.*, 1998; Flint *et al.*, 1999; Matsuura *et al.*, 2001; Buonocore *et al.*, 2002). However, such modifications may interfere with the multiple and complex roles of the E1 and E2 transmembrane domains (Op De Beeck *et al.*, 2001) and may perturb the conformation and functions of E1-E2 complexes.

On this background, the establishment of infectious retroviral pseudotypes displaying functional HCV glycoproteins as a robust model system for the study of viral entry represented a major breakthrough (Bartosch *et al.*, 2003a; Hsu *et al.*, 2003). HCV pseudoparticle infectivity is restricted primarily to human hepatocytes and hepatocyte-derived cell lines, and entry is pH-dependent. Thus, HCV entry likely involves transit through an endosomal, low pH compartment and fusion with the endosomal membrane.

The structural basis for low pH-induced membrane fusion has recently been elucidated for the dengue, TBE and Semliki Forest viruses (Bressanelli *et al.*, 2004; Gibbons *et al.*, 2004; Modis *et al.*, 2004). The envelope proteins of these related flavi- and alphaviruses possess an internal fusion peptide that is exposed during low pH-mediated domain rearrangement and trimerization of the protein. The scaffolds of these so called class II fusion proteins are remarkably similar, suggesting that all members of the Flaviviridae family, including HCV, could behave similarly.

Replication complex

Formation of a membrane-associated replication complex, composed of viral proteins, replicating RNA, and altered cellular membranes, is a hallmark of all positive-strand RNA viruses investigated thus far (see Egger *et al.*, 2002a, Ahlquist *et al.*, 2003 and Moradpour *et al.*, 2003a for reviews). Depending on the virus, replication may occur on altered membranes derived from the ER, Golgi apparatus, mitochondria or even lysosomes. The role of membranes in viral RNA synthesis is not well understood. It may include (1) the physical support and organization of the RNA replication complex (Lyle *et al.*, 2002), (2) the compartimentalization and local concentration of viral products (Schwartz *et al.*, 2002), (3) tethering of the viral RNA during unwinding (Egger *et al.*, 2002a), (4) provision of lipid constituents important for replication, and (5) protection of the viral RNA from double-strand RNA-mediated host defenses or RNA interference.

In the case of HCV, protein–protein interactions among HCV nonstructural proteins have been described (Shirota *et al.*, 2002; Dimitrova *et al.*, 2003) and determinants for membrane association of the HCV proteins have been mapped. For a more comprehensive review on the interactions of HCV proteins with host cell membranes the reader is referred to Dubuisson *et al.* (2002) and Moradpour *et al.* (2003a).

A specific membrane alteration, designated the membranous web, was recently identified as the site of RNA replication in HuH-7 cells harboring subgenomic HCV replicons (Gosert *et al.*, 2003) (Figure 11.6). Formation of the membranous web could be induced by NS4B alone (see above) and was very similar to the "sponge-like inclusions" previously found by electron microscopy in the liver of HCV-infected chimpanzees. The membranous web was often found in close association or even continuity with the rough ER. Based on this observation, the colocalization of individually expressed HCV proteins with membranes of the ER (Wölk *et al.*, 2000; Hügle *et al.*, 2001; Schmidt-Mende *et al.*, 2001; Brass *et al.*, 2002) and biochemical data (Ivashkina *et al.*, 2002), it is currently believed that the membranous web is derived from membranes of the ER. Ongoing studies are aimed at isolating and further characterizing this complex and at defining the viral and cellular processes involved in formation of the membranous web.

Given that numerous core protein molecules are needed to form a nucleocapsid for incorporation of one genomic RNA, a large excess of nonstructural proteins are produced in an infected cell. Thus, one can conceive that these proteins have additional functions as arrays or lattices on membranes. In fact, it has been shown that only a small proportion of HCV nonstructural proteins expressed in cells harboring HCV replicons are actively en-

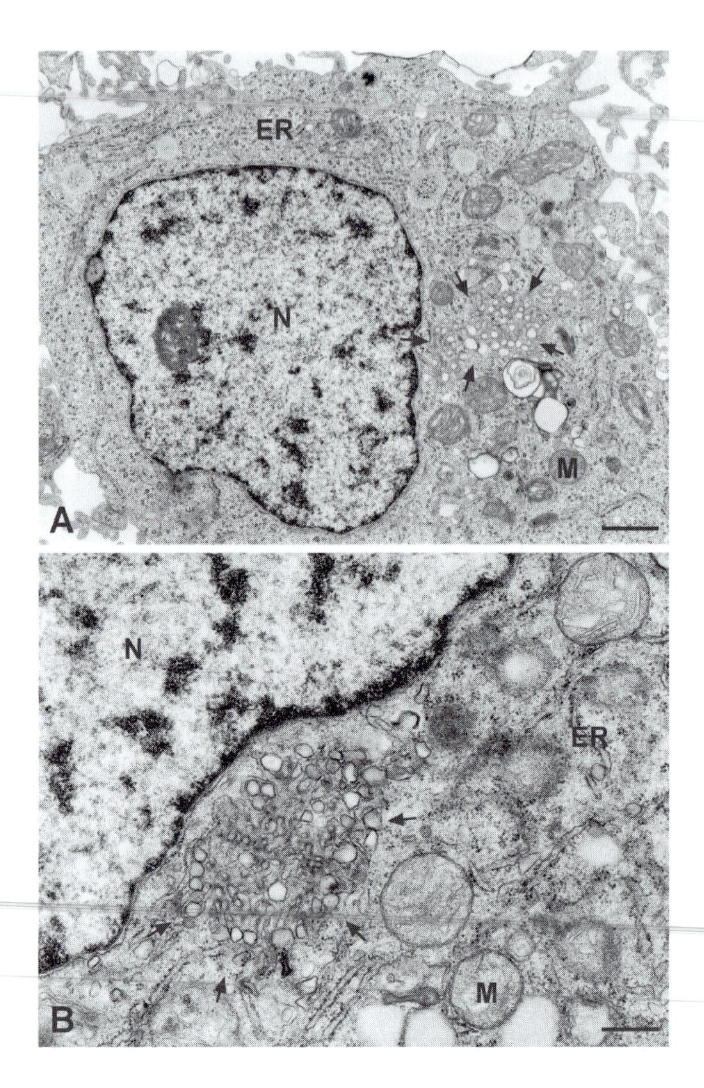

Figure 11.6 HCV replication complex (Gosert *et al.*, 2003). (A) Low-power overview of a Huh-7 cell harboring a subgenomic HCV replicon. A distinct membrane alteration, named membranous web (arrows), is found in the juxtanuclear region. Note the circumscript nature of this specific membrane alteration and the otherwise unaltered cellular organelles. Bar, 1 μm. (B) Higher magnification of a membranous web (arrows) composed of small vesicles embedded in a membrane matrix. Note the close association of the membranous web with the rough endoplasmic reticulum. Bar, 500 nm. The membranous web harbors all HCV nonstructural proteins and nascent viral RNA in HuH-7 cells harboring subgenomic replicons, and therefore represents the HCV RNA replication complex. N, nucleus; ER, endoplasmic reticulum; M, mitochondria.

gaged in RNA replication (Miyanari *et al.*, 2003; Quinkert *et al.*, 2005) and that viral RNA is protected not only by membranes but also by a protease-sensitive component (Yang *et al.*, 2004).

Recent studies demonstrate a complex interaction between HCV RNA replication and the cellular lipid metabolism. It was found, for example, that geranylgeranylation of one or more host proteins plays an important role (Ye *et al.*, 2003; Kapadia and Chisari, 2005). More recently, FBL2 was identified as geranylgeranylated cellular protein required for HCV RNA replication (Wang *et al.*, 2005). Such observations suggest that pharmacologic manipulation of lipid metabolism may have therapeutic potential in hepatitis C.

Pathogenesis

HCV infection is a highly dynamic process with a viral half-life of only a few hours and an average daily virion production and clearance of up to more than 10^{12} (Neumann *et al.*, 1998). This high replicative activity, together with the lack of a proof-reading function of the viral RdRp, provides the basis for the genetic variability of HCV. In addition, these findings are similar to the dynamics of HIV infection and provide a rationale for the development and implementation of combination antiviral therapies.

The mechanisms responsible for liver injury in acute and chronic hepatitis C are complex and incompletely understood. For comprehensive reviews, the reader is referred to Cerny and Chisari (1999); Shoukry *et al.* (2004); Rehermann and Nascimbeni (2005). In acute HCV infection, liver cell damage coincides with the development of the host immune response and not with infection and viral replication. In addition, persistent viral replication often occurs without evidence of liver cell damage, suggesting that HCV is not directly cytopathic. The immune response against HCV, therefore, plays a central role in the HCV pathogenesis of hepatitis C.

Patients who clear HCV infection have a more vigorous CD4^{+} (Diepolder *et al.*, 1995; Missale *et al.*, 1996) and CD8^{+} T cell response early on (Lechner *et al.*, 2000). The role of specific CD4^{+} and CD8^{+} T cell responses in control of HCV infection was elegantly illustrated by *in vivo* depletion studies in chimpanzees (Grakoui *et al.*, 2003; Shoukry *et al.*, 2003). Despite the presence of an immune response, however, HCV is rarely eliminated. Thus, HCV may overwhelm, not induce, or evade antiviral immune responses.

Most patients develop chronic infection with relatively stable viral titers, about 2–3 logs lower than in the acute phase. Only a small proportion of patients recover and test negative for HCV RNA using standard assays. Whether HCV is completely cleared after recovery or whether trace amounts of virus persist, similar to hepatitis B virus, is debated (Pham *et al.*, 2004). HCV-specific antibodies may disappear completely 10–20 years after recovery (Takaki *et al.*, 2000).

Microarray analyses of serial liver biopsy samples in experimentally infected chimpanzees revealed that HCV induces the intrahepatic expression of many genes, including type I IFN, i.e. IFN-α and -β, responses (Su *et al.*, 2002).

However, even if HCV RNA replication *in vitro* is efficiently inhibited by type I IFNs (Frese *et al.*, 2001), HCV seems to be resistant to these responses and frequently succeeds in establishing chronic hepatitis. Indeed, HCV may have evolved numerous mechanisms to counteract the innate immune response, including interference with the IFN system at the induction (Foy *et al.*, 2003; Foy *et al.*, 2005; Li *et al.*, 2005; Meylan *et al.*, 2005), signaling (Heim *et al.*, 1999; Blindenbacher *et al.*, 2003; Duong *et al.*, 2004) and effector levels (Gale *et al.*, 1997 and 1998; Taylor *et al.*, 1999; reviewed in Gale and Foy, 2005).

In addition, HCV may interfere with natural killer (NK) cell functions (Crotta *et al.*, 2002; Tseng and Klimpel, 2002). In this context, a recent large immunogenetic study revealed an association between a NK cell receptor (*KIR2DL3* allele)-HLA compound genotype and HCV clearance and clinical recovery, pointing toward a role of NK cells in early HCV infection (Khakoo *et al.*, 2004).

Additional mechanisms of viral persistence may include the induction of peripheral tolerance or exhaustion of the T cell response, infection of immunologically privileged sites, inhibition of antigen presentation, downregulation of viral gene expression, and viral mutations that abrogate, anergize or antagonize antigen recognition by virus-specific T cells (Cerny and Chisari, 1999). An impairment of dendritic cell function has been proposed (Bain *et al.*, 2001), but this is controversial (Longman *et al.*, 2004). There is some evidence that privileged sites may play a role since HCV may infect extrahepatic cells and tissues. As mentioned above, the role of viral escape mutations and the quasispecies nature of HCV as a cause of viral persistence has attracted considerable interest. In this context, HCV escape to antibodies (Shimizu *et al.*, 1994; Farci *et al.*, 2000) and T cells (Chang *et al.*, 1997; Tsai *et al.*, 1998; Erickson *et al.*, 2001; Seifert *et al.*, 2004) has been demonstrated both in humans and chimpanzees.

The role of the humoral immune response in the natural course and pathogenesis of hepatitis C is not well understood. Recent studies using HCV pseudoparticles (see above) validated earlier studies demonstrating neutralizing of antibodies (Logvinoff *et al.*, 2004; Yu *et al.*, 2004). However, the highest antibody titers are found in patients with chronic hepatitis C and the role of antibodies capable of neutralizing a minor fraction of the HCV population is unknown.

Conclusions and perspectives

The development of powerful model systems has allowed to systematically dissect important steps of the HCV life cycle. These efforts have translated into the identification of novel antiviral targets and the development of new therapeutic strategies, some of which are already in early phase clinical evaluation. Much work remains to be done with respect to the virion structure, the early and late steps of the HCV life cycle, the mechanism and regulation of RNA replication, and the pathogenesis of HCV-induced liver disease. Ultimately, an improved understanding of the viral life cycle should result in innovative therapeutic and preventive strategies for one of the most common causes of chronic hepatitis, liver cirrhosis, and HCC worldwide.

References

Agnello, V., Abel, G., Elfahal, M., Knight, G.B., and Zhang, Q.X. (1999). Hepatitis C virus and other flaviviridae viruses enter cells via low density lipoprotein receptor. Proc. Natl. Acad. Sci. USA *96*, 12766–12771.

Ago, H., Adachi, T., Yoshida, A., Yamamoto, M., Habuka, N., Yatsunami, K., and Miyano, M. (1999). Crystal structure of the RNA-dependent RNA polymerase of hepatitis C virus. Structure Fold. Des. *7*, 1417–1426.

Ahlquist, P., Noueiry, A.O., Lee, W.M., Kushner, D.B., and Dye, B.T. (2003). Host factors in positive-strand RNA virus genome replication. J. Virol. *77*, 8181–8186.

André, P., Komurian-Pradel, F., Deforges, S., Perret, M., Berland, J.L., Sodoyer, M., Pol, S., Bréchot, C., Paranhos-Baccalà, G., and Lotteau, V. (2002). Characterization of low- and very-low-density hepatitis C virus RNA-containing particles. J. Virol. *76*, 6919–6928.

Appel, N., Pietschmann, T., and Bartenschlager, R. (2005). Mutational analysis of hepatitis C virus nonstructural protein 5A: potential role of differential phosphorylation in RNA replication and identification of a genetically flexible domain. J. Virol. *79*, 3187–3194.

Bain, C., Fatmi, A., Zoulim, F., Zarski, J.P., Trépo, C., and Inchauspé, G. (2001). Impaired al-

lostimulatory function of dendritic cells in chronic hepatitis C infection. Gastroenterology *120*, 512–524.
Bain, C., Parroche, P., Lavergne, J.P., Duverger, B., Vieux, C., Dubois, V., Komurian-Pradel, F., Trépo, C., Gebuhrer, L., Paranhos-Baccala, G., Penin, F., and Inchauspé, G. (2004). Memory T-cell-mediated immune responses specific to an alternative core protein in hepatitis C virus infection. J. Virol. *78*, 10460–10469.
Barba, G., Harper, F., Harada, T., Kohara, M., Goulinet, S., Matsuura, Z., Eder, G., Schaff, Z., Chapman, M.J., Miyamura, T., and Bréchot, C. (1997). Hepatitis C virus core protein shows a cytoplasmic localization and associates to cellular lipid storage droplets. Proc. Natl. Acad. Sci. USA *94*, 1200–1205.
Bartenschlager, R., Frese, M., and Pietschmann, T. (2004). Novel insights into hepatitis C virus replication and persistence. Adv. Virus Res. *63*, 71–180.
Bartenschlager, R. (2005). The hepatitis C virus replicon system: from basic research to clinical application. J. Hepatol. *43*, 210–216.
Bartosch, B., Dubuisson, J., and Cosset, F.L. (2003a). Infectious hepatitis C virus pseudo-particles containing functional E1-E2 envelope protein complexes. J. Exp. Med. *197*, 633–642.
Bartosch, B., Vitelli, A., Granier, C., Goujon, C., Dubuisson, J., Pascale, S., Scarselli, E., Cortese, R., Nicosia, A., and Cosset, F.L. (2003b). Cell entry of hepatitis C virus requires a set of co-receptors that include the CD81 tetraspanin and the SR-B1 scavenger receptor. J. Biol. Chem. *278*, 41624–41630.
Beames, B., Chavez, D., Guerra, B., Notvall, L., Brasky, K.M., and Lanford, R.E. (2000). Development of a primary tamarin hepatocyte culture system for GB virus-B: a surrogate model for hepatitis C virus. J. Virol. *74*, 11764–11772.
Behrens, S.-E., Tomei, L., and De Francesco, R. (1996). Identification and properties of the RNA-dependent RNA polymerase of hepatitis C virus. EMBO J. *15*, 12–22.
Berke, J.M., and Moradpour, D. (2005). Hepatitis V virus comes full circle: production of recombinant infections virus in tissue culture. Hepatology *42*, 1264–1269.
Blanchard, E., Brand, D., Trassard, S., Goudeau, A., and Roingeard, P. (2002). Hepatitis C virus-like particle morphogenesis. J. Virol. *76*, 4073–4079.
Blight, K.J., Kolykhalov, A.A., and Rice, C.M. (2000). Efficient initiation of HCV RNA replication in cell culture. Science *290*, 1972–1974.
Blight, K.J., McKeating, J.A., and Rice, C.M. (2002). Highly permissive cell lines for subgenomic and genomic hepatitis C virus RNA replication. J. Virol. *76*, 13001–13014.
Blight, K.J., McKeating, J.A., Marcotrigiano, J., and Rice, C.M. (2003). Efficient replication of hepatitis C virus genotype 1a RNAs in cell culture. J. Virol. *77*, 3181–3190.
Blindenbacher, A., Duong, F.H., Hunziker, L., Stutvoet, S.T., Wang, X., Terracciano, L., Moradpour, D., Blum, H.E., Alonzi, T., Tripodi, M., La Monica, N., and Heim, M.H. (2003). Expression of hepatitis C virus proteins inhibits interferon alpha signaling in the liver of transgenic mice. Gastroenterology *124*, 1465–1475.
Boulant, S., Becchi, M., Penin, F., and Lavergne, J.P. (2003). Unusual multiple recoding events leading to alternative forms of hepatitis C virus core protein from genotype 1b. J. Biol. Chem. *278*, 45785–45792.
Branch, A.D., Stump, D.D., Gutierrez, J.A., Eng, F., and Walewski, J.L. (2005). The hepatitis C virus alternate reading frame (ARF) and its family of novel products: the alternate reading frame protein/F-protein, the double-frameshift protein, and others. Semin. Liver Dis. *25*, 105–117.
Brass, V., Bieck, E., Montserret, R., Wölk, B., Hellings, J.A., Blum, H.E., Penin, F., and Moradpour, D. (2002). An aminoterminal amphipathic alpha-helix mediates membrane association of the hepatitis C virus nonstructural protein 5A. J. Biol. Chem. *277*, 8130–8139.
Bressanelli, S., Tomei, L., Roussel, A., Incitti, I., Vitale, R.L., Mathieu, M., De Francesco, R., and Rey, F.A. (1999). Crystal structure of the RNA-dependent RNA polymerase of hepatitis C virus. Proc. Natl. Acad. Sci. USA *96*, 13034–13039.
Bressanelli, S., Tomei, L., Rey, F.A., and De Francesco, R. (2002). Structural analysis of the hepatitis C virus RNA polymerase in complex with ribonucleotides. J. Virol. *76*, 3482–3492.
Bressanelli, S., Stiasny, K., Allison, S.L., Stura, E.A., Duquerroy, S., Lescar, J., Heinz, F.X., and Rey, F.A. (2004). Structure of a flavivirus envelope glycoprotein in its low-pH-induced membrane fusion conformation. EMBO J. *23*, 728–738.
Bukh, J., Apgar, C.L., and Yanagi, M. (1999). Toward a surrogate model for hepatitis C virus: an infectious molecular clone of the GB

virus-B hepatitis agent. Virology *262*, 470–478.

Bukh, J., Pietschmann, T., Lohmann, V., Krieger, N., Faulk, K., Engle, R.E., Govindarajan, S., Shapiro, M., St Claire, M., and Bartenschlager, R. (2002). Mutations that permit efficient replication of hepatitis C virus RNA in Huh-7 cells prevent productive replication in chimpanzees. Proc. Natl. Acad. Sci. USA *99*, 14416–14421.

Bukh, J. (2004). A critical role for the chimpanzee model in the study of hepatitis C. Hepatology *39*, 1469–1475.

Buonocore, L., Blight, K.J., Rice, C.M., and Rose, J.K. (2002). Characterization of vesicular stomatitis virus recombinants that express and incorporate high levels of hepatitis C virus glycoproteins. J. Virol. *76*, 6865–6872.

Butcher, S.J., Grimes, J.M., Makeyev, E.V., Bamford, D.H., and Stuart, D.I. (2001). A mechanism for initiating RNA-dependent RNA polymerization. Nature *410*, 235–240.

Carrère-Kremer, S., Montpellier-Pala, C., Cocquerel, L., Wychowski, C., Penin, F., and Dubuisson, J. (2002). Subcellular localization and topology of the p7 polypeptide of hepatitis C virus. J. Virol. *76*, 3720–3730.

Castet, V., Fournier, C., Soulier, A., Brillet, R., Coste, J., Larrey, D., Dhumeaux, D., Maurel, P., and Pawlotsky, J.M. (2002). Alpha interferon inhibits hepatitis C virus replication in primary human hepatocytes infected *in vitro*. J. Virol. *76*, 8189–8199.

Cerny, A., and Chisari, F.V. (1999). Pathogenesis of chronic hepatitis C: immunological features of hepatic injury and viral persistence. Hepatology *30*, 595–601.

Chang, K.-M., Rehermann, B., McHutchison, J.G., Pasquinelli, C., Southwood, S., Sette, A., and Chisari, F.V. (1997). Immunological significance of cytotoxic T lymphocyte epitope variants in patients chronically infected by the hepatitis C virus. J. Clin. Invest. *100*, 2376–2385.

Choi, J., Xu, Z., and Ou, J.H. (2003). Triple decoding of hepatitis C virus RNA by programmed translational frameshifting. Mol. Cell. Biol. *23*, 1489–1497.

Choi, K.H., Groarke, J.M., Young, D.C., Kuhn, R.J., Smith, J.L., Pevear, D.C., and Rossmann, M.G. (2004). The structure of the RNA-dependent RNA polymerase from bovine viral diarrhea virus establishes the role of GTP in de novo initiation. Proc. Natl. Acad. Sci. USA *101*, 4425–4430.

Choo, Q.-L., Kuo, G., Weiner, A.J., Overby, L.R., Bradley, D.W., and Houghton, M. (1989). Isolation of a cDNA clone derived from a blood-borne non-A, non-B viral hepatitis genome. Science *244*, 359–362.

Cocquerel, L., Op de Beeck, A., Lambot, M., Roussel, J., Delgrange, D., Pillez, A., Wychowski, C., Penin, F., and Dubuisson, J. (2002). Topologic changes in the transmembrane domains of hepatitis C virus envelope glycoproteins. EMBO J. *21*, 2893–2902.

Collier, A.J., Gallego, J., Klinck, R., Cole, P.T., Harris, S.J., Harrison, G.P., Aboul-Ela, F., Varani, G., and Walker, S. (2002). A conserved RNA structure within the HCV IRES eIF3-binding site. Nat. Struct. Biol. *9*, 375–380.

Cooper, S., Erickson, A.L., Adams, E.J., Kansopon, J., Weiner, A.J., Chien, D.Y., Houghton, M., Parham, P., and Walker, C.M. (1999). Analysis of a successful immune response against hepatitis C virus. Immunity *10*, 439–449.

Cormier, E.G. Durso, R.J., Tsamis, F., Boussemart, L., Manix, C., Olson, W.C., Gardner, J.P., and Dragic, T. (2004a). L-SIGN (CD209L) and DC-SIGN (CD209) mediate transinfection of liver cells by hepatitis C virus. Proc. Natl. Acad. Sci. USA *101*, 14067–14072.

Cormier, E.G. Tsamis, F., Kajumo, F., Durso, R.J., Gardner, J.P., and Dragic, T. (2004b). CD81 is an entry coreceptor for hepatitis C virus. Proc. Natl. Acad. Sci. USA *101*, 7270–7274.

Crotta, S., Stilla, A., Wack, A., D'Andrea, A., Nuti, S., D'Oro, U., Mosca, M., Filliponi, F., Brunetto, R.M., Bonino, F., Abrignani, S., and Valiante, N.M. (2002). Inhibition of natural killer cells through engagement of CD81 by the major hepatitis C virus envelope protein. J. Exp. Med. *195*, 35–41.

Date, T., Kato, T., Miyamoto, M., Zhao, Z., Yasui, K., Mizokami, M., and Wakita, T. (2004). Genotype 2a hepatitis C virus subgenomic replicon can replicate in HepG2 and IMY-N9 cells. J. Biol. Chem. *279*, 22371–22376.

Davis, G.L., Albright, J.E., Cook, S.F., and Rosenberg, D.M. (2003). Projecting future complications of chronic hepatitis C in the United States. Liver Transpl. *9*, 331–338.

De Francesco, R., and Steinkühler, C. (2000). Structure and function of the hepatitis C virus NS3-NS4A serine protease. Curr. Top. Microbiol. Immunol. *242*, 149–169.

De Francesco, R., and Migliaccio, G. (2005) Challenges and successes in developing new therapies for hepatitis C. Nature *436*, 953–960.

De Tomassi, A., Pizzuti, M., Graziani, R., Sbardellati, A., Altamura, S., Paonessa, G., and Traboni, C. (2002). Cell clones selected from the Huh7 human hepatoma cell line support efficient replication of a subgenomic GB virus B replicon. J. Virol. *76*, 7736–7746.

Diepolder, H.M., Zachoval, R., Hoffmann, R.M., Wierenga, E.A., Santantonio, T., Jung, M.C., Eichenlaub, D., and Pape, G.R. (1995). Possible mechanism involving T lymphocyte response to non structural protein 3 in viral clearance in acute hepatitis C virus infection. Lancet *346*, 1006–1007.

Dimitrova, M., Imbert, I., Kieny, M.P., and Schuster, C. (2003). Protein–protein interactions between hepatitis C virus nonstructural proteins. J. Virol. *77*, 5401–5414.

Dubuisson, J., Penin, F., and Moradpour, D. (2002). Interaction of hepatitis C virus proteins with host cell membranes and lipids. Trends Cell Biol. *12*, 517–523.

Duong, F.H., Filipowicz, M., Tripodi, M., La Monica, N., and Heim, M.H. (2004). Hepatitis C virus inhibits interferon signaling through up-regulation of protein phosphatase 2A. Gastroenterology *126*, 263–277.

Egger, D., Gosert, R., and Bienz, K. (2002a). Role of cellular structures in viral RNA replication. In Molecular biology of picornaviruses, B. Semler and E. Wimmer, eds. (Washington DC, ASM Press), pp. 247–253.

Egger, D., Wölk, B., Gosert, R., Bianchi, L., Blum, H.E., Moradpour, D., and Bienz, K. (2002b). Expression of hepatitis C virus proteins induces distinct membrane alterations including a candidate viral replication complex. J. Virol. *76*, 5974–5984.

Elazar, M., Cheong, K.H., Liu, P., Greenberg, H.B., Rice, C.M., and Glenn, J.S. (2003). Amphipathic helix-dependent localization of NS5A mediates hepatitis C virus RNA replication. J. Virol. *77*, 6055–6061.

Enomoto, N., Sakuma, I., Asahina, Y., Kurosaki, M., Murakami, T., Yamamoto, C., Ogura, Y., Izumi, N., Marumo, F., and Sato, C. (1996). Mutations in the nonstructural protein 5A gene and response to interferon in patients with chronic hepatitis C virus 1b infection. N. Engl. J. Med. *334*, 77–81.

Erickson, A.L., Kimura, Y., Igarashi, S., Eichelberger, J., Houghton, M., Sidney, J., McKinney, D., Sette, A., Hughes, A.L., and Walker, C.M. (2001). The outcome of hepatitis C virus infection is predicted by escape mutations in epitopes targeted by cytotoxic T lymphocytes. Immunity *15*, 883–895.

Evans, M.J., Rice, C.M., and Goff, S.P. (2004). Phosphorylation of hepatitis C virus nonstructural protein 5A modulates its protein interactions and viral RNA replication. Proc. Natl. Acad. Sci. USA *101*, 13038–13043.

Falck-Ytter, Y., Kale, H., Mullen, K.D., Sarbah, S.A., Sorescu, L., and McCullough, A.J. (2002). Surprisingly small effect of antiviral treatment in patients with hepatitis C. Ann. Intern. Med. *136*, 288–292.

Farci, P., Shimoda, A., Wong, D., Cabezon, T., De Gioannis, D., Strazzera, A., Shimizu, Y., Shapiro, M., Alter, H.J., and Purcell, R.H. (1996). Prevention of hepatitis C virus infection in chimpanzees by hyperimmune serum against the hypervariable region 1 of the envelope 2 protein. Proc. Natl. Acad. Sci. USA *93*, 15394–15399.

Farci, P., Shimoda, A., Coiana, A., Diaz, G., Peddis, G., Melpolder, J.C., Strazzera, A., Chien, D.Y., Munoz, S.J., Balestrieri, A., Purcell, R.H., and Alter, H.J. (2000). The outcome of acute hepatitis C predicted by the evolution of the viral quasispecies. Science *288*, 339–344.

Ferlenghi, I., Clarke, M., Ruttan, T., Allison, S.L., Schalich, J., Heinz, F.X., Harrison, S.C., Rey, F.A., and Fuller, S.D. (2001). Molecular organization of a recombinant subviral particle from tick-borne encephalitis virus. Mol. Cell *7*, 593–602.

Ferrari, E., Wright-Minogue, J., Fang, J.W., Baroudy, B.M., Lau, J.Y., and Hong, Z. (1999). Characterization of soluble hepatitis C virus RNA-dependent RNA polymerase expressed in *Escherichia coli*. J. Virol. *73*, 1649–1654.

Filocamo, G., Pacini, L., Nardi, C., Bartholomew, L., Scaturro, M., Delmastro, P., Tramontano, A., de Francesco, R., and Migliaccio, G. (1999). Selection of functional variants of the NS3-NS4A protease of hepatitis C virus by using chimeric Sindbis viruses. J. Virol. *73*, 561–575.

Flint, M., Thomas, J.M., Maidens, C.M., Shotton, C., Levy, S., Barclay, W.S., and McKeating, J.A. (1999). Functional analysis of cell surface-expressed hepatitis C virus E2 glycoprotein. J. Virol. *73*, 6782–6790.

Forns, X., Bukh, J., and Purcell, R.H. (2002). The challenge of developing a vaccine against hepatitis C virus. J. Hepatol. *37*, 684–695.

Foy, E., Li, K., Wang, C., Sumpter, R., Jr., Ikeda, M., Lemon, S.M., and Gale, M., Jr. (2003). Regulation of interferon regulatory factor-3 by the hepatitis C virus serine protease. Science *300*, 1145–1148.

Foy, E., Li, K., Sumpter, R., Jr., Loo, Y.M., Johnson, C.L., Wang, C., Fish, P.M., Yoneyama, M., Fujita, T., Lemon, S.M., and Gale, M., Jr. (2005). Control of antiviral defenses through hepatitis C virus disruption of retinoic acid-inducible gene-I signaling. Proc. Natl. Acad. Sci. USA *102*, 2986–2991.

Frese, M., Pietschmann, T., Moradpour, D., Haller, O., and Bartenschlager, R. (2001). Interferon-α inhibits hepatitis C virus subgenomic RNA replication by an MxA-independent pathway. J. Gen. Virol. *82*, 723–733.

Frese, M., Schwärzle, V., Barth, K., Krieger, N., Lohmann, V., Mihm, S., Haller, O., and Bartenschlager, R. (2002). Interferon-γ inhibits replication of subgenomic and genomic hepatitis C virus RNAs. Hepatology *35*, 694–703.

Friebe, P., Lohmann, V., Krieger, N., and Bartenschlager, R. (2001). Sequences in the 5′ nontranslated region of hepatitis C virus required for RNA replication. J. Virol. *75*, 12047–12057.

Friebe, P., and Bartenschlager, R. (2002). Genetic analysis of sequences in the 3′ nontranslated region of hepatitis C virus that are important for RNA replication. J. Virol. *76*, 5326–5338.

Friebe, P., Boudet, J., Simorre, J.-P., and Bartenschlager, R. (2005). A kissing loop interaction in the 3′ end of the hepatitis C virus genome essential for RNA replication. J. Virol. *79*, 380–392.

Fried, M., Shiffman, M., Reddy, K., Smith, C., Marinos, G., Goncales, F.J., Häussinger, D., Diago, M., Carosi, G., Dhumeaux, D., Craxi, A., Lin, A., Hoffman, J., and Yu, J. (2002). Peginterferon alfa-2a plus ribavirin for chronic hepatitis C virus infection. N. Engl. J. Med. *347*, 975–982.

Gale, M., Jr, Korth, M.J., Tang, N.M., Tan, S.-L., Hopkins, D.A., Dever, T.E., Polyak, S.J., Gretch, D.R., and Katze, M.G. (1997). Evidence that hepatitis C virus resistance to interferon is mediated through repression of the PKR protein kinase by the nonstructural 5A protein. Virology *230*, 217–227.

Gale, M., Jr., Blakely, C.M., Kwieciszewski, B., Tan, S.L., Dossett, M., Tang, N.M., Korth, M.J., Polyak, S.J., Gretch, D.R., and Katze, M.G. (1998). Control of PKR protein kinase by hepatitis C virus nonstructural 5A protein: molecular mechanisms of kinase regulation. Mol. Cell. Biol. *18*, 5208–5218.

Gale, M., Jr., and Foy, E.M. (2005) Evasion of intracellular host defence by hepatitis C virus. Nature *436*, 939–945.

Gardner, J.P., Durso, R.J., Arrigale, R.R., Donovan, G.P., Maddon, P.J., Dragic, T., and Olson, W.C. (2003). L-SIGN (CD 209L) is a liver-specific capture receptor for hepatitis C virus. Proc. Natl. Acad. Sci. USA *100*, 4498–4503.

Gibbons, D.L., Vaney, M.C., Roussel, A., Vigouroux, A., Reilly, B., Lepault, J., Kielian, M., and Rey, F.A. (2004). Conformational change and protein–protein interactions of the fusion protein of Semliki Forest virus. Nature *427*, 320–325.

Gosert, R., Egger, D., Lohmann, V., Bartenschlager, R., Blum, H.E., Bienz, K., and Moradpour, D. (2003). Identification of the hepatitis C virus RNA replication complex in Huh-7 cells harboring subgenomic replicons. J. Virol. *77*, 5487–5492.

Grakoui, A., McCourt, D.W., Wychowski, C., Feinstone, S.M., and Rice, C.M. (1993). A second hepatitis C virus-encoded proteinase. Proc. Natl. Acad. Sci. USA *90*, 10583–10587.

Grakoui, A., Shoukry, N.H., Woollard, D.J., Han, J.H., Hanson, H.L., Ghrayeb, J., Murthy, K.K., Rice, C.M., and Walker, C.M. (2003). HCV persistence and immune evasion in the absence of memory T cell help. Science *302*, 659–662.

Griffin, S.D., Beales, L.P., Clarke, D.S., Worsfold, O., Evans, S.D., Jaeger, J., Harris, M.P., and Rowlands, D.J. (2003). The p7 protein of hepatitis C virus forms an ion channel that is blocked by the antiviral drug, Amantadine. FEBS Lett. *535*, 34–38.

Hadziyannis, S.J., Sette, H., Jr., Morgan, T.R., Balan, V., Diago, M., Marcellin, P., Ramadori, G., Bodenheimer, H., Jr., Bernstein, D., Rizzetto, M., Zeuzem, S., Pockros, P.J., Lin, A., and Ackrill, A.M. (2004). Peginterferon-alpha2a and ribavirin combination therapy in chronic hepatitis C: a randomized study of treatment duration and ribavirin dose. Ann. Intern. Med. *140*, 346–355.

Harada, T., Tautz, N., and Thiel, H.J. (2000). E2-p7 region of the bovine viral diarrhea virus polyprotein: processing and functional studies. J. Virol. *74*, 9498–9506.

Heim, M.H., Moradpour, D., and Blum, H.E. (1999). Expression of hepatitis C virus proteins inhibits signal transduction through the Jak-STAT pathway. J. Virol. *73*, 8469–8475.

Hijikata, M., Kato, N., Ootsuyama, Y., Nakagawa, M., Ohkoshi, S., and Shimotohno, K. (1991). Hypervariable regions in the putative glycoprotein of hepatitis C virus. Biochem. Biophys. Res. Commun. *175*, 220–228.

Hijikata, M., Mizushima, H., Akagi, T., Mori, S., Kakiuchi, N., Kato, N., Tanaka, T., Kimura, K., and Shimotohno, K. (1993). Two distinct proteinase activities required for the processing of a putative nonstructural precursor protein of hepatitis C virus. J. Virol. *67*, 4665–4675.

Hinrichsen, H., Benhamou, Y., Wedemeyer, H., Reiser, M., Sentjens, R.E., Calleja, J.L., Forns, X., Erhardt, A., Cronlein, J., Chaves, R.L.,

Yong, C.L., Nehmiz, G., and Steinmann, G.G. (2004). Short-term antiviral efficacy of BILN 2061, a hepatitis C virus serine protease inhibitor, in hepatitis C genotype 1 patients. Gastroenterology *127*, 1347–1355.

Hope, R.G., Murphy, D.J., and McLauchlan, J. (2002). The domains required to direct core proteins of hepatitis C virus and GB virus-B to lipid droplets share common features with plant oleosin proteins. J. Biol. Chem. *277*, 4261–4270.

Houghton, M., and Abrignani, S. (2005). Prospects for a vaccine against the hepatitis C virus. Nature *436*, 961–966.

Hsu, M., Zhang, J., Flint, M., Logvinoff, C., Cheng-Mayer, C., Rice, C.M., and McKeating, J.A. (2003). Hepatitis C virus glycoproteins mediate pH-dependent cell entry of pseudotyped retroviral particles. Proc. Natl. Acad. Sci. USA *100*, 7271–7276.

Hügle, T., Fehrmann, F., Bieck, E., Kohara, M., Kräusslich, H.-G., Rice, C.M., Blum, H.E., and Moradpour, D. (2001). The hepatitis C virus nonstructural protein 4B is an integral endoplasmic reticulum membrane protein. Virology *284*, 70–81.

Ikeda, M., Yi, M., Li, K., and Lemon, S.M. (2002). Selectable subgenomic and genome-length dicistronic RNAs derived from an infectious molecular clone of the HCV-N strain of hepatitis C virus replicate efficiently in cultured Huh7 cells. J. Virol. *76*, 2997–3006.

Ivashkina, N., Wölk, B., Lohmann, V., Bartenschlager, R., Blum, H.E., Penin, F., and Moradpour, D. (2002). The hepatitis C virus RNA-dependent RNA polymerase membrane insertion sequence is a transmembrane segment. J. Virol. *76*, 13088–13093.

Kaito, M., Watanabe, S., Tsukiyama-Kohara, K., Yamaguchi, K., Kobayashi, Y., Konishi, M., Yokoi, M., Ishida, S., Suzuki, S., and Kohara, M. (1994). Hepatitis C virus particle detected by immunoelectron microscopy study. J. Gen. Virol. *75*, 1755–1760.

Kapadia, S.B., and Chisari, F.V. (2005). Hepatitis C virus RNA replication is regulated by host geranylgeranylation and fatty acids. Proc. Natl. Acad. Sci. USA *102*, 2561–2566.

Kato, T., Date, T., Miyamoto, M., Furusaka, A., Tokushige, K., Mizokami, M., and Wakita, T. (2003). Efficient replication of the genotype 2a hepatitis C virus subgenomic replicon. Gastroenterology *125*, 1808–1817.

Kato, T., Date, T., Miyamoto, M., Zhao, Z., Mizokami, M., and Wakita, T. (2005). Nonhepatic cell lines HeLa and 293 support efficient replication of the hepatitis C virus genotype 2a subgenomic replicon. J. Virol. *79*, 592–596.

Katze, M.G., Kwieciszewski, B., Goodlett, D.R., Blakely, C.M., Neddermann, P., Tan, S.L., and Aebersold, R. (2000). Ser(2194) is a highly conserved major phosphorylation site of the hepatitis C virus nonstructural protein NS5A. Virology *278*, 501–513.

Khakoo, S.I., Thio, C.L., Martin, M.P., Brooks, C.R., Gao, X., Astemborski, J., Cheng, J., Goedert, J.J., Vlahov, D., Hilgartner, M., Cox, S., Little, A.M., Alexander, G.J., Cramp, M.E., O'Brien, S.J., Rosenberg, W.M., Thomas, D.L., and Carrington, M. (2004). HLA and NK cell inhibitory receptor genes in resolving hepatitis C virus infection. Science *305*, 872–874.

Kieft, J.S., Zhou, K., Grech, A., Jubin, R., and Duodna, J.A. (2002). Crystal structure of an RNA tertiary domain essential to HCV IRES-mediated translation initiation. Nat. Struct. Biol. *9*, 370–374.

Kim, J.L., Morgenstern, K.A., Lin, C., Fox, T., Dwyer, M.D., Landro, J.A., Chambers, S.P., Markland, W., Lepre, C.A., O'Malley, E.T., Harbeson, S.L., Rice, C.M., Murcko, M.A., Caron, P.R., and Thomson, J.A. (1996). Crystal structure of the hepatitis C virus NS3 protease domain complexed with a synthetic NS4A cofactor peptide. Cell *87*, 343–355.

Kim, J.L., Morgenstern, K.A., Griffith, J.P., Dwyer, M.D., Thomson, J.A., Murcko, M.A., Lin, C., and Caron, P.R. (1998). Hepatitis virus NS3 RNA helicase domain with a bound oligonucleotide: the crystal structure provides insights into the mode of unwinding. Structure *6*, 89–100.

Kolykhalov, A.A., Feinstone, S.M., and Rice, C.M. (1996). Identification of a highly conserved sequence element at the 3′ terminus of hepatitis C virus genome RNA. J. Virol. *70*, 3363–3371.

Kolykhalov, A.A., Agapov, E.V., Blight, K.J., Mihalik, K., Feinstone, S.M., and Rice, C.M. (1997). Transmission of hepatitis C by intrahepatic inoculation with transcribed RNA. Science *277*, 570–574.

Kolykhalov, A.A., Mihalik, K., Feinstone, S.M., and Rice, C.M. (2000). Hepatitis C virus-encoded enzymatic activities and conserved RNA elements in the 3′ nontranslated region are essential for virus replication *in vivo*. J. Virol. *74*, 2046–2051.

Komurian-Pradel, F., Rajoharison, A., Berland, J.L., Khouri, V., Perret, M., Van Roosmalen, M., Pol, S., Negro, F., and Paranhos-Baccala, G. (2004). Antigenic relevance of F protein in chronic hepatitis C virus infection. Hepatology *40*, 900–909.

Krieger, N., Lohmann, V., and Bartenschlager, R. (2001). Enhancement of hepatitis C virus RNA replication by cell culture-adaptive mutations. J. Virol. *75*, 4614–4624.

Kuhn, R.J., Zhang, W., Rossmann, M.G., Pletnev, S.V., Corver, J., Lenches, E., Jones, C.T., Mukhopadhyay, S., Chipman, P.R., Strauss, E.G. Baker, T.S., and Strauss, J.H. (2002). Structure of dengue virus: implications for flavivirus organization, maturation, and fusion. Cell *108*, 717–725.

Kunkel, M., Lorinczi, M., Rijnbrand, R., Lemon, S.M., and Watowich, S.J. (2001). Self-assembly of nucleocapsid-like particles from recombinant hepatitis C virus core protein. J. Virol. *75*, 2119–2129.

Kwong, A.D., Kim, J.L., and Lin, C. (2000). Structure and function of hepatitis C virus NS3 helicase. Curr. Top. Microbiol. Immunol. *242*, 171–196.

Lagging, L.M., Meyer, K., Owens, R.J., and Ray, R. (1998). Functional role of hepatitis C virus chimeric glycoproteins in the infectivity of pseudotyped virus. J. Virol. *72*, 3539–3546.

Lamarre, D., Anderson, P.C., Bailey, M., Beaulieu, P., Bolger, G., Bonneau, P., Bos, M., Cameron, D.R., Cartier, M., Cordingley, M.G., Faucher, A.M., Goudreau, N., Kawai, S.H., Kukolj, G., Lagace, L., LaPlante, S.R., Narjes, H., Poupart, M.A., Rancourt, J., Sentjens, R.E., St George, R., Simoneau, B., Steinmann, G., Thibeault, D., Tsantrizos, Y.S., Weldon, S.M., Yong, C.L., and Llinas-Brunet, M. (2003). An NS3 protease inhibitor with antiviral effects in humans infected with hepatitis C virus. Nature *426*, 186–189.

Lambot, M., Fretier, S., Op De Beeck, A., Quatannens, B., Lestavel, S., Clavey, V., and Dubuisson, J. (2002). Reconstitution of hepatitis C virus envelope glycoproteins into liposomes as a surrogate model to study virus attachment. J. Biol. Chem. *277*, 20625–20630.

Lanford, R.E., and Bigger, C. (2002). Advances in model systems for hepatitis C virus research. Virology *293*, 1–9.

Lechner, F., Wong, D.K., Dunbar, P.R., Chapman, R., Chung, R.T., Dohrenwend, P., Robbins, G., Phillips, R., Klenerman, P., and Walker, B.D. (2000). Analysis of successful immune responses in persons infected with hepatitis C virus. J. Exp. Med. *191*, 1499–1512.

Lerat, H., Honda, M., Beard, M.R., Loesch, K., Sun, J., Yang, Y., Okuda, M., Gosert, R., Xiao, S.Y., Weinman, S.A., and Lemon, S.M. (2002). Steatosis and liver cancer in transgenic mice expressing the structural and nonstructural proteins of hepatitis C virus. Gastroenterology *122*, 352–365.

Lesburg, C.A., Cable, M.B., Ferrari, E., Hong, Z., Mannarino, A.F., and Weber, P.C. (1999). Crystal structure of the RNA-dependent RNA polymerase from hepatitis C virus reveals a fully encircled active site. Nat. Struct. Biol. *6*, 937–943.

Lescar, J., Roussel, A., Wien, M.W., Navaza, J., Fuller, S.D., Wengler, G., Wengler, G., and Rey, F.A. (2001). The fusion glycoprotein shell of Semliki Forest virus: an icosahedral assembly primed for fusogenic activation at endosomal pH. Cell *105*, 137–148.

Li, K., Foy, E., Ferreon, J.C., Nakamura, M., Ferreon, A.C., Ikeda, M., Ray, S.C., Gale, M., Jr., and Lemon, S.M. (2005). Immune evasion by hepatitis C virus NS3/4A protease-mediated cleavage of the Toll-like receptor 3 adaptor protein TRIF. Proc. Natl. Acad. Sci. USA *102*, 2992–2997.

Lindenbach, B.D., Evans, M.J., Syder, A.J., Wölk, B., Tellinghuisen, T.L., Liu, C.C., Maruyama, T., Hynes, R.O., Burton, D.R., McKeating, J.A., and Rice, C.M. (2005a). Complete replication of hepatitis C virus in cell culture. Science *309*, 623–626.

Lindenbach, B.D., and Rice, C.M. (2005b). Unravelling hepatitis C virus replication from genome to function. Nature *436*, 933–938.

Logvinoff, C., Major, M.E., Oldach, D., Heyward, S., Talal, A., Balfe, P., Feinstone, S.M., Alter, H., Rice, C.M., and McKeating, J.A. (2004). Neutralizing antibody response during acute and chronic hepatitis C virus infection. Proc. Natl. Acad. Sci. USA *101*, 10149–10154.

Lohmann, V., Körner, F., Herian, U., and Bartenschlager, R. (1997). Biochemical properties of hepatitis C virus NS5B RNA-dependent RNA polymerase and identification of amino acid sequence motifs essential for enzymatic activity. J. Virol. *71*, 8416–8428.

Lohmann, V., Körner, F., Koch, J.-O., Herian, U., Theilmann, L., and Bartenschlager, R. (1999). Replication of subgenomic hepatitis C virus RNAs in a hepatoma cell line. Science *285*, 110–113.

Lohmann, V., Körner, F., Dobierzewska, A., and Bartenschlager, R. (2001). Mutations in hepatitis C virus RNAs conferring cell culture adaptation. J. Virol. *75*, 1437–1449.

Lohmann, V., Hoffmann, S., Herian, U., Penin, F., and Bartenschlager, R. (2003). Viral and cellular determinants of hepatitis C virus RNA replication in cell culture. J. Virol. *77*, 3007–3019.

Longman, R.S., Talal, A.H., Jacobson, I.M., Albert, M.L., and Rice, C.M. (2004). Presence of functional dendritic cells in patients chronically infected with hepatitis C virus. Blood *103*, 1026–1029.

Love, R.A., Parge, H.E., Wickersham, J.A., Hostomsky, Z., Habuka, N., Moomaw, E.W., Adachi, T., and Hostomska, Z. (1996). The crystal structure of hepatitis C virus NS3 proteinase reveals a trypsin-like fold and a structural zinc binding site. Cell *87*, 331–342.

Lozach, P.Y., Amara, A., Bartosch, B., Virelizier, J.L., Arenzana-Seisdedos, F., Cosset, F.L., and Altmeyer, R. (2004). C-type lectins L-SIGN and DC-SIGN capture and transmit infectious hepatitis C virus pseudotype particles. J. Biol. Chem. *279*, 32035–32045.

Lukavsky, P.J., Otto, G.A., Lancaster, A.M., Sarnow, P., and Puglisi, J.D. (2000). Structures of two RNA domains essential for hepatitis C virus internal ribosome entry site function. Nat. Struct. Biol. *7*, 1105–1110.

Lukavsky, P.J., Kim, I., Otto, G.A., and Puglisi, J.D. (2003). Structure of HCV IRES domain II determined by NMR. Nat. Struct. Biol. *10*, 1033–1038.

Lundin, M., Monne, M., Widell, A., Von Heijne, G., and Persson, M.A. (2003). Topology of the membrane-associated hepatitis C virus protein NS4B. J. Virol. *77*, 5428–5438.

Lyle, J.M., Bullitt, E., Bienz, K., and Kirkegaard, K. (2002). Visualization and functional analysis of RNA-dependent RNA polymerase lattices. Science *296*, 2218–2222.

Macdonald, A., and Harris, M. (2004). Hepatitis C virus NS5A: tales of a promiscuous protein. J. Gen. Virol. *85*, 2485–2502.

Manns, M.P., McHutchison, J.G., Gordon, S.C., Rustgi, V.K., Shiffman, M., Reindollar, R., Goodman, Z.D., Koury, K., Ling, M., Albrecht, J.K., and the International Hepatitis Interventional Therapy Group (2001). Peginterferon alfa-2b plus ribavirin compared with interferon alfa-2b plus ribavirin for initial treatment of chronic hepatitis C: a randomised trial. Lancet *358*, 958–965.

Matsuura, Y., Tani, H., Suzuki, K., Kimura-Someya, T., Suzuki, R., Aizaki, H., Ishii, K., Moriishi, K., Robison, C.S., Whitt, M.A., and Miyamura, T. (2001). Characterization of pseudotype VSV possessing HCV envelope proteins. Virology *286*, 263–275.

McLauchlan, J., Lemberg, M.K., Hope, G., and Martoglio, B. (2002). Intramembrane proteolysis promotes trafficking of hepatitis C virus core protein to lipid droplets. EMBO J. *21*, 3980–3988.

Mercer, D.F., Schiller, D.E., Elliott, J.F., Douglas, D.N., Hao, C., Rinfret, A., Addison, W.R., Fischer, K.P., Churchill, T.A., Lakey, J.R.T., Tyrrell, D.L.J., and Kneteman, N.M. (2001). Hepatitis C virus replication in mice with chimeric human livers. Nat. Med. *7*, 927–933.

Meylan, E., Curran, J., Hofmann, K., Moradpour, D., Binder, M., Bartenschlager, R., and Tschopp, J. (2005). Cardif is a novel adaptor protein in RIG-I-mediated antiviral responses targeted by hepatitis C virus. Nature *437*, 1167–1172.

Missale, G., Bertoni, R., Lamonaca, V., Valli, A., Massari, M., Mori, C., Rumi, M.G., Houghton, M., Fiaccadori, F., and Ferrari, C. (1996). Different clinical behaviors of acute hepatitis C virus infection are associated with different vigor of the anti-viral cell-mediated immune response. J. Clin. Invest. *98*, 706–714.

Miyanari, Y., Hijikata, M., Yamaji, M., Hosaka, M., Takahashi, H., and Shimotohno, K. (2003). Hepatitis C virus non-structural proteins in the probable membranous compartment function in viral RNA replication. J. Biol. Chem. *278*, 50301–50308.

Modis, Y., Ogata, S., Clements, D., and Harrison, S.C. (2003). A ligand-binding pocket in the dengue virus envelope glycoprotein. Proc. Natl. Acad. Sci. USA *100*, 6986–6991.

Modis, Y., Ogata, S., Clements, D., and Harrison, S.C. (2004). Structure of the dengue virus envelope protein after membrane fusion. Nature *427*, 313–319.

Moradpour, D., Englert, C., Wakita, T., and Wands, J.R. (1996). Characterization of cell lines allowing tightly regulated expression of hepatitis C virus core protein. Virology *222*, 51–63.

Moradpour, D., Gosert, R., Egger, D., Penin, F., Blum, H.E., and Bienz, K. (2003a). Membrane association of hepatitis C virus nonstructural proteins and identification of the membrane alteration that harbors the viral replication complex. Antiviral Res. *60*, 103–109.

Moradpour, D., Heim, M.H., Cerny, A., Rice, C.M., and Blum, H.E. (2003b). Cell lines that allow regulated expression of HCV proteins: principles and applications. In Frontiers in viral hepatitis, R. F. Schinazi, J.-P. Sommadossi, and C. M. Rice, eds. (Amsterdam, Elsevier Science), pp. 175–186.

Moradpour, D., Brass, V., Bieck, E., Friebe, P., Gosert, R., Blum, H.E., Bartenschlager, R., Penin, F., and Lohmann, V. (2004a). Membrane association of the RNA-dependent RNA polymerase is essential for hepatitis C virus RNA replication. J. Virol. *78*, 13278–13284.

Moradpour, D., Evans, M.J., Gosert, R., Yuan, Z.H., Blum, H.E., Goff, S.P., Lindenbach, B.D., and Rice, C.M. (2004b). Insertion of green fluorescent protein into nonstructural protein 5A allows direct visualization of functional hepatitis C virus replication complexes. J. Virol. *78*, 7400–7409.

Moradpour, D., Brass, V., and Penin, F. (2005). Function follows form: the structure of the N-terminal domain of HCV NS5A. Hepatology *42*, 732–735.

Moriya, K., Fujie, H., Shintani, Y., Yotsuyanagi, H., Tsutsumi, T., Ishibashi, K., Matsuura, Y., Kimura, S., Miyamura, T., and Koike, K. (1998). The core protein of hepatitis C virus induces hepatocellular carcinoma in transgenic mice. Nat. Med. *4*, 1065–1067.

Muir, A.J., Bornstein, J.D., and Killenberg, P.G. (2004). Peginterferon alfa-2b and ribavirin for the treatment of chronic hepatitis C in blacks and non-Hispanic whites. N. Engl. J. Med. *350*, 2265–2271.

Mukhopadhyay, S., Kim, B.S., Chipman, P.R., Rossmann, M.G., and Kuhn, R.J. (2003). Structure of West Nile virus. Science *302*, 248.

Murray, E.M., Grobler, J.A., Markel, E.J., Pagnoni, M.F., Paonessa, G., Simon, A.J., and Flores, O.A. (2003). Persistent replication of hepatitis C virus replicons expressing the beta-lactamase reporter in subpopulations of highly permissive Huh7 cells. J. Virol. *77*, 2928–2935.

Neddermann, P., Quintavalle, M., Di Pietro, C., Clementi, A., Cerretani, M., Altamura, S., Bartholomew, L., and De Francesco, R. (2004). Reduction of hepatitis C virus NS5A hyperphosphorylation by selective inhibition of cellular kinases activates viral RNA replication in cell culture. J. Virol. *78*, 13306–13314.

Neumann, A.U., Lam, N.P., Dahari, H., Gretch, D.R., Wiley, T.E., Layden, T.J., and Perelson, A.S. (1998). Hepatitis C viral dynamics *in vivo* and the antiviral efficacy of interferon-alpha therapy. Science *282*, 103–107.

National Institutes of Health (2002). Consensus Development Conference Statement: Management of Hepatitis C: (2002). Hepatology *36 (Suppl 1)*, S2-S20.

Op De Beeck, A., Cocquerel, L., and Dubuisson, J. (2001). Biogenesis of hepatitis C virus envelope glycoproteins. J. Gen. Virol. *82*, 2589–2595.

Otto, G.A., and Puglisi, J.D. (2004). The pathway of HCV IRES-mediated translation initiation. Cell *119*, 369–380.

Pallaoro, M., Lahm, A., Biasiol, G., Brunetti, M., Nardella, C., Orsatti, L., Bonelli, F., Orrù, S., Narjes, F., and Steinkühler, C. (2001). Characterization of the hepatitis C virus NS2/3 processing reaction by using a purified precursor protein. J. Virol. *75*, 9939–9946.

Pavlovic, D., Neville, D.C., Argaud, O., Blumberg, B., Dwek, R.A., Fischer, W.B., and Zitzmann, N. (2003). The hepatitis C virus p7 protein forms an ion channel that is inhibited by long-alkyl-chain iminosugar derivatives. Proc. Natl. Acad. Sci. USA *100*, 6104–6108.

Penin, F., Combet, C., Germanidis, G., Frainais, P.O., Deleage, G., and Pawlotsky, J.M. (2001). Conservation of the conformation and positive charges of hepatitis C virus E2 envelope glycoprotein hypervariable region 1 points to a role in cell attachment. J. Virol. *75*, 5703–5710.

Penin, F., Brass, V., Appel, N., Ramboarina, S., Montserret, R., Ficheux, D., Blum, H.E., Bartenschlager, R., and Moradpour, D. (2004a). Structure and function of the membrane anchor domain of hepatitis C virus nonstructural protein 5A. J. Biol. Chem. *279*, 40835–40843.

Penin, F., Dubuisson, J., Rey, F.A., Moradpour, D., and Pawlotsky, J.M. (2004b). Structural biology of hepatitis C virus. Hepatology *39*, 5–19.

Pham, T.N., MacParland, S.A., Mulrooney, P.M., Cooksley, H., Naoumov, N.V., and Michalak, T.I. (2004). Hepatitis C virus persistence after spontaneous or treatment-induced resolution of hepatitis C. J. Virol. *78*, 5867–5874.

Pietschmann, T., Lohmann, V., Kaul, A., Krieger, N., Rinck, G., Rutter, G., Strand, D., and Bartenschlager, R. (2002). Persistent and transient replication of full-length hepatitis C virus genomes in cell culture. J. Virol. *76*, 4008–4021.

Pileri, P., Uematsu, Y., Campagnoli, S., Galli, G., Falugi, F., Petracca, R., Weiner, A.J., Houghton, M., Rosa, D., Grandi, G., and Abrignani, S. (1998). Binding of hepatitis C virus to CD81. Science *282*, 938–941.

Pöhlmann, S., Zhang, J., Baribaud, F., Chen, Z., Leslie, G., Lin, G., Granelli-Piperno, A., Doms, R.W., Rice, C.M., and McKeating, J.A. (2003). Hepatitis C virus glycoproteins interact with DC-SIGN and DC-SIGNR. J. Virol. *77*, 4070–4080.

Quinkert, D., Bartenschlager, R., and Lohmann, V. (2005). Quantitative analysis of the hepatitis C virus replication complex. J. Virol. *in press*.

Reed, K., and Rice, C.M. (1999). Identification of the major phosphorylation site of the hepatitis C virus H strain NS5A protein as serine (2321). J. Biol. Chem. *274*, 28011–28018.

Rehermann, B., and Nascimbeni, M. (2005). Immunology of hepatitis B virus and hepatitis C virus infection. Nat. Rev. Immunol. *5*, 215–229.

Rey, F.A., Heinz, F.X., Mandl, C., Kunz, C., and Harrison, S.C. (1995). The envelope glycoprotein from tick-borne encephalitis virus at 2 Å resolution. Nature *375*, 291–298.

Rijnbrand, R.C., and Lemon, S.M. (2000). Internal ribosome entry site-mediated translation in hepatitis C virus replication. Curr. Top. Microbiol. Immunol. *242*, 85–116.

Sakai, A., Claire, M.S., Faulk, K., Govindarajan, S., Emerson, S.U., Purcell, R.H., and Bukh, J. (2003). The p7 polypeptide of hepatitis C virus is critical for infectivity and contains functionally important genotype-specific sequences. Proc. Natl. Acad. Sci. USA *100*, 11646–11651.

Sarnow, P. (2003). Viral internal ribosome entry site elements: novel ribosome-RNA complexes and roles in viral pathogenesis. J. Virol. *77*, 2801–2806.

Scarselli, E., Ansuini, H., Cerino, R., Roccasecca, R.M., Acali, S., Filocamo, G., Traboni, C., Nicosia, A., Cortese, R., and Vitelli, A. (2002). The human scavenger receptor class B type I is a novel candidate receptor for the hepatitis C virus. EMBO J. *21*, 5017–5025.

Schmidt-Mende, J., Bieck, E., Hügle, T., Penin, F., Rice, C.M., Blum, H.E., and Moradpour, D. (2001). Determinants for membrane association of the hepatitis C virus RNA-dependent RNA polymerase. J. Biol. Chem. *276*, 44052–44063.

Schwartz, M., Chen, J., Janda, M., Sullivan, M., den Boon, J., and Ahlquist, P. (2002). A positive-strand RNA virus replication complex parallels form and function of retrovirus capsids. Mol. Cell *9*, 505–514.

Seifert, U., Liermann, H., Racanelli, V., Halenius, A., Wiese, M., Wedemeyer, H., Ruppert, T., Rispeter, K., Henklein, P., Sijts, A., Hengel, H., Kloetzel, P.M., and Rehermann, B. (2004). Hepatitis C virus mutation affects proteasomal epitope processing. J. Clin. Invest. *114*, 250–259.

Serebrov, V., and Pyle, A.M. (2004). Periodic cycles of RNA unwinding and pausing by hepatitis C virus NS3 helicase. Nature *430*, 476–480.

Shimizu, Y.K., Hijikata, M., Iwamoto, A., Alter, H.J., Purcell, R.H., and Yoshikura, H. (1994). Neutralizing antibodies against hepatitis C virus and the emergence of neutralization escape mutant viruses. J. Virol. *68*, 1494–1500.

Shimizu, Y.K., Feinstone, S.M., Kohara, M., Purcell, R.H., and Yoshikura, H. (1996a). Hepatitis C virus: detection of intracellular virus particles by electron microscopy. Hepatology *23*, 205–209.

Shimizu, Y.K., Igarashi, H., Kiyohara, T., Cabezon, T., Farci, P., Purcell, R.H., and Yoshikura, H. (1996b). A hyperimmune serum against a synthetic peptide corresponding to the hypervariable region 1 of hepatitis C virus can prevent viral infection in cell cultures. Virology *223*, 409–412.

Shirota, Y., Luo, H., Qin, W., Kaneko, S., Yamashita, T., Kobayashi, K., and Murakami, S. (2002). Hepatitis C virus NS5A binds RNA-dependent RNA polymerase NS5B and modulates RNA-dependent RNA polymerase activity. J. Biol. Chem. *277*, 11149–11155.

Shoukry, N.H., Grakoui, A., Houghton, M., Chien, D.Y., Ghrayeb, J., Reimann, K.A., and Walker, C.M. (2003). Memory CD8+ T cells are required for protection from persistent hepatitis C virus infection. J. Exp. Med. *197*, 1645–1655.

Shoukry, N.H., Cawthon, A.G., and Walker, C.M. (2004). Cell-mediated immunity and the outcome of hepatitis C virus infection. Annu. Rev. Microbiol. *58*, 391–424.

Simmonds, P. (2004). Genetic diversity and evolution of hepatitis C virus—15 years on. J. Gen. Virol. *85*, 3173–3188.

Spahn, C.M., Kieft, J.S., Grassucci, R.A., Penczek, P.A., Zhou, K., Doudna, J.A., and Frank, J. (2001). Hepatitis C virus IRES RNA-induced changes in the conformation of the 40S ribosomal subunit. Science *291*, 1959–1962.

Su, A.I., Pezacki, J.P., Wodicka, L., Brideau, A.D., Supekova, L., Thimme, R., Wieland, S., Bukh, J., Purcell, R.H., Schultz, P.G., and Chisari, F.V. (2002). Genomic analysis of the host response to hepatitis C virus infection. Proc. Natl. Acad. Sci. USA *99*, 15669–15674.

Sumpter, R., Jr., Loo, Y.M., Foy, E., Li, K., Yoneyama, M., Fujita, T., Lemon, S.M., and Gale, M., Jr. (2005). Regulating intracellular antiviral defense and permissiveness to hepatitis C virus RNA replication through a cellular RNA helicase, RIG-I. J. Virol. *79*, 2689–2699.

Takaki, A., Wiese, M., Maertens, G., Depla, E., Seifert, U., Liebetrau, A., Miller, J.L., Manns, M.P., and Rehermann, B. (2000). Cellular immune responses persist and humoral responses decrease two decades after recovery from a single-source outbreak of hepatitis C. Nat. Med. *6*, 578–582.

Tan, S.-L., and Katze, M.G. (2001). How hepatitis C virus counteracts the interferon response: the jury is still out on NS5A. Virology *284*, 1–12.

Tanaka, T., Kato, N., Cho, M.-J., Sugiyama, K., and Shimotohno, K. (1996). Structure of the 3′ terminus of the hepatitis C virus genome. J. Virol. *70*, 3307–3312.

Taylor, D.R., Shi, S.T., Romano, P.R., Barber, G.N., and Lai, M.M.C. (1999). Inhibition of the interferon-inducible protein kinase PKR by HCV E2 protein. Science *285*, 107–110.

Tellinghuisen, T.L., and Rice, C.M. (2002). Interaction between hepatitis C virus proteins and host cell factors. Curr. Opin. Microbiol. *5*, 419–427.

Tellinghuisen, T.L., Marcotrigiano, J., Gorbalenya, A.E., and Rice, C.M. (2004). The NS5A protein of hepatitis C virus is a zinc metalloprotein. J. Biol. Chem. *279*, 48576–48587.

Tellinghuisen, T.L., Marcotrigiano, J., and Rice, C.M. (2005). Structure of the zinc-binding domain of an essential replicase component of hepatitis C virus reveals a novel fold. Nature *435*, 375–379.

Thibeault, D., Maurice, R., Pilote, L., Lamarre, D., and Pause, A. (2001). *In vitro* characterization of a purified NS2/3 protease variant of hepatitis C virus. J. Biol. Chem. *276*, 46678–46684.

Triyatni, M., Saunier, B., Maruvada, P., Davis, A.R., Ulianich, L., Heller, T., Patel, A., Kohn, L.D., and Liang, T.J. (2002). Interaction of hepatitis C virus-like particles and cells: a model system for studying viral binding and entry. J. Virol. *76*, 9335–9344.

Tsai, S.L., Chen, Y.M., Chen, M.H., Huang, C.Y., Sheen, I.S., Yeh, C.T., Huang, J.H., Kuo, G.C., and Liaw, Y.F. (1998). Hepatitis C virus variants circumventing cytotoxic T lymphocyte activity as a mechanism of chronicity. Gastroenterology *115*, 954–965.

Tseng, C.T., and Klimpel, G.R. (2002). Binding of the hepatitis C virus envelope protein E2 to CD81 inhibits natural killer cell functions. J. Exp. Med. *195*, 43–49.

van Regenmortel, M.H.V., Fauquet, C.M., Bishop, D.H.L., Carstens, E.B., Estes, M.K., Lemon, S.M., Maniloff, J., Mayo, M.A., McGeoch, D.J., Pringle, C.R., and Wickner, R.B., eds. (2000). Virus Taxonomy. The VIIth Report of the International Committee on Taxonomy of Viruses (San Diego, Academic Press).

Varaklioti, A., Vassilaki, N., Georgopoulou, U., and Mavromara, P. (2002). Alternate translation occurs within the core coding region of the hepatitis C viral genome. J. Biol. Chem. *277*, 17713–17721.

Wakita, T., Pietschmann, T., Kato, T., Date, T., Miyamoto, M., Zhao, Z., Murthy, K., Habermann, A., Kräusslich, H.G., Mizokami, M., Bartenschlager, R., and Liang, T.J. (2005). Production of infectious hepatitis C virus in tissue culture from a cloned viral genome. Nat. Med. *11*, 791–796.

Walewski, J.L., Keller, T.R., Stump, D.D., and Branch, A.D. (2001). Evidence for a new hepatitis C virus antigen encoded in an overlapping reading frame. RNA *7*, 710–721.

Wang, C., Gale, M., Jr., Keller, B.C., Huang, H., Brown, M.S., Goldstein, J.L., and Ye, J. (2005). Identification of FBL2 as a geranylgeranylated cellular protein required for hepatitis C virus RNA replication. Mol. Cell *18*, 425–434.

Watashi, K., Ishii, N., Hijikata, M., Inoue, D., Murata, T., Miyanari, Y., and Shimotohno, K. (2005). Cyclophilin B is a functional regulator of hepatitis C virus RNA polymerase. Mol. Cell *19*, 111–122.

Weiner, A.J., Brauer, M.J., Rosenblatt, J., Richman, K.H., Tung, J., Crawford, K., Bonino, F., Saracco, G., Choo, Q.L., Houghton, M., and Han, J.H. (1991). Variable and hypervariable domains are found in the regions of HCV corresponding to the flavivirus envelope and NS1 proteins and the pestivirus envelope glycoproteins. Virology *180*, 842–848.

Wellnitz, S., Klumpp, B., Barth, H., Ito, S., Depla, E., Dubuisson, J., Blum, H.E., and Baumert, T.F. (2002). Binding of hepatitis C virus-like particles derived from infectious clone H77C to defined human cell lines. J. Virol. *76*, 1181–1193.

Wölk, B., Sansonno, D., Kräusslich, H.-G., Dammacco, F., Rice, C.M., Blum, H.E., and Moradpour, D. (2000). Subcellular localization, stability and *trans*-cleavage competence of the hepatitis C virus NS3-NS4A complex expressed in tetracycline-regulated cell lines. J. Virol. *74*, 2293–2304.

Xie, Z.C., Riezu-Boj, J.I., Lasarte, J.J., Guillen, J., Su, J.H., Civeira, M.P., and Prieto, J. (1998). Transmission of hepatitis C virus infection to tree shrews. Virology *244*, 513–520.

Xu, Z., Choi, J., Yen, T.S., Lu, W., Strohecker, A., Govindarajan, S., Chien, D., Selby, M.J., and Ou, J. (2001). Synthesis of a novel hepatitis C virus protein by ribosomal frameshift. EMBO J. *20*, 3840–3848.

Yagnik, A.T., Lahm, A., Meola, A., Roccasecca, R.M., Ercole, B.B., Nicosia, A., and Tramontano, A. (2000). A model for the hepatitis C virus envelope glycoprotein E2. Proteins *40*, 355–366.

Yamashita, T., Kaneko, S., Shirota, Y., Qin, W., Nomura, T., Kobayashi, K., and Murakami, S. (1998). RNA-dependent RNA polymerase activity of the soluble recombinant hepatitis C virus NS5B protein truncated at the C-terminal region. J. Biol. Chem. *273*, 15479–15486.

Yan, Y., Li, Y., Munshi, S., Sardana, V., Cole, J.L., Sardana, M., Steinkuehler, C., Tomei, L., De Francesco, R., Kuo, L.C., and Chen, Z. (1998). Complex of NS3 protease and NS4A peptide of BK strain of hepatitis C virus: a 2.2 Å resolution structure in a hexagonal crystal form. Protein Sci. *7*, 837–847.

Yanagi, M., St Claire, M., Emerson, S.U., Purcell, R.H., and Bukh, J. (1999). In vivo analysis of the 3′ untranslated region of the hepatitis C virus after *in vitro* mutagenesis of an infectious cDNA clone. Proc. Natl. Acad. Sci. USA *96*, 2291–2295.

Yang, G., Pevear, D.C., Collett, M.S., Chunduru, S., Young, D.C., Benetatos, C., and Jordan, R. (2004). Newly synthesized hepatitis C virus replicon RNA is protected from nuclease activity by a protease-sensitive factor(s). J. Virol. *78*, 10202–10205.

Yao, N., Hesson, T., Cable, M., Hong, Z., Kwong, A.D., Le, H.V., and Weber, P.C. (1997). Structure of the hepatitis C virus RNA helicase domain. Nat. Struct. Biol. *4*, 463–467.

Yao, N., Reichert, P., Taremi, S.S., Prosise, W.W., and Weber, P.C. (1999). Molecular views of viral polyprotein processing revealed by the crystal structure of the hepatitis C virus bifunctional protease-helicase. Structure Fold. Des. *7*, 1353–1363.

Ye, J., Wang, C., Sumpter, R., Jr., Brown, M.S., Goldstein, J.L., and Gale, M., Jr. (2003). Disruption of hepatitis C virus RNA replication through inhibition of host protein geranylgeranylation. Proc. Natl. Acad. Sci. USA *100*, 15865–15870.

Yi, M., Bodola, F., and Lemon, S.M. (2002). Subgenomic hepatitis C virus replicons inducing expression of a secreted enzymatic reporter protein. Virology *304*, 197–210.

Yi, M., and Lemon, S.M. (2003a). 3′ nontranslated RNA signals required for replication of hepatitis C virus RNA. J. Virol. *77*, 3557–3568.

Yi, M., and Lemon, S.M. (2003b). Structure-function analysis of the 3′ stem–loop of hepatitis C virus genomic RNA and its role in viral RNA replication. RNA *9*, 331–345.

You, S., Stump, D.D., Branch, A.D., and Rice, C.M. (2004). A cis-acting replication element in the sequence encoding the NS5B RNA-dependent RNA polymerase is required for hepatitis C virus RNA replication. J. Virol. *78*, 1352–1366.

Yu, M.Y., Bartosch, B., Zhang, P., Guo, Z.P., Renzi, P.M., Shen, L.M., Granier, C., Feinstone, S.M., Cosset, F.L., and Purcell, R.H. (2004). Neutralizing antibodies to hepatitis C virus (HCV) in immune globulins derived from anti-HCV-positive plasma. Proc. Natl. Acad. Sci. USA *101*, 7705–7710.

Zhang, J., Randall, G., Higginbottom, A., Monk, P., Rice, C.M., and McKeating, J.A. (2004). CD81 is required for hepatitis C virus glycoprotein-mediated viral infection. J. Virol. *78*, 1448–1455.

Zhao, W.D., Wimmer, E., and Lahser, F.C. (1999). Poliovirus/hepatitis C virus (internal ribosomal entry site-core) chimeric viruses: improved growth properties through modification of proteolytic cleavage site and requirement for core RNA sequences but not for core-related polypeptides. J. Virol. *73*, 1546–1554.

Zhong, J., Gastaminza, P., Cheng, G., Kapadia, S., Kato, T., Burton, D.R., Wieland, S.F., Uprichard, S.L., Wakita, T., and Chisari, F.V. (2005). Robust hepatitis C virus infection *in vitro*. Proc. Natl. Acad. Sci. USA *102*, 9294–9299.

Zhu, Q., Guo, J.T., and Seeger, C. (2003). Replication of hepatitis C virus subgenomes in nonhepatic epithelial and mouse hepatoma cells. J. Virol. *77*, 9204–9210.

Treatment of HCV-related Liver Diseases and Prevention of Hepatocellular Carcinoma

12

Kyuichi Tanikawa

Abstract

HCV-related liver diseases are characterized by persistent inflammation and the frequent development of hepatocellular carcinoma (HCC). These characteristics are due to immunological alterations caused by dysfunction of the dendritic cells and by oxidative stress in the hepatocytes. Immunological alterations brought about by persistent HCV infection and oxidative stress are among the most important risk factors for the development of HCC. Therefore, the most important clinical targets for the treatment of HCV-related liver diseases are the elimination of HCV from the body and the prevention of HCC. For these purposes Interferon treatment has been carried out with some success. Reduction of oxidative stress is also an important target in the treatment of HCV-related liver diseases and prevention of HCC.

Introduction

It is presently estimated that there are approximately 170 million HCV (hepatitis C virus) carriers in the world (WHO, 1999). Unlike hepatitis B virus (HBV) carriers, HCV carriers are mainly found in developed countries, because HCV is primarily transmitted by parenteral routes involving medical treatment or intravenous drug abuse.

Japan has a relatively high prevalence of HCV infection. HCV infection rates in Japan rose sharply soon after the end of the Second World War. Intravenous drug abuse became very popular during this period because of social disorder, while at the same time all citizens had access to medical treatment such as intravenous injections or blood transfusions because of the well-established national medical security system. As a result, many people were infected with HCV soon after the war through parenteral transmission, and we estimate that there are presently about 1.5 million HCV carriers in Japan. We have seen many HCV-infected patients during the last 60 years, and we have had a chance to follow many patients with HCV-related liver diseases for a long period of time. Our experience indicates that once chronic hepatitis C is established, in most cases chronic inflammation persists with no decrease in viral load throughout the clinical course, resulting in no clinical cure and eventual progression of the liver injuries to hepatocellular carcinoma (HCC).

The annual number of deaths due to liver diseases in Japan is approximately 50 000, and HCC accounts for more than two-thirds or 35 000 of these deaths. Additionally, more than 80 percent of the deaths due to HCC are HCV-related. At present, complications of cirrhosis such as esophageal variceal rupture, ascites or hepatic encephalopathy are clinically well controlled and few patients die of these complications, especially in HCV-related liver cirrhosis.

Therefore the attention of hepatologists in Japan today is focused primarily on early detection of small HCC lesions, and on preventing the development of HCC.

The most important challenge from a clinical point of view is to identify the reasons why in most cases, once HCV infection is established, there is persistent inflammation with progression to cirrhosis and finally to HCC with no clearance of the infected virus. It is extremely important that the diagnosis and treatment of HCV-related liver diseases, and efforts to prevent HCC, should be based on a thorough understanding of the pathogenesis of HCV-related liver diseases.

This chapter will focus on the treatment of HCV-related liver injuries and the prevention of HCC.

Natural course of HCV infection

For the most part, HCV infection (unlike HBV infection) occurs only through parenteral routes, because the viral load in HCV carriers is about one-hundredth that of HBV carriers. Thus, mother to infant or sexual transmission is only observed in rare cases with an unusually high viral load (Vandelli *et al.*, 2004).

Once HCV transmission occurs, most patients become persistent carriers. Persistent infection is observed in at least 60–80% of persons infected with HCV (Gerlach *et al.*, 2003; Omata *et al.*, 1991). In HCV-related acute hepatitis, which is generally characterized by symptoms such as general malaise, jaundice and others, the transition to persistent infection is often observed in the clinical course following the acute stage. However, the vast majority of chronic hepatitis C cases never experience an acute episode of hepatitis.

In acute hepatitis, immunological reactions to HCV infection are relatively strong and thus general malaise, jaundice and other symptoms of marked liver injuries are induced (Lechner *et al.*, 2000). However, as mentioned above, in most cases of HCV infection acute episodes are not recognized, probably due to weak immunological responses at the time of infection. As a result, once HCV infections occurs, most cases seem to become persistent carriers, gradually develop chronic hepatitis, and eventually progress to liver cirrhosis without a natural cure. However, approximately 30% of HCV carriers show minimal liver changes with normal serum ALT (Bruno *et al.*, 1994; Jamal *et al.*, 1999: Mathurin *et al.*, 1998). These "healthy carriers" have mostly mild liver injuries for a long period with no development of HCC, and only a few of them progress to an advanced stage of chronic liver disease (Brillanti *et al.*, 1993).

Once chronic hepatitis is established, most patients progress to liver cirrhosis without natural cure, and among cirrhotic patients the annual incidence of HCC is about 7%. This means that all cirrhotic patients will finally progress to HCC after about 15 years (Ikeda *et al.*, 1993) (Figure 12.1). A recent Japanese study showed that the annual occurrence rates of HCC in F_2 and F_3 chronic hepatitis are 2 and 5.3%, and in F_4 (cirrhosis) the rate is 7.9% (Yoshida *et al.*, 1999). In addition, in cases completely cured of HCC by surgical resection or ablation therapy, the annual occurrence of newly developed second primary HCC lesions is 20% (Tanikawa and Majima, 1993; Ikeda *et al.*, 2000) (Figure 12.2). These patients are at super high risk

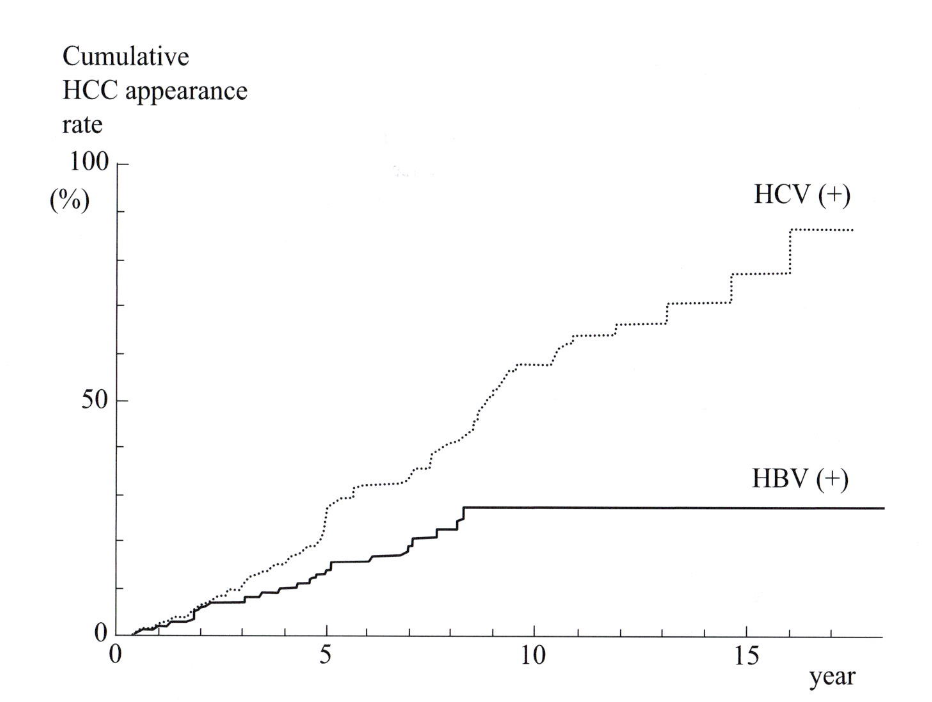

Figure 12.1 Cumulative HCC appearance rates of the HBsAg-positive subgroup and the anti-HCV-positive subgroup. HCC occurred less frequently in the patients with HBsAg than in those with anti-HCV (Ikeda *et al*., 1993).

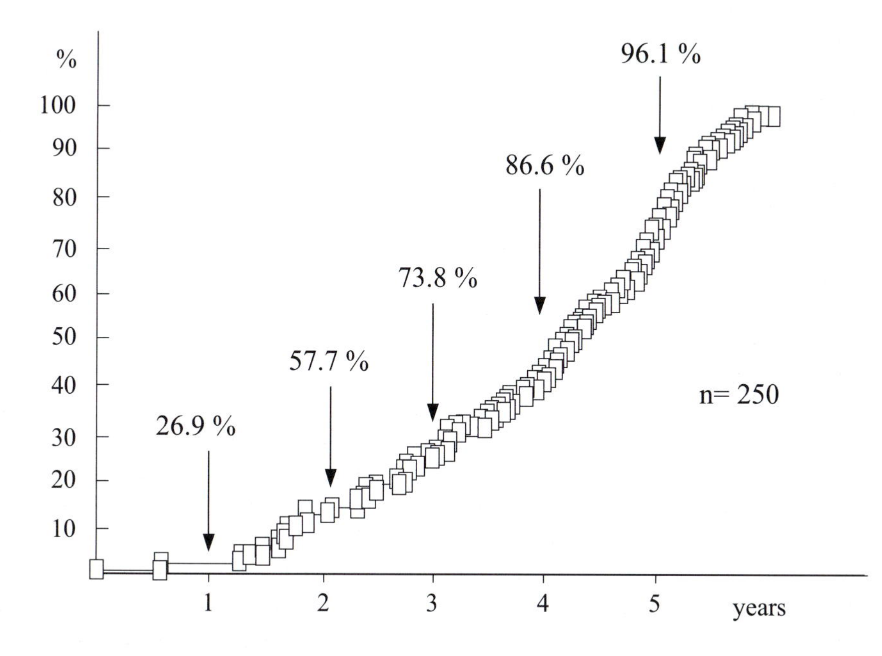

Figure 12.2 Incidence of recurrence after successful percutaneous ethanol injection therapy (PEIT) in cases of small HCC (less than 3cm in diameter). Most cases of small HCC develop recurrence five years after first PEIT (Tanikawa *et al*., 1993).

of HCC and should be followed very carefully after the first treatment.

HCC generally develops in most HCV-related cirrhotic patients before the appearance of severe end-stage cirrhotic complications, and thus the cause of death in these patients is usually HCC.

Taken these findings into account, it becomes clear that the most important factors for successful treatment are detection of the HCC at an early stage, and prevention of the progression to HCC.

Considering the above facts, we should carefully examine the reasons why, in HCV infection, persistent infection occurs which results in chronic hepatitis and leads finally to the development of HCC. A full understanding of the pathogenesis of HCV infection is essential in order to achieve the best diagnosis and treatment for HCV-related liver injuries.

Pathogenesis of persistent infection and development of HCC

As described earlier, once chronic hepatitis is established, inflammation in the liver persists without a decrease of viral load and the hepatic injuries progress gradually. In HBV-related chronic hepatitis, on the other hand, more than half of all patients achieve a clinically cured state with a decrease of viral load. These differences are reflected in the types of pathological changes seen in HBV- and HCV-infected livers, respectively. Histological changes accompanying liver cirrhosis differ markedly depending on the type of virus involved. Namely in HCV cirrhosis, regenerative nodules appear smaller and have active inflammatory changes. In HBV cirrhosis, regenerative nodules are generally larger with minimal inflammation (Shimamatsu *et al.*, 1994).

One of the explanations for the persistence of HCV infection is the infection of dendritic cells by HCV (Kaimori *et al.*, 2004), causing functional disorders in these cells (Kanto *et al.*, 1999; Jinushi *et al.*, 2003a; Muratori *et al.*, 1996). For example, the impaired function of HCV-specific cytotoxic T cells seen in HCV-chronic liver injuries (Giuggio *et al.*, 1998; Gruener *et al.*, 2001) is probably due to their infection by HCV, which would explain the persistent infection. Impaired NK cell function is also seen in HCV-related liver injuries (Corado *et al.*, 1997). This dysfunction is also probably due to impaired dendritic cell functions (Jinushi *et al.*, 2003b). It is important to remember that the liver has the highest number and highest activity of NK cells among all organs. One reason for this finding might be the high number of foreign antigens that reach the liver by the portal blood stream. One of the important functions of the NK cells is to recognize cells infected with viruses or malignant cells and to eliminate them nonspecifically. Therefore, impaired NK functions can induce an environment in the liver where malignant cell development can take place more readily.

Recent studies have indicated that the core protein of HCV in hepatocytes induces a remarkable increase in reactive oxygen species (ROS) (Yamamoto et el., 1998; Jain *et al.*, 2002; Weinman *et al.*, 2003)). In fact, malondialdehyde (MDA) and 4-hydroxynonenal (HNE) are detected by immunological staining as lipid peroxidation products due to oxidative stress in the hepatocytes of HCV patients (Paradis *et al.*, 1997). Similarly, 8-OhdG staining is observed in the nucleus of the hepatocytes. Thioredoxin, which is produced as a result of oxidative stress, is elevated in the blood and also is stained histochemically in the hepatocytes according to light and electron microscopic observations (Mahmood *et al.*, 2004). These studies indicate that ROS production is enhanced in hepatocytes infected with HCV, resulting in increased oxidative stress.

The mechanisms of the oxidative stress in HCV infection have not yet been elucidated. Some reports suggest that it is due to mitochondrial changes induced by HCV core protein (Okuda *et al.*, 2002). In fact, characteristic changes in these organelles have been observed in hepatocytes (Barbaro *et al.*, 1999).

In addition, recent studies have shown that NF-κB (Yoshida *et al.*, 2002), and other signal factors are remarkably activated by HCV core protein. These activations are probably due to oxidative stress in the hepatocytes and are favorable for malignant transformation.

At present, the most frequently mentioned cause of cancer development is oxidative stress, which induces oxidative nuclear DNA injuries as the first step toward malignant transformation. Thus, it is conceivable that HCV infection induces HCC development, not by direct effect of the virus on the DNA, but by induced oxidative stress in the hepatocyte. In other words, it is easy to imagine that oxidative stress in the hepatocyte, occurring over a period of 20–30 years of persistent HCV infection, may be the main causative factor of HCV-induced HCC. In addition, the impaired immunological tumor surveillance in the liver due to dysfunction of NK cells as a result of infection of dendritic cells by HCV makes the liver more susceptible to tumor development. Figure 12.3 shows the pathogenesis of HCV-related liver injuries that finally lead to development of HCC.

Promoting factors for development of HCC

Two types of mechanisms of cancer development have been identified, namely hereditary and environmental. However, no cases of hereditary tumors have been identified in human HCC. Two main environmental factors have been known to promote human tumor development, namely direct DNA injuries by environmental factors, and oxidative stress caused

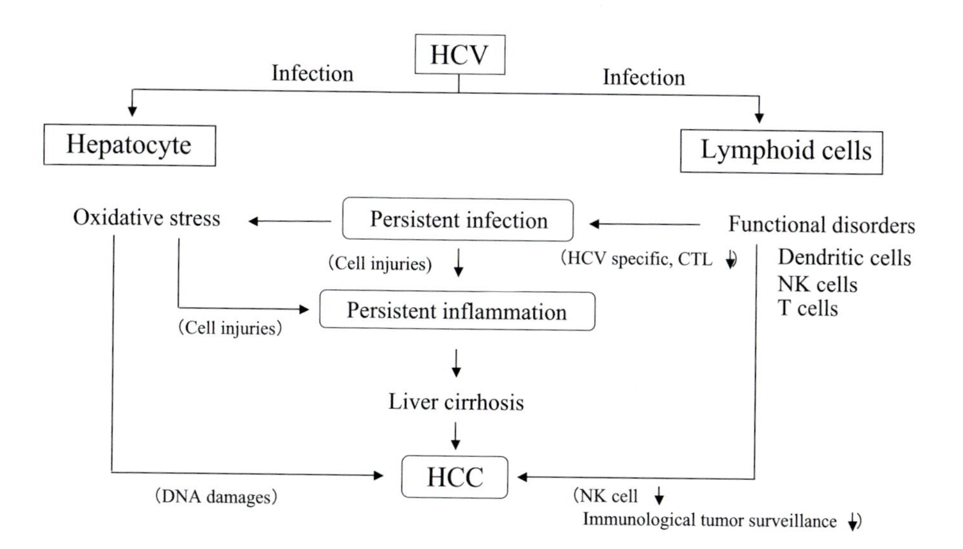

Figure 12.3 Mechanism of HCV-related liver injuries.

by ROS (Dreher and Junod, 1996). In human HCC, direct DNA injuries are considered the causative factors in cases of HBV infection (Koike and Takada, 1995), and aflatoxin intoxication (Hertzog *et al.*, 1982). In contrast, HCC caused by HCV infection, alcoholic liver injuries, non-alcoholic steatohepatitis (NASH) and hemochromatosis are all probably caused by oxidative stress. Therefore, oxidative stress seems to be, at present, the most important factor in HCC development.

The liver is a center of metabolism in the body. Among the organelles in the hepatocyte, mitochondria carry out a considerable number of metabolic activities. One of their representative metabolic activities is the production of ATP by β oxidation of fatty acids. In this oxidation process, a large amount of ROS is produced. Moreover, ROS are produced in the smooth ER whenever a large amount of drugs or alcohol is ingested. The liver has a strong antioxidant system to cope with such ROS production in the hepatocytes, including SOD, GSH, catalase, etc. Of all the various cells and organs in the body, the hepatocyte has the strongest antioxidant system.

We see on one hand that ROS are produced more abundantly in the hepatocyte than in any other type of cell in the body, and on the other, that hepatocytes have the most powerful antioxidant system of any cell type. Thus, when additional oxidative stress is induced by environmental factors such as viral infection, alcohol intake or others, the hepatocyte's powerful antioxidant system is impaired, eventually resulting in HCC development.

As a result, we may go so far as to say that cancer development is more readily induced in hepatocytes than in any other cell type or organ in the body. In addition, the factors described below are also considered to play important roles in the development of HCV-related HCC.

Association with liver cirrhosis

In HCV-related HCC, most of the patients have liver cirrhosis. Interestingly, the reported annual frequency of HCC development in patients with cirrhosis is almost the same (around 7%) at medical institutions throughout Japan. In other words, the frequency of HCC development in 10 years is 70% and 100% in 15 years. The reasons why the frequency is so high in cirrhotic patients are not clear. One possible explanation is the poor supply of oxygen and nutrients to the hepatocytes due to the capillarization of the sinusoidal wall in the cirrhotic liver (Schaffner and Popper, 1963), which might induce oxygen stress in the hepatocyte. In the processes involved in hepatic fibrosis, the stellate cells transform to fibroblasts with a loss of fat droplets containing vitamin A. In fact, a remarkable loss of vitamin A is noted in cirrhotic livers. A recent study showed that steatohepatitis and HCC develop at a high frequency in mice expressing retinotic acid receptor dominant negative forms (Yanagitani *et al.*, 2004). This result suggested that oxidative stress in the hepatocytes of mice with dominant negative retinotic acid receptors probably induced HCC. It appears, therefore, that hepatocytes need a large supply of vitamin A to prevent the induction of oxidative stress in liver cirrhosis. For this reason, the hepatic stellate cells play an important role in the metabolic activities of hepatocytes.

Zinc deficiency is also often noted in the cirrhotic liver (Boyett and Sullivan, 1970). Zinc is an important component of various enzymes. For instance Zn-SOD is an important enzyme in the antioxidant system and therefore the antioxidant activities in the hepatocyte could be impaired by zinc deficiency associated with liver cirrhosis. Hepatic NK cell count and its activity reduced in cirrhosis would be also favorable in development of HCC and IFN administration shows to improve it activity (Shirachi *et al.*, 1998)

Persistent inflammation

Persistent inflammation without a decreased viral load is one of the characteristic features of HCV-related chronic liver injuries. The persistent inflammation is probably caused both by impaired activities of HCV-specific cytopathic T cells and by oxidative stress induced in the hepatocyte. The incidence of HCC is significantly higher in patients with serum ALT over 80 units than in patients with lower ALT levels in HCV-related liver cirrhosis (Tarao *et al.*, 1999). Persistent inflammation would cause instability of genes, which is favorable for development of HCC (Hino *et al.*, 1998).

Unlike HBV-related liver cirrhosis, persistent inflammation with elevated ALT is always noted in HCV-related liver cirrhosis. This is one of the reasons why a higher incidence of HCC development is noted in HCV-related liver cirrhosis.

Age

It seems that age is an important factor in the development of HCC. In HBV-related HCC, age is not always related with HCC development because this type of HCC is occasionally found in children. On the other hand, our experience with 865 cases of HCV-related HCC revealed that 97% of patients developed HCC after 50 years of age and that two thirds of the patients were over 60 at the time of discovery of HCC. The development of HCC was not related with the time of HCV infection, but rather with age at the time of detection, which was generally when the patient was over 60 (Hamada *et al.*, 2002).

Thus, HCC appears to develop more readily in the livers of cirrhotic patients with HCV infection who are over 60 years of age. Electron microscopic observation of liver biopsy specimens taken from elderly persons reveals numerous lipofuscin granules, which are thought to be a complex of peroxidative lipid and degenerated lipoprotein in the hepatocyte. Such deposits of granules in the hepatocyte suggest that protein or lipid in the hepatocytes of older people is affected by oxidative stress. The levels of antioxidants such as SOD in the hepatocyte in older people could be low, and these lower antioxidant activities in the elderly could help to promote the development of HCC. More extensive studies are required on the induction of ROS and the changes in the antioxidant system in the hepatocytes of older people.

Iron overload

An excess of iron accumulation is often noted in the hepatocytes of chronic hepatitis C (Hézode *et al.*, 1999). Generally, iron accumulation is clinically evaluated by the elevation of serum ferritin (Chino *et al.*, 2002). Excess iron in the hepatocyte produces ROS, and this leads to additional oxidative stress on top of the HCV-induced oxidative stress. HCC is often found in patients with hemochromatosis (Niederau *et al.*, 1985), probably because of the oxidative stress induced by ROS due to the iron deposits. Therefore, iron accumulation could promote HCC development in HCV-related liver injuries (Chapoutot *et al.*, 2000). In general, iron deposits are frequently observed in hepatocytes of alcoholic liver diseases (Ito *et al.*, 1999), NASH (George *et al.*, 1998) and HCV-related liver injuries. Those hepatic diseases are all characterized etiologically by oxidative stress as one of the main pathogenetic factors.

However, it is not clear why such iron accumulation occurs in these hepatic diseases. Enhanced transferrin receptor expression has been suggested as one of the factors contributing to the excess iron accumulation in alcoholic liver injuries (Suzuki *et al.*, 2002). Hepcidin (Robson, 2004; Walker *et al.*, 2004)), a recently recognized hepatic factor controlling iron absorption in the duodenum, could be related to such iron accumulation.

Obesity and diabetes

Obesity has become a major health problem all over the world, including Japan, and is associated with an increasing number of diabetic patients. Visceral obesity in particular is etiologically related with insulin resistance and type 2 diabetes. This association is supported by the fact that adipocytes in visceral fatty tissue excrete more TNFα and free fatty acid, both known as causative factors of the insulin resistance, than those in subcutaneous fatty tissue (Shibasaki *et al.*, 2002).

The most common cause of fatty liver is obesity, and the fatty liver in obesity is generally related to visceral fat accumulation because the main sources of the fat in the hepatocyte are dietary along with free fatty acid released from the visceral fat via the portal vein. One of the important etiological factors in NASH developing in the fatty liver is reported to be oxidative stress (Younossi *et al.*, 2002; Weltman *et al.*, 1998; Sumida *et al.*, 2003), and it has been noted that HCC frequently occurs during the clinical course of NASH (Bugianesi *et al.*, 2002). Thus, the combination of oxidative stress related to HCV infection and that from fatty liver due to obesity is expected to enhance the development of HCC. In fact, obesity is reported to be the most serious risk factor related to disease progression in HCV-related liver injuries (Hourigan *et al.*, 1999). As described before, insulin resistance is closely related with visceral fat accumulation. TNFα and free fatty acid released from lipocytes in the visceral fat tissue directly affect the hepatocytes via the portal vein. It is thought that the portal blood flow passes through the hepatic sinusoidal and affects numerous hepatocytes in livers weighing about 1500 g and induces insulin resistance in large numbers of hepatocytes by TNFα and free fatty acid. Expression of insulin resistance has been compared between mice with knockout of the insulin receptor in the liver and those with knockout of the insulin receptor in muscle tissue, and it was found that insulin resistance is much more severe in the case of insulin receptor knockout in the liver (Brüning *et al.*, 1998; Michael *et al.*, 2000).

Thus, clinical and experimental studies indicate that the liver seems to be the most important organ for the induction of insulin resistance. In fact, insulin resistance and type 2 diabetes are often noted clinically in HCV-related liver diseases (Allison *et al.*, 1994; Mason *et al.*, 1999; Mehta *et al.*, 2000; Petit *et al.*, 2001). Our recent study showed an inhibited expression of IRS-1 in liver biopsy specimens taken from patients with HCV-chronic hepatitis (Kawaguchi *et al.*, 2004). Furthermore, a remarkable insulin resistance was noted in HCV core gene transgenic mice, and type 2 diabetes developed in the transgenic mice given a high fat diet (Shintani *et al.*, 2004). From these findings, it is clear that the liver plays an important role in the development of insulin resistance. Another recent study showed that insulin resistance is closely related to hepatic iron overload (Mendler *et al.*, 1999; Furutani *et al.*, 2003), which is commonly seen in chronic hepatitis C, NASH and alcoholic liver injuries. It is not clear that the oxidative stress is the main causative factor of insulin resistance. However, the liver seems to play the most important role in the development of insulin resistance and type 2 diabetes. In chronic hepatitis C, the high blood insulin level caused by insulin resistance also induces fat accumulation in the liver and promotes oxidative stress. In fact, obesity and obesity-related diabetes are the most serious risk factors in the prognosis of HCV-related chronic liver diseases (Hu *et al.*, 2004; Oritz *et al.*, 2002).

Alcohol

Alcohol consumption is a common habit and quite a number of people drink alcohol every day. ROS are produced in the process of alcohol metabolization in the liver, especially in cases of heavy drinking (Hoek *et al.*, 2002). ROS are produced in large amounts along with induction of P450 2E1. It is believed that one of the main pathogenetic pathways in alcohol liver injuries is the oxidative stress induced by ROS overproduction in the hepatocytes (Kurose *et al.*, 1997). Thus, it is reasonable to assume that alcohol intake can exacerbate chronic hepatitis C by producing additional oxidative stress in the hepatocytes.

The incidence of HCC in HCV-related chronic liver diseases caused by blood transfusion is higher in alcohol drinkers than in nondrinkers (Miyakawa *et al.*, 1993). A recent study reported that alcohol activates an intrinsic opioid system and enhances HCV proliferation, and that as a result alcohol intake promotes disease progression and resistance to IFN therapy (Zhang *et al.*, 2003). Thus, heavy drinking is a significant risk factor for HCC in HCV-related liver cirrhosis (Chiba *et al.*, 1996; Hassan *et al.*, 2002).

Occult HBV infection

Once HBV infection has occurred, a very small amount of virus remains in the liver in spite of clinical recovery in adult cases (Loriot *et al.*, 1997; Rehermann *et al.*, 1996). In fact, most recipients receiving liver transplants from donors positive to HBc antibody alone develop type B hepatitis (Dickson *et al.*, 1997). Generally, however, people positive for HBc antibody alone have no liver diseases and do not develop HCC in most cases. In areas of endemic HBV infection, many residents are expected to be positive for HBc-Ab alone without liver diseases. My own experience has shown that 52% of Japanese persons over 80 years old were positive for HBc-Ab alone. These people had no history of acute infection and thus at some point they probably had an unapparent infection of HBV and still had a small amount of the virus in their livers at the time of testing. It is reported that HBV DNA is often detected in HCC or its surrounding tissues in HCV-related HCC (Yotsuyanagi *et al.*, 1998). Furthermore, eight of 11 patients with HCV-related HCC (73.3%) showed positive HBV DNA in the liver tissue, while only 24% of patients without HCC were positive for HBV DNA (Fujioka *et al.*, 2003). In addition, HCC develops in most cases of HCV-related liver diseases with positive HBc-Ab before they progress to liver cirrhosis, and the result of their surgical resection is poor (Kubo *et al.*, 1999).

These facts suggest some contribution of occult HBV infection to the development of HCC in HCV-related liver diseases (Tamori *et al.*, 1999; Urashima *et al.*, 1997). Prospective studies of a large number of cases are needed to further clarify the significance of occult HBV infection.

Difference in HCC development between the USA and Japan

About 2.7 million HCV carriers are estimated to live in the United States (Alter *et al.*, 1999), thus the carrier rate is approximately the same in Japan and the United States. However, a great difference is noted in the age of HCV carriers. In Japan, a high percentage of carriers are in their 60s and 70s (Yoshizawa, 2002). On the other hand, most HCV carriers in the USA are in their 30s to 40s (1988–1994) (Alter *et al.*, 1999). This is explained by the fact that HCV infection occurred in large numbers of people soon after the Second World War in Japan. On the other hand, HCV infection in the USA mainly occurred after the Vietnam War. As mentioned before, the development of HCC is mainly seen after the age of 60 in HCV-related chronic liver disease. For this reason, the incidence of HCC in chronic liver diseases associated with HCV infection is not yet as high in the USA as it is in Japan.

In Japan, the incidence of death due to HCC began to reach high levels in 1975 (Yoshizawa, 2002) and is now at its peak. In the USA, the incidence of deaths due to HCV-related liver disease began rising in 1993 (Vong and Bell, 2004) and HCC is presently on the increase. The time difference in the development of HCC is about 20 years between the two countries. It is possible that genetic factors such as those affecting the antioxidant system in the hepatocytes could affect HCC development differently in the two countries. However, there is little doubt that HCV-related HCC will increase remarkably in the USA in the very near future.

In conclusion, as oxidative stress is one of the most important factors in hepatic carcinogenesis, the liver, in which a large amount of ROS are produced in the course of various metabolic activities in the hepatocyte, is the organ at greatest risk of developing malignancy. In other words, hepatocytes are very likely to transform into malignant cells under the conditions described above. Therefore, in the presence of oxidative stress induced by HCV infection, additional oxidative stress due to iron

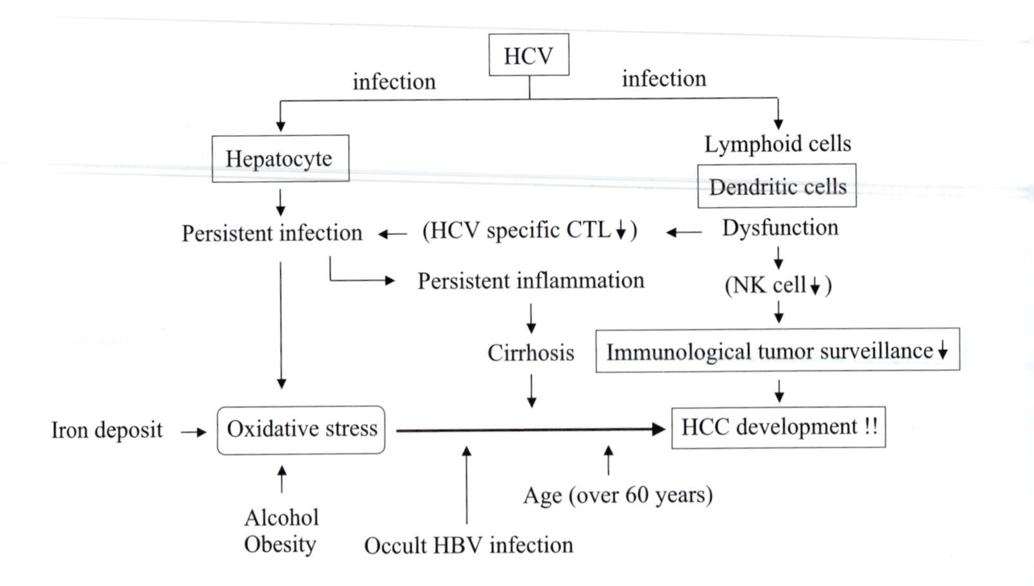

Figure 12.4 Promoting factors of HCC development in HCV-related liver injuries.

overload, alcohol intake, or fat accumulation due to obesity may also contribute to the malignant transformation of hepatocytes.

In addition, cirrhosis and age are important promoting factors in this process. Figure 12.4 summarizes multiple promoting factors in the development of HCC in HCV-related liver disease. It is easy to see why the development of HCC occurs so frequently in chronic HCV liver diseases in Japan. For the diagnosis and treatment of HCV-related liver diseases, it is clearly essential to fully understand their pathogenesis, especially the mechanisms of HCC development.

Prevention of HCC in HCV-related liver diseases

Once chronic hepatitis C is established, persistent inflammation continues with no decreased viral load. Very few patients experience a natural cure. Most eventually, progress to liver cirrhosis and death due to HCC usually occurs before the development of complications of cirrhosis.

Therefore, based upon this understanding of the clinical course of HCV-related liver diseases, at present our treatment goal is to achieve the complete clearance of infected virus from the body with IFN. However, IFN treatment succeeds in eliminating HCV from only half of all chronic hepatitis C patients, and therefore we must also focus our efforts on those patients who do not respond to IFN.

Interferon (IFN) treatment for the elimination of HCV

Various predictive factors for sustained response to IFN have been identified, and the following factors tend to be associated with poor responders: patients with (1) high viral load (Kanai *et al.*, 1992; Yamada *et al.*, 1995), (2) virus genotype I (Tsubota *et al.*, 1994), (3) cirrhosis (Jouet *et al.*, 1994), and (4) few amino acid mutations in the interferon sensitivity determining region (ISDR) of the NS5A (Enomoto *et al.*, 1996).

However, at present over 50% of patients with high viral load and genotype I respond to IFN (Peg IFN) + ribavirin combination treatment for 1 year (McHutchison *et al.*, 1998). One urgent area of study is to clarify which patients do not respond to the combination treatment. It is not clear why HCV of genotype I responds poorly to IFN, and why only a limited number of cirrhotic patients respond to IFN. Our research on experimental cirrhotic rats has shown that the amount of IFN reaching the hepatocytes is inadequate because of sinusoidal capillarization (Miyajima *et al.*, 1996). IFN receptors should also be targets for further study. In any case, Japanese studies have clearly demonstrated that among complete responders to IFN, few develop HCC (Yoshida *et al.*, 1999; Tanaka *et al.*, 1998; Kasahara *et al.*, 1998) (Figure 12.5). Thus, at present, all patients with chronic hepatitis C-related liver diseases are, in principle, candidates for IFN treatment and therefore it is very important to improve IFN therapies.

IFN treatment for prevention of HCC

Japanese studies have shown that even in nonresponders to IFN, HCC incidence is lower than in non-IFN treated patients (Yoshida *et al.*, 1999). In fact, the response to IFN is only 16% in cirrhotic hepatitis C patients. Nevertheless, later in their clinical courses HCC developed much less often in cirrhotic hepatitis C patients that had received IFN treatment than in those who received no IFN treatment (Nishiguchi, *et al.*, 1995). The super high-risk group of HCC consists of those who received curative treatments such as surgical resection for small HCC. The annual incidence of

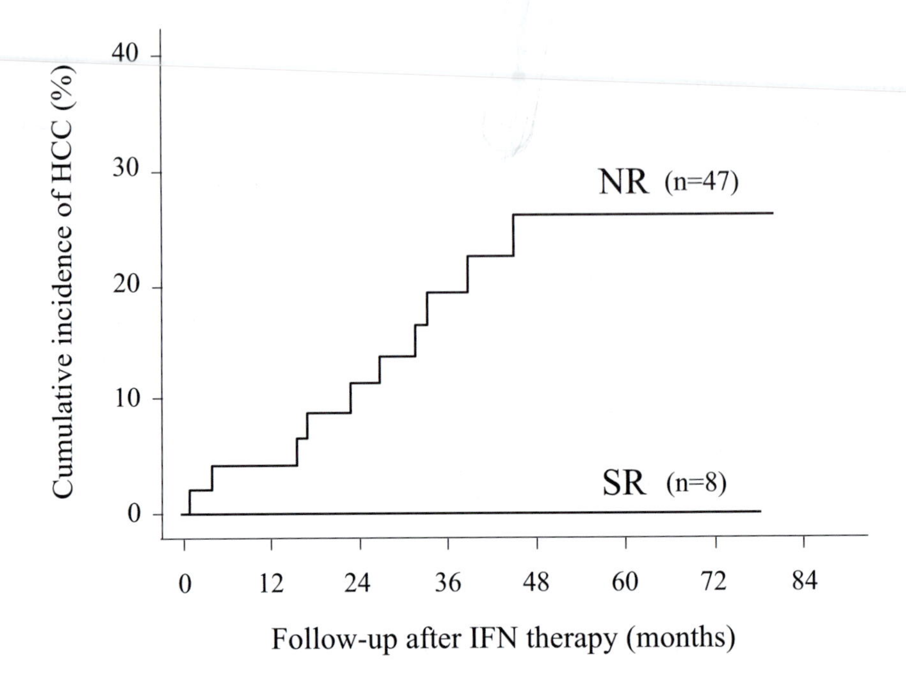

Figure 12.5 Cumulative incidence of hepatocellular carcinoma (HCC). The incidence of HCC in the sustained responder to IFN group (SR) was significantly lower than that in nonresponder group (NR) (*P*<0.01) (Tanaka *et al*., 1998).

new HCC development in different parts of the liver is around 20% in these patients (Tanikawa and Majima, 1993). A significantly decreased incidence of secondary primary HCCs was reported in patients who received IFN treatment for 6 months to 3 years after the initial curative treatment (Suou, *et al.*, 2001; Ikeda, *et al.*, 2000; Kubo, *et al.*, 2001). These findings suggest the possibility that the antitumor or immunological actions of IFN could affect small HCC nodules or preneoplastic lesions not detected by imaging. In fact, such antitumor effects of IFN have been shown by experimental and clinical studies. Almost all HCC cells have receptors for IFN, and IFN has shown antitumor effects on HCC cell lines (Yano, *et al.*, 1999). It is also certain from a clinical viewpoint that IFN has some antitumor effects because of the fact that, at present, chemotherapy combined with IFN is the best treatment for advanced HCC (Sakon, *et al.*, 2002). In addition, the effects of IFN on preneoplastic nodules in experimental animals have been reported (Alvarez, *et al.*, 2002).

Thus, IFN seems to be effective in preventing HCC even among nonresponders to IFN. It is possible that IFN at small doses and administered for short periods of time may also be effective in preventing HCC in patients who are unable to tolerate the standard dose IFN treatment because of its severe side effects. The same is true for elderly patients over 70 years of age. It must be emphasized that IFN treatment is essential for those patients receiving a complete curative therapy for an initial HCC tumor in order to prevent development of a second primary HCC.

Anti-inflammatory and antioxidant preparations

Chronic hepatitis C patients with higher serum ATL levels show a higher incidence of HCC and tend to progress more rapidly to an advanced stage disease. Therefore, in Japan, UDCA and glycyrrhizin preparations have been used in patients with elevated serum ALT, and one study has shown that long-term glycyrrhizin administration prevented HCC development (Arase *et al.*, 1997). Although the mechanisms of the anti-inflammatory effects of these two agents have not been clarified, they do appear to be useful antioxidants (Mitsuyoshi *et al.*, 1999).

Vitamin E, already well known as an antioxidative agent, can lower the serum ALT level in chronic hepatitis C (Herbay *et al.*, 1997; Mahmood *et al.*, 2003)). Thus, vitamin E could be recommended in chronic liver disease with HCV infection, as oxidative stress plays an important role in the development of HCC. However, the indications for vitamin E administration in these patients require more study.

At present, no good markers of oxidative stress have been identified. It is conceivable that a remarkable elevation of oxidative stress markers may be present in the blood of chronic HCV-related liver disease patients because the oxidative stress occurs in the liver, the largest organ of the body.

Thioredoxin, a product induced by oxidative stress, could be one candidate marker in the serum (Sumida *et al.*, 2000; Nakashima *et al.*, 2000). Upon the examination of such markers, antioxidative agents such as vitamin E could be more readily given to patients.

Phlebotomy, iron-restricted diet

There is no doubt that iron deposits in the hepatocytes are a considerable risk factor in exacerbation of HCV-related liver diseases, although the mechanism of this iron accumulation is not known. A decline in serum ALT and ferritin is often observed in patients with iron deposits in the liver after phlebotomy (Yano *et al.*, 2004), and a drop in the 8-OhdG level of these hepatocytes has also been observed (Kato *et al.*, 2001). It is suggested that phlebotomy should be carried out in cases with high serum ferritin levels, and should be followed by an iron-restricted diet.

Retinoid

Retinoid is essential for maintaining metabolic activities in the hepatocytes, and HCC occurs frequently in cirrhotic liver in which retinoid is deficient. Thus, the constant supply of retinoid is of great importance in treating liver cirrhosis. In fact, synthetic retinoid has been shown to prevent a second primary HCC occurrence in cirrhotic HCV patients after a complete curative treatment for an initial HCC tumor (Muto *et al.*, 1996). Additionally, carotinoid + vitamin E is reported to be effective in preventing the development of HCC in cirrhotic HCV patients (Nishino *et al.*, 2002). Further large-scale trials will be required to confirm these effects.

Vitamin K

Vitamin K has been reported to be effective in inhibiting growth, inducing apoptosis and cell differentiation, etc. in various kinds of malignant cells (Wu *et al.*, 1993). In addition, its inhibitory effect on cell growth has been reported on HCC cell lines (Nishikawa *et al.*, 1995). The exact mechanism of such effects has not yet been clarified. However, one study recently showed that vitamin K inhibits the growth and invasiveness of HCC cells via protein kinase A activation (Otsuka *et al.*, 2004).

Recently, two interesting findings have been reported from clinical trials in Japan. One trial reported a preventive effect of Vitamin K_2 on recurrence of HCC after treatment of a primary HCC, and the other showed an inhibitory effect on the invasion of HCC into the portal vein. No

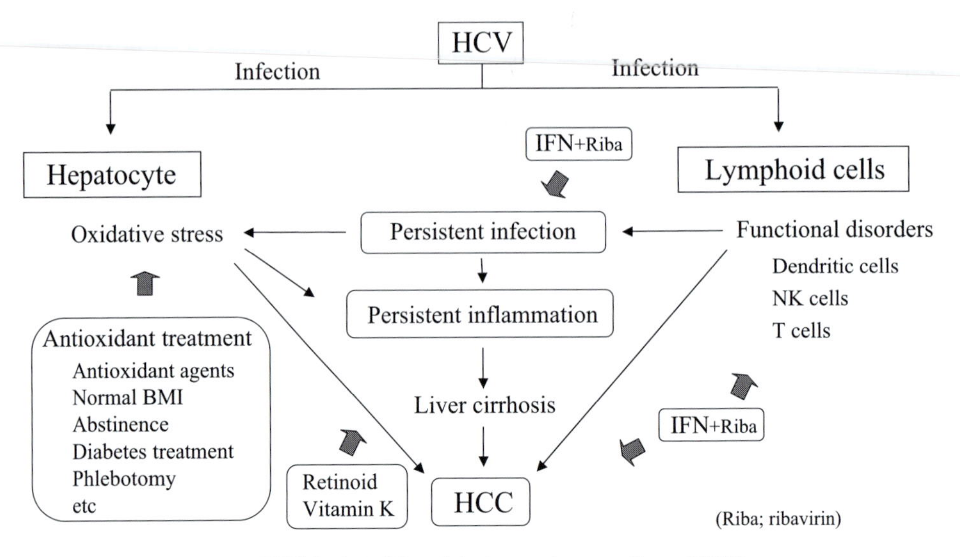

Figure 12.6 Treatment of HCV-related liver injuries and prevention of HCC.

side effects were noted for long-term administration of vitamin K_2 at a daily dose of 45 mg. In addition, a very recent clinical trial revealed preventive effects of HCC in women with viral cirrhosis (Habu *et al.*, 2004).

Control of obesity and diabetes

Obesity, especially due to visceral fat accumulation, induces fatty liver, in which ROS are overproduced in the process of fatty acid oxidation in the hepatocyte, resulting in increased oxidative stress. Therefore, in theory obesity should be expected to exacerbate the oxidative changes induced by HCV infection. In fact, obesity is reported in quite a number of clinical studies to be one of the most serious risk factors for progression of HCV-related liver diseases.

A decline of serum ALT is noted in obese patients with HCV infection after weight reduction (Hickman *et al.*, 2002). Type 2 diabetes is often associated with HCV type liver diseases, and 10% of all diabetic patients are complicated with HCV-related liver diseases (Arao *et al.*, 2003). Complications of diabetes further promote oxidative stress in the hepatocyte. Thus, diabetic patients not improved by weight reduction should receive alternative treatments for diabetes, keeping in mind that IFN therapy may have an adverse effect on diabetes. Figure 12.6 shows the whole spectrum of HCC prevention modalities in HCV-related liver injuries. Among them IFN treatment is the most important. However, in addition to non-responders to IFN, there are numerous patients who cannot have IFN therapy because of side effects or age. For these patients we should make an effort to reduce oxidative stress, which is the most important risk factor for development of HCC.

In conclusion, a full understanding of the pathogenesis of HCV-related liver diseases is essential if we are to provide the best treatment possible and if we are to succeed in preventing HCC (Tanikawa, 2004).

References

Allison, M.E.D., Wreghitt, T., Palmer, C.R., and Alexander, G.J.M. (1994). Evidence for a link between hepatitis C virus infection and diabetes mellitus in a cirrhotic population. J. Hepatol. *21*, 1135–1139.

Alter, M.J., Kruszon-Moran, D., Nainan, O.V., McQuillan, G.M., Gao, F., Moyer, L.A., Kaslow, R.A., and Margolis, H.S. (1999). The prevalence of hepatitis C virus infection in the United States, 1988 through 1994. N. Engl. J. Med. *341*, 556–562.

Alvarez, M.L., Cerliani, J.P., Monti, J., Carnovale, C., Ronco, M.T., Pisani, G., Lugano, M.C., and Carrillo, M.C. (2002). The *in vivo* apoptotic effect of interferon alfa-2b on rat preneoplastic liver involves Bax protein. Hepatology *35*, 824–833.

Arao, M., Murase, K., Kusakabe, A., Yoshioka, K., Fukuzawa, Y., Ishikawa, T., Tagaya, T., Yamanouchi, K., Ichimiya, H., Sameshima, Y., *et al.* (2003). Prevalence of diabetes mellitus in Japanese patients infected chronically with hepatitis C virus. J. Gastroenterol. *38*, 355–360.

Arase, Y., Ikeda, K., Murashima, N., Chayama, K., Tsubota, A., Koida, I., Suzuki, Y., Saitoh, S., Kobayashi, M., and Kumada, H. (1997). The long term efficacy of glycyrrhizin in chronic hepatitis C patients. Cancer *79*, 1494–1500.

Barbaro, G., Lorenzo, G., Asti, A., Ribersani, M., Belloni, G., Grisorio, B., Filice, G., and Barbarini, G. (1999). Hepatocellular mitochondrial alterations in patients with chronic hepatitis C: Ultrastructural and biochemical findings. Am. J. Gastroenterol. *94*, 2198–2205.

Boyett, J.D., and Sullivan, J.F. (1970). Zinc and collagen content of cirrhotic liver. Am. J. Dig. Dis. *15*, 797–802.

Brillanti, S., Foli, M., Gaiani, S., Masci, C., Miglioli, M., Barbara, L. (1993). Persistent hepatitis C viraemia without liver disease. Lancet *341*, 464–465.

Brüing, J.C., Michael, M.D., Winnay, J.N., Hayashi, T., Hösch, D., Accili, D., Goodyear, L.J., and Kahn, R. (1998). A muscle-specific insulin receptor knockout exhibits features of the metabolic syndrome of NIDDM without altering glucose tolerance. Mol. Cell *2*, 559–569.

Bruno, S., Rossi, S., Petroni, M.L., Villa, E., Zuin, M., and Podda, M. (1994). Normal aminotransferase concentrations in patients with antibodies to hepatitis C virus. Br. Med. J. *308*, 697.

Bugianesi, E., Leone, N., Vanni, E., Marchesini, G., Brunello, F., Carucci, P., Musso, A., Paolis, P., Capussotti, L, Salizzoni, M., *et al.* (2002). Expanding the natural history of nonalcoholic steatohepatitis: From cryptogenic cirrhosis to hepatocellular carcinoma. Gastroenterology *123*, 134–140.

Chapoutot, C., Esslimani, M., Joomaye, Z., Ramos, J., Perney, P., Laurent, C., Fabbro-Peray, P., Larrey, D., Domergue, J., and Blanc, F. (2000). Liver iron excess in patients with hepatocellular carcinoma developed on viral C cirrhosis. Gut *46*, 711–714.

Chiba, T., Matsuzaki, Y., Abei, M., Shoda, J., Aikawa, T., Tanaka, N., and Osuga, T. (1996). Multivariate analysis of risk factors for hepatocellular carcinoma in patients with hepatitis C virus-related liver cirrhosis. J. Gastroenterol. *31*, 552–558.

Chino, S., Moriyama, M., Matsumura, H., Ono, Y., and Arakawa, Y. (2002). Clinical pathological significance of iron metabolism with chronic hepatitis C patients. Hepatol. Res. *24*, 245–255.

Corado, J., Toro, F., Rivera, H., Bianco, N.E., Deibis, L., and Sanctis, J.B. (1997). Impairment of natural killer (NK) cytotoxic activity in hepatitis C virus (HCV) infection. Clin. Exp. Immunol. *109*, 451–457.

Dickson, R.C., Everhart, J.E., Lake, J.R., Wei, Y., Seaberg, E.C., Wiesner, R.H., Zetterman, R.K., Pruett, T.L., Ishitani, M.B., Hoofnagle, J.H., *et al.* (1997). Transmission of hepatitis B by transplantation of livers from donors positive for antibody to hepatitis B core antigen. Gastroenterology *113*, 1668–1674.

Dreher, D., and Junod, A.F. (1996). Role of oxygen free radicals in cancer development. Eur. J. Cancer *32A*, 30–38.

Enomoto, N., Sakuma, I., Asahina, Y., Kurosaki, M., Murakami, T., Yamamoto, C., Ogura, Y., Izumi, N., Marumo, F., and Sato, C. (1996). Mutations in the nonstructural protein 5A gene and response to interferon in patients with chronic hepatitis C virus 1b infection. N. Engl. J. Med. *334*, 77–81.

Fujioka, S., Shimomura, H., Iwasaki, Y., Fujio, K., Nakagawa, H., Onishi, Y., Takagi, S., Taniguchi, H., Umeoka, F., Nakajima, H., *et al.* (2003). Hepatitis B virus gene in liver tissue promotes hepatocellular carcinoma development in chronic hepatitis C patients. Dig. Dis. Sci. *48*, 1920–1924.

Furutani, M., Nakashima, T., Sumida, Y., Hirohama, A., Yoh, T., Kakisaka, Y., Mitsuyoshi, H., Senmaru, H., and Okanoue, T. (2003). Insulin resistance/β-cell function and serum ferritin level in non-diabetic patients with hepatitis C virus infection. Liver Int. *23*, 294–299.

George, D.K., Goldwurm, S., MacDonald, G.A., Cowley, L.L., Walker, N.I., Ward, P.J.,

Jazwinska, E.C., and Powell, L.W. (1998). Increased hepatic iron concentration in non-alcoholic steatohepatitis is associated with increased fibrosis. Gastroenterology *114*, 311–318.

Gerlach, J.T., Diepolder, H.M., Zachoval, R., Gruener, N.H., Jung, M.C., Ulsenheimer, A., Schraut, W.W., Schirren, C.A., Waechtler, M., Backmund, M., *et al.* (2003). Acute hepatitis C: High rate of both spontaneous and treatment-induced viral clearance. Gastroenterology *125*, 80–88.

Giuggio, V.M., Bonkovsky, H.L., Smith, J., and Rothman, A.L. (1998). Inefficient recognition of autologous viral sequences by intrahepatic hepatitis C virus-specific cytotoxic T lymphocytes in chronically infected subjects. Virology *251*, 132–140.

Gruener, N.H., Lechner, F., Jung, M.C., Diepolder, H., Gerlach, T., Lauer, G., Walker, B., Sullivan, J., Phillips, R., Pape, G.R., and Klenerman, P. (2001). Sustained dysfunction of antiviral $CD8^+$ T lymphocytes after infection with hepatitis C virus. J. Virol. *75*, 5550–5558.

Habu, D., Shiomi, S., Tamori, A., Takeda, T., Tanaka, T., Kubo, S., and Nishiguchi, S. (2004). Role of vitamin K_2 in the development of hepatocellular carcinoma in women with viral cirrhosis of the liver. J. Am. Med. Assoc. *292*, 358–361.

Hamada, H., Yatsuhashi, H., Yano, K., Daikoku, M., Arisawa, K., Inoue, O., Koga, M., Nakata, K., Eguchi, K., Yano, M. (2002). Impact of aging on the development of hepatocellular carcinoma in patients with posttransfusion chronic hepatitis C. Cancer 95, 331–339.

Hassan, M.M., Hwang, L.Y., Hatten, C.J., Swaim, M., Li, D., Abbruzzese, J.L., Beasley, P., and Patt, Y.Z. (2002). Risk factors for hepatocellular carcinoma: Synergism of alcohol with viral hepatitis and diabetes mellitus. Hepatology *36*, 1206–1213.

Herbay, A., Stahl, W., Niederau C., and Sies, H. (1997). Vitamin E improves the aminotransferase status of patients suffering from viral hepatitis C: A randomized, double-blind, placebo-controlled study. Free Rad. Res. *27*, 599–605.

Hertzog, P.J., Smith, J.R.L., and Garner, R.C. (1982). Production of monoclonal antibodies to guanine imidazole ring-opened aflatoxin B_1DNA, the persistent DNA adduct *in vivo*. Carcinogenesis 3, 825–828.

Hézode, C., Cazeneuve, C., Coué, O., Roudot-Thoraval, F., Lonjon, I., Bastie, A., Duvoux, C., Pawlotsky, J.M., Zafrani, E.S., Amselem, S., *et al.* (1999). Liver iron accumulation in patients with chronic active hepatitis C: prevalence and role of hemochromatosis gene mutations and relationship with hepatic histological lesions. J. Hepatol. *31*, 979–984.

Hickman, I.J., Clouston, A.D., Macdonald, G.A., Purdie, D.M., Prins, J.B., Ash, S., Jonsson, J.R., and Powell, E.E. (2002). Effect of weight reduction on liver histology and biochemistry in patients with chronic hepatitis C. Gut *51*, 89–94.

Hino, O., Yamamoto, T., and Kajino, K. (1998). Inflammation mediated hepatocarcinogenesis; hypercarcinogenic state of chronic liver disease. In Progress in Hepatology Volume 5, M. Yamanaka, *et al.*, eds. (Amsterdam: Elsevier Science), pp. 35–39.

Hoek, J.B., Cahill, A., and Pastorino, J.G. (2002) Alcohol and mitochondria: A dysfunctional relationship. Gastroenterology *122*, 2049–2063.

Hourigan, L.F., MacDonald, G.A., Purdie, D., Whitehall, V.H., Shorthouse, C., Clouston, A., and Powell, E.E. (1999). Fibrosis in chronic hepatitis C correlates significantly with body mass index and steatosis. Hepatology *29*, 1215–1219.

Hu, K.Q., Kyulo, N.L., Esrailian, E., Thompson, K., Chase, R., Hillebrand, D.J., and Runyon, B.A. (2004). Overweight and obesity, hepatic steatosis, and progression of chronic hepatitis C: a retrospective study on a large cohort of patients in the United States. J. Hepatol. *40*, 147–154.

Ikeda, K., Arase, Y., Saitoh, S., Kobayashi, M., Suzuki, Y., Suzuki, F., Tsubota, A., Chayama, K., Murashima, N., and Kumada, H. (2000). Interferon beta prevents recurrence of hepatocellular carcinoma after complete resection of ablation of the primary tumor-A prospective randomized study of hepatitis C virus-related liver cancer. Hepatology 32, 228–232.

Ikeda, K., Saitoh, S., Koida, I., Arase, Y., Tsubota, A., Chayama, K., Kumada, H., and Kawanishi, M. (1993) A multivariate analysis of risk factors for hepatocellular carcinogenesis: A prospective observation of 795 patients with viral and alcoholic cirrhosis. Hepatology *18*, 47–53.

Ito, K., Mitchel, D.G., Gabata, T., Hann, H.W.L., Kim, P.N., Fujita, T., Awaya, H., Honjo, K., Matsunaga, N. (1999). Hepatocellular carcinoma: Association with increased iron deposition in the cirrhotic liver at MR Imaging. Radiology *212*, 235–240.

Jain, S.K., Pemberton, P.W., Smith, A., McMahon, R.F.T., Burrows, P.C., Aboutwerat, A., and Warnes, T.W. (2002). Oxidative stress in chronic hepatitis C: not just a feature of late stage disease. J. Hepatol. *36*, 805–811.

Jamal, M.M., Soni, A., Quinn, P.G., Wheeler, D.E., Arora, S., and Johnston, D.E. (1999). Clinical features of hepatitis C-Infected patients with persistently normal alanine transaminase levels in the southwestern United States. Hepatology *30*, 1307–1311.

Jinushi, M., Takehara, T., Tatsumi, T., Kanto, T., Groh, V., Spies, T., Suzuki, T., Miyagi, T., and Hayashi, N. (2003a). Autocrine/paracrine IL-15 that is required for type I IFN-mediated dendritic cell expression of MHC class I-related chain A and B is impaired in hepatitis C virus infection. J. Immunol. *171*, 5423–5429.

Jinushi, M., Takehara, T., Kanto, T., Tatsumi, T., Groh, V., Spies, T., Miyagi, T., Suzuki, T., Sasaki, Y., and Hayashi, N. (2003b). Critical role of MHC class I-related chain A and B expression on IFN-α-stimulated dendritic cells in NK cell activation: Impairment in chronic hepatitis C virus infection. J. Immunol. *170*, 1249–1256.

Jouet, P., Roudot-Thoraval, F., Dhumeaux, D., Metreau, J.M. (1994). Comparative efficacy of interferon alpha in cirrhotic and noncirrhotic patients with non-A, non-B, C hepatitis. Gastroenterology *106*, 686–690.

Kaimori, A., Kanto, T., Limn, C.K., Komoda, Y., Oki, C., Inoue, M., Miyatake, H., Itose, I., Sakakibara, M., Yakushijin, T., *et al.* (2004). Pseudotype hepatitis C virus enters immature myeloid dendritic cells through the interaction with lectin. Virology *324*, 74–83.

Kanai, K., Kako, M., Okamoto, H. (1992). HCV genotypes in chronic hepatitis C and response to interferon. Lancet 339, 1543.

Kanto, T., Hayashi, N., Takehara, T., Tatsumi, T., Kuzushita, N., Ito, A., Sasaki, Y., Kasahara, A., and Hori, M. (1999). Impaired allostimulatory capacity of peripheral blood dendritic cells recovered from hepatitis C virus-infected individuals. J. Immunol. *162*, 5584–5591.

Kasahara, A., Hayashi, N., Mochizuki, K., Takayanagi, M., Yoshioka, K., Kakumu, S., Iijima, A., Urushihara, A., Kiyosawa, K., Okuda, M., *et al.* (1998). Risk factors for hepatocellular carcinoma and its incidence after interferon treatment in patients with chronic hepatitis C. Hepatology *27*, 1394–1402.

Kato, J., Kobune, M., Nakamura, T., Kuroiwa, G., Takada, K., Takimoto, R., Sato, Y., Fujikawa, K., Takahashi, M., Takayama, T., *et al.* (2001). Normalization of elevated hepatic 8-hydroxy-2'-deoxyguanosine levels in chronic hepatitis C patients by phlebotomy and low iron diet. Cancer Res. *61*, 8697–8702.

Kato, N., Yoshida, H., Ono-Nita, S.K., Kato, J., Goto, T., Otsuka, M., Lan, K.H., Matsushima, K., Shiratori, Y., and Omata, M. (2000). Activation of intracellular signaling by hepatitis B and C viruses: C-viral core is the most potent signal inducer. Hepatology *32*, 405–412.

Kawaguchi, T., Harada, M., Yoshida, T., Taniguchi, E., Kumemura, H., Hanada, S., Baba, S., Maeyama, M., Koga, H., Kumashiro, R., *et al.* (2004). Hepatitis C virus down-regulates insulin receptor substrate 1 and 2 through up-regulation of suppressor of cytokine signaling 3. Am. J. Pathol. *165*, 1499–1508.

Koike, K., and Takada, S. (1995). Biochemistry and functions of hepatitis B virus X protein. Intervirology *38*, 89–99.

Kubo, S., Nishiguchi, S., Hirohashi, K., Tanaka, H., Shuto, T., Yamazaki, O., Shiomi, S., Tamori, A., Oka, H., Igawa, S., *et al.* (2001). Effects of long-term postoperative interferon-α therapy on intrahepatic recurrence after resection of hepatitis C virus-related hepatocellular carcinoma. Ann. Intern. Med. *134*, 963–967.

Kubo, S., Nishiguchi, S., Hirohashi, K., Tanaka, H., Tsukamoto, T., Hamba, H., Shuto, T., Yamamoto, T., Ikebe, T., and Kinoshita, H. (1999). Clinical significance of prior hepatitis B virus infection in patients with hepatitis C virus-related hepatocellular carcinoma. Cancer *86*, 793–798.

Kurose, I., Higuchi, H., Kato, S., Miura, S., Watanabe, N., Kamegaya, Y., Tomita, K., Takaishi, M., Horie, Y., Fukuda, M., *et al.* (1997). Oxidative stress on mitochondria and cell membrane of cultured rat hepatocytes and perfused liver exposed to ethanol. Gastroenterology *112*, 1331–1343.

Lechner, F., Wong, D.K.H., Dunbar, P.R., Chapmen, R., Chung, R.T., Dohrenwend, P., Robbins, G., Phillips, R., Klenerman, P., and Walker, B.D. (2000). Analysis of successful immune responses in persons infected with hepatitis C virus. J. Exp. Med. *191*, 1499–1512.

Loriot, M.A., Marcellin, P., Walker, F., Boyer, N., Degott, C., Randrianatoavina, I., Benhamou, J.P., and Erlinger, S. (1997). Persistence of hepatitis B virus DNA in serum and liver from patients with chronic hepatitis B after loss of HBsAg. J. Hepatol. *27*, 251–258.

Mahmood, S., Kawanaka, M., Kamei, A., Izumi, A., Nakata, K., Niiyama, G., Ikeda, H., Hanano, S., Suehiro, M., Togawa, K., *et al.* (2004). Immunohistochemical evaluation of oxidative stress markers in chronic hepatitis C. Antioxid. Redox Signal. 6, 19–24.

Mahmood, S., Yamada, G., Niiyama, G., Kawanaka, M., Togawa, K., Sho, M., Ito, T., Sasagawa, T., Okita, M., Nakamura, H., *et al.* (2003). Effect of Vitamin E on serum aminotransferase and thioredoxin levels in patients with viral hepatitis C. Free Rad. Res. *37*, 781–785.

Mason, A.L., Lau, J.Y.N., Hoang, N., Qian K., Alexander, G.J.M., Xu, L., Guo, L., Jacob, S., Regenstein, F.G., Zimmerman, R., *et al.* (1999). Association of diabetes mellitus and chronic hepatitis C virus infection. Hepatology *29*, 328–333.

Mathurin, P., Moussalli, J., Cadranel, J.F., Thibault, V., Charlotte, F., Dumouchel, P., Cazier, A., Huraux, J.M., Devergie, B., Vidaud, M., *et al.* (1998). Slow progression rate of fibrosis in hepatitis C virus patients with persistently normal alanine transaminase activity. Hepatology 27, 868–872.

McHutchison, J.G., Gordon, S.C., Schiff, E.R., Shiffman, M.L., Lee, W.M., Rustge, V.K., Goodman, Z.D., Ling, M.H, Cort, S., and Albrecht, J.K. (1998). Interferon alfa-2b alone or in combination with ribavirin as initial treatment for chronic hepatitis C. N. Engl. J. Med. *339*, 1485–1492.

Mehta, S.H., Brancati, F.L., Sulkowski, M.S., Strathdee, S.A., Szklo, M., and Thomas, D.L. (2000). Prevalence of type 2 diabetes mellitus among persons with hepatitic C virus infection in the United States. Ann. Intern. Med. *133*, 592–599.

Mendler, M.H., Turlin, B., Moirand, R., Jouanolle, A.M., Sapey, T., Guyader, D., Gall, J.Y., Brissot, P., David, V., and Deugnier, Y. (1999). Insulin resistance-Associated hepatic iron overload. Gastroenterology *117*, 1155–1163.

Michael, M.D., Kulkarni, R.N., Postic, C., Previs, S.F., Shulman, G.I., Magnuson, M.A., and Kahn, C.R. (2000). Loss of insulin signaling in hepatocytes leads to severe insulin resistance and progressive hepatic dysfunction. Mol. Cell *6*, 87–97.

Mitsuyoshi, H., Nakashima, T., Sumida, Y., Yoh, T., Nakajima, Y., Ishikawa, H., Inaba, K., Sakamoto, Y., Okanoue, T., and Kashima, K. (1999). Ursodeoxycholic acid protects hepatocytes against oxidative injury via induction of antioxidants. Biochem. Biophys. Res. Commun. *263*, 537–542.

Miyajima, I., Sata, M., Gondo, K., Suzuki, H., and Tanikawa, K. (1996). Interferon accumulation in cirrhotic rat liver. J. Viral Hepatitis 3, 23–27.

Miyakawa, H., Sato, C., Izumi, N., Tazawa, J., Ebata, A., Hattori, K., Sakai, H., Ikeda, T., Hirata, R., Sakai, Y., *et al.* (1993). Hepatitis C virus infection in alcoholic liver cirrhosis in Japan: Its contribution to the development of hepatocellular carcinoma. Alcohol & Alcoholism *28* (S1A): 85–90.

Muratori, L., Gibellini, D., Lenzi, M., Cataleta, M., Muratori, P., Morelli, M.C., and Bianchi, F.B. (1996). Quantification of hepatitis C virus-Infected peripheral blood mononuclear cells by in situ reverse transcriptase-polymerase chain reaction. Blood *88*, 2768–2774.

Muto, Y., Moriwaki, H., Ninomiya, M., Adachi, S., Saito, A., Takasaki, K.T., Tanaka, T., Tsurumi, K., Okuno, M., Tomita, E., *et al.* (1996). Prevention of second primary tumors by an acyclic retinoid, polyprenoic acid, in patients with hepatocellular carcinoma. N. Engl. J. Med. *334*, 1561–1567.

Nakashima, T., Sumida Y., Yoh, T., Kakisaka, Y., Nakajima, Y., Ishikawa, H., Mitsuyoshi, H., Kashima, K., Nakamura, H., and Yodoi, J. (2000). Thioredoxin levels in the sera of untreated viral hepatitis patients and those treated with glycyrrhizin or ursodeoxycholic acid. Antiox. Redox Signal. 2, 687–694

Niederau, C., Fischer, R., Sonnenberg, A., Stremmel, W., Trampisch, H.J., and Strohmeyer, G. (1985). Survival and causes of death in cirrhotic and in noncirrhotic patients with primary hemochromatosis. N. Engl. J. Med. *313*, 1256–1262.

Nishiguchi, S., Kuroki, T., Nakatani, S., Morimoto, H., Takeda, T., Nakajima, S., Shiomi, S., Seki, S., Kobayashi, K., and Otani, S. (1995). Randomised trial of effects of interferon-α on incidence of hepatocellular carcinoma in chronic active hepatitis C with cirrhosis. Lancet *346*, 1051–1055.

Nishikawa, Y., Carr, B.I., Wang, M., Kar, S., Finn, F., Dowd, P., Zheng, Z.B., Kerns, J., and Naganathan, S. (1995). Growth inhibition of hepatoma cells induced by vitamin K and its analogs. J. Biol. Chem. *270*, 28304–28310.

Nishino, H., Murakoshi, M., Ii, T., Takemura, M., Kuchide, M., Kanazawa, M., Mou, X.Y., Wada, S., Masuda, M., Ohsaka, Y., *et al.* (2002). Carotenoids in cancer chemoprevention. Cancer and Metastasis Rev. *21*, 257–264.

Okuda, M., Li, K., Beard, M.R., Showalter, L.A., Scholle, F., Lemon, S.M., and Weinman, S.A. (2002). Mitochondrial injury, oxidative stress, and antioxidant gene expression are induced by hepatitis C virus core protein. Gastroenterology *122*, 366–375.

Omata, M., Yokosuka, O., Takano, S., Kato, N., Hosoda, K., Imazeki, F., Tada, M., Ito, Y., and Ohto, M. (1991). Resolution of acute hepatitis

C after therapy with natural beta interferon. Lancet *338*, 914–915.

Ortiz, V., Berenguer, M., Rayón, J.M., Carrasco, D., and Berenguer (2002). Contribution of obesity to hepatitis C-related fibrosis progression. Am. J. Gastroenterol. *97*, 2408–2414.

Otsuka, M., Kato, N., Shao, R.X., Hoshida, Y., Ijichi, H., Koike, Y., Taniguchi, H., Moriyama, M., Shiratori, Y., Kawabe, T., *et al.* (2004). Vitamin K_2 inhibits the growth and invasiveness of hepatocellular carcinoma cells via protein kinase A activation. Hepatology *40*, 243–251.

Paradis, V., Mathurin, P., Kollinger, M., Imbert-Bismut, F., Charlotte, F., Piton, A., Opolon, P., Holstege, A., Poynard, T., and Bedossa, P. (1997). In situ detection of lipid peroxidation in chronic hepatitis C: correlation with pathological features. J. Clin. Pathol. *50*, 401–406.

Petit, J.M., Bour, J.B., Galland-Jos, C., Minello, A., Verges, B., Guiguet, M., Brun, J.M., and Hillon, P. (2001). Risk factors for diabetes mellitus and early insulin resistance in chronic hepatitis C. J. Hepatol. *35*, 279–283.

Rehermann, B., Ferrari, C., Pasquinelli, C., and Chisari, F.V. (1996). The hepatitis B virus persists for decades after patients' recovery from acute viral hepatitis despite active maintenance of a cytotoxic T-lymphocyte response. Nature Med. *2*, 1104–1108.

Robson, K.J. (2004). Hepcidin and its role in iron absorption. Gut *53*, 617–619.

Sakon, M., Nagano, H., Dono, K., Nakamori, S., Umeshita, K., Yamada, A., Kawata, S., Imai, Y., Iijima, S., and Monden, M. (2002). Combined intraarterial 5-Fluorouracil and subcutaneous interferon-α therapy for advanced hepatocellular carcinoma with tumor thrombi in the major portal branches. Cancer *94*, 435–442.

Schaffner, F., and Popper, H. (1963). Capillarization of hepatic sinusoids in man. Gastroenterology *44*, 239–242.

Shibasaki, M., Takahashi, K., Itou, T., Miyazawa, S., Ito, M., Kobayashi, J., Bujo, H., and Saito, Y. (2002). Alterations of insulin sensitivity by the implantation of 3T3-L1 cells in nude mice. A role for TNF-alpha? Diabetologia *45*, 518–526.

Shimamatsu, K., Kage, M., Nakashima, O., and Kojiro, M. (1994). Pathomorphological study of HCV antibody-positive liver cirrhosis. J. Gastroenterol. Hepatol. *9*, 624–630.

Shintani, Y., Fujie, H., Miyoshi, H., Tsutsumi, T., Tsukamoto, K., Kimura, S., Moriya, K., and Koike, K. (2004). Hepatitis C virus infection and diabetes: Direct involvement of the virus in the development of insulin resistance. Gastroenterology *126*, 840–848.

Shirachi, M., Sata, M., Miyajima, I., Tanaka, M., and Tanikawa, K. (1998). Liver-associated natural killer activity in cirrhotic rats. Microbiol. Immunol. *42*, 117–124.

Sumida, Y., Nakashima, T., Yoh, T., Furutani, M., Hirohama, A., Kakisaka, Y., Nakajima, Y., Ishikawa, H., Mitsuyoshi, H., Okanoue, T., *et al.* (2003). Serum thioredoxin levels as a predictor of steatohepatitis in patients with nonalcoholic fatty liver disease. J. Hepatol. *38*, 32–38.

Sumida, Y., Nakashima, T., Yoh, T., Nakajima, Y., Ishikawa, H., Mitsuyoshi, H., Sakamoto, Y., Okanoue, T., Kashima, K., Nakamura, H., *et al.* (2000). Serum thioredoxin levels as an indicator of oxidative stress in patients with hepatitis C virus infection. J. Hepatol. *33*, 616–622.

Suou, T., Mitsuda, A., Koda, M., Matsuda, H., Maruyama, S., Tanaka, H., Kishimoto, Y., Kohono, M., Hirooka, Y., and Kawasaki, H. (2001). Interferon alpha inhibits intrahepatic recurrence in hepatocellular carcinoma with chronic hepatitis C: a pilot study. Hepatol. Res. *20*, 301–311.

Suzuki, Y., Saito, H., Suzuki, M., Hosoki, Y., Sakurai, S., Fujimoto, Y., and Kohgo, Y. (2002). Up-regulation of transferrin receptor expression in hepatocytes by habitual alcohol drinking is implicated in hepatic iron overload in alcoholic liver disease. Alcohol Clin. Exp. Res. *26*, 26S–31S.

Tamori, A., Nishiguchi, S., Kubo S., Koh, N., Moriyama, Y., Fujimoto, S., Takeda, T., Shiomi, S., Hirohashi, K., Kinoshita, H., *et al.* (1999). Possible contribution to hepatocarcinogenesis of X transcript of hepatitis B virus in Japanese patients with hepatitis C virus. Hepatology *29*, 1429–1434.

Tanaka, K., Sata, M., Uchimura, Y., Suzuki, H., and Tanikawa, K. (1998). Long-term evaluation of interferon therapy in hepatitis C virus-associated cirrhosis: Does IFN prevent development of hepatocellular carcinoma? Oncol. Rep. *5*, 205–208.

Tanikawa, K. (2004) Pathogenesis and treatment of hepatitis C virus-related liver diseases. Hepatobiliary Pancreat. Dis. Int. *3*, 17–20.

Tanikawa, K., and Majima, Y. (1993). Percutaneous ethanol injection therapy for recurrent hepatocellular carcinoma. Hepatogastroenterol. *40*, 324–327.

Tarao, K., Rino, Y., Ohkawa, S., Shimizu, A., Tamai, S., Miyakawa, K., Aoki, H., Imada, T., Shindo, K., Okamoto, N., and Totsuka, S.

(1999). Association between high serum alanine aminotransferase levels and more rapid development and higher rate of incidence of hepatocellular carcinoma in patients with hepatitis C virus-associated cirrhosis. Cancer *86*, 589–595.

Tsubota, A., Chayama, K., Ikeda, K., Yasuji, A., Koida, I., Saitoh, S., Hashimoto, M., Iwasaki, S., Kobayashi, M., and Kumada, H. (1994). Factors predictive of response to interferon-α therapy in hepatitis C virus infection. Hepatology *19*, 1088–1094.

Urashima, T., Saigo, K., Kobayashi, S., Imaseki, H., Matsubara, H., Koide, Y., Asano, T., Kondo, Y., Koike, K., and Isono, K. (1997). Identification of hepatitis B virus integration in hepatitis C virus-infected hepatocellular carcinoma tissues. J. Hepatol. *26*, 771–778.

Vandelli, C., Renzo, F., Romanó, L., Tisminetzky, S., Palma, M., Stroffolini, T., Ventura, E., and Zanetti, A. (2004). Lack of evidence of sexual transmission of hepatitis C among monogamous couples: Results of a 10-year prospective follow-up study. Am. J. Gasroenterol. *99*, 855–859

Vong, S., and Bell, B.P. (2004). Chronic liver disease mortality in the United States, 1990–1998. Hepatology *39*, 476–483.

W.H.O. (1999). Global surveillance and control of hepatitis C. J. Viral Hepatitis *6*, 35–47.

Walker, A.P., Partridge, J., Srai, S.K., and Dooley, J.S. (2004). Hepcidin: what every gastroenterologist should know. Gut 53, 624–627.

Weinman, S.A., Okuda, M., Li, K., Showalter, L.A., Otani, K., Lemon, S.M., and Beard, M.R. (2003). Role of core protein-induced oxidative stress in the pathogenesis of hepatitis C. In HCV/Oxidative Stress and Liver Disease, K. Okita, ed. (Tokyo: Springer-Verlag), pp. 8–18.

Weltman, M.D., Farrell, G.C., Hall, P., Ingelman-Sundberg, M., and Liddle, C. (1998). Hepatic cytochrome P450 2E1 is increased in patients with nonalcoholic steatohepatitis. Hepatology *27*, 128–133.

Wu, F.Y.H., Chang, N.T., Chen, W.J., and Juan, C.C. (1993). Vitamin K_3-induced cell cycle arrest and apoptotic cell death are accompanied by altered expression of *c-fos* and *c-myc* in nasopharyngeal carcinoma cells. Oncogene *8*, 2237–2244.

Yamada, G., Takatani, M., Kishi, F., Takahashi, M., Doi, T., Tsuji, T., Shin, S., Tanno, M., Urdea, M.S., and Kolberg, J.A. (1995). Efficacy of interferon alpha therapy in chronic hepatitis C patients depends primarily on hepatitis C virus RNA level. Hepatology 22, 1351–1354.

Yamamoto, Y., Yamashita, S., Fujisawa, A., Kokura, S., and Yoshikawa, T. (1998). Oxidative stress in patients with hepatitis, cirrhosis, and hepatoma evaluated by plasma antioxidants. Biochem. Biophys. Res. Commun. *247*, 166–170.

Yanagitani, A., Yamada, S., Yasui, S., Shimomura, T., Murai, R., Murawaki, Y., Hashiguchi, K., Kanbe, T., Saeki, T., Ichiba, M., *et al.* (2004). Retinoic acid receptor α dominant negative form causes steatohepatitis and liver tumors in transgenic mice. Hepatology *40*, 366–375.

Yano, H., Iemura, A., Haramaki, M., Ogasawara, S., Takayama, A., Akiba, J., and Kojiro, M. (1999). Interferon alpha receptor expression and growth inhibition by interferon alpha in human liver cancer cell lines. Hepatology *29*, 1708–1717.

Yano, M., Hayashi, H., Yoshioka, K., Kohgo, Y., Saito, H., Niitsu, Y., Kato, J., Iino, S., Yotsuyanagi, H., Kobayashi, Y., *et al.* (2004). A significant reduction in serum alanine aminotransferase levels after 3-month iron reduction therapy for chronic hepatitis C: a multicenter, prospective, randomized, controlled trial in Japan. J. Gastroenterol. 39, 570–574.

Yoshida, H., Shiratori, Y., Moriyama, M., Arakawa, Y., Ide, T., Sata, M., Inoue, O., Yano, M., Tanaka, M., Fujiyama, S., *et al.* (1999). Interferon therapy reduces the risk for hepatocellular carcinoma: National surveillance program of cirrhotic and noncirrhotic patients with chronic hepatitis C in Japan. Ann. Intern. Med. *131*, 174–181.

Yoshida, T., Hanada, T., Tokuhisa, T., Kosai, K., Sata, M., Kohara, M., and Yoshimura, A. (2002). Activation of STAT3 by the hepatitis C virus core protein leads to cellular transformation. J. Exp. Med. *196*, 641–653.

Yoshizawa, H. (2002). Hepatocellular carcinoma associated with hepatitis C virus infection in Japan: Projection to other countries in the foreseeable future. Oncology *62* Suppl 1, 8–17.

Yotsuyanagi, H., Yasuda, K., Iino, S., Moriya, K., Shintani, Y., Fujie, H., Tsutsumi, T., Kimura, S., and Koike, K. (1998). Persistent viremia after recovery from self-limited acute hepatitis B. Hepatology *27*, 1377–1382.

Younossi, Z.M., Diehl, A.M., and Ong, J.P. (2002). Nonalcoholic fatty liver disease: An agenda for clinical research. Hepatology 35, 746–752.

Zhang, T., Li, Y., Lai, J.P., Douglas, S.D., Metzger, D.S., O'Brien, C.P., and Ho, W.Z. (2003). Alcohol potentiates hepatitis C virus replicon expression. Hepatology 38, 57–65.

A New Perspective in the Pathophysiology of Hepatitis C Virus (HCV) Infection: Interaction of the NS3 Protein of HCV With Serine-/Threonine-specific Protein Kinases

13

Matthias Kalitzky and Peter Borowski

Abstract

Hepatitis C virus (HCV) causes aggressive infections of the liver with a high percentage of chronic courses. The pathophysiology of this finding has not been clarified on a molecular level. The present work describes a new aspect of the pathophysiology of this HCV related disease: the interference of viral antigens in intracellular signal transduction pathways.

The analysis of the amino acid sequence of the product of the HCV genome revealed an arginine-rich motif localized between the amino acids 1487 and 1500 of the HCV polyprotein. This sequence shows close similarity to arginine-rich motifs localized in the regulatory domains of the cAMP-dependent protein kinase A (PKA) and in the phospholipid/Ca^{2+}-dependent protein kinase C (PKC). These motifs determine the recognition of the substrates and the kinetic activity of the protein kinases.

A synthetic peptide with the amino acid sequence of the arginine-rich motif of HCV inhibits the phosphorylation reaction mediated by the two enzymes by a competitive modus. Recombinant HCV proteins with the sequence of the arginine-rich motif do not only block the catalytic activity of the protein kinases, but also inhibit further functions of these enzymes like their translocation between cell compartments and association with specific receptor proteins.

Considering the important role of these protein kinases in cell growth, differentiation, carcinogenesis, and tumor promotion, it can be assumed that chronic impairment of the functions of these signal proteins by viral antigens can represent an essential part in the pathogenesis of HCV infection and in the carcinogenesis associated with HCV. The specific inhibition of interactions of these domains with cellular proteins can be used as a basis for the development of new therapeutic strategies.

Review

The plus-strand RNA of HCV encodes a 3010 amino acid polyprotein which is being cleaved by cellular and viral proteases into structural proteins (core (C) protein, envelope proteins (E1 and E2)) and non-structural proteins (NS2, NS3, NS4A, NS4B, NS5A, and NS5B) and a so-called P7 polypeptide (Miller and Purcell RH, 1990; Choo *et al.*, 1991).

The envelope proteins represent the outer spikes of the viral membrane. It is assumed that these proteins are important for association of the virus with the target cell (van Doorn, 1994; Major and Feinstone, 1997). Hypervariability of the epitopes of these envelope proteins inhibits an effective immune response (Tomei *et al.*, 1993). The core protein contains several basic amino acids mediating the binding to RNA. In cooperation with parts of NS3, NS2 exerts the function of a metalloprotease mediating the autoproteolytic cleavage of the NS2/NS3 region. The terminal region of NS3 functions as a serine protease generating the five other nonstructural proteins. Complex formation with a defined region of NS4A is necessary for the development of the complete protease activity of NS3 (Kim *et al.*, 1996). Besides its protease activity, the protein exerts an ATP-dependent helicase function (Kim *et al.*, 1995; Tai *et al.*, 1996). NS5B has the function of an RNA-dependent RNA polymerase, necessary for viral replication (Behrens *et al.*, 1996). The functions of the P7 polypeptide and of NS1, NS4B, and NS5A still remain to be elucidated. Until present, it has not been possible to establish a stable cell culture system for HCV. Still, direct pathogenic and carcinogenic activity of isolated structural and nonstructural proteins of HCV could be detected. Among other HCV proteins, the core protein was shown to have biologic activity in cell lines associated with development of tumor phenotype.

Among the structural proteins, NS3 seems to play an important role in experimental carcinogenesis. Several studies show malignant transformation of NIH 3T3 cells transfected with NS3 c-DNA (Sakamuro *et al.*, 1995). Stable expression of NS3 fragments in the same cells leads to augmented resistance to actinomycin-D-induced apoptosis (Fujita *et al.*, 1996).

Disturbances of intracellular signal transduction are of great importance for the pathogenesis of viral diseases, especially for the development of tumors of viral etiology. Modulation of concentration or activity of cellular protein kinases mediated by viral genome products is responsible for the oncogenic potential of hepatitis B virus (HBV), bovine papillomavirus (BPV) and Rous sarcoma virus (RSV) (Uehara *et al.*, 1989; Kekule *et al.*, 1993; Meyer *et al.*, 1994).

Small segments of the viral antigen structure can be sufficient to exert a pathogenic activity. These segments or motifs consist of short parts of the primary structure with close similarity to peptide transmitters or functional domains of intracellular proteins. The function of these segments can be imitated experimentally by synthetic peptides (Meyer *et al.*, 1994).

Analysis of the HCV genome product revealed and arginine-rich amino acid sequence within the region of the NTPase/helicase of the NS3 protein between the amino acids 1487 and 1500.

This sequence bears a distinct homology with arginine-rich sequences present in regulatory domains of AMP-dependent protein kinase A (PKA) and phospholipid/Ca^{2+}-dependent protein kinase C (PKC). These sequences are responsible for the regulation of the catalytic activity of the enzymes (Figure 13.1).

PKA mediates most of the effects contributed to the "second-messenger" cAMP in eukaryotic cells. Occupation of membrane associated receptors with ligands using this cAMP-dependent pathway results in activation of adenylate cyclase and consequently in increase of intracellular cAMP level. cAMP binds to the regulatory subunit (R) of PKA and dissociates into a dimer consisting of two R-subunits and two fully active C-subunits.

Fluorescence and immunocytochemical studies showed that the holoenzyme is situated in the cytosol at low cAMP concentrations. The tetrameric form of the holoenzyme of PKA is inactive. The inactive conformation is being stabilized by a short arginine-rich motif, the so-called autoinhibitory pseudosubstrate domain. This domain is localized in the R-subunit and blocks the active center of PKA (Knighton *et al.*, 1991). This autoinhibitory pseudosubstrate domain exerts high similarity with characteristic sequences of the substrates of PKA (consensus sequence) (Knighton *et al.*, 1991).

At elevated cAMP concentrations, the subunit C dislocates and translocates into the nucleus where this subunit participates in the regulation of protein transcription by direct protein–protein interaction and by phosphorylation of target proteins. Translocation of the enzyme is a transient process. When the intracellular cAMP concentration decreases, the C-subunit leaves the nucleus and binds to the cytosolic R-subunit again. It remains to be elucidated which molecular mechanisms are responsible for the translocation process.

It is still unclear whether this "shuttle" function of the C-subunit is being realized by passive diffusion, active transport or by an "anchorage release model," involving specific receptors for the C-subunit (so-called "A-kinase anchoring proteins").

PKA represents a heterogeneous family of proteins. In most cells, two types of protein kinases exist, differing in their regulatory subunits (RI and RII). As cAMP causes the release of an identical subunit from both types of holoenzymes, the existence of several isoenzymes with different substrate spectrums cannot be explained. Therefore, the question of the isoform-specific functions has to remain unanswered to date.

PKC shows a similar autoregulatory modus. This enzyme was first characterized as a protein kinase activated by limited Ca^{2+}/Calpain-dependent proteolysis. Physiologically, the enzyme is being activated by a complex consisting of the "second messengers" phosphatidylserine, diacylglycerol (DAG) and Ca^{2+}. PKC could be identified as a specific receptor for the most potent tumor promoter 12-O-tetradecanoylphorbol-13-acetate (TPA). TPA has a chemical structure similar to DAG. It binds like DAG to the regulatory subunit of the enzyme and increases its affinity to Ca^{2+} and its kinetic activity.

Furthermore, TPA (like DAG) is able to induce the translocation of the enzyme to the "particular fraction" of the cell. It is assumed that most of the processes effectuated by TPA are mediated by PKC.

This protein kinase consists of two structural subunits with distinct functions: an NH_2-terminally localized regulatory domain and a catalytic domain localized at the COOH-terminus (Newton, 1995). These subunits are connected to one another by a relatively short and flexible amino acid sequence, the so-called "hinge region," allowing a high degree of mobility of the subunits.

Beside the binding sites for allosteric activators like PS, DAG and Ca^{2+}, the regulatory subunit contains a short arginine-rich motif, the so-called autoinhibitory pseudosubstrate domain. In the absence of allosteric activators, the autoinhibitory pseudosubstrate domain blocks the catalytic center of the protein kinase and propagates its inactive state (House and Kemp, 1987). Similarly to PKA, the autoinhibitory pseudosubstrate domain of PKC shows a high degree of homology to the consensus sequences of the substrates (Hanks *et al.*, 1988). Contrary to the consensus sequence of the substrates, the autoinhibitory pseudosubstrate domain of PKC does not possess a phosphate acceptor (serine or threonine), but a nonphosphorylizable amino acid residue (House and Kemp, 1987) (Figure 13.1A). In the presence of the activators a conformational shift takes place, dislocating the autoinhibitory pseudosubstrate domain from the catalytic center and thereby allowing recognition and binding of the substrates.

In brief, activation of PKC is mediated by the dislocation of the regulatory subunit from the catalytic center, unlike PKA which is activated by dissociation of the holoenzyme.

Synthetic peptides containing the amino acid sequence of the autoinhibitory pseudosubstrate domain, the so-called pseudosubstrate (like PKC-(19–31) peptide), inhibited the enzymatic activity of PKC competitively in regard to the substrates (House and Kemp, 1987).

The translocation to the particular fraction of the cell is closely connected with activation of the enzyme. There the protein kinase forms stable complexes with specific receptors, the so-called "receptors for activated C kinase" (RACK) (Ron and Mochly-Rosen, 1995). The bound RACK conserves the active status of the protein kinase and thereby allows the phosphorylation of the enzyme.

PKC also represents a heterogeneous family of proteins (Mellor and Parker, 1998). Although all isoforms of PKC present a similar structure, they differ in their response to allosteric activators, in their affinity for substrates (Borowski *et al.*, 1998; Borowski *et al.*, 1999c), and in their translocalization to cell compartments. Each tissue and cell expresses an individual pattern of PKC isoforms (Borowski *et al.*, 1999d). Therefore, it is assumed that the different isoforms control different processes like cell growth and other biologic processes.

Like in most of the serine and threonine-specific protein kinases, the presence of basic and hydrophobic amino acids within the consensus sequence or within the autoinhibitory pseudosubstrate domain determines the specificity of the interaction in PKA and PKC (Lee *et al.*, 1994).

Studies performed with peptides containing amino acid residues of the consensus sequences of substrates and/or autoinhibitory pseudosubstrate domains showed high specificity of the catalytic center of the protein kinases in regard to the substrates but relatively low specificity to the autoinhibitory pseudosubstrate domain (Smith *et al.*, 1990; Borowski *et al.*, 1999c).

This could be an explanation for the observation that arginine or lysine-rich amino acid sequences of some nonphosphorylizable proteins are being recognized as inhibitors of serine/threonine-specific protein kinases. Nevertheless, consensus sequences of the substrates as well as inhibitory sequences of other proteins compete with autoinhibitory pseudosubstrate domains for the catalytic center of the protein kinases (House and Kemp, 1987; Smith *et al.*, 1990). This mechanism of autoregulation could experimentally be found in many other serine/threonine-specific protein kinases.

Taken together, basic amino acids like arginine determine the recognition and binding of the substrates, the degree of activity of the enzymes and the recognition and binding of the "anchor proteins." Therefore, the presence of a viral protein bearing an arginine-rich amino acid sequence, similar to the consensus sequence, may disturb the balance between the cellular receptors and the protein kinases and could thereby lead to impairment of its functions.

Results

The arginine-rich motif of NS3

Our studies have revealed evidence for formerly not described biologic functions of NS3 and of the translocation of PKA and PKC. These functions seem to be associated with an arginine-rich motif Arg-Arg-Gly-Arg-Thr-Gly-Arg-Gly-Arg-Arg-Gly-Ile-Tyr-Arg localized between the amino acids 1487 and 1500 of the HCV polyprotein in the region of the NS3-associated NTPase/helicase (Figure 13.1A). The resolution of the tertiary structure of the three domains containing HCV NTPase/helicase allowed the precise localization of the motif on the surface of domain 2 (Yao *et al.*, 1997; Kim *et al.*, 1998).

Comparison of the distribution of basic and hydrophobic amino acids within this HCV sequence with functional domains of PKA and the PKC revealed a high degree of similarity with (1) the autoinhibitory pseudosubstrate domains of PKA and PKC, localized within the regulatory domains of the enzymes, (2) the consensus sequences of the substrates of PKA (Arg-Xaa-Xaa-Ser/Thr-Xaa) and PKC (Ser/Thr-Xaa-Arg, Arg-Xaa-Xaa-Ser/Thr, Arg-Xaa-Ser/Thr, Arg-Xaa-Ser/Thr-Xaa-Arg), and (3) the inhibitory domain of the physiologic heat-stable protein kinase A inhibitor (PKI) (Scott *et al.*, 1985).

Such small segments of the viral antigen structure can be sufficient for protein–protein interaction by which the viral proteins interfere with intracellular signal cascades by competition with physiological molecules.

Inhibition of the enzymatic activity of PKA by a synthetic peptide in vitro

It is known that peptides containing the amino acid sequence of the autoinhibitory pseudosubstrate domain of the subunit R (pseudosubstrate) and the peptide bearing the amino acid sequence of the PKA-inhibiting domain of PKI (PKI-tid), act as competitive inhibitors of PKA in regard to the substrates of PKA (Scott *et al.*, 1985; Smith *et al.*, 1990; Poteet-Smith *et al.*, 1997). We have synthesized a peptide and tested it for its inhibitory potential as a substrate antagonist *in vitro*.

In order to study the role of the amino-terminally and carboxy-terminally localized basic groups in recognition of the

(A)

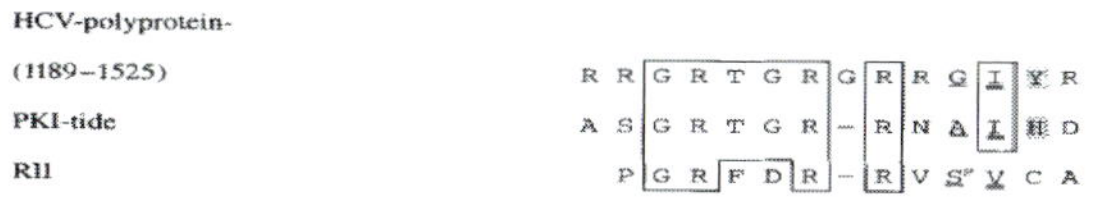

HCV-polyprotein-(1189–1525)	R R G R T G R G R R G I Y R
PKI-tide	A S G R T G R – R N A I H D
RII	P G R F D R – R V S^P V C A

(B)

Peptide	Sequence	Substrate			
		protamine	MBP	Histone H4	Histone H2B
		IC_{50} (μM)			
HCV-polyprotein-(1487-1500)-peptide	R R G R T G R G R R G I Y R	5.0	70	4.5	70
HCV-polyprotein-(1490-1500)-peptide	R T G R G R R G I Y R	30	>100	6.7	>100
HCV-polyprotein-(1493-1500)-peptide	R G R R G I Y R	0.6	90	10	95
HCV-polyprotein-(1496-1500)-peptide	R G I Y R	1.2	>100	2.8	>100
HCV-polyprotein-(1489-1497)-peptide	G R T G R G R R G	>100	40	77	75

Figure 13.1 The alignment of amino acid sequences of residues 1487–1500 of HCV polyprotein, PKI-tide, and the autophosphorylation site of RII (A) and the amino acid sequence of the synthetic peptides used in this study and their inhibition of PKA-mediated phosphorylation of protein substrates (B). A: identical residues are demonstrated in boxes. Underlined residues, amino acids with small and neutral residues; double underlined residues, hydrophobic residues; shadowed residues, ring-containing residues. S^P represents the autophosphorylation site of RII. B: phosphorylation of protein substrates (at concentrations equal to K_m values) was performed and the inhibitory constants were determined by nonlinear regression analysis and are reported as averages of 2 or 3 independent assays.

NS3 sequence by PKA, we synthesized several deleted analogues of this peptide (Figure 13.2).

The investigations of the inhibitory potential of the peptides were carried out with protein substrates showing different affinities to the substrate binding site of the C-subunit of PKA. Despite the similarity of the amino acid sequences of the PKI-tid and the HCV (1487–1500) peptide some differences in their inhibitory properties could be observed.

Contrary to the previously characterized truncated PKI-tid derivates (Scott *et al.*, 1985), all synthesized HCV-peptides inhibited the PKA-mediated phosphorylation. The inhibitory potential of these peptides was dependent on the nature of the substrate. It did neither correlate with the length of the peptide, nor with the K_m of the substrates. Figure 13.1B shows the concentrations at which half maximal inhibition was measured (IC_{50}). The graphic analysis of the modus of inhibition revealed a purely competitive type of inhibition (Borowski *et al.*, 1996).

The existence of this arginine-rich motif within NS3 let us expect that interactions between the viral protein and PKA could be observed. Therefore, a sequence of 337 amino acids of the HCV-polyprotein (HCV-polyprotein 1189–1525) was expressed as a fusion protein in *E. coli*, isolated by affinity chromatography and tested for its interactions with PKA.

Studies of the inhibition of PKA-activity showed a potent inhibition of phosphorylation of the substrates by the HCV polyprotein (1189–1525). Similar inhibition parameters were measured with the isolated C-subunit of the protein kinase. Yet, the modus of inhibition was not purely competitive but a mixed type of inhibition (Borowski *et al.*, 1996).

The homology of the arginine-rich motif of NS3 with functional domains of PKA suggests that the binding of the HCV polyprotein (1189–1525) to the C-subunit is mediated by substrate binding site of the protein kinase. This hypothesis could be confirmed by the inhibition of this binding by the HCV peptide (1487–1500), the PKI-tid, or the PKI.

These latter peptides were able to inhibit the formation of a complex of the C-subunit and the HCV polyprotein (1189–

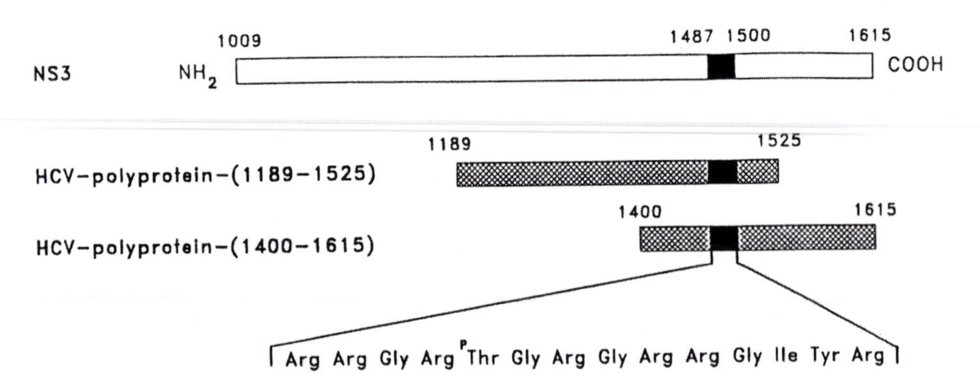

Figure 13.2 Schematic representation of the NS3 fragments used in this study. The top bar represents the entire NS3 molecule. The shadowed bars below represent the expressed fragments of NS3: HCV polyprotein (1189–1525) and HCV polyprotein (1400–1615). The position of the arginine-rich sequence within the NS3 is shown with a solid box. PThr represents the phosphorylatable threonine residue Thr^{1491} of the synthetic peptide that reproduces the arginine-rich domain of NS3.

1525) significantly (Borowski *et al.*, 1997), showing that the binding of the C-subunit to NS3 takes place via the catalytic center of the protein kinase.

Yet, the inhibition of the complex formation by the peptides was not complete and could not even be totally inhibited by the complete PKI or by poly-L-arginine. This finding suggests that the interactions between the two molecules are mediated by additional binding sites.

Interference of NS3 with functions of PKA in vivo

The effects of the interactions of the NS3 protein with PKA on the functions of the protein kinase that was observed *in vitro*, was studied *in vivo*. The HCV polyprotein (1189–1525) was introduced in Hep-2 cells after proteolytic permeabilization, and intracellular processing of the protein was characterized. Immunofluorescence studies performed with an anti-NS3 antibody showed accumulation of the introduced protein in the cytosolic compartment. The half-life of the protein was determined by Western blot analysis of the introduced cells.

The introduction of the protein did not affect the subcellular localization of the C-subunit. When not-introduced Hep-2 cells were treated with Forskolin (an activator of PKA (Witters and Blackshear, 1987)), rapid translocalization of the C-subunit to the nucleus was observed. In Hep-2 cells with the introduced HCV polyprotein (1189–1525), the translocalization process after Forskolin treatment was inhibited. Control experiments with introduced bovine serum albumin, GST or GST-HCV-polyprotein (1923–2043) revealed no inhibition of the Forskolin-induced translocalization of the C-subunit. Competition assays and kinetic *in vitro* analyses of the protein–protein interaction demonstrated that the arginine-rich motif of the NS3 protein mediated not only the binding of PKA but also the inhibition of phosphorylation of the substrates. The biologic relevance of the NS3-mediated inhibition of PKA activity could be verified in *in vivo* phosphorylation studies.

It was investigated whether the impairment of the "shuttle function" of the PKA was associated with reduction of the phosphorylation of the substrates. It was shown that the HCV polyprotein (1189–1525) present in Hep-2 cells inhibited the PKA-mediated Forskolin-stimulated phosphorylation of the histone H2B and other nucleoproteins *in vivo*. Control studies with other proteins introduced into Hep-2 cells confirmed the specificity of the inhibition of the functions of PKA mediated by the NS3 protein (Borowski *et al.*, 1997).

Inhibition of the enzymatic activity of PKC by a synthetic peptide in vitro

The arginine-rich motif of NS3 also plays an essential role in the interaction of the viral protein with protein kinase C (PKC). The NS3 amino acid sequence bears similarities with the autoinhibitory substrate domain and the consensus sequence of the PKC substrates. The synthetic peptide with the amino acid sequence of the arginine-rich motif of NS3 (HCV 1487–1500) was first tested in *in vitro* studies for its properties as an inhibitor of PKC with the protein substrates (histones H1, H2B, MBP and p80). Inhibition of the protein kinase activity was shown with all substrates, and graphic analysis revealed a competitive type of inhibition of the phosphorylation activity in regard to the substrates.

Several analogues with substituted NH_2-terminal arginine residues and shorter derivatives of the HCV peptide 1487–1500 were synthesized in order to study the role of NH_2-terminal and COOH-terminal basic groups of the arginine-rich motif of NS3 for its recognition by the catalytic center of PKC (Figure 13.3). In contrast to PKA, inhibition of the PKC-mediated phosphorylation was only observed with the complete peptide HCV (1487–1500), shorter derivatives only showed a minimal inhibitory effect (Borowski *et al.*, 2000). Although not included in the autoinhibitory pseudosubstrate domain of PKC, NH_2-terminal arginine residues seem to play an important role for the recognition of the peptide as an inhibitor and for the recognition of the threonine residue as a phosphate acceptor.

Modulation of PKC-mediated phosphorylation by NS3

Phosphorylation experiments showed that the HCV polyprotein (1189–1525) modulated the PKC-mediated protein phosphorylation. The extent of this modulating effect of the NS3 protein was dependent on the nature and the concentration of the substrate.

A kinetic analysis of PKC-mediated phosphorylation demonstrated that the maximal activity of the enzyme was not measurable at saturating concentrations of the substrate, but at optimal substrate concentrations (Borowski *et al.*, 1999a). Consequently, a biphasic and not a saturating experimental curve was determined for the phosphorylation reaction of PKC. A similar substrate-induced inhibition of enzymatic activity was observed with PKA.

The HCV polyprotein (1189–1525) was an effective inhibitor of PKC at low concentrations of the substrates. Unlike for the inhibition mediated by the HCV peptide (1487–1500), graphic analysis of the type of inhibition mediated by the HCV polyprotein (1189–1525) did not reveal a purely competitive inhibition modus, but a mixed type of inhibition. The inhibitory effect of the substrates could be reduced or completely antagonized by the HCV polyprotein (1189–1525) (Figure 13.4).

The match-marker effect of NS3

The influence of NS3 on the properties of the interacting proteins can be compared with the action of protein modulators, so-called "match markers." Complex formation of the target protein and the match marker

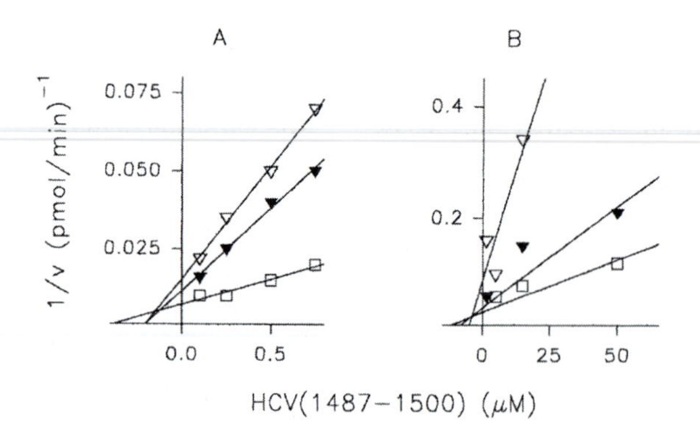

Figure 13.3 Plots demonstrating competitive inhibition of PKC by HCV (1487–1500) compared to [Ser^{25}]PKC-(19–31) and histone III-S. Rat brain PKC activity was determined with [Ser^{25}]PKC-(19–31) (A) or histone III-S (B) as substrates in the presence of HCV peptide (1487–1500) as inhibitor. The substrates were phosphorylated at concentrations corresponding to 3- (open squares), 1- (filled triangles), or 1/3-fold (open triangles) of the K_m value: 600 nM, 200 nM, or 60 nM for [Ser^{25}]PKC(19–31) and 13.5 μM, 4.5 μM, or 1.5 μM for histone phosphorylation. The data obtained were plotted according to the method of Dixon.

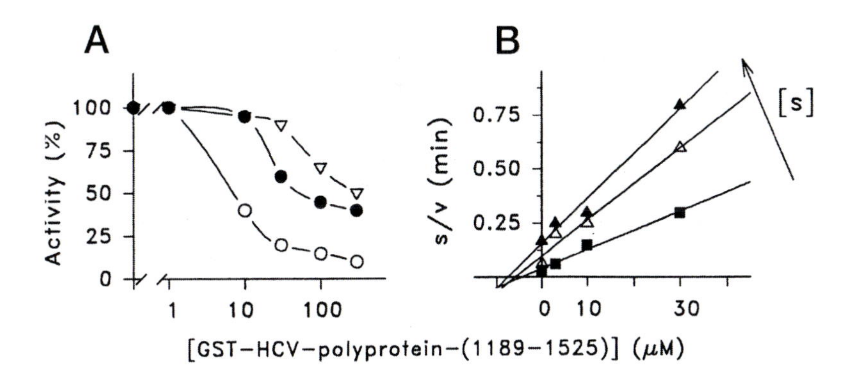

Figure 13.4 Inhibition of the PKC-mediated phosphorylation by GST-HCV polyprotein (1189–1525) and a plot demonstrating the mixed type of PKC inhibition relative to histone III-S. A: the phosphorylation of protein substrates by PKC was performed in the presence of increasing concentrations of GST-HCV polyprotein (1189–1525) as inhibitor. The substrates: histone III-S (open circles), histone H2B (filled circles), and myelin basic protein (triangles) were phosphorylated at concentrations corresponding to their K_m values (4.5 μM for histone III-S, 2.5 μM for histone H2B, and 25 μM for myelin basic protein). The kinase activity toward each substrate in the presence of the inhibitor was referred to as 100%. The ^{32}P incorporation into substrates was determined. B: PKC-activity in the presence of GST-HCV polyprotein (1189–1525) as inhibitor was investigated with histone III-S. The substrate was phosphorylated at concentrations of 13.5; 4.5; and 1.5 μM corresponding to 3-, 1-, or 1/3-fold of the K_m of histone phosphorylation (filled triangles, open triangles, and squares, respectively). The PKC activity was determined and the data obtained were plotted according to Cornish-Bowden. The results are representative for three independent experiments.

molecule leads to a conformational change of the interacting components and results in increased accessibility of the phosphorylation sites for the protein kinase.

Several studies investigating the binding of DNA to histones and other basic control proteins confirmed the conception of a match marker activity of NS3. Immobilized histones bound significant amounts of radioactive DNA in the presence of HCV polyprotein (1189–1525). The viral protein inhibited the interaction of DNA and histones in a concentration-dependent manner, and the inhibition resulted from conformational change of the histones caused by NS3. The HCV polyprotein (1189–1525) was subjected to limited proteolysis, and NH_2-terminal sequencing was performed with the resulting fragments which contained an intact, conserved, histone-binding site. This way, it was possible to identify the domain of NS3 NTPase/helicase interacting with histones, the so-called hinge region, joining the domains 1 and 2 of the enzyme (Borowski *et al.*, 1999b).

Inhibition of PKC activity by a synthetic peptide in vivo

The biologic significance of the inhibition of PKC activity by the arginine-rich domain of NS3 *in vitro* was first investigated in cells into which the HCV peptide 1487–1500 or its analogs were introduced after permeabilization of the cell membrane. By introduction of a radioactively labeled peptide it was possible to determine its half-life and subcellular distribution. Additionally to the Hep-2 cells, with which the interaction of the HCV-polyprotein (1189–1525) with PKA was investigated,

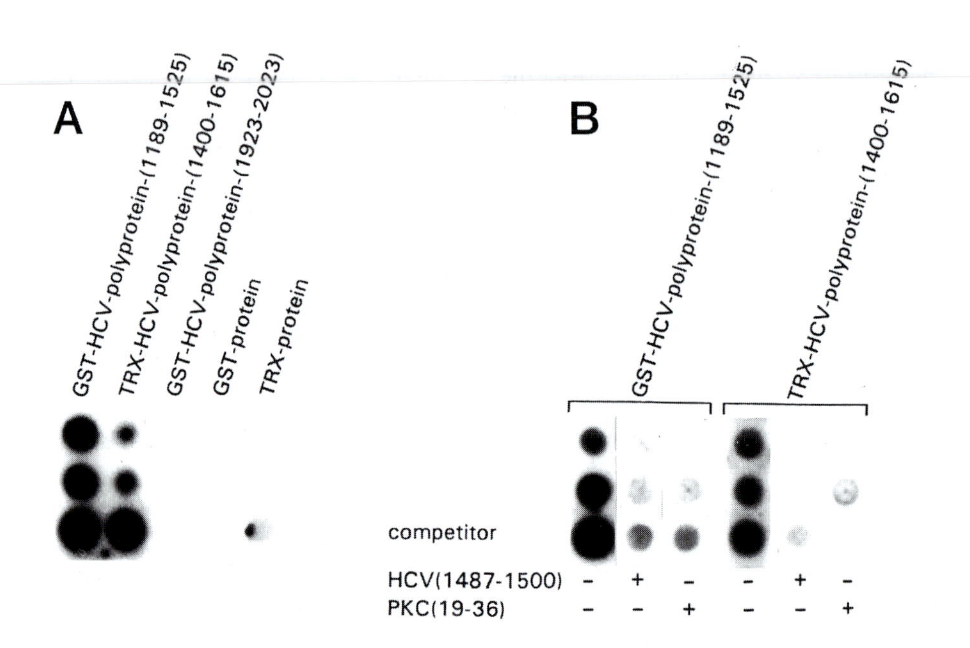

Figure 13.5 Binding of PKC to recombinant fragments of NS3 and inhibition of the binding by synthetic peptides that correspond to the arginine-rich domain of NS3 and to the autoregulatory domain of PKC. A: increasing amounts (0.1; 0.3; and 1 nmol) of GST-HCV polyprotein (1189–1525), TRX-HCV polyprotein (1400–1615), GST-HCV polyprotein (1913–2023), GST, and TRX proteins were immobilized on nitrocellulose and overlaid with ^{125}I-labeled PKC (50 nM). The nitrocellulose was washed, dried, and exposed to Kodak film for 4 h. B: increasing amounts of immobilized GST-HCV polyprotein (1189–1525) and TRX-HCV polyprotein (1400–1615) were overlaid with ^{125}I-labeled PKC in the presence of a 5-fold molar excess of the peptides HCV (1487–1500) of PKC (1936) in the overlaying buffer. The nitrocellulose was washed, dried, and autoradiographed for 4 h.

we have used human neutrophil granulocytes. These cells represent a potential reservoir for virus replication.

The HCV peptide 1487–1500 and its analogues retained their properties as inhibitors of PKC after introduction to the cell. Similarly to the *in vitro* experiments, the analog of the HCV peptide (1487–1500) bearing alanine instead of threonine inhibited the phosphorylation of p80 in Hep2 cells and neutrophils much more effectively than the original peptide. Analogs with different substitutions of the threonine showing weaker inhibition *in vitro*, also exerted a weaker inhibitory potential *in vivo* (Borowski *et al.*, 1999e).

The *in vitro* experiments presented above demonstrate that the HCV peptide (1487–1500) and its analogues inhibited the enzymatic activity of PKC by blockade of the catalytic center of the enzyme. The inhibition coincides with the formation of stable complexes of the peptide with PKC.

NS3 blocks functions of PKC by an arginine-rich motif

Besides the HCV-polyprotein (1189–1525), another fragment of the NS3 protein was produced in *E. coli* as a fusion protein with thioredoxin and was included in our studies: the HCV-polyprotein (1400–1615). The NH_2-terminus of this protein containing the domains 2 and 3 of the NTPase/helicase was constructed carboxyterminally of the hinge region joining the domains 1 and 2 (Figure 13.2). This way, the protein contains no binding site for the substrates of PKC. As a consequence, the HCV-polyprotein (1400–1615) shows no phosphorylation modulating properties, in contrast to the HCV polyprotein (1189–1525). Yet, the inhibitory potential of the two peptides in regard to the PKC-mediated protein phosphorylation was similar when tested in *in vitro* studies.

Experiments investigating the interaction of PKC and NS3 fragments in the presence of the HCV (1487–1500) polypeptide revealed a significant reduction of the binding affinity between PKC and NS3 in the presence of HCV (1487–1500).This way, it was possible to identify the catalytic center as the domain of the protein kinase mediating the interaction with NS3 (Figure 13.5).

In vivo studies of the interaction of the viral proteins with PKC were carried out as described for the study of PKA. After introduction of FITC-labeled HCV-polyprotein (1189–1525) and HCV polyprotein (1400–1615) into the cells, the influence of the viral proteins on the distribution and catalytic functions of PKC was studied in cells treated with TPA and control cells. Quantification of the amount of PKC in subcellular fractions showed that TPA induced translocalization and association of PKC with the particulate fraction of the cell was inhibited in the presence of the NS3 fragments. This finding was confirmed by the introduction of control proteins (HCV-polyprotein (1923–2043), GST and TXT) into the cells.

In order to study the specificity of the interaction of PKC and the NS3 fragments, we tested weather the introduced NS3 proteins were able to inhibit the TPA induced "oxidative burst" of the cells, another reaction mediated by PKC. Experiments showed an effective inhibition of the "oxidative burst" by the introduced HCV polyproteins (1189–1525) and (1400–1615).

Comparison of NS3-mediated inhibition with the effect of established inhibitors of PKC demonstrated that the viral protein inhibited the functions of the protein kinase very effectively (Figure 13.6).

Conclusions

The PKA signal pathway

PKA is one of the most important post-transcriptional regulators of carbohydrate and lipid metabolism (Gomez *et al.*, 1999). Most important intracellular signal transduction pathways are being regulated by PKA-mediated phosphorylation of enzymes such as glycogen synthase, glycogen phosphorylase, fructose-1,6-bis-phosphatase and proteins of the Ras-signal way (Baron-Delage and Cherqui, 1997). Thus, blockade of PKA functions may have an influence on cell growth, division, differentiation, and regeneration.

Activation of PKA by cAMP leads to interruption of the interaction of the C-subunit with the R-subunit of the protein kinase and thereby makes the catalytic center of the enzyme susceptible for interactions with the NS3 protein. This way, the viral protein can compete with the physiologic target proteins of the protein kinase. The presence of such "pseudo-anchor proteins" like NS3 can affect the PKA-mediated transcriptional regulation of "cAMP-responsive genes" (Riabowol *et al.*, 1988; Baouz *et al.*, 2001).

The role of PKC in carcinogenesis and tumor promotion

Whereas the role of PKA remains still unclear in this respect, the essential role of PKC in the development of virus associated tumors could be clearly demonstrated

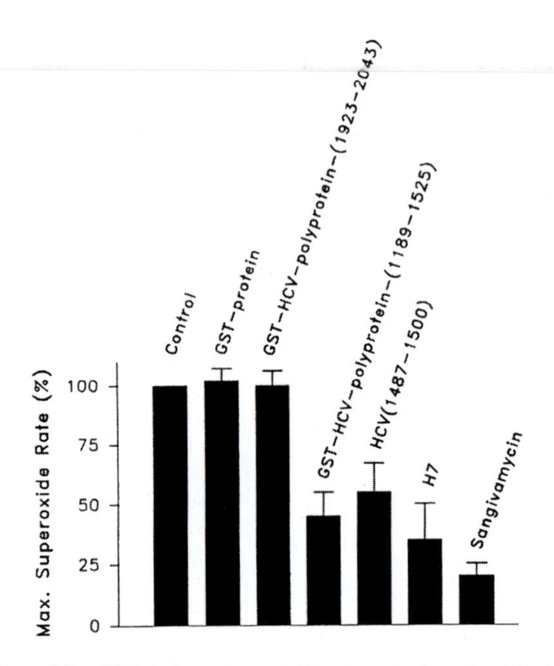

Figure 13.6 Inhibition of the TPA-induced respiratory burst of neutrophils by HCV polyprotein (1189–1525) and HCV (1487–1500) peptide. HCV polyprotein (1189–1525), HCV polyprotein (1923–2043) and HCV (1487–1500) peptide were induced into neutrophils. The cells were suspended in RPMI 1640 medium for 4 hours. Respiratory burst was measured by visible spectroscopy of superoxide dismutase-inhibitable reduction of cytochrome *c*. Data are normalized to the maximum rate of superoxide production by the control cells (not introduced neutrophils). Further control stimulations were performed on not introduced cells pretreated with the PKC inhibitors 1-(5-isoquinolinylsulfonyl)-2-methylpiperszine (150 μM) or sangivamycin (30 nM). The results are representative for five independent experiments.

(Gomez *et al.*, 1999; Baouz *et al.*, 2001). Although it is generally accepted that activation of the enzyme is a causative factor in PKC-mediated carcinogenesis (e.g. by an increase of intracellular DAG concentration mediated by a transactivator-protein in HBV-associated hepatocellular carcinoma (Kekule *et al.*, 1993)), induction of the tumor phenotype can also be effectuated by reduction of PKC activity (Harbers *et al.*, 1992). This paradox role of PKC in cell transformation can be explained by the complex interaction between the protein kinase and the tumor promoter. It remains to be elucidated whether activation or inhibition of PKC activity, or the sequential occurrence of both effects leads to transformation of the cell. Yet it is proven that induction of the tumor phenotype requires the chronic interference of the tumor promoter (Dion *et al.*, 1988). Consequently, reduction rather than increase of PKC activity should be regarded as a factor of tumor promotion. In this context, chronic inhibition of PKC functions by NS3 seems to be a potential carcinogenic co-factor. Further studies of the interaction of NS3 with the PKC signal pathway could be of interest for the study of the pathogenesis of hepatocellular carcinoma.

Future perspectives

In this chapter we report a novel possible pathomechanism of HCV-infection: the blockade of the functions of PKA and PKC mediated by a small segment of NS3. A possibility to neutralize or reduce this inhibiting activity of NS3 could be to change the conformation of the protein in a way that its binding site becomes inaccessible for the protein kinases. Allosteric modulators of the helicase activity like N7 or N9 chlorethylguanines might be able to fulfill this task. These substances and their derivatives change the conformation of the HCV NTPase/helicase and thereby the NTPase and helicase activities of the enzyme. Studies are presently carried out in order to investigate whether the protein kinase inhibiting function of NS3 can be antagonized *in vivo*.

References

Baouz, S., Jacquet, E., Accorsi, K., Hountondij, C., Balestrini, M., Zippel, R., Struani, E., and Parmeggiani, A. (2001). Sites of phosphorylation by protein kinase A in CDC25Mm/GRF1, a guanine nucleotide exchange factor for Ras. J. Biol. Chem. *276*, 1742–1749.

Baron-Delage, S. and Cherqui, G. (1997). Review: Proteine kinase C et potentiel tumoral. Bull Cancer *84*, 829–832.

Behrens, S., Tomei, L., and DeFrancesco, R. (1996). Identification and properties of the RNA-dependent RNA polymerase of hepatitis C virus. EMBO *15*, 12–22.

Borowski, P., Heiland, M., Feucht, H., and Laufs, R. (1999a). Characterization of non-structural protein 3 of hepatitis C virus as modulator of protein phosphorylation mediated by PKA and PKC; Evidences for action on the level of substrate and enzyme. Arch. Virol. *144*, 687–701.

Borowski, P., Heiland, M., Kornetzky, L., Medem, S., and Laufs, R. (1998). Purification of catalytical domain of rat spleen $p72^{syk}$ kinase and demonstration of evidences for its phosphorylation and activation by protein kinase C. Biochem J *331*, 649–657.

Borowski, P., Heiland, M., Oehlmann, K., Becker, B., Kornetzky, L., Feucht, H., and Laufs, R. (1996). Non-structural protein 3 of hepatitis C virus inhibits phosphorylation mediated by cAMP dependent protein kinase. Eur J Biochem *237*, 611–618.

Borowski, P., Kuehl, R., Laufs, R., Schulze zur Wiesch, J., and Heiland, M. (1999b). Identification and characterization of a histone binding site of the non-structural protein 3 of hepatitis C virus. J. Clin. Virol. *13*, 61–69.

Borowski, P., Kühl, R., Mueller, O., Hwang, L., Schulze zur Wiesch, J., and Schmitz, H. (1999c). Biochemical properties of a minimal functional domain with ATP-binding activity of the NTPase/helcase of hepatitis C virus. Eur J Biochem *266*, 715–723.

Borowski, P., Oehlmann, K., Heiland, M., and Laufs, R. (1997). Nonstructural protein 3 of hepatitis C virus blocks the distribution of free catalytic subunit of cyclic AMP-dependent protein kinase. J. Virol. *71*, 2838–2843.

Borowski, P., Resch, K., Schmitz, H., and Heiland, M. (2000). Synthetic peptide derived from the non-structural protein of hepatitis C virus serves as a substrate specific for protein kinase C. Biol. Chem. *381*, 19–27.

Borowski, P., Roloff, S., Medem, S., Kühl, R., and Laufs, R. (1999d). Protein kinase C produces reciprocal effects on the phorbol ester stimulated tyrosine phosphorylation of a 50-kDa kinase in jurkat cells. Biol. Chem. *380*, 403–412.

Borowski, P., Schulze zur Wiesch, J., Resch, K., Feucht, H., Laufs, R., and Schmitz, H. (1999e). Protein kinase C recognizes the protein kinase A-binding motif of nonstructural protein 3 of hepatitis C virus. J. Biol. Chem. *274*, 30722–30728.

Choo, Q., Richman, K., Han, J., Berger, K., Lee, C., Domg, C., and Gallegos, C. (1991). Genetic organisation and diversity of the Hepatitis C virus. Proc. Natl. Acad. Sci. USA *88*, 2451–2455.

Dion, L., Gindhart, T., and Colburn, N. (1988). Four-day duration of tumor promoter exposure required to transform JB6 promotion-sensitive cells to anchorage independence. Cancer Res. *48*, 7126–7131.

Fujita, T., Ishido, S., Muramatsu, S., Itoh, M., and Hotta, H. (1996). Suppression of actinomycin D-induced apoptosis by the NS3 protein of hepatitis C virus. Biochem. Biophys. Res. Commun. *229*, 825–831.

Gomez, D., Skilton, G., Alonso, D., and Kazanietz, M. (1999). Review: The role of protein kinase C and novel phorbol ester receptors in tumor cell invasion and metastasis. Oncol. Rep. *6*, 1363–1370.

Hanks, S., Quinn, A., and Hunter, T. (1988). The protein kinase family: Conserved features and deduced phylogeny of the catalytic domains. Nature *241*, 42–52.

Harbers, M., Borowski, P., Fanick, W., Lengyel, H., Buck, F., Hinsch, K., and Hilz, H. (1992). Epigenetic activation of Gi-2 protein, the product of a putative protooncogene, mediates tumor promotion *in vivo*. Carcinogenesis *13*, 2403–2406.

House, C. and Kemp, B. (1987). Protein kinase C contains a pseudosubstrate prototype in its regulatory domain. Science *238*, 1726–1728.

Kekule, A., Lauer, U., Weiss, L., Luber, B., and Hofschneider, P.H. (1993). The hepatitis B virus transactivator HBx uses a tumour promoter signalling pathway. Nature *361*, 742–745.

Kim, J., Morgenstern, K., Griffith, J.P., Dwyer, M., and Thomson, J.A. (1998). Hepatitis C virus NS3 RNA helicase domain with a bound oligonucleotide: the crystal structure provides insights into the mode of unwinding. Structure *6*, 89–100.

Kim, J., Morgenstern, K., Lin, C., Fox, T., Dwyer, M., Landro, J., Chambers, S., Markland, W., Lepre, C., O'Malley, E., Harbeson, S., Rice, C., Murcko, M., Caron, P., and Thomson, J. (1996). Crystal structure of the Hepatitis C virus NS3 protease domain complexed with a synthetic NS4A cofactor peptide. Cell *87*, 343–355.

Kim, W., Gwack, Y., Han, J., and Choe, J. (1995). C-terminal domain of the hepatitis C virus NS3 protein contains an RNA helicase activity. Biochem. Biophys. Res. Commun. *215*, 160–166.

Knighton, D., Zheng, J., Eyck, F., Xuong, N., Taylor, S., and Sowadski, J. (1991). Structure of a peptide inhibitor bound to the catalytic subunit of cyclic adenosine monophosphate-dependent protein kinase. Science *253*, 414–420.

Lee, J., Kwon, Y., Lawrence, S., and Edelman, A. (1994). A requirement of hydrophobic and basic amino acid residues for substrate recognition by Ca^{2+}/calmodulin-dependent protein kinase Ia. Proc. Natl. Acad. Sci. USA *91*, 6413–6417.

Major, M. and Feinstone, S. (1997). The Molecular Virology of Hepatitis C Virus. Hepatology *25*, 1527–1538.

Mellor, H. and Parker, P. (1998). The extended protein kinase C superfamily. Biochem J *332*, 281–292.

Meyer, A., Xu, Y., Webster, M., Smith, A., and Donoghue, D. (1994). Cellular transformation by a transmembrane peptide: Structural requirements for the bovine papilloma virus E5 oncoprotein. Proc. Natl. Acad. Sci. USA 4634–4638.

Miller, R. and Purcell RH (1990). Hepatitis C virus shares amino acid sequence similarity with pestiviruses and flaviviruses as well as members of two plant virus supergroups. Proc. Natl. Acad. Sci. USA *87*, 2057–2061.

Newton, A. (1995). Protein kinase C. Seeing two domains. Curr Biol *5*, 973–976.

Poteet-Smith, C., Shabb, J., Francis, S., and Corbin, J. (1997). Identification of critical determinants for autoinhibition in the pseudosubstrate region of type I alpha cAMP-dependent protein kinase. J. Biol. Chem. 379–388.

Riabowol, K., Fink, J., Gilman, M., Walsh, D., Goodman, R., and Feramisco, J. (1988). The catalytic subunit of cAMP-dependent protein kinase induces expression of genes containing cAMP-responsive enhances elements. Nature *336*, 83–86.

Ron, D. and Mochly-Rosen, D. (1995). An autoregulatory region in the protein kinase C: The pseudoanchoring site. Proc. Natl. Acad. Sci. USA 92, 3997–4000.

Sakamuro, D., Furukawa, T., and Takegami, T. (1995). Hepatitis C virus nonstructural protein NS3 transforms NIH 3T3 cells. J. Virol. *69*, 3893–3896.

Scott, J., Fisher, E., Demaille, J., and Krebs, E. (1985). Identification of an inhibitory region of the heat-stable protein inhibitor of the cAMP-dependent protein kinase. Proc. Natl. Acad. Sci. USA 4379–4383.

Smith, K., Colbran, R., and Soderling, T. (1990). Specificities of autoinhibitory domain peptides for four protein kinases. Implications for intact cell studies of protein kinase function. J. Biol. Chem. *265*, 1837–1840.

Tai, C., Chi, W., Chen, D., and Hwang, L. (1996). The helicase activity associated with Hepatitis C virus nonstructural Protein 3 (NS3). J. Virol. *70*, 8477–8484.

Tomei, L., Failla, C., Santolini, E., DeFrancesco, R., and LaMonica, N. (1993). NS3 is a serine protease required for processing of Hepatitis C virus polyprotein. J. Virol. *67*, 4017–4026.

Uehara, Y., Murukami, Y., Sugimoto, Y., and Mizuno, S. (1989). Mechanism of reversion of Rous sarcoma virus transformation by herbimycin A: reduction of total phosphotyrosine levels due to educed kinase activity and increased turnover of $p60^{v\text{-}scr}$. Cancer Res. *49*, 780–785.

van Doorn, L. (1994). Molecular Biology of the Hepatitis C Virus. J. Med. Virol. *67*, 989–996.

Witters, L. and Blackshear, P. (1987). Protein kinase C-mediated phosphorylation in intact cells. Methods Enzymol *141*, 412–424.

Yao, N., Hesson, T., Cable, N., Hong, Z., and Kwong, A. (1997). Structure of the hepatitis C virus RNA helicase domain. Nat. Struct. Biol. *4*, 463–467.

Index

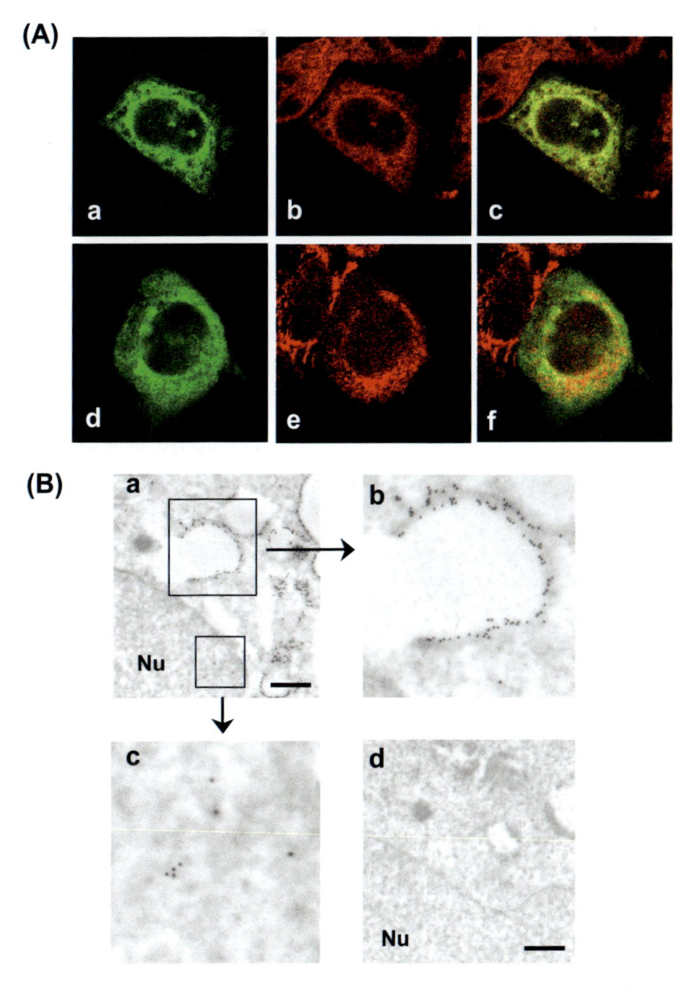

Figure 10.2 (A) Confocal microscopic images of HepG2 cells transfected with the HCV core (aa 1–191) cDNA. Cells were double stained with anti-core and anti-calregulin, ER marker, or Mitotracker, mitochondrial marker. a, d: core protein, b: calregulin, e: Mitotracker, c, f: overlay. (B) Immunoelectron microscopy. Cells expressing the core protein (a, b, c) or nonexpressing cells (d) were fixed and immunogold labeled with anti-core. Gold particles were found at cytoplasmic membranes (a, b) and in the nucleus (a, c). Bars, 0.5 μm.